COMPACT STARS IN BINARIES

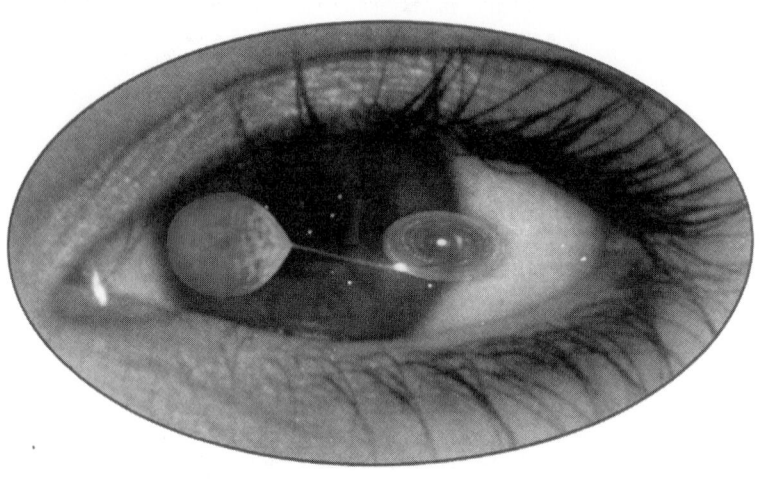

INTERNATIONAL ASTRONOMICAL UNION

UNION ASTRONOMIQUE INTERNATIONALE

COMPACT STARS IN BINARIES

PROCEEDINGS OF THE 165TH SYMPOSIUM OF THE
INTERNATIONAL ASTRONOMICAL UNION,
HELD IN THE HAGUE, THE NETHERLANDS,
AUGUST 15–19, 1994

EDITED BY

JAN VAN PARADIJS

*Astronomical Institute 'Anton Pannekoek',
University of Amsterdam, and
Center for High Energy Astrophysics (CHEAF),
Amsterdam, The Netherlands, and
Physics Department, University of Alabama in Huntsville,
Huntsville, Alabama, U.S.A.*

EDWARD P. J. VAN DEN HEUVEL

*Astronomical Institute 'Anton Pannekoek',
University of Amsterdam, and
Center for High Energy Astrophysics (CHEAF),
Amsterdam, The Netherlands*

and

ERIK KUULKERS

*ESA/ESTEC, Astrophysics Division (SA),
Noordwijk, The Netherlands, and
Astronomical Institute 'Anton Pannekoek',
University of Amsterdam, and
Center for High Energy Astrophysics (CHEAF),
Amsterdam, The Netherlands*

KLUWER ACADEMIC PUBLISHERS
DORDRECHT / BOSTON / LONDON

Library of Congress Cataloging-in-Publication Data

```
Compact stars in binaries / edited by J. van Paradijs, E.P.J. van den
  Heuvel, [and] E. Kuulkers.
       p.   cm.
    "IAU symposium 165 'Compact stars in binaries' was held from 15
  through 19 August 1994, as part of the 22nd General Assembly of the
  IAU in The Hague"--Pref.
    Includes indexes.
    ISBN 0-7923-3845-6 (hb : alk. paper). -- ISBN 0-7923-3846-4 (pb :
  alk. paper)
    1. Double stars--Congresses.  2. Supernovae--Congresses.
  I. Paradijs, J. van.  II. Heuvel, Edward Peter Jacobus van den,
  1948-     .  III. Kuulkers, E. (Erik)  IV. International Astronomical
  Union.  Symposium (165 : 1994 : Hague, Netherlands)
  QB821.C65  1996
  523.8'41--dc20                                              95-46313
```

ISBN 0-7923-3845-6

Published on behalf of
the International Astronomical Union
by
Kluwer Academic Publishers, P.O. Box 17, 3300 AA Dordrecht, The Netherlands.

Kluwer Academic Publishers incorporates
the publishing programmes of
D. Reidel, Martinus Nijhoff, Dr W. Junk and MTP Press.

Sold and distributed in the U.S.A. and Canada
by Kluwer Academic Publishers,
101 Philip Drive, Norwell, MA 02061, U.S.A.

In all other countries, sold and distributed
by Kluwer Academic Publishers Group,
P.O. Box 322, 3300 AH Dordrecht, The Netherlands.

Printed on acid-free paper

All Rights Reserved
©1996 International Astronomical Union

No part of the material protected by this copyright notice may be reproduced or utilized in any form or by any means, electronic or mechanical including photocopying, recording or by any information storage and retrieval system, without written permission from the publisher.

Printed in the Netherlands

– Jacob Shaham –

1942 – 1995

Photograph courtesy Mrs. Meira Shaham

Table of Contents

Preface   xi

1 Binary Evolution   1

R.E. Taam – *Common-Envelope Evolution, the Formation of CVs, LMXBs, and the Fate of HMXBs* 3

F.A. Rasio & S.L. Shapiro – *Hydrodynamic Evolution of Coalescing Compact Binaries*. 17

Ph. Podsiadlowski – *The Structure and Evolution of Thorne-Żytkow Objects*. 29

J.P. Lasota – *Mechanisms for Dwarf Nova Outbursts and Soft X-ray Transients (A Critical Review)* 43

P. Ghosh – *Spin Evolution of the Progenitors of Binary and Millisecond Pulsars* 57

H. Ritter, Z. Zhang & U. Kolb – *The Reaction of Low-Mass Stars to Anisotropic Irradiation and its Implications for the Secular Evolution of Cataclysmic Binaries* 65

E. Ergma & M.J. Sarna – *The Evolutionary Status of PSR 1718–19* 73

V.M. Lipunov – *The Ecology of Magnetic Rotators* 81

R.W. Romani – *The Formation and Evolution of Black-Hole Binaries*. 93

2 Supernovae in Binaries   105

B. Leibundgut – *Type Ib/c supernovae and their Relation to Binary Stars* 107

K. Nomoto, K. Iwamoto, T. Suzuki, O.R. Pols, H. Yamaoka, M. Hashimoto, P. Höflich & E.P.J. van den Heuvel – *The Origin of Type Ib-Ic-IIb-IIL Supernovae and Binary Star Evolution* . 119

S.D. van Dyk, A.J. Barth & A.V. Filippenko – *The Environments of Type Ib/c Supernovae* 135

P.C. Joss – *Type II Supernovae in Binary Systems* 141

3 Gravitational waves from Binaries   151

K.S. Thorne – *Gravitational Waves from Compact Bodies* 153

4 Radio Pulsars — 185

- A. Wolszczan – *Planets around Pulsars* 187
- D.C. Backer – *Timing of Millisecond Pulsars* 197
- M. Bailes – *Pulsar Velocities* 213
- A.G. Lyne – *A Review of Galactic Millisecond Pulsar Searches* 225
- D.R. Lorimer – *The Local Low-Mass Binary Pulsar Population* 235
- D. Bhattacharya – *Models for the Formation of Binary and Millisecond Pulsars* 243
- D.A. Frail – *Pulsar/Supernova Remnant Associations* 257
- S. Johnston – *Periastron Observations of the PSR B1259−63/SS 2883 Binary System* 263
- V.M. Kaspi, R.N. Manchester, M. Bailes & J.F. Bell – *Timing Observations of the SMC Binary PSR J0045−7319* 271
- W.T.S. Deich & S.R. Kulkarni – *The Masses of the Neutron Stars in M15C* 279

5 X-ray Binaries — 287

- F. Nagase – *High-Mass X-ray Binaries: Recent Developments* 289
- M. van der Klis – *Low-Mass X-ray Binaries — Recent Developments* 301
- M.H. Finger, R.B. Wilson, B.A. Harmon & W.S. Paciesas – *Aperiodic Flux Variability in A 0535+262* 313
- H. Inoue – *ASCA Observations of White Dwarfs, Neutron Stars and Black Holes* 321
- F. Verbunt – *ROSAT Observations of Soft X-ray Transients in Quiescence* 333
- P.A. Charles – *Black-Hole Systems: Optical Spectroscopy and IR Photometry* 341
- C.A. Haswell – *Optical Photometry of Black-Hole Candidates* 351
- W.S. Paciesas, S.N. Zhang, B.C. Rubin, B.A. Harmon, C.A. Wilson & G.J. Fishman – *Discovery of a New X-ray Transient in Scorpius (GRO J1655−40 ≡ X-ray Nova Scorpii 1994)* 363
- R.W. Hunstead, D. Campbell-Wilson & T. Ye – *The Radio Outburst from GRO J1655−40* 369

6 Binaries in Globular Clusters — 375

- P. Hut – *Dynamics and Binary (Trans)formation in Globular Clusters* 377
- H.M. Johnston, F. Verbunt, G. Hasinger & W. Bunk – *ROSAT Observations of Globular Clusters in the Galaxy and in M31* 389

7 Cataclysmic Variables — 401

K. Beuermann – *AM Herculis Binaries* 403

S. Rappaport & R. Di Stefano – *Luminous Supersoft X-ray Sources* 415

P. Kahabka & J. Trümper – *Supersoft ROSAT Sources in the Galaxies* 425

A.P. Cowley, P.C. Schmidtke, D. Crampton & J.B. Hutchings – *Supersoft X-ray Sources in the LMC* 439

A. van Teeseling, J. Heise & P. Kahabka – *Are Supersoft X-ray Sources Consistent with White Dwarfs?* 445

M. Mikołajewski, J. Mikołajewska & T. Tomov – *Propellors—A New Class of Interacting Binaries* 451

B.D. Oppenheimer & J.A. Mattei – *Analysis of Long-Term AAVSO Observations of RS Ophiuchi* 457

8 Gamma-ray Bursts — 465

G.J. Fishman – *Gamma-ray Bursts: Observational Overview* . . . 467

C. Kouveliotou – *Soft Gamma Repeaters Revisited with BATSE* . 477

T. Piran – *Gamma-ray Bursts and Binary Neutron Star Mergers* 489

Author index — 503

Subject index — 515

Object index — 531

Listing of Poster Papers — 537

Preface

IAU symposium 165 'Compact Stars in Binaries' was held from 15 through 19 August 1994, as part of the 22nd General Assembly of the IAU in The Hague. The symposium, supported by IAU Commissions 35, 37, 44 and 48, and co-sponsored by Commission 42, was attended by about 400 to 500 participants.

This symposium received support from:
- The International Astronomical Union;
- The Royal Netherlands Academy of Sciences;
- The Netherlands Ministry of Education and Science;
- The Leids Kerkhoven Bosscha Fonds;
- The Stichting Fysica.

The field of compact stars in binaries is one of the most active areas of present-day astrophysics. An absolute highlight of the last few years was the 1993 Nobel Prize of physics, awarded to Taylor and Hulse for their discovery of the binary pulsar PSR 1913+16, and the measurement of the orbital decay of this system due to the emission of gravitational waves.

The aim of the organizers of the symposium was to present an overview of the most significant observational discoveries of the past decade, in combination with a review of the most important theoretical developments. We were very happy that most of the world's leading experts in observation and theory were present at the symposium to review the various aspects of the subject. The contents of their oral presentations are now published in the form of these proceedings, which we expect to become an important source of reference for the coming years.

Among the highlight discoveries of the past several years we mention here just a few:

– The discovery by Backer, Lyne, Kulkarni, Taylor and their co-workers of several dozen millisecond pulsars, many of them in binary systems. These objects and their possible formation mechanisms are reviewed here by Backer, Lyne, Lorimer, Bailes, Kulkarni, Deich, Ergma, Bhattacharya, H. Johnston and Hut.

– The discovery in 1991 of a planetary system around a millisecond radiopulsar, reviewed here by its discoverer Wolszczan, while possible formation mechanisms of such systems are reviewed by Podsialowski.

– The discovery of a class of black-hole X-ray binaries, the so-called "soft X-ray transients", consisting of a stellar-mass black hole and a low-mass K- or G-star (notably: an overabundance of lithium). The amazing X-ray and optical properties of these systems, their structure and their possible formation and evolution are reviewed here by Charles, Haswell, Romani, Verbunt and Lasota (unfortunately, the manuscripts of the important contributions by Sunyaev and Grebenev – presented at the symposium by Sunyaev – had not reached us at the moment when the manuscript of these proceedings was sent to the printers).

– The discovery in 1990 by the ROSAT team of a new class of luminous X-ray sources with a very soft spectrum, the so-called "super soft sources". These are reviewed here by two of the discoverers: Trümper and Kahabka. Some 30 of these sources have been found now in external galaxies and another dozen in our own Galaxy.

It has become clear that they are white dwarfs in binaries which are steadily nuclearly burning on their surface the hydrogen which they receive from a companion star. The physics, formation and possible fate (Type Ia supernova?) of these systems are reviewed at this symposium by Van Teeseling and Rappaport, and their optical characteristics by Cowley.

– The discovery by Mirabel of superluminal expansion in the radio source associated with the bright transient galactic gamma-ray source GRS 1915 +105.

During the symposium the discovery with the Molonglo Telescope of a second superluminal source took place and was reported by Campbell-Wilson and Paciesas: the X-ray transient "Nova Scorpii" (GRO J1655−40) which flared up in X-rays on 27 July 1994, and in radio on 15–20 August.

– The discovery of two radio pulsars in very eccentric orbits around B-type stars, reviewed here by S. Johnston and Kaspi.

– The gamma-ray burst sources (now widely believed to be associated with merging double neutron stars or neutron-star black hole binaries), reviewed here by Fishman and Piran, and the Soft Gamma Repeaters reviewed here by Kouveliotou.

– The discovery of quasi-periodic oscillations in the pulsating X-ray binary A0535−26, by Finger, which appears to confirm the Alpar-Shaham "beat-frequency" model for this system.

Other highlights of the meeting and these proceedings are:
– Thorne's review of the expected sources of gravitational radiation in the Universe, in relation to presenty planned gravitational wave observatories. Merging close neutron star and black-hole binaries are the most certain sources among those expected to be detectable on Earth.
– The presentation by Oppenheimer and Mattei of long-term AAVSO observations of the recurrent nova RS Oph. Ben Oppenheimer (16 yr) is the youngest author of which a paper was presented at this symposium (and in these proceedings), and at the entire 22nd General Assembly.

Apart from the oral presentations, there were 205 posters presented at this meeting, many of them of such excellent quality that we regret very much that space did not allow us to publish them.

August 1995

J. van Paradijs
E.P.J. van den Heuvel
E. Kuulkers

Dedication: These proceedings are dedicated to the memory of our dear friend and colleague Jacob Shaham, one of the foremost workers in the field of compact stars in binaries, who died on April 20, 1995 in New York.

1

Binary Evolution

COMMON-ENVELOPE EVOLUTION, THE FORMATION OF CVS, LMXBS, AND THE FATE OF HMXBS

R.E. TAAM
Northwestern University
Department of Physics and Astronomy
2145 Sheridan Road
Evanston, IL 60208, U.S.A.

Abstract. Recent three-dimensional studies of the common-envelope phase of binary evolution have provided important insights into its theoretical description. The role of non-axisymmetric effects associated with gravitational torques is essential for understanding all aspects of the evolution. For successful ejection of the common envelope and survival of the remnant compact binary it is required that the orbital period of the progenitor system is long, so that one of the components of the system is in the red giant or red supergiant stage of evolution. Not only must there be sufficient energy released from the orbit to unbind the common envelope, but it is also necessary that a sufficiently steep density gradient exist above the evolved core of the giant. If these conditions are satisfied, the time scale for orbital decay in the region above the core exceeds the time scale for mass loss from the common envelope and merger is avoided. The implications of these results for the formation of cataclysmic variables (CVs), low-mass X-ray binaries (LMXBs), and the descendants of high-mass X-ray binaries (HMXBs) are discussed.

1. Introduction

The common-envelope phase of binary evolution has long been recognized as important for understanding the formation of many classes of binaries containing compact objects in short-period systems. Originally suggested nearly two decades ago by Ostriker (1975) and Paczynski (1976) as an evolutionary path for the origin of cataclysmic variables, the paradigm has been extended to include the formation of neutron star systems (such as LMXBs and binary radio pulsars) and black-hole candidate X-ray transient systems. In this phase of evolution, the two stars of the progenitor system gravitationally interact within a differentially rotating common envelope. This interaction leads to a significant shrinkage of the orbit, and the energy released in the process facilitates the ejection of the envelope. A significant fraction of the initial orbital angular momentum of the system is converted into spin angular momentum of the common envelope which is lost as well. This mass ejection results either from the hydrodynamical expansion induced by the high energy deposition rate into the common envelope or by processes responsible for mass loss in red giant stars.

The binary systems which evolve into the common-envelope phase are, generally, systems in which the mass ratio of the two components significantly differs from unity. For systems consisting of a red giant and a dwarf, the dwarf-like component can plunge into the envelope of its giant companion as a result of a tidal instability (Counselman 1973; Kopal 1978). In this case, there is insufficient angular momentum in the orbit for the dwarf to spin up the red giant to a state of corotation. Alternatively, the system can evolve into a common-envelope phase as a result of a mass transfer instability in which the mass is transferred from the giant to its companion at such a rapid rate that the matter cannot be assimilated by the mass gainer in a state of corotation (Webbink 1979).

A quantitative description of the common-envelope phase is lacking since the computational resources required for the calculation of the hydrodynamical interaction of the two stars in three spatial dimensions over a 100–1000 fold shrinkage of the orbit is prohibitive. As a consequence, all investigations directed toward a population synthesis of these compact binaries are based on crude energy arguments with a parameterization introduced for the efficiency of the mass ejection process. Although sufficient energy may be lost from the orbit to unbind the common envelope, it is still possible that the two cores coalesce. Survival of a remnant binary also requires that the orbital decay time scale increases sufficiently rapidly that the common envelope is lost before the cores can merge. The major goal of theoretical studies of the common-envelope phase is to (1) identify the distinguishing characteristics of the binary system parameters which lead to the survival of a remnant binary and to (2) determine the relationship between the parameters of the progenitor and post-common-envelope system. In this paper, I will highlight recent results from multi-dimensional simulations of the common-envelope phase. Particular reference will be made to recent studies in three spatial dimensions. The implications of these results for the formation of CVs, LMXBs, and the fate of HMXBs as well as the key issues that remain to be resolved will be discussed.

2. Numerical Results

Three dimensional simulations of the common envelope phase of evolution have been carried out by De Kool (1987), Livio & Soker (1988), and Terman, Taam & Hernquist (1994). These studies demonstrated that non-axisymmetric effects are important in determining the evolution especially during the initial phases when the orbital period of the system is longer than the time scale for the orbital decay. Their results confirmed the earlier two dimensional studies of Bodenheimer & Taam (1984) and Taam & Bodenheimer (1989, 1991) in showing that the mass ejection occurs primarily in the orbital plane of the binary. These three dimensional studies, however, were limited in resolution and in the time over which the calculations were followed. In particular, it has not yet been demonstrated that the entire common envelope is ejected. For recent reviews of the observational and theoretical status see Iben & Livio (1993) and Taam (1994).

In this paper we present a brief summary of some recent results from the study by Terman, Taam & Hernquist (1995) in which it has been demonstrated that nearly the entire common envelope is ejected as a result of its hydrodynamical interaction with the two cores. Although we primarily concentrate on the evolution of a neutron star with a massive companion, the results are of a general nature and can be applied qualitatively to the evolution of a main-sequence star with its red giant companion.

As an example, we consider the common-envelope evolution of a binary system consisting of a $16\,M_\odot$ red supergiant and a $1.4\,M_\odot$ neutron star in orbit about their center of mass with a period of $\sim 1.3\,\mathrm{yrs}$. In this illustration the supergiant is in its late core helium burning phase. Its luminosity and effective temperature are $7.12 \times 10^4\,L_\odot$ and $3{,}690\,\mathrm{K}$, respectively. The evolution is numerically simulated using the smoothed particle hydrodynamics technique (Lucy 1977; Gingold & Monaghan 1977; Monaghan 1985, 1992) which is adaptive in both space and time (Hernquist & Katz 1989). The equation of state includes contributions from both gas and radiation, and the gravitational potential of the binary system is calculated with a hierarchical tree algorithm (Appel 1985; Barnes & Hut 1986; Greengard & Rokhlin 1987; Hernquist 1987). The $11\,M_\odot$ envelope of the $16\,M_\odot$ star is represented by 10,000 particles.

We assume that the supergiant is not rotating and, hence, the neutron star rapidly plunges into the envelope (see Fig. 1) on a time scale of $0.7\,\mathrm{yrs}$. The action of the gravitational torques are effective in the outer layers of the star since a significant fraction of the mass of the star ($\sim 25\%$) is located in its outer half. As the binary orbit shrinks, the orbit tends to circularize. The time scale of the orbital decay is shown in Fig. 2, and it can be seen that after reaching a minimum of ~ 95 days after the first ~ 1000 days of

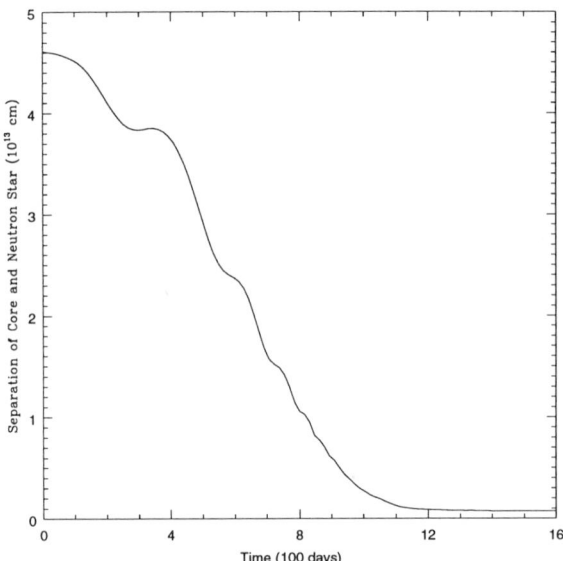

Figure 1. The variation of the orbital separation between the core of the red supergiant and the neutron star companion. The separation is in units of 10^{13} cm and the evolution time is in units of 100 days.

evolution it increased dramatically by more than a factor of 40, to ~11 yr by the end of the calculation.

The rate at which energy is deposited into the common envelope generally increased and was maintained at a level of ~$1.3\ 10^{40}$ ergs s^{-1} for ~50 days before decreasing within ~200 days by nearly a factor of 3 (see Fig. 3). The latter trend reflects the spin up and ejection of matter in the vicinity of the core. In particular, the gas is spun up to within ~50% of corotation with nearly all of the initial orbital angular momentum from the binary converted into rotational angular momentum of the common envelope. We note that the use of a Bondi-Hoyle prescription for the energy deposition rate overestimates that calculated in the three dimensional simulations by several orders of magnitude. This is due to the subsonic nature of the flow associated with the reduction in relative velocity of the two cores with respect to the common envelope due to spin up, to the overestimate of the accretion radius when the density scale heights are much smaller, and to the higher sound speeds achieved when radiation pressure effects are included in the description of the accretion process (see also Shankar, Kley & Burkert 1994).

The spatial distribution of the ejecta is similar to that found by Terman *et al.* (1994) in their calculation of a common-envelope configuration

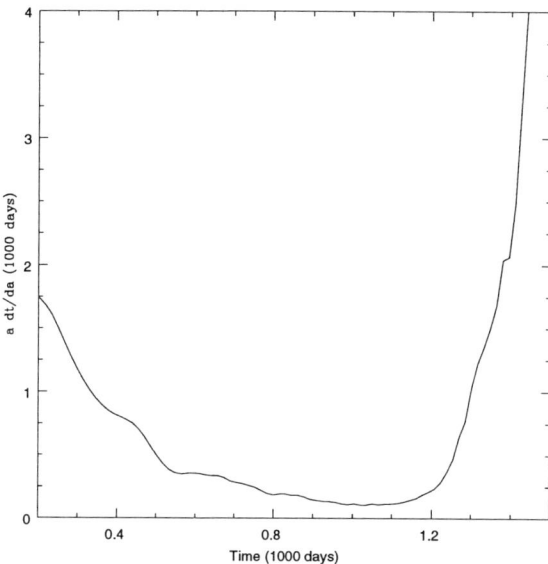

Figure 2. The time scale of the orbital decay $(a/\dot{a})$ as a function of time. Both times are in units of 1000 days.

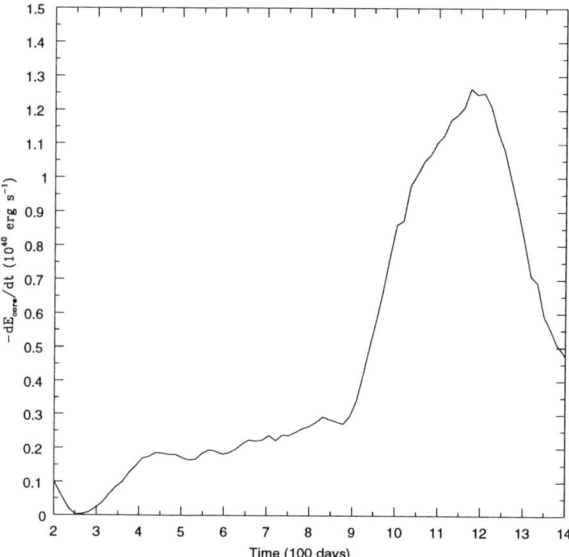

Figure 3. The energy deposition rate into the common envelope resulting from the interaction of the two cores as a function of time. The deposition rate is in units of 10^{40} ergs s^{-1} and the time is in units of 100 days.

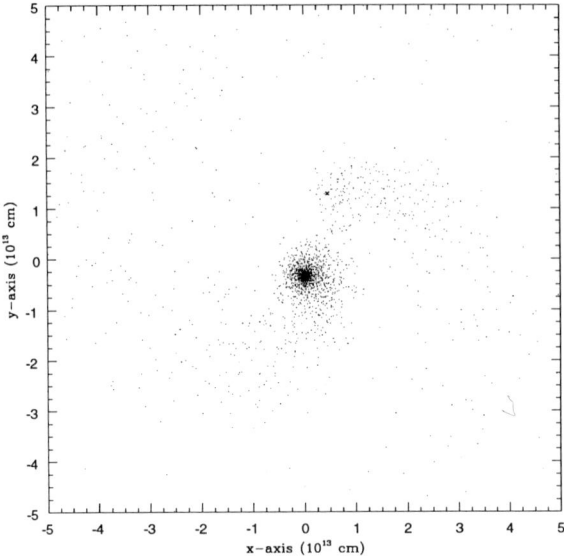

Figure 4. The distribution of the unbound matter projected onto the orbital plane of the binary system at an evolution time of 1.89 yrs. The spatial dimensions are in units of 10^{13} cm.

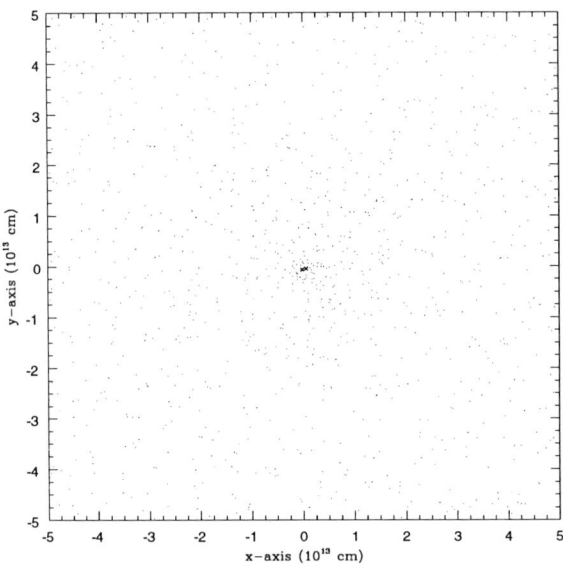

Figure 5. The distribution of the unbound matter projected onto the orbital plane of the binary system at an evolution time of 3.82 yrs. The spatial dimensions are in units of 10^{13} cm.

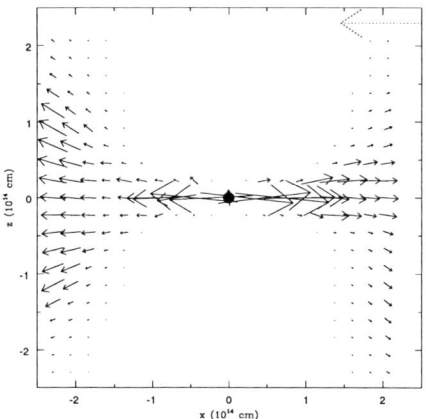

Figure 6. The velocity field in the plane perpendicular to the orbital plane of the binary system containing the two cores during the last stage of evolution after a time of 3.8 yrs. The dotted arrow in the upper right hand corner of the figure corresponds to a velocity of 68 km s^{-1}. The spatial dimensions are in units of 10^{14} cm.

consisting of a 4.67 M$_\odot$ red giant with a 0.94 M$_\odot$ dwarf. During the initial stage the matter is ejected in the form of spirals (see Fig. 4) whereas, at later stages, it is ejected more uniformly in a circular distribution (see Fig. 5) for matter projected onto the orbital plane. Additional insight into the morphology of the mass ejection process is obtained upon inspection of the velocity field in the plane perpendicular to the equatorial plane (x, z) described by the two cores illustrated in Fig. 6. It can be seen that most of the matter that is ejected from the red giant is the result of the action of an equatorial wind.

The fraction of unbound mass with respect to the common envelope rapidly rises within ~100 days to ~85% of the envelope at a time (~1050 days) when the neutron star enters the dense gas surrounding the core of the red supergiant. The entire common envelope is likely to be ejected as a result of tidal spin up (Taam & Bodenheimer 1991; Taam, Bodenheimer & Rozyczka 1994) since the density distribution of the red supergiant is sufficiently steep, and the mass contained above the helium core is small. The efficiency of the mass ejection process, defined to be given by the ratio of the actual binding energy of the ejected mass (which is the sum of the internal and gravitational potential energies) to the amount of energy lost from the orbit of the binary, is found to be ~42%. Since the ratio of the time scale for orbital decay to the time scale for ejection of the remaining matter in the common envelope rapidly increases (~5 at the end of the calculation), the orbital decay will eventually cease with the mass inside the orbit contracting away from the neutron star to the helium core (see Taam & Bodenheimer 1991). The end product of this initial long period binary system will be a neutron star orbiting about the helium core at a distance $\lesssim 6\,R_\odot$ in a period $\lesssim 12\,\text{hrs}$.

3. Discussion

The numerical results of SPH simulations for a range of masses and orbital periods (Terman et al. 1995) indicate that survival of a remnant binary is likely for progenitor systems containing a red giant or supergiant component. For binary systems containing an unevolved star or slightly evolved star, the engulfed companion will merge with the core. In all cases the majority of the mass is ejected when the two cores evolve in a tight orbit. The cores strongly torque up gas in its vicinity and produce a strong equatorial wind that ejects mass primarily in the orbital plane of the binary. The initial orbital angular momentum is distributed throughout the common envelope and, for the cases in which successful envelope ejection occurs, the gas surrounding the red giant core is spun up to within ~50%-60% of the corotational value. Since most of the energy and angular momentum is imparted to matter in the equatorial plane of the common envelope the efficiency for the mass ejection process is less than 100%, ranging from ~30%-50%. The final stage of the common envelope phase ends with a slow material outflow and with the time scale for the decay of the orbit rapidly increasing as the gravitational torques become ineffectual in response to the formation of a low density region about the two cores. Thus, the qualitative aspects gleaned from numerical simulations in two spatial dimensions in the equatorial plane (Taam, Bodenheimer & Rozyczka 1994) and in the meridional plane (Taam & Bodenheimer 1991) are confirmed. Taken together the results of the multi-dimensional simulations indicate that the presence of a steep density gradient above the core in the progenitor star is essential for the successful ejection of the common envelope and the production of a short-period compact binary system.

Based on these results, the progenitor systems of cataclysmic variables with white dwarfs more massive than about $0.6\,M_\odot$ must have had orbital periods $\gtrsim 1\,\text{yr}$. Although Terman et al. (1994) found very low efficiencies for the mass ejection process (~15%), more recent unpublished calculations in which the evolution was followed for a greater orbital shrinkage and in which an equation of state including radiation pressure was included indicate that an efficiency factor ~50% is more representative. The evolutionary calculations for the progenitors of CVs ($M \sim 2$–$8\,M_\odot$) indicate that the steep density gradients necessary for survival of the remnant binary depend on the mass of the star on the asymptotic red-giant branch (see Taam &

Bodenheimer 1992). For example, the gradients are sufficiently steep and extend to orbital separations of several solar radii for carbon-oxygen degenerate cores more massive than 0.65, 0.9, and 1.06 $M_\odot$ for stars of 3, 5, and 7 $M_\odot$ respectively. On the other hand, for stars less massive than 2.25 $M_\odot$, the steep density gradients that exist in stars on the red-giant branch with helium degenerate cores more massive than 0.35 $M_\odot$ (corresponding to orbital periods greater than several months) make these stars more favorable for successful ejection of the common envelope with regard to the formation of cataclysmic variables containing low-mass white-dwarf components ($\lesssim 0.5\,M_\odot$) with orbital separations of several solar radii.

For systems which contain white dwarfs outside these mass ranges successful ejection of the common envelope is difficult to understand within the common-envelope framework since the above conditions cannot be simultaneously satisfied. A possible solution for the formation of systems of this type is to tap the additional energy source associated with nuclear fusion (Taam & Bodenheimer 1989). If the material circulations induced in the inner region of the common envelope lead to significant compositional mixing in the hydrogen and helium rich regions, then the ejection of the remainder of the common envelope could be facilitated. The absence of steep density and pressure gradients in less evolved giant stars may facilitate this mixing to enhance the energy generation rates in the nuclear burning shells. Such mixing may render the hydrogen and helium burning shells unstable and lead to the ejection of the common envelope (see Taam 1994). If the masses of the two cores are comparable, such a process might be effective since significant differential rotation is induced in the inner regions surrounding the red-giant core, in this case, as a result of its motion about the center of mass. The degree of differential rotation and presumably the turbulence induced by hydrodynamical instabilities (see Fujimoto 1987) may be sufficient to produce the required mixing.

The numerical results of the common envelope phase of systems containing a massive star, as described in the previous section, are directly relevant to the formation of LMXBs and to the fate of HMXBs. In particular, the critical orbital periods of progenitor binary systems, $P_{\rm crit}$, which favor the survival of the binary through the common-envelope phase can be estimated on energetic grounds if an efficiency for the mass ejection process is assumed. Provided that no other energy sources can be tapped besides that derived from the orbit, and that an efficiency for the mass ejection process corresponding to the average found from the numerical calculations ($\sim 38\%$) is assumed, we find that $P_{\rm crit}$ increases with increasing mass ranging from ~ 80 days to 2 yrs for systems consisting of a 1.4 $M_\odot$ neutron star and massive companions ranging in mass from 12 $M_\odot$ to 24 $M_\odot$. The estimate for the critical period at the lower end of the mass range is likely to

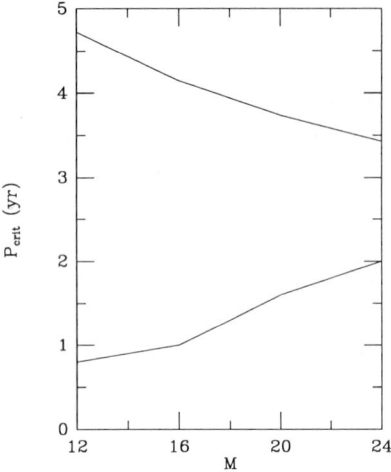

Figure 7. The critical orbital periods as a function of the mass of the massive star (in units of $M_\odot$). For periods below the lower curve a $1.4\,M_\odot$ neutron star is expected to merge with the companion. Above the upper curve the binary system does not enter into the common-envelope phase. Within the boundaries described by the two curves the formation of a short period system consisting of a neutron star and a helium core is possible.

be an underestimate since the density gradients above the nuclear burning shells for these stars are not as steep as those found for the more massive stars at longer orbital periods. A critical period of about 0.8 yr is a better estimate for a star of $12\,M_\odot$. We display this critical period relation as well as the period relation above which the system would not evolve into the common-envelope stage (for an assumed maximum stellar radius of 10^{14} cm) in Fig. 7. Although these curves have been derived for progenitor systems with characteristics similar to HMXBs, they are also relevant for systems consisting of a low-mass main-sequence star and a massive companion (for application to the formation of LMXBs) provided that the engulfed star's mass is not significantly different from $1.4\,M_\odot$. In order for the system to survive the common-envelope phase, the orbital period of the system must be enclosed within the boundaries defined by these two curves. For such long orbital period systems the companion to the neutron star must be in its red-supergiant phase. For $P < P_{\rm crit}$, the system will merge to form a single star. In the case of a neutron star companion, a recycled pulsar of low space velocity could form (see Bhattacharya & Van den Heuvel 1991) following its intermediate phase as a Thorne-Zytkow object (Thorne & Zytkow 1977; Biehle 1991; Cannon et al. 1992; Cannon 1993). We point out that stars with $M \gtrsim 30\,M_\odot$ do not expand to the red-supergiant phase due to the action of strong stellar winds. Hence, these stars would not contribute to the population of compact systems formed via the common-envelope phase (see Van den Heuvel 1994).

The progenitor systems which survive the common-envelope phase as a compact binary are a class of systems known as Be/X-ray binaries (see Rappaport & Van den Heuvel 1982; Van den Heuvel & Rappaport 1987; Van den Heuvel 1992). The remnant binary will consist of the helium core of the red supergiant and its neutron star companion. Systems with low-mass main-sequence companions are also expected to survive, however their progenitor systems may not be Be binaries. The future evolution of these products can either lead to the merger of the two components or to the production of close compact binary systems such as PSR 1913+16 and PSR 1534+12 (see Flannery & Van den Heuvel 1975; De Loore et al. 1975) for a neutron star companion, and to LMXBs for a main-sequence like companion.

Coalescence of the two components may take place if the remnant system undergoes a second common-envelope phase as a result of the expansion of the helium star beyond the orbit of its companion. This is a distinct possibility since helium stars less massive than about $4\,M_\odot$ (corresponding to hydrogen-rich stars less massive than $\sim 16\,M_\odot$) expand to $\gtrsim 3\,R_\odot$ with stars in the mass range of 2–$2.7\,M_\odot$ developing a red-giant structure after central helium burning (Habets 1985, 1986). In fact, for sufficiently large mass ratios, the mass transfer process will not be conservative and a second common-envelope phase will result. We remark that our calculations require that the progenitor star has a red-giant like structure for successful ejection of the common envelope. Assuming that this requirement is also necessary for the survival of the system through the second spiral-in phase as well, then the small orbital separation characteristic of the products of the first spiral-in phase ($\lesssim 6\,R_\odot$) is likely to lead to merger of the helium core and the neutron star and to the formation of a neutron star or a black hole (see below) with a surrounding remnant accretion disk in the second spiral-in stage unless sufficiently steep density gradients are produced as a result of the mass loss process itself. Hence, the survival probability of a binary system through two common-envelope phases in massive star systems may be expected to be low.

On the other hand, for remnant binary systems where the mass ratio is not large the system may avoid a common-envelope phase and quasi-conservative mass transfer can result. In this case, the helium star transfers mass to its neutron star companion at rates ($\sim 10^{-4}$ to $10^{-3}\,M_\odot\,\text{yr}^{-1}$) significantly in excess of the Eddington limit. For such high mass transfer rates and spherically symmetric accretion by the neutron star, photon trapping in the flow can be effective with the gravitational potential energy lost directly in the form of neutrinos (Chevalier 1993). As a result, the neutron star may accrete sufficient mass to exceed its mass limit and, thereby, produce a black hole. Further evolution of the helium star may lead to the formation of a neutron star, and to the existence of a short-period binary consisting of a black hole and a neutron star. Alternatively, if the mass transfer process leads to jet formation as in SS 433, the neutron star in the remnant binary may not accrete appreciably and the system likely evolves to the close double neutron star stage.

In contrast, systems containing helium stars more massive than $\sim 4\,M_\odot$ are not expected to merge with their companion provided that the orbital decay time scale is longer than the nuclear burning evolutionary time scale. In these systems, the helium star will evolve to the supernova stage either producing a high-velocity pulsar or a neutron star remnant in a bound orbit. It is likely that a kick velocity in the retrograde sense relative to the orbit will be necessary to bind the system because the system would be unbound for a helium core more massive than $4.2\,M_\odot$ for a $1.4\,M_\odot$ neutron star companion, or more massive than $3.8\,M_\odot$ for a $1\,M_\odot$ main sequence-like companion in the absence of a kick. This suggests that, for a given helium core remnant, the mass loss associated with the supernova event is more likely to disrupt a system composed of a low-mass main-sequence star than a neutron star. Additional effects related to the sizes of the stellar objects further restricts the survival probability of main-sequence stars with a neutron star companions since the orbit cannot be so highly eccentric that the newly born neutron star collides with its companion during periastron passage.

A number of key issues remain for detailed study. Among them I include the investigation of the dependence of the efficiency of the mass ejection process on the mass of each of the components of the system, the evolutionary state of the stars, and the degree to which the components are out of synchronous rotation at the onset of the common-envelope phase. A better understanding of the final stage of this phase of evolution with regard to the possibility of forming steep density gradients in the structure of the common envelope as a result of the mass ejection process itself will be necessary in order to place constraints on the fraction of binaries containing compact stars which form short-period systems, and at the same time, to establish the relationship between the orbital parameters of the post common-envelope binary to that of the pre common-envelope binary. Finally, studies of the nuclear assisted hypothesis are necessary to determine its viability for the formation of low-mass white dwarfs in close binary systems within the framework of the common-envelope paradigm.

Acknowledgements. This research has been supported in part by the Pittsburgh Supercomputing Center and the NSF under grant AST 91-13150.

References

Appel, A.W. 1985, SIAM, J. Sci. Stat. Comput. 6, 85
Barnes, J.E. & Hut, P. 1986, Nat 324, 446
Bhattacharya, D. & Van den Heuvel, E.P.J. 1991, Physics Reports 203, 1
Biehle, G.T. 1991, ApJ 380, 167
Bodenheimer, P. & Taam, R.E. 1984, ApJ 280, 771
Cannon, R.C. 1993, MNRAS 263, 817
Cannon, R.C. et al. 1992, ApJ 386, 206
Chevalier, R.A. 1993, ApJ 411, L33
Counselman, C.C. 1973, ApJ 180, 307
De Kool, M. 1987, Ph.D. thesis, University of Amsterdam
De Loore, C., de Greve, J.P. & De Cuyper, J.P. 1975, Ap&SS 36, 219
Flannery, B.P. & Van den Heuvel, E.P.J. 1975, A&A 39, 61
Fujimoto, M.Y. 1987, A&A 176, 53
Gingold, R.A. & Monaghan, J.J. 1977, MNRAS 181, 375
Greengard, L. & Rokhlin, V. 1987, J. Comp. Phys., 73, 325
Habets, G.M.H.J. 1985, Ph.D. Thesis, University of Amsterdam
Habets, G.M.H.J. 1986, A&A 167, 61
Hernquist, L. 1987, ApJS 64, 715
Hernquist, L. & Katz, N. 1989, ApJS 70, 419
Iben Jr, I. & Livio, M. 1993, PASP 105, 1373
Kopal, Z. 1978, *Dynamics of Close Binary Systems*, Reidel Publ. Cy
Livio, M. & Soker, N. 1988, ApJ 329, 764
Lucy, L. 1977, AJ 82, 1013
Monaghan, J.J. 1985, Computer Phys. Rep. 3, 71
Monaghan, J.J. 1992, ARA&A 30, 543
Ostriker J.P. 1975, talk presented at IAU Symposium 73
Paczynski, B. 1976, in *The Structure and Evolution of Close Binary Systems*, IAU Symposium 73, P. Eggleton, S. Mitton & J. Whelan (Eds.), Reidel Publ. Cy, p. 75
Rappaport, S. & Van den Heuvel, E.P.J. 1982, in *Be Stars*, IAU Symposium 98, M. Jaschek & H.G. Groth (Eds.), Reidel Publ. Cy, p. 327
Shankar, A., Kley, W. & Burkert, A. 1994, in *Interacting Binary Stars*, A.W. Shafter (Ed.), ASP Conf. Proc. Vol. 56, p. 436
Taam, R.E. 1994, in *Interacting Binary Stars*, A.W. Shafter (Ed.), ASP Conf. Proc. Vol. 56, p. 208
Taam, R.E. & Bodenheimer, P. 1989, ApJ 337, 849
Taam, R.E. & Bodenheimer, P. 1991, ApJ 373, 246
Taam, R.E. & Bodenheimer, P. 1992, in *X-Ray Binaries and Recycled Pulsars*, E.P.J. van den Heuvel & S.A. Rappaport (Eds.), Kluwer Academic Publishers, p. 281
Taam, R.E., Bodenheimer, P. & Rozyczka, M. 1994, ApJ 431, 247
Terman, J.L., Taam, R.E. & Hernquist, L. 1994, ApJ 422, 729
Terman, J.L., Taam, R.E. & Hernquist, L. 1995, ApJ (submitted)
Thorne, K.S. & Zytkow, A.N. 1977, ApJ 212, 832
Van den Heuvel, E.P.J. 1992, in *X-Ray Binaries and Recycled Pulsars*, E.P.J. van den Heuvel & S.A. Rappaport (Eds.), Kluwer Academic Publishers, p. 233
Van den Heuvel, E.P.J. 1994, Space Sci. Rev., 66, 309
Van den Heuvel, E.P.J. & Rappaport, S. 1987, in *Physics of Be Stars*, A. Slettebak & T.P. Snow (Eds.), Cambridge University Press, p. 291
Webbink, R.F. 1979, in *Changing Trends in Variable Star Research*, IAU Coll. No. 46, F.M. Bateson, J. Smak & I.H. Urch (Eds.), Univ. Waikato, p. 102

HYDRODYNAMIC EVOLUTION OF COALESCING COMPACT BINARIES

FREDERIC A. RASIO
Institute for Advanced Study
Princeton, NJ 08540, USA

AND

STUART L. SHAPIRO
Center for Radiophysics and Space Research
Cornell University, Ithaca, NY 14853, USA

Abstract. In addition to their possible relevance to gamma-ray bursts, coalescing binary neutron stars have long been recognized as important sources of gravitational radiation that should become detectable with the new generation of laser interferometers such as LIGO. Hydrodynamics plays an essential role near the end of the coalescence when the two stars finally merge into a single object. The shape of the corresponding burst of gravitational waves provides a direct probe into the interior structure of a neutron star and the nuclear equation of state. The interpretation of the gravitational waveform data will require detailed theoretical models of the complicated three-dimensional hydrodynamic processes involved. Here we review the results of our recent work on this problem, using both approximate quasi-analytic methods and large-scale numerical hydrodynamics calculations on supercomputers. We also discuss briefly the coalescence of white-dwarf binaries, which are also associated with a variety of interesting astrophysical phenomena.

1. Introduction

The coalescence and merging of two stars into a single object is the almost inevitable end point of compact binary evolution. Dissipation mechanisms such as the emission of gravitational radiation are always present and cause the binary orbit to decay. The terminal stage of this decay is always hydrodynamic in nature, with the final merging of the two stars taking place on a time scale comparable to the rotation period. In some systems, this is because mass transfer from one component to the other can become dynamically unstable, with the mass donor undergoing complete tidal disruption eventually. In addition, it was realized recently that, even if the mass transfer is stable or does not occur, *global hydrodynamic instabilities* can drive the binary system to rapid coalescence once the tidal interaction between the two stars becomes sufficiently strong (Rasio & Shapiro 1992, 1994, 1995a,b hereafter RS1–4; Lai, Rasio & Shapiro 1993a,b, 1994a,b,c, hereafter LRS1–5 or collectively LRS). Using numerical hydrodynamic calculations, we demonstrated for the first time in RS1 the existence of these global instabilities for binary systems containing a compressible fluid. In addition, the classical *analytic* work for binaries containing an *incompressible* fluid (Chandrasekhar 1969) was extended to compressible fluids in the work of LRS. This new analytic study confirmed the existence of dynamical and secular instabilities for sufficiently close binary systems containing polytropes. However, numerical calculations remain essential for establishing the stability limits of close binaries accurately and for following the non-linear evolution of unstable systems all the way to complete coalescence.

This review will concentrate on the coalescence of compact binaries, containing either two neutron stars (Section 2) or two white dwarfs (Section 3). Many of the results for white dwarfs, however, are also relevant to low-mass main-sequence stars in contact systems and the problem of blue-straggler formation through binary coalescence (RS3).

2. Coalescing Neutron Star Binaries

2.1. ASTROPHYSICAL MOTIVATION

Coalescing neutron star binaries are the most promising known sources of gravitational radiation that could be detected by the new generation of laser interferometers such as the Caltech-MIT LIGO (Thorne 1987; Abramovici *et al.* 1992; Cutler *et al.* 1992) and the European VIRGO (Bradaschia *et al.* 1990). Statistical arguments based on the observed local population of binary pulsars with probable neutron star companions lead to an estimate of the rate of neutron star binary coalescence in the Universe of order $10^{-7}\,\mathrm{yr}^{-1}\,\mathrm{Mpc}^{-3}$ (Narayan, Piran & Shemi 1991; Phinney 1991). Using this estimate, Finn & Chernoff (1993) predict that an advanced LIGO detector could observe about 70 events per year[1]. In addition to providing a major

---

[1] Theoretical models of the binary star population in our Galaxy suggest that the neutron star binary coalescence rate may be even higher, perhaps as high as

new confirmation of Einstein's theory of general relativity, the detection of gravitational waves from coalescing binaries at cosmological distances could provide the first accurate measurement of the Hubble constant and mean density of the Universe (Schutz 1986; Chernoff & Finn 1993; Marković 1993).

Recent calculations of the gravitational radiation waveforms from coalescing binaries have focused on the signal emitted during the last few thousand orbits, as the frequency sweeps upward from about 10 Hz to 1000 Hz. The waveforms in this regime can be calculated fairly accurately by performing high-order post-Newtonian expansions of the equations of motion for two *point masses*[2] (Lincoln & Will 1990; Junker & Schäfer 1992; Kidder, Will & Wiseman 1992; Wiseman 1993; Cutler & Flanagan 1994). However, at the end of the inspiral, when the binary separation becomes comparable to the stellar radii, hydrodynamic effects become important and the character of the waveforms will change. Special purpose narrow-band detectors that can sweep up frequency in real time will be used to try to catch the corresponding final few cycles of gravitational waves (Meers 1988; Strain & Meers 1991). In this terminal phase of the coalescence, the waveforms contain information not just about the effects of general relativity, but mostly about the internal structure of the stars and the nuclear equation of state at high density. Extracting this information from observed waveforms, however, requires detailed theoretical knowledge about all relevant hydrodynamic processes.

Coalescing neutron star binaries are also at the basis of numerous models of gamma-ray bursters at cosmological distances (Paczyński 1986; Eichler *et al.* 1989; Narayan, Paczyński, & Piran 1992; Nemiroff 1994). The isotropic angular distribution of the bursts detected with the BATSE experiment on the Compton GRO satellite (Meegan *et al.* 1992) strongly suggests a cosmological origin, and the rate of gamma-ray bursts detected with BATSE, of order one per day, is in rough agreement with theoretical predictions for the rate of neutron star binary coalescence in the Universe (cf., above). The complete hydrodynamic evolution during final merging, especially in the outermost, low-density regions of the system, must be understood in detail before realistic, three-dimensional models can be constructed for the gamma-ray emission (Davies *et al.* 1994; Piran, these Proceedings).

$\gtrsim 10^{-6}$ yr^{-1} Mpc^{-3} (Tutukov & Yungelson 1993).

[2] High accuracy is essential here because the observed signals will be matched against theoretical templates. Since the templates must cover $\gtrsim 10^3$ orbits, a fractional error as small as 10^{-3} can prevent detection.

2.2. DYNAMICAL COALESCENCE

Hydrostatic equilibrium configurations for binary systems with sufficiently close components can become *dynamically unstable* (Chandrasekhar 1975; Tassoul 1975). The physical nature of this instability is common to all binary interaction potentials that are sufficiently steeper than $1/r$ (see, e.g., Goldstein 1980, Section 3.6). It is analogous to the familiar instability of circular orbits sufficiently close to a black hole (Shapiro & Teukolsky 1983, Section 12.4). Here, however, it is the *Newtonian tidal interaction* that is responsible for the steepening of the effective interaction potential between the two stars and for the destabilization of the circular orbit (LRS3).

Close binaries containing neutron stars with stiff equations of state (adiabatic exponent $\Gamma \gtrsim 2$) are particularly susceptible to this instability. This is because tidal effects are stronger for stars containing a less compressible fluid. As the dynamical stability limit is approached, the secular orbital decay driven by gravitational wave emission can be dramatically accelerated (LRS2, LRS3). The two stars then plunge rapidly toward each other, and merge together into a single object in just a few rotation periods. This dynamical instability was first identified in RS1, where we calculated the evolution of equilibrium configurations containing two identical polytropes with $\Gamma = 2$. It was found that when $r \lesssim 3R$ (r is the binary separation and R the radius of an unperturbed neutron star), the orbit becomes unstable to radial perturbations and the two stars undergo rapid coalescence. For $r \gtrsim 3R$, the system could be evolved dynamically for many orbital periods without showing any sign of orbital evolution (in the absence of dissipation).

The dynamical evolution of an unstable, initially synchronized (i.e., rigidly rotating) binary can be described typically as follows (RS1, RS2). During the initial, linear stage of the instability, the two stars approach each other and come into contact after about one orbital revolution. In the corotating frame of the binary, the relative velocity remains very subsonic, so that the evolution is adiabatic at this stage. This is in sharp contrast to the case of a head-on collision between two stars on a free-fall, radial orbit, where shocks are very important for the dynamics (RS1). Here the stars are constantly being held back by a (slowly receding) centrifugal barrier, and the merging, although dynamical, is much more gentle. After typically two orbital revolutions the innermost cores of the two stars have merged and the system resembles a single, very elongated ellipsoid. At this point a secondary instability occurs: *mass shedding* sets in rather abruptly. Material is ejected through the outer Lagrange points of the effective potential and spirals out rapidly. In the final stage, the spiral arms widen and merge together. The relative radial velocities of neighboring arms as they merge are supersonic, leading to some shock-heating and dissipation. As a result, a hot, nearly axisymmetric rotating halo forms around the central dense core. No measurable amount of mass escapes from the system. The halo contains about 20% of the total mass and has a pseudo-barotropic structure (Tassoul 1978, Section 4.3), with the angular velocity decreasing as a power-law $\Omega \propto \varpi^{-\nu}$ where $\nu \lesssim 2$ and ϖ is the distance to the rotation axis (RS1). The core is rotating uniformly near break-up speed and contains about 80% of the mass still in a cold, degenerate state.

We calculate the emission of gravitational radiation during dynamical coalescence using the quadrupole approximation (RS1). Both the frequency and amplitude of the emission peak somewhere during the final dynamical coalescence, typically just before the onset of mass shedding. Immediately after the peak, the amplitude drops abruptly as the system evolves towards a more axially symmetric state. For an initially synchronized binary containing two identical polytropes, the properties of the waves near the end of the coalescence depend very sensitively on the stiffness of the equation of state. When $\Gamma < \Gamma_{crit}$, with $\Gamma_{crit} \approx 2.3$, the final merged configuration is perfectly axisymmetric[3] and the amplitude of the waves drops to zero in just a few periods (RS1). In contrast, when $\Gamma > \Gamma_{crit}$, the dense central core of the final configuration remains *triaxial* (its structure is basically that of a compressible Jacobi ellipsoid; cf. LRS1) and therefore it continues to radiate gravitational waves. The amplitude of the waves first drops quickly to a non-zero value and then decays more slowly as gravitational waves continue to carry angular momentum away from the central core (RS2). Because realistic neutron star models give effective Γ values precisely in the range 2—3 (LRS3), i.e., close to $\Gamma_{crit} \approx 2.3$, a simple determination of the absence or presence of persisting gravitational radiation after the coalescence (i.e., after the peak in the emission) could place a strong constraint on the stiffness of the equation of state.

2.3. MASS TRANSFER AND THE DEPENDENCE ON THE MASS RATIO

Clark & Eardley (1977) suggested that secular, *stable* mass transfer from one neutron star to another could last for hundreds of orbital revolutions before the lighter star is tidally disrupted. Such an episode of stable mass transfer would be accompanied by a secular *increase* of the orbital separation. Thus, if stable mass transfer could indeed occur, a characteristic "reversed chirp" would be observed in the gravitational wave signal at the end of the inspiral phase (Jaranowski & Krolak 1992).

---

[3] A polytropic fluid with $\Gamma < 2.3$ (polytropic index $n > 0.8$) cannot sustain a non-axisymmetric, uniformly rotating configuration in equilibrium (see, e.g., Tassoul 1978, Section 10.3).

The question was reexamined recently by Kochanek (1992) and Bildsten & Cutler (1992), who both argued against the possibility of stable mass transfer on the basis that very large mass transfer rates and extreme mass ratios would be required. Moreover, in LRS3 it was pointed out that mass transfer has in fact little importance for most neutron star binaries (except perhaps those containing a very low-mass neutron star). This is because for $\Gamma \gtrsim 2$, *dynamical instability always arises before the Roche limit* along a sequence of binary configurations with decreasing r. Therefore, by the time mass transfer begins, the system is already in a state of dynamical coalescence and it can no longer remain in a nearly circular orbit. Thus stable mass transfer from one neutron star to another appears impossible.

In RS2 we presented a complete dynamical calculation for a system containing two polytropes with $\Gamma = 3$ and a mass ratio $q = 0.85$[4]. For this system we found that the dynamical stability limit is at $r/R \approx 2.95$, whereas the Roche limit is at $r/R \approx 2.85$. The dynamical evolution turns out to be quite different from that of a system with $q = 1$. The Roche limit is quickly reached while the system is still in the linear stage of growth of the instability. Dynamical mass transfer from the less massive to the more massive star begins within the first orbital revolution. Because of the proximity of the two components, the fluid acquires very little velocity as it slides down from the inner Lagrange point to the surface of the other star. As a result, relative velocities of fluid particles remain largely subsonic and the coalescence proceeds quasi-adiabatically, just as in the $q = 1$ case. In fact, the mass transfer appears to have essentially no effect on the dynamical evolution. After about two orbital revolutions the smaller-mass star undergoes complete tidal disruption. Most of its material is quickly spread on top of the more massive star, while a small fraction of the mass is ejected from the outermost Lagrange point and forms a single-arm spiral outflow. The more massive star, however, remains little perturbed during the entire evolution and simply becomes the inner core of the merged configuration.

The dependence of the peak amplitude $h_{\max}$ of gravitational waves on the mass ratio q appears to be very strong, and nontrivial. In RS2 we obtained an approximate scaling $h_{\max} \propto q^2$. This is very different from the scaling obtained for a detached binary system with a given binary separation. In particular, for two point masses in a circular orbit with separation r we have $h \propto \Omega^2 \mu r^2$, where $\Omega^2 = G(M + M')/r^3$ and $\mu = MM'/(M+M')$. At constant r, this gives $h \propto q$. This linear scaling is obeyed (only approximately, because of finite-size effects) by the wave amplitudes of the

---

[4]This is the most probable value of the mass ratio in the binary pulsar PSR 2303+46 (Thorsett et al. 1993) and represents the largest observed departure from $q = 1$ in any observed binary pulsar with likely neutron star companion. For comparison, $q = 1.386/1.442 = 0.96$ in PSR 1913+16 (Taylor & Weisberg 1989) and $q = 1.32/1.36 = 0.97$ in PSR 1534+12 (Wolszczan 1991).

various systems at the *onset* of dynamical instability. For determining the *maximum* amplitude, however, hydrodynamics plays an essential role. In a system with $q \neq 1$, the more massive star tends to play a far less active role in the hydrodynamics and, as a result, *there is a rapid suppression of the radiation efficiency as q departs even slightly from unity*. For the peak luminosity of gravitational radiation we found approximately $L_{\max} \propto q^6$. Again, this is a much steeper dependence than one would expect based on a simple point-mass estimate, which gives $L \propto q^2(1+q)$ at constant r.

2.4. MEASURING THE RADIUS OF A NEUTRON STAR WITH LIGO

The most important parameter that enters into quantitative estimates of the gravitational wave emission during the final coalescence is the relativistic parameter M/R for a neutron star (we take $G = c = 1$). In particular, for two identical point masses we know that the wave amplitude obeys $(r_O/M)h \propto (M/R)$, where r_O is the distance to the observer, and the total luminosity $L \propto (M/R)^5$. Thus one expects that any quantitative measurement of the emission near maximum should lead to a direct determination of the radius R, assuming that the mass M has already been determined from the low-frequency inspiral waveform (Cutler & Flanagan 1994). Most current nuclear equations of state for neutron stars give $M/R \sim 0.1$, with $R \sim 10$ km nearly independent of the mass in the range $0.8 \, M_\odot \lesssim M \lesssim 1.5 \, M_\odot$ (see, e.g., Baym 1991; Cook et al. 1994; LRS3).

However, the details of the hydrodynamics also enter into this determination. The importance of hydrodynamic effects introduces an explicit dependence of all wave properties on the internal structure of the stars (which we represent here by a single dimensionless parameter Γ), and on the mass ratio q. If relativistic effects were taken into account for the hydrodynamics itself, an additional, non-trivial dependence on M/R would also be present. This can be written conceptually as

$$\left(\frac{r_O}{M}\right) h_{\max} \equiv \mathcal{H}(q, \Gamma, M/R) \times \left(\frac{M}{R}\right) \quad (1)$$

$$\frac{L_{\max}}{L_o} \equiv \mathcal{L}(q, \Gamma, M/R) \times \left(\frac{M}{R}\right)^5 \quad (2)$$

Combining all the results of RS, we can write, in the limit where $M/R \to 0$ and for q not too far from unity,

$$\mathcal{H}(q, \Gamma, M/R) \approx 2.2 \, q^2 \quad \mathcal{L}(q, \Gamma, M/R) \approx 0.5 \, q^6, \quad (3)$$

essentially independent of Γ in the range $\Gamma \approx 2\text{-}3$ (RS2). This is in the case of synchronized spins. For non-synchronized configurations, the spin frequency of the stars must be considered as additional parameters.

2.5. NON-SYNCHRONIZED BINARIES

Recent theoretical work suggests that the synchronization time in close neutron star binaries remains always longer than the orbital decay time due to gravitational radiation (Kochanek 1992; Bildsten & Cutler 1992). In particular, Bildsten & Cutler (1992) show with simple dimensional arguments that one would need an implausibly small value of the effective viscous time, $t_{\rm visc} \sim R/c$, in order to reach complete synchronization just before final merging. In the opposite limiting regime where viscosity is completely negligible, the fluid circulation in the binary system is conserved during the orbital decay and the stars behave approximately as Darwin-Riemann ellipsoids (Kochanek 1992; LRS3). Of particular importance are the irrotational Darwin-Riemann configurations, obtained when two initially non-spinning (or, in practice, slowly spinning) neutron stars evolve in the absence of significant viscosity. Compared to synchronized systems, these irrotational configurations exhibit smaller deviations from point-mass Keplerian behavior at small r. However, as shown in LRS3 and RS4, irrotational configurations for binary neutron stars with $\Gamma \gtrsim 2$ can nevertheless become dynamically unstable near contact. Thus the final coalescence of two neutron stars in a non-synchronized binary system must still be driven by hydrodynamic instabilities.

The details of the hydrodynamics are very different, however (RS4). Because the two stars appear to be counter-spinning in the corotating frame of the binary, a vortex sheet with $\Delta v = |v_+ - v_-| \approx \Omega r$ appears when the surfaces come into contact. Such a vortex sheet is Kelvin-Helmholtz unstable on all wavelengths and the hydrodynamics is therefore rather difficult to model accurately given the limited spatial resolution of three-dimensional calculations. The breaking of the vortex sheet generates a large turbulent viscosity so that the final configuration may no longer be irrotational. In numerical simulations, however, vorticity is generated mostly through spurious shear viscosity introduced by the spatial discretization. An additional difficulty is that non-synchronized configurations evolving rapidly by gravitational radiation emission tend to develop significant tidal lags, with the long axes of the two components becoming misaligned (LRS5). This is a purely dynamical effect, present even if the viscosity is zero, but its magnitude depends on the entire previous evolution of the system. Thus the construction of initial conditions for hydrodynamic calculations of non-synchronized binary coalescence must incorporate the gravitational radiation reaction *self-consistently*. Instead, previous studies of non-synchronized, equal-mass binary coalescence by Shibata, Nakamura & Oohara (1992), Davies *et al.* (1994), and Zughe, Centrella & McMillan (1994) used very approximate initial conditions consisting of two identical *spheres* (polytropes with $\Gamma \approx 2$) placed on an inspiral trajectory calculated for two point masses.

3. Coalescing White-Dwarf Binaries

3.1. ASTROPHYSICAL MOTIVATION

Coalescing white-dwarf binaries are thought to be likely progenitors for type Ia supernovae (Iben & Tutukov 1984; Webbink 1984; Paczyński 1985; Mochkovitch & Livio 1989; Yungelson et al. 1994). To produce a supernova, the total mass of the system must be above the Chandrasekhar mass. Given evolutionary considerations, this requires two C-O or O-Ne-Mg white dwarfs. Yungelson et al. (1994) show that the expected merger rate for close pairs of white dwarfs with total mass exceeding the Chandrasekhar mass is consistent with the rate of type Ia supernovae deduced from observations. Alternatively, a massive enough merger may collapse to form a rapidly rotating neutron star (Nomoto & Iben 1985; Colgate 1990). Chen & Leonard (1993) have discussed the possibility that most millisecond pulsars in globular clusters may have formed in this way. In some cases planets may form in the disk of material ejected during the coalescence and left in orbit around the central pulsar (Podsiadlowski, Pringle & Rees 1991). Indeed the first extrasolar planets have been discovered in orbit around a millisecond pulsar, PSR B1257+12 (Wolszczan 1994). A merger of two highly magnetized white dwarfs might lead to the formation of a neutron star with extremely high magnetic field, and this scenario has been proposed as a source of gamma-ray bursts (Usov 1992).

Close white-dwarf binaries are expected to be extremely abundant in our Galaxy. Iben & Tutukov (1984, 1986) predict that $\sim 20\%$ of all binary stars produce close pairs of white dwarfs at the end of their stellar evolution. The most common systems should be those containing two low-mass helium white dwarfs. Their final coalescence can produce an object massive enough to start helium burning. Bailyn (1993) suggests that extreme horizontal-branch stars in globular clusters may be such helium-burning stars formed by the coalescence of two white dwarfs. Paczyński (1990) has proposed that the peculiar X-ray pulsar 1E 2259+586 may be the product of a recent white-dwarf merger. Planets in orbit around a massive white dwarf may also form following a merger (Livio, Pringle & Saffer 1992).

Coalescing white-dwarf binaries are also important sources of low-frequency gravitational waves that should be easily detectable by future space-based interferometers. Recent proposals for space-based interferometers include the LAGOS experiment (Stebbins et al. 1989), which should have an extremely high sensitivity (down to an amplitude $h \sim 10^{-23}$–10^{-24}) to sources with frequencies in the range ~ 0.1–$100\,\text{mHz}$. Evans, Iben & Smarr (1987) estimate a white-dwarf merger rate of order one every 5 yr in our own Galaxy. Coalescing systems closest to Earth should produce quasi-periodic gravitational waves of amplitude $h \sim 10^{-21}$ in the frequency range ~ 10–100 mHz. In addition, the total number ($\sim 10^4$) of close white-dwarf binaries in our Galaxy emitting at lower frequencies ~ 0.1–1 mHz (the emission lasting for $\sim 10^2$–10^4 yr before final coalescence) should provide a continuum background signal of amplitude $h_c \sim 10^{-20}$–10^{-21}. Individual sources should be detectable by LAGOS above this background when their frequency becomes $\gtrsim 10$ mHz. The detection of the final burst of gravitational waves emitted during the actual merging would provide a unique opportunity to observe in "real time" the hydrodynamic interaction between the two white dwarfs, possibly followed immediately by a supernova explosion, nuclear outburst, or some other type of electromagnetic signal.

3.2. HYDRODYNAMICS OF COALESCING WHITE-DWARF BINARIES

The results of RS3 for polytropes with $\Gamma = 5/3$ show that hydrodynamics also plays an important role in the coalescence of two white dwarfs, either because of dynamical instabilities of the equilibrium configuration, or following the onset of dynamically unstable mass transfer. Systems with $q \approx 1$ must evolve into deep contact before they become dynamically unstable and merge. Instead, equilibrium configurations for binaries with q sufficiently far from unity never become dynamically unstable, but once these binaries reach their Roche limit, we find that dynamically unstable mass transfer occurs and that the less massive star is completely disrupted after a small number (< 10) of orbital periods (see also Benz et al. 1990). In both cases, the final merged configuration is an axisymmetric, rapidly rotating object with a core-halo structure similar to that obtained for coalescing neutron stars (RS2, RS3; see also Mochkovitch & Livio 1989).

For two massive enough white dwarfs, the merger product may be well above the Chandrasekhar mass M_{Ch}. The object may therefore explode as a (type Ia) supernova, or perhaps collapse to a neutron star. The rapid rotation and possibly high mass (up to $2M_{Ch}$) of the object must be taken into account for determining its final fate. Unfortunately, this is not done in current theoretical calculations of accretion induced collapse (AIC), which always consider a non-rotating white dwarf just below the Chandrasekhar limit accreting matter slowly and quasi-spherically (Canal et al. 1990; Nomoto & Kondo 1991; Isern 1994). Under these assumptions it is found that collapse to a neutron star is possible only for a narrow range of initial conditions. In most cases, a supernova explosion follows the ignition of the nuclear fuel in the degenerate core. However, the fate of a much more massive object with substantial rotational support and large deviations from spherical symmetry (as would be formed by dynamical coalescence) may be very different.

Acknowledgements. Support for this work was provided by NSF Grant AST 91–19475 and NASA Grant NAGW–2364. F.A.R. was supported by a Hubble Fellowship, funded by NASA through Grant HF-1037.01-92A from the Space Telescope Science Institute, which is operated by AURA, Inc., for NASA, under contract NAS5-26555. Computations were performed at the Cornell Theory Center, which receives major funding from the NSF and IBM, with additional support from the New York State Science and Technology Foundation and members of the Corporate Research Institute.

References

Abramovici, A. et al. 1992, Science 256, 325
Bailyn, C.D. 1993, in *Structure and Dynamics of Globular Clusters*, S.G. Djorgovski & G. Meylan (Eds.), ASP Conf. Proc. Vol. 50, p. 191
Baym, G. 1991, in *Neutron Stars: Theory and Observation*, J. Ventura & D. Pines (Eds.), Kluwer Academic Publishers, p. 21
Benz, W. et al. 1990, ApJ 348, 647
Bildsten, L. & Cutler, C. 1992, ApJ 400, 175
Bradaschia, C. et al. 1990, Nucl. Instr. Methods A289, 518
Canal, R. et al. 1990, ApJ 356, L51
Chandrasekhar, S. 1969, *Ellipsoidal Figures of Equilibrium*, Yale University Press; revised Dover edition 1987
Chandrasekhar, S. 1975, ApJ 202, 809
Chen, K. & Leonard, P.J.T. 1993, ApJ 411, L75
Chernoff, D.F. & Finn, L.S. 1993, ApJ 411, L5
Clark, J.P.A. & Eardley, D.M. 1977, ApJ 251, 311
Colgate, S.A. 1990, in *Supernovae*, S.E. Woosley (Ed.), Springer-Verlag, p. 585
Cook, G.B., Shapiro, S.L. & Teukolsky, S.L. 1994, ApJ 424, 823
Cutler, C. & Flanagan, E.E. 1994, Phys. Rev. D49, 2658
Cutler, C. et al. 1993, Phys. Rev. Lett. 70, 2984
Davies, M.B. et al. 1994, ApJ 431, 742
Eichler, D. et al. 1989, Nat 340, 126
Evans, C.R., Iben, I. & Smarr, L. 1987, ApJ 323, 129
Finn, L.S. & Chernoff, D. 1993, Phys. Rev. D47, 2198
Goldstein, H. 1980, *Classical Mechanics*, Addison-Wesley
Iben Jr, I. & Tutukov, A.V. 1984, ApJS 54, 335
Iben Jr, I. & Tutukov, A.V. 1986, ApJ 311, 753
Isern, P. 1994, in *Evolutionary Links in the Zoo of Interacting Binaries*, F. D'Antona (Ed.), Mem. Soc. Astron. Ital. (in press)
Jaranowski, P. & Krolak, A. 1992, ApJ 394, 586
Junker, W. & Schäfer, G. 1992, MNRAS 254, 146
Kidder, L.E., Will, C.M. & Wiseman, A.G. 1992, Class. Quantum Grav. 9, L125
Kochanek, C.S. 1992, ApJ 398, 234
Lai, D., Rasio, F.A. & Shapiro, S.L. 1993a, ApJS 88, 205 [LRS1]
Lai, D., Rasio, F.A. & Shapiro, S.L. 1993b, ApJ 406, L63 [LRS2]
Lai, D., Rasio, F.A. & Shapiro, S.L. 1994a, ApJ 420, 811 [LRS3]
Lai, D., Rasio, F.A. & Shapiro, S.L. 1994b, ApJ 423, 344 [LRS4]

Lai, D., Rasio, F.A. & Shapiro, S.L. 1994c, ApJ (in press) [LRS5]
Lincoln, C.W. & Will, C.M. 1990, Phys. Rev. D42, 1123
Livio, M., Pringle, J.E. & Saffer, R.A. 1992, MNRAS 257, 15P
Marković, D. 1993, Phys. Rev. D48, 4738
Meegan, C.A. et al. 1992, Nat 355, 143
Meers, B.J. 1988, Phys. Rev. D38, 2317
Mochkovitch, R. & Livio, M. 1989, A&A 209, 111
Narayan, R., Paczyński, B. & Piran, T. 1992, ApJ 395, L83
Narayan, R., Piran, T. & Shemi, A. 1991, ApJ 379, L17
Nemiroff, R.J. 1994, Comments on Astrophysics (in press)
Nomoto, K. & Iben Jr, I. 1985, ApJ 297, 531
Nomoto, K. & Kondo, Y. 1991, ApJ 367, L19
Paczyński, B. 1985, in *Cataclysmic Variables and Low-mass X-ray Binaries*, D.Q. Lamb & J. Patterson (Eds.), Reidel, p. 1
Paczyński, B. 1986, ApJ 308, L43
Paczyński, B. 1990, ApJ 365, L9
Phinney, E.S. 1991, ApJ 380, L17
Podsiadlowski, P., Pringle, J.E. & Rees, M.J. 1991, Nat 352, 783
Rasio, F.A. & Shapiro, S.L. 1992, ApJ 401, 226 [RS1]
Rasio, F.A. & Shapiro, S.L. 1994, ApJ 432, 242 [RS2]
Rasio, F.A. & Shapiro, S.L. 1995a, ApJ (in press) [RS3]
Rasio, F.A. & Shapiro, S.L. 1995b, ApJ (in preparation) [RS4]
Schutz, B.F. 1986, Nat 323, 310
Shapiro, S.L. & Teukolsky, S.A. 1983, *Black Holes, White Dwarfs, and Neutron Stars*, Wiley
Shibata, M., Nakamura, T. & Oohara, K. 1992, Prog. Theor. Phys. 88, 1079
Stebbins, R.T. et al. 1989, in *Proc. 5th Marcel Grossman Meeting*, D.G. Blair & M.J. Buckingham (Eds.), Cambridge Univ. Press, p. 179.
Strain, K.A. & Meers, B.J. 1991, Phys. Rev. Lett. 66, 1391
Tassoul, M. 1975, ApJ 202, 803
Tassoul, J.-L. 1978, *Theory of Rotating Stars*, Princeton University Press
Taylor, J.H. & Weisberg, J.M. 1989, ApJ 345, 434
Thorne, K.S. 1987, in *300 Years of Gravitation*, S.W. Hawking & W. Israel (Eds.), Cambridge University Press, p. 330
Thorsett, S.E. et al. 1993, ApJ 405, L29
Tutukov, A.V. & Yungelson, L.R. 1993, MNRAS 260, 675
Usov, V.V. 1992, Nat 357, 472
Webbink, R.F. 1984, ApJ 277, 355
Wiseman, A.G. 1993, Phys. Rev. D48, 4757
Wolszczan, A. 1991, Nat 350, 688
Wolszczan, A. 1994, Science 264, 538
Yungelson, L.R. et al. 1994, ApJ 420, 336
Zughe, X., Centrella, J.M. & McMillan, S.L.W. 1994, Phys. Rev. D (in press)

THE STRUCTURE AND EVOLUTION OF THORNE-ŻYTKOW OBJECTS

PHILIPP PODSIADLOWSKI

Institute of Astronomy
Cambridge, CB3 0HA, United Kingdom

Abstract. Thorne-Żytkow objects (TŻOs) are red supergiants with neutron cores. The energy source in TŻOs with low-mass envelopes ($\lesssim 8\,M_\odot$) is accretion onto the neutron core, while for TŻOs with massive envelopes ($\gtrsim 14\,M_\odot$) it is nuclear burning via the exotic rp process. TŻOs are expected to form as a result of unstable mass transfer in high-mass X-ray binaries, the direct collision of a neutron star with a massive companion after a supernova or the collision of a neutron star with a low-mass star in a globular cluster. We estimate a birth rate of massive TŻOs in the Galaxy of $\sim 2\,10^{-4}\,\mathrm{yr}^{-1}$. Thus, for a characteristic TŻO lifetime of 10^5–10^6 yr there should be 20–200 TŻOs in the Galaxy at present. These can be distinguished from ordinary red supergiants because of anomalously high surface abundances of lithium and rp-process elements, produced in the TŻO interior. The TŻO phase ends when either the star has exhausted its rp-process seed elements or the envelope mass decreases below a critical mass ($\sim 14\,M_\odot$). Then nuclear burning becomes inefficient and a neutrino runaway ensues, leading to the dynamical accretion of matter near the core onto the neutron star and its spin up to spin frequencies of up to ~ 100 Hz. The fate of the massive envelope is not entirely clear. If a significant fraction can be accreted onto the core, the formation of a black hole becomes likely. Part of the envelope may collapse into a massive disk which may ultimately become gravitationally unstable and lead to the formation of planets or even low-mass stars. We discuss the various possible outcomes and suggest a possible link between massive TŻOs and soft X-ray transients.

1. Introduction

The study of stars with neutron cores has a history as long as the history of modern stellar-evolution theory. In the 1930s, Gamow (1937) and Landau (1938) speculated that the Sun might have a neutron core to solve the solar-energy problem. The modern study of stars with neutron cores starts with the work of Thorne and Żytkow (Thorne & Żytkow 1975, 1977). However, Thorne and Żytkow did not solve the problem of the nuclear energy source in massive supergiants with neutron cores, now referred to as Thorne-Żytkow objects (TŻO). This was done independently by Biehle (1991, 1994) and Cannon (1993), who showed that nuclear burning in massive TŻOs occurs via the exotic rapid proton process (rp-process).

In this review we first summarize the aspects of TŻOs that are reasonably well understood, such as their internal structure and the chemical signatures by which they may be detected observationally (Section 2). We then discuss the more uncertain aspects of their evolution, in particular their formation (Section 3) and their final fate (Section 4). In Section 5 we speculate that soft X-ray transients may be descendants of TŻOs.

2. The Structure of Thorne-Żytkow Objects

The outer appearance of a TŻO is that of a more-or-less normal red supergiant (with temperature $T_{\text{eff}} \sim 3200\,\text{K}$ and radius $R \sim 1400\,\text{R}_\odot$ for a TŻO of mass $M_{\text{TZO}} = 15\,\text{M}_\odot$). TŻOs obey a simple mass-luminosity relation (Cannon 1993)

$$L \simeq (1.6\,10^5\,\text{L}_\odot) \left(\frac{\alpha}{1.5}\right)^2 \left(\frac{M_{\text{TZO}}}{15\,\text{M}_\odot}\right)^{2/3},$$

where α is the mixing-length parameter (i.e., the ratio of the mixing length to the pressure scale height). The strong dependence of the luminosity on the mixing-length parameter already indicates that the internal structure of a TŻO must be fundamentally different from that of an ordinary red supergiant (whose luminosity is independent of α).

In general, two types of TŻOs can be distinguished, based on their central energy source. In low-mass models (with envelope masses $\lesssim 8\,\text{M}_\odot$), the main energy source is gravitational energy, released by the Eddington-limited accretion of matter onto the central neutron core. In massive models (with envelope masses $\gtrsim 14\,\text{M}_\odot$), gravitational energy release is relatively unimportant because of the generation of e^+-e^- pairs near the core which reduces the Eddington accretion rate by a factor of 10–100. Thus, the energy source has to be nuclear energy. However, the region near the core which is hot enough for nuclear reactions to take place contains very little mass ($\sim 10^{-11}\,M_\odot$), only sufficient to produce the required surface luminosity ($\sim 10^5\,L_\odot$) for a few minutes. Therefore, fresh fuel has to be continually injected from the envelope, which in effect serves as a fuel reservoir. This implies that the burning region has to be linked with the envelope by convection, i.e., the burning zone has to be at the base of the convective supergiant envelope. This is the reason why the TŻO luminosity depends on the mixing length. A further complication is that the convective turnover time in the burning region is only ~ 0.01 s, much shorter than the β^+-decay times of many weak interactions in the CNO-Ne cycle, which have lifetimes of up to ~ 100 s. As a result, the CNO-Ne reaction chain gets hung up (by the time the β^+-decays have occurred, the matter has moved out of the hot burning region) and cannot provide the necessary energy to support a TŻO envelope.

The problem of the missing energy source has recently been solved by Biehle (1991) and, in most detail, by Cannon (1993). They showed that the energy is provided by the rapid proton process (rp-process). The rp-process (strictly speaking the irp-process, which stands for interrupted rp-process; see Cannon 1993) consists of sequences of proton-capture reactions, which are terminated when the time scale for the next proton capture exceeds the β^+-decay time; a typical reaction chain is:

$$^{23}\mathrm{Na}\,(p,\gamma)\,^{24}\mathrm{Mg}\,(p,\gamma)\,^{25}\mathrm{Al}\,(p,\gamma)\,^{26}\mathrm{Si}\,(\beta^+\nu)\,^{26}\mathrm{Al}.$$

Note that, in many respects, the rp-process is analogous to the rapid-neutron process or r-process for neutron captures.

The fact that TŻOs generate their luminosities by the exotic rp-process has two important implications. One, it is possible that TŻOs provide the main site for the generation of proton-rich elements, for which no site has been unambiguously identified in the past (Cannon 1993). Two, the rp-process provides an opportunity for distinguishing TŻOs from ordinary red supergiants. Biehle (1994) estimates that the surface abundance of many rp-process elements should be enhanced by several orders of magnitude (relative to solar) in TŻOs (e.g. Mo, Br, Rb, Y, Nb) and suggests several spectral lines of these elements which should be detectable with modern spectrographs.

In addition to the rp-process, TŻOs provide an ideal environment for the production of ^{7}Li by the ^{7}Be transport mechanism (Cameron 1955). In ordinary hydrogen burning stars, ^{7}Li is continually produced as part of the PPII reaction chain (see, e.g., Clayton 1968) by the reactions

$$^3\mathrm{He} + {}^4\mathrm{He} \longrightarrow {}^7\mathrm{Be} + \gamma \qquad (1)$$
$$^7\mathrm{Be} + e^- \longrightarrow {}^7\mathrm{Li} + \nu, \qquad (2)$$

but is also continually being destroyed by the reaction

$$^7\text{Li} + \text{p} \longrightarrow 2\,^4\text{He}. \tag{3}$$

The equilibrium abundance in the centers of stars is determined by the balancing of the production and the destruction rates and is generally very low. However, in lithium stars (e.g., Sackmann & Boothroyd 1992) and TŻOs a blob of material spends very little time in the hot burning region and, before ^{7}Be can capture an electron, the material has moved out of the region hot enough for the ^{7}Li-destruction reaction (Eq. 3) to occur. This is possible since reaction (2) does not involve a Coulomb barrier and therefore also occurs at much lower temperatures than reaction (3) which involves a Coulomb barrier. ^{7}Li will only be destroyed when the blob of material passes again through the burning region. Because the ^{7}Li-destruction process is so inefficient, lithium stars and TŻOs are able to build up a large ^{7}Li abundance.

To illustrate this, we have performed a nucleosynthesis calculation with a full pp-reaction network for a typical $16\,M_\odot$ TŻO model (Podsiadlowski, Cannon & Rees 1994; the numerical procedure and the TŻO models are described in detail in Cannon 1993). We assume a solar-type initial abundance for ^{7}Li and ^{3}He. Note that this means that we make the plausible assumption that the ^{3}He abundance has not been significantly reduced in the progenitor system. In Fig. 1 we show the time evolution of the ^{7}Li and ^{3}He surface abundances. Very rapidly, the ^{7}Li abundance reaches the value seen in V404 Cyg and V616 Mon (see Section 5), which is of order the primordial ^{7}Li abundance (e.g. Boesgaard & Steigman 1985). After $\sim 10^5$ yr, the ^{7}Li abundance reaches a maximum some three orders of magnitude larger than the primordial value and subsequently decreases, because ^{3}He is destroyed in the TŻO envelope on this time scale. But even after 10^6 yr, which is the maximum expected TŻO lifetime, the ^{7}Li is still larger then the primordial value. This demonstrate that TŻOs should have anomalously high ^{7}Li abundances and this can be used as a secondary indicator for identifying TŻOs, although detection of a lithium anomaly does not provide conclusive evidence for a TŻO.

3. The Formation of Thorne-Żytkow Objects

While no TŻO has yet been identified, they have been predicted to form by a variety of mechanisms: (1) the direct disruptive collision of a low-mass main-sequence star with a neutron star in a globular cluster (e.g., Ray, Kembhavi & Antia 1987); (2) the complete coalescence of a neutron star with a massive companion following a high-mass X-ray binary phase (Taam, Bodenheimer & Ostriker 1978); and (3) the disruption of a companion star by a newly formed neutron star which received a supernova kick in the

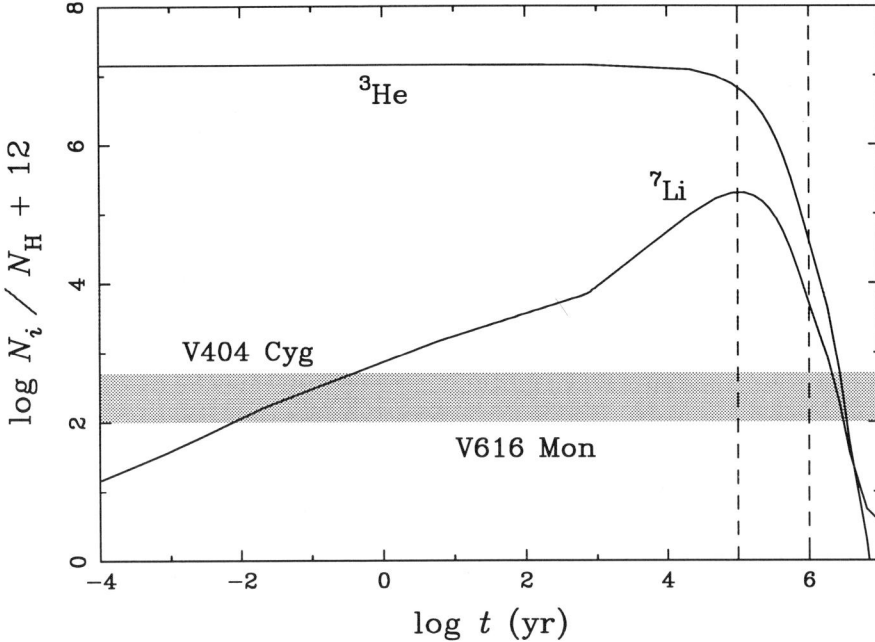

Figure 1. The time evolution of the ^{7}Li and ^{3}He surface abundances (by number) in a typical Thorne-Żytkow object (TŻO) since its formation. The shaded area shows the range of ^{7}Li abundances measured in V404 Cyg and V616 Mon (Charles *et al.* 1994). The dashed lines indicate the range of expected TŻO lifetimes.

direction of the companion (Leonard, Hills & Dewey 1994). In this review we will concentrate on the evolution and final fate of massive TŻOs, i.e. those formed via the second and third route.

The "classical" formation scenario for TŻOs is that they are the descendants of high-mass X-ray binaries (HMXBs) (for a detailed review of the evolution of HMXBs, see Bhattacharya & Van den Heuvel 1991). When in these systems the massive component completely fills its Roche lobe, the mass transfer rate ($\dot{M} \sim 10^{-3}$–$10^{-5}\,M_\odot\,\mathrm{yr}^{-1}$) will exceed the Eddington accretion rate of the neutron star ($\sim 10^{-8}\,M_\odot\,\mathrm{yr}^{-1}$) by several orders of magnitude. As a result, the neutron star will not be able to accrete most of the transferred mass, and the excess mass will form an extended envelope around the neutron star, ultimately (over)filling the neutron star's Roche lobe. The system will now have entered into a common-envelope phase, where the neutron star is completely engulfed within the massive star's envelope. Due to gas drag, the neutron star will then spiral towards the center of the system. If the orbital energy released in the process is sufficient to eject the common envelope, the end product will be a close binary, consisting of a neutron star and a helium star (i.e. the core of the massive star). If the energy is not sufficient to eject the envelope, the neutron star will settle at the center and the system will become a TŻO. Taam et al. (1978) estimated that TŻOs form from HMXBs with initial periods less than ~ 100 d. The birth rate of TŻOs, $\nu_{\rm TZO}$, can be estimated from the observed number of HMXBs with periods less than 100 d and their expected lifetimes before they fill their Roche lobes. Assuming that there are at least 10 HMXBs with the correct properties (e.g., Biehle 1994) in the Galaxy and that the HMXB phase lasts less than 10^5 yr, one obtains a *conservative* birth rate

$$\nu_{\rm TZO} \gtrsim 10^{-4} \, {\rm yr}^{-1}.$$

The third route for forming TŻOs, immediately after the supernova as a result of a kick which leads to the immediate merger of the binary (Leonard et al. 1994), has become significantly more important with the recent realization that supernova birth velocities of pulsars have been systematically underestimated in the past by a factor of about 3 (Lyne & Lorimer 1994). Lyne & Lorimer (1994) find an average pulsar birth velocity of $450 \, {\rm km \, s}^{-1}$ with a standard deviation of $290 \, {\rm km \, s}^{-1}$. A consequence of these high kick velocities is that only $\sim 1/4$ of potential HMXB progenitors survive the supernova as bound systems and that $\sim 1/4$ of those receive kicks in which the neutron star spirals into the companion star immediately after the supernova (Brandt & Podsiadlowski 1994). Assuming that the neutron star birth rate in the Galaxy is 1 every 50 yr, that $\sim 25\%$ of neutron stars are born in close binaries (Podsiadlowski, Joss & Hsu 1992) and that $\sim 1/2$ of the resulting systems are massive enough for rp-processing (see Cannon 1993), we estimate a TŻO birth rate for this channel of

$$\nu_{\rm TZO} \sim 10^{-4} \, {\rm yr}^{-1}.$$

Thus, both of these formation channels should be of comparable importance. This should be compared to the birth rate of low-mass X-ray binaries (LMXBs), $\nu_{\rm LMXB}$. Assuming that there are ~ 100 LMXBs in the Galaxy and that the lifetime of the LMXB phase is 10^7–10^8 yr, one obtains an estimate for the LMXB birth rate,

$$\nu_{\rm LMXB} \sim 10^{-6} - 10^{-5} \, {\rm yr}^{-1}.$$

Thus, the expected birth rate of TŻOs *exceeds* the birth rate of LMXBs. In other words, while TŻOs may be exotic objects, they may not be as exotic as LMXBs!

The characteristic lifetime of a TŻO is in the range of 10^5–10^6 yr (see Cannon 1993; Biehle 1994, and Section 4.1). Combining this lifetime with the above birth rates predicts that there should be between 20 and 200

TŻOs at any given time in the Galaxy. Since there are only a few thousand normal red supergiants of comparable luminosities (as estimated from Humphreys 1984), the fraction of TŻOs among the red-supergiant population should be between a few per cent up to about 10 per cent.

Chevalier (1994) has recently argued that TŻOs may not be able to form since, during the spiral-in phase, accretion onto the neutron star may occur in the neutrino-dominated rather than the Eddington-limited regime and may lead to the collapse of the neutron star into a black hole. However, his argument assumes spherically symmetric accretion at a rate determined by the orbital motion of the neutron star. This is unlikely to be a realistic assumption (Taam 1994). It appears more likely that the spiraling-in neutron star is surrounded by a quasi-hydrostatic structure (in effect, the inner envelope of a TŻO), which acts like a "hard" obstacle and deflects the large-scale flow. In addition, observationally a number of neutron stars are known which are very likely to have experienced and survived a spiral-in phase (see, e.g., Bhattacharya & Van den Heuvel 1991). Nevertheless, we note as a *caveat* that no fully realistic calculation of the formation of a TŻO has yet been performed and that it has not been demonstrated how a forming TŻO bridges the mass gap in which there are no known hydrostatic solutions (Thorne & Żytkow 1977; Cannon 1993).

4. The Fate of Thorne-Żytkow Objects

4.1. THE NEUTRINO RUNAWAY

The steady-burning phase of a massive TŻO is terminated by either of two events. One is that the supply of rp-process seed elements is exhausted and that the rp-process becomes inefficient (after $\sim 10^6$ yr). The second is that, due to the expected strong stellar wind, the envelope mass decreases below the minimum mass for nuclear burning. The minimum envelope mass is $\simeq 14\,M_\odot$, but depends strongly on the efficiency of convection (Cannon 1993). For typical red-supergiant wind mass loss rates, $\dot{M} \sim 10^{-5} M_\odot\,\mathrm{yr}^{-1}$ (Kudritzki & Reimers 1978), this will also happen after $\sim 10^6$ yr. At this point, a radiative zone develops somewhere between the burning region and the outer envelope, and the supply of fresh fuel to the burning region is choked off. (Note that there is a mass gap between low-mass models in which gravity provides the energy source and high-mass models in which the rp-process does; there are no steady-state solutions in between.) In either of these two cases, the star effectively runs out of fuel. As a result, the region just above the core will heat up until neutrino losses become the dominant energy-loss mechanism (at $\log T \gtrsim 9.4$), and accretion onto the neutron core will be no longer Eddington-limited. At this point, a neutrino runaway becomes unavoidable (see also Bisnovatyi-Kogan & Lamzin

1984). We have been able to follow the initial phase of this runaway with the code described by Cannon et al. (1992) and Cannon (1993). However, very quickly the evolution becomes dynamical and our calculation becomes invalid. We typically stop the calculations when the central accretion rate has reached $\sim 10^{-6}\,M_\odot\,\text{yr}^{-1}$, i.e., about two orders of magnitude larger than the "standard" Eddington accretion rate.

In the neutrino runaway, all of the material that has less specific angular momentum than the maximum specific angular momentum allowed for a neutron star will fall onto the neutron star on a dynamical time scale. We can estimate the total amount of accreted mass by assuming that the total angular momentum of the envelope is of order the initial orbital angular momentum of the progenitor HMXB and that the TŻO envelope is in solid-body rotation. The latter is likely to be a good approximation for our purposes, since convection is very efficient in redistributing angular momentum and will prevent any strong differential rotation in the envelope. For an initial HMXB with an orbital period P, consisting of a $1.4\,M_\odot$ neutron star and a $15\,M_\odot$ companion, the angular velocity, ω, of the resulting TŻO (with a radius $R \simeq 1400\,R_\odot$ and envelope moment of inertia $I_\text{env} \simeq 8\ 10^{61}\,\text{g cm}^2$) is

$$\omega \sim (3\ 10^{-9}\,\text{s}^{-1})\,P_{10\,\text{d}}^{1/3},$$

where $P_{10\,\text{d}} \equiv P/10\,\text{d}$. This is much less than the maximum possible (breakup) angular velocity at the surface of a typical TŻO, $\omega_\text{max} \sim 5\ 10^{-8}\,\text{s}^{-1}$. Thus, TŻOs are expected to be relatively slow rotators. This also implies that most of the material that is accreted onto the neutron star during the TŻO phase ($\lesssim 10^{-4}\,M_\odot$ for a lifetime of 10^6 yr and a reduced Eddington accretion rate of $10^{-10}\,M_\odot$) has very little specific angular momentum and that the neutron star will be spun down rather than spun up in this phase. The total mass, ΔM_acc, that can be accreted directly during the neutrino runaway is the mass within a cylinder of radius R_max given by the conservation of specific angular momentum

$$R_\text{max}^2\,\omega \simeq \sqrt{GM_\text{NS}R_\text{NS}},$$

where the right-hand side gives the maximum (Newtonian) specific angular momentum of a neutron star of mass M_NS and radius R_NS. For $M_\text{NS} = 1.4\,M_\odot$ and $R_\text{NS} = 10^6$ cm, we obtain from our TŻO models

$$\Delta M_\text{acc} \sim (3\ 10^{-2}\,M_\odot)\,P_{10\,\text{d}}^{-1/3}.$$

Similarly, we can estimate the total angular momentum accreted,

$$\Delta J_\text{acc} \sim (6\ 10^{47}\,\text{g cm}^2\,\text{s}^{-1})\,P_{10\,\text{d}}^{-1/3}.$$

For a typical neutron star model (with a moment of inertia $I_{\rm NS} \simeq 10^{45}\,{\rm g\,cm^2}$), a slowly rotating neutron star can thereby be spun up to a spin period

$$P_{\rm NS} \sim (10\,{\rm ms})\,P_{10\,{\rm d}}^{1/3}.$$

This period is typical of many "recycled" pulsars, but not as short as the period of the shortest millisecond pulsars.

4.2. THE ENVELOPE COLLAPSE

The fate of the large TŻO envelope is less obvious and is mainly governed by the competition of its cooling time scale and the various (uncertain) viscous time scales (Krolik 1984). After the initial neutrino runaway, the inner part of the envelope will be centrifugally supported. If the viscous time scale of the bulk of the envelope is shorter than the cooling time, it can be supported by viscous energy transport from the gravitational energy released in the inner contracting part. The initial cooling time will be of order the Kelvin-Helmholtz time of the TŻO ($t_{\rm KH} \sim 10\text{-}100\,{\rm yr}$). If convection provides the dominant angular-momentum transport mechanism, the viscous time scale, $t_{\rm visc}$, will be of order the convective turnover time times the square of the number of pressure scale heights in the envelope. For the bulk of the envelope (which contains most of the mass), this time scale is only a few times longer than the dynamical time scale ($t_{\rm visc} \sim 1\text{-}10\,{\rm yr}$). This is not much shorter than the Kelvin-Helmholtz time. Thus, initially the envelope can probably be supported by viscous energy dissipation. However, the mass in the inner centrifugally supported disk increases and ultimately most of the envelope is likely to form a disk-like structure with a characteristic radius $r_{\rm centr}$, (where the Keplerian specific angular momentum equals the initial specific orbital angular momentum in the outer envelope), of order, but somewhat smaller, than the initial binary separation,

$$r_{\rm centr} \sim (3\,10^{11}\,{\rm cm})\,P_{10\,{\rm d}}^{2/3}.$$

The further evolution of this massive, disk-like structure depends on when gravitational instabilities start to develop and whether these lead to bar instabilities or disk fragmentation. This also depends on the amount of material accreted by the central object, which in turn is governed by the viscous time scale of the inner envelope. As a result, a variety of different outcomes are possible.

The central object may be become either a slightly spun up neutron star or, perhaps more likely, a stellar-mass black hole. If the massive disk fragments into self-gravitating objects, these will ultimately form one or more low-mass objects. These may be planet-mass objects ("giant planets") or, in view of the large available mass in the disk, more probably, low-mass stars. Whichever the outcome, the result is bound to be of some interest.

(1) The final system may be a spun up pulsar, possibly surrounded by one or more planets. This possibility requires that the system can expel some $10\,M_\odot$, for example, by a pulsar ejection mechanism (see, e.g., Ostriker & Gunn 1971). We note that, since most of the envelope mass is at a large distance from the neutron star ($\sim 10^{14}$ cm) and hence has very low binding energy, very little mass has to be accreted onto the neutron star to generate enough energy to, in principle, eject the whole envelope. A distinctive feature of the resulting single, recycled pulsars would be that they would have relatively low space velocities, since HMXBs, unlike LMXBs, only receive a system kick velocity of $50 \pm 20\,\mathrm{km\,s^{-1}}$ as a result of the supernova in which the neutron stars formed (Brandt & Podsiadlowski 1994). Indeed, a significant fraction of such systems could remain bound in young globular clusters and possibly contribute to their neutron star (millisecond pulsar?) populations.

(2) If the pulsar is surrounded by low-mass stars (with a characteristic separation $\gtrsim r_\mathrm{centr}$), the system would be a potential progenitor for a low-mass X-ray binary. While only a fraction of all TŻOs would be sufficient to produce all LMXBs in the Galaxy (see Section 3), the spatial distribution of LMXBs in the Galaxy suggests that their progenitors are population I systems that received a substantial supernova kick (e.g., Bailes 1989; Naylor & Podsiadlowski 1993; Brandt & Podsiadlowski 1994). Since this is inconsistent with the distribution of HMXBs and hence TŻOs, this possibility is not favoured.

(3) The most likely outcome may be a stellar-mass black hole. The central neutron star can be converted into a black hole provided that the viscosity in the envelope can transfer angular momentum fast enough to maintain accretion in the (high $\dot{M}$) neutrino-dominated regime rather than the Eddington-limited regime after the initial neutrino runaway. Once a black hole has formed, subsequent accretion from the envelope becomes easier since photons are trapped into a flow and are accreted by the black hole (Begelman & Meier 1982). If the black hole is single or surrounded by planets, this would perhaps be the least interesting possible outcome, since it would be extremely difficult to discover such systems.

(4) If the black hole is surrounded by one or more stellar-mass objects, such systems would be excellent candidates for the progenitors of soft X-ray transients like V404 Cygni and V616 Mon (A 0620–00).

5. Soft X-Ray Transients

V404 Cyg and V616 Mon (A 0620–00) are members of a class of X-ray binaries, known as soft X-ray transients, which are widely believed to consist of a stellar-mass black hole and a low-mass companion. In V404 Cyg, the minimum mass of the compact object is $6.3\,M_\odot$ (Casares, Charles & Naylor 1992). In addition, Tanaka (e.g., Tanaka 1992) has argued from an observational point of view that soft X-ray transients are relatively common. This is rather puzzling since the formation of such systems in a scenario analogous to the formation of cataclysmic variables (see, e.g., Romani 1992) requires a number of assumptions, some of which may be difficult to fulfill. Most importantly such a scenario requires close binaries with an extreme initial mass ratio. This is not favoured by observations which suggest that the components of massive close binaries are typically of comparable mass (e.g., Garmany, Conti & Massey 1980). In addition, a model in which the progenitor of a soft X-ray transient experienced a common-envelope phase requires that the orbital energy released by the spiral-in of a low-mass star is sufficient to eject a very massive common envelope. This may be possible for binaries with very large initial separations, for which the common envelope would be only loosely bound, but even then only if the common-envelope ejection process is very efficient. While this may be the case, it is clear that the formation of soft X-ray transients by this route would be a very rare event, with a birth rate much lower than the already low birth rate of LMXBs (also see Romani 1992).

On the other hand, the formation of a black-hole binary with a low-mass companion may be a rather natural outcome of the evolution of a TŻO. Although a TŻO scenario involves more uncertainties than a more *conservative* common-envelope scenario, the birth rate could be quite high, even substantially higher than the birth rate of LMXBs (see Section 3).

A TŻO scenario allows several possible tests. First, one might expect that, in a significant fraction of systems, more than one companion is formed (i.e. a hierarchical multiple system). Second, such systems should have relatively low space velocities and should therefore have a very small Galactic disk scale height, as is indeed observed for soft X-ray transients (White 1994). Third and most importantly, the companion should show signatures of rp-processing and an enhanced lithium abundance (see Section 2). The latter has been seen in V404 Cyg and V616 Mon (Martin *et al.* 1992; Charles *et al.* 1994). The observed abundances are consistent with our calculations, although they seem to be somewhat too low (see Section 2 and Fig. 1), in particular if the TŻO lifetime is only a few 10^5 yr. A likely explanation for this discrepancy is that the ^{7}Li abundance has been reduced during the "main-sequence" phase of the newly formed companion stars. For G and

K dwarfs, main-sequence ^{7}Li burning can reduce the ^{7}Li abundance by up to three orders of magnitude (for a review of lithium burning in stars, see Michaud & Charbonneau 1991).

V822 Cen (Cen X-4) is another X-ray binary in which the stellar companion has an anomalously high ^{7}Li abundance (Charles et al. 1994). Unlike the other two systems, the compact component is known to be neutron star since it exhibits X-ray bursts. While this may argue somewhat against a TŻO scenario as a generic explanation of lithium anomalies in these systems, we emphasize that it does not rule it out, since, as discussed in Section 4, the TŻO collapse may lead to the formation of black-hole as well as neutron star binaries. Whether the neutron star is transformed into a black hole might, for example, depend on the total angular momentum in the TŻO envelope, which in turn depends on the initial orbital period of the TŻO progenitor and/or the TŻO formation scenario (Section 3).

Fortunately, detection or non-detection of the predicted rp-process anomalies should conclusively verify or refute our suggested TŻO connection for the formation of soft X-ray transients. If it were indeed confirmed, it would provide an indirect proof of the existence of TŻO.

References

Bailes, M. 1989, ApJ 342, 917
Begelman, M.L. & Meier, D.L. 1982, ApJ 253, 873
Bhattacharya, D. & Van den Heuvel, E.P.J. 1991, Phys. Rep. 203, 1
Biehle, G.T. 1991, ApJ 380, 167
Biehle, G.T. 1994, ApJ 420, 364
Bisnovatyi-Kogan, G.S. & Lamzin, S.A. 1984, SvA 28, 187
Boesgaard, A.M. & Steigman, G. 1985, ARA&A 23, 319
Brandt, W.N. & Podsiadlowski, Ph. 1994, MNRAS (submitted)
Cameron, A.G.W. 1955, ApJ 121, 144
Cannon, R.C. 1993, MNRAS 263, 817
Cannon, R.C. et al. 1992, ApJ 386, 206
Casares, J., Charles, P.A. & Naylor, T. 1992, Nat 355, 614
Charles, P.A. et al. 1994, in *The Evolution of X-Ray Binaries*, S.S. Holt & C.S. Day (Eds.), AIP Press (New York), p. 371
Chevalier R.A. 1994, ApJ 411, L33
Clayton, D.D., 1968, *Principles of Stellar Evolution and Nucleosynthesis*, University of Chicago Press (Chicago)
Garmany, C.D., Conti, P.S. & Massey, P. 1980, ApJ 242, 1063
Gamow, G. 1937, *Structure of Atomic Nuclei and Nuclear Transformations*, Oxford University Press (Oxford)
Humphreys, R.M. 1984, in *Observational Tests of the Stellar Evolution Theory*, A. Maeder & A. Renzini (Eds.), Reidel (Dordrecht), p. 279
Krolik, J.H. 1984, ApJ 282, 452
Kudritzki, R.P. & Reimers, D. 1978, A&A 70, 277
Landau, L.D. 1938, Nat 141, 333
Leonard, P.J.T., Hills, J.G. & Dewey, R.J. 1994, ApJ 423, L19
Lyne, A.G. & Lorimer, D.R. 1994, Nat 369, 127
Martin, E.L. et al. 1992, Nat 358, 129

Michaud, G. & Charbonneau, P. 1991, Sp. Sci. Rev. 57, 1
Naylor, T. & Podsiadlowski, Ph. 1993, MNRAS 262, 929
Ostriker, J.P. & Gunn, J.E. 1971, ApJ 164, L95
Podsiadlowski, Ph., Cannon, R.C. & Rees, M.J. 1994, MNRAS (submitted)
Podsiadlowski, Ph., Joss, P.C. & Hsu, J.J.L. 1992, ApJ 391, 246
Ray, A., Kembhavi, A.K. & Antia, H.M. 1987, A&A 184, 164
Romani, R.W. 1992, ApJ 399, 621
Sackmann, I.-J. & Boothroyd, A.I. 1992, ApJ 392, L71
Taam, R. E. 1994, private communication
Taam, R.E., Bodenheimer, P. & Ostriker, J.P. 1978, ApJ 222, 269
Tanaka, Y. 1992, in *Evolutionary Processes in Binary Stars*, Y. Kondo, R.F. Sisteró & R.S. Polidan (Eds.), Kluwer (Dordrecht), p. 215
Thorne, K.S. & Żytkow, A.N. 1975, ApJ 199, L19
Thorne, K.S. & Żytkow, A.N. 1977, ApJ 212, 832
White, N.E. 1994, in *The Evolution of X-Ray Binaries*, S.S. Holt & C.S. Day (Eds.), AIP Press (New York), p. 53

MECHANISMS FOR DWARF NOVA OUTBURSTS AND SOFT X-RAY TRANSIENTS

A critical review

J.P. LASOTA
UPR 176 du CNRS; DARC
Observatoire de Paris, Section de Meudon
92190 Meudon Cedex, France

Abstract. I review models that try to explain dwarf nova outbursts and soft X-ray transients. The disc instability model for dwarf novae is still in a preliminary state of development: its predictions depend very strongly on the unknown viscosity mechanism. It is also doubtful that a *pure* disc instability phenomenon will be able to describe *all* types of dwarf nova outbursts, in particular superoutbursts. The disc instability model for SXTs suffers from the same difficulties but in addition its predictions are contradicted by observations of transient sources in quiescence. The mass transfer model, incorporating illumination of the secondary, cannot describe correctly the time scales of SXT events for main-sequence secondaries with masses less than 1 $M_\odot$. The existence of at least three systems with $P_{\rm orb} < 10$ hr seems to rule it out as an explanation of the SXT phenomenon.

1. Introduction

Dwarf novae and soft X-ray transients (SXTs) are subclasses of cataclysmic variables (CVs) and low-mass X-ray binaries (LMXBs), respectively, which undergo, more or less regularly, high-amplitude outbursts. CVs and LMXBs are close binary systems in which a Roche lobe filling low-mass (secondary) star transfers matter to a compact (primary) object: a white dwarf in CVs, a neutron star or a black hole in LMXBs. The primary accretes matter through an accretion disc, except if it is a strongly magnetized white dwarf. There is no doubt that outbursts are accretion events and that their site is the accretion disc around the compact objects. The uncertainty concerns the origin and the cause of the outbursts. Are they due to an increased mass transfer rate from the secondary or to some instability in the disc itself?

In the case of dwarf nova outbursts the second hypothesis – the disc instability model – gained popularity after is was realised that *locally* accretion disc equilibria become thermally and viscously unstable due to strong opacity variations caused by partial ionization of hydrogen (see Cannizzo 1993a, for a short history of the disc instability model). In the mass transfer model the cause of an increased mass transfer rate was rather obscure. One should realise, however, that the disc instability is local. To obtain a global "limit cycle" behaviour corresponding to observations one has to "tune" the viscosity prescription and/or the vertical disc structure as shown first by Smak (1984). The α-disc approximation (Shakura & Sunyaev 1973) fails to give a consistent physical mechanism for dwarf novae outbursts. At present, as I discuss below, it provides mostly phenomenological descriptions of various processes. It is clear that further progress in this domain will require a substantial improvement of the description of viscosity in accretion discs. This is not a very optimistic perspective.

In Section 2 I summarize the main elements of the disc instability model and some recent work on the subject. Section 3 will be devoted to the origin of so-called superoutbursts in SU UMa type systems. Particular attention will be given to WZ Sge, a system which seems to be a close cousin of SXTs.

The disc instability model possesses the undeniable merit of offering a description of the main properties of normal (U Gem type) dwarf novae outbursts. The same model extended to SXTs does not function very well since it is unable, at least in its present version, to reproduce basic observed properties such as recurrence times and accretion rates in quiescence.

In the case of SXTs, illumination of the secondary by X rays emitted in the vicinity of the compact object provides a mechanism for an increased mass transfer. Unfortunately, time scales obtained in this type of model cannot apply to SXTs in which the companion star is on the main sequence. The exact cause of SXT phenomena is therefore still uncertain. This problem as well as some recent attempts to combine mass-transfer and disc-instability mechanisms will be reviewed in Section 4.

2. The Disc Instability Model

2.1. WORKING PRINCIPLES

The local (i.e., at a given radial distance) equilibria of accretion discs are conveniently represented as curves on the accretion rate $\dot{M}$ (or effective temperature T_{eff}) versus surface density Σ plane. For $2500 \lesssim T_{\text{eff}} \lesssim 25000\,\text{K}$ the equilibrium curves have a characteristic "S" shape. The upper and lower branches of the S correspond to stable, respectively "hot" and "cold" equilibria. The middle branch with a negative slope corresponds to disc equilibria which are thermally and viscously unstable. If the accretion rate at a certain radius is fixed in the unstable regime the system will undergo "limit cycle" oscillations between the lower and the upper branches. The S-shaped curve implies only a *local* instability. To obtain a *global* "limit cycle" the whole disc (in the simplest case) has to behave in a certain way. From the local $T_{\rm eff} - \Sigma$ relation it follows that the surface density in the "cold" (quiescent) state has to be lower than a certain $\Sigma_{\max}$, which has the general form

$$\Sigma_{\max} = \Sigma_0 r^{b_r} M^{-b_r/3} \alpha^{b_\alpha} \qquad (1)$$

where M is the mass of the central object, r the distance from its center and α the Shakura-Sunyaev viscosity parameter, and $b_r \sim 1$ so that $\Sigma_{\max}$ is an increasing function of r whereas in an equilibrium solution Σ is a decreasing function of r in the same regime of physical parameters. An accretion disc configuration which is globally on the lower branch will be in a non-equilibrium state when the mass-transfer corresponds to a unstable equilibrium.

For a constant mass transfer rate the type of outburst will depend on two characteristic times: the time $t_{\rm accum}$ it takes the matter accumulating at the outer disc to build up a surface density larger than the critical one, compared to the viscous time $t_{\rm vis}$ it takes the matter to diffuse inward and cross the $\Sigma_{\max}$ barrier somewhere nearer to the inner disc boundary. If $t_{\rm vis} > t_{\rm accum}$ one obtains a so-called "outside-in" outburst beginning at the outer disc edge; in the opposite case the outburst is of the "inside-out" type (Smak 1984).

When, locally, the disc becomes thermally unstable its system point on the $\Sigma - T_{\rm eff}$ plane jumps to the upper hot branch. This increases the temperature and viscosity and an abrupt temperature (and density) gradient forms in a local thermal time. This gradient propagates towards the cold disc regions, igniting them on its passage. This gives rise to a global outburst. (In practice, one has to increase α by hand or to use a suitable viscosity prescription.) The density profile in outburst is a decreasing function of r, so that outer disc regions are always closer to the $\Sigma_{\max}$ barrier. Therefore the cooling wave that brings the disc back to the lower state (below the $\Sigma_{\max}$ line) always proceeds from outside (see e.g., Cannizzo 1993a). It is interesting to notice that, according to most model calculations, only a small fraction (10–20%) of the matter stored in quiescence actually accretes on to the central body. The main reason for this rather surprising result is the speed at which the cooling front brings the disc back to the cold state, before a significant amount of matter has had time to accrete (Cannizzo 1993a).

2.2. THE DISC INSTABILITY MODEL OF DWARF NOVA OUTBURSTS

As has been shown already in early papers (Cannizzo, Wheeler & Ghosh 1985; Papaloizou, Faulkner & Lin 1983; Meyer & Meyer-Hofmeister 1984; Mineshige & Osaki 1983; Smak 1984), time-dependent calculations in the framework of the disc instability model are able to reproduce some properties of dwarf nova outbursts. The excellent review of the subject by Cannizzo (1993a) ends with a list of eight arguments in favour of the "limit cycle model". Here I would like to stress the weak points of the model. In the next paragraphs I will discuss the claim that *all* categories of dwarf nova outbursts are due to "pure disc phenomena" in the sense that all outbursts can be modelled assuming a constant mass transfer rate.

Ludwig, Meyer-Hofmeister & Ritter (1994) have made a systematic study of how the observed features of dwarf nova outbursts in various systems can be described by the disc instability model. They used a two-alpha viscosity prescription in which viscosity in quiescence and in outbursts is parameterised by two constants $\alpha_{\rm cold}$ and $\alpha_{\rm hot}$, respectively. They concluded that no single viscosity prescription using an $(\alpha_{\rm cold}, \alpha_{\rm hot})$ pair can account for all observed properties of dwarf nova outbursts. In particular, it is difficult to get "outside-in" outbursts, whereas in such systems as U Gem and VW Hyi observations show that outbursts start in the outer disc region (Smak 1984).

One possible conclusion is that one should modify the viscosity prescription. This can be done either *ad hoc*, by assuming that, for example, α is a function of r, or by finding inspiration in the theoretical work on viscosity in accretion discs. The result in both cases is the same: an α depending on r, T, $\dot{M}_T$, the binary system mass ratio, etc. Among the most popular are: $\alpha = \alpha_0 (H/r)^n$, where H is the disc half-thickness, or just $\alpha \sim r^{\epsilon}$. The formula $\nu = \nu_0 (r/r_0)^a (\Sigma/\Sigma_0)^b$ was proposed in a study that is supposed to demonstrate the superiority of the disc instability model over the mass transfer enhancement model (Mineshige, Yamasaki & Ishizaka 1993).

Another difficulty is the so called UV lag: in some dwarf novae the rise to outburst of the optical flux precedes that of the UV flux by 0.5 to 1 day. The disc instability model in its standard form is unable to account for this effect because the heating front arrives too fast at the inner, UV emitting, disc regions (obviously the outburst has to be of the "outside-in" type). This difficulty could be avoided if the inner disc regions were absent during quiescence. The heating front propagating inward would then end its trip at some $r_{\rm in}(\text{quiescence}) > r_{\rm in}(\text{outburst})$. The increased viscosity would rebuild the inner UV emitting region in a viscous time. In this way one obtains the required UV lag. Two models removing the inner disc have been proposed. Livio & Pringle (1992) show that a weak magnetic field of the white dwarf may have the required effect. Meyer & Meyer-Hofmeister (1994) show that the inner disc may be evaporated through a coronal flow.

The second model has the advantage of solving another difficulty: to keep $\Sigma < \Sigma_{\max}$ for the duration of low states the disc instability model requires very low accretion rates in quiescence (compare with Eq. 4 below): $\dot{M} \lesssim 5\ 10^{12}$–$10^{13}\,\mathrm{g\,s^{-1}}$ whereas observations of UV and hard X-ray radiation in quiescence suggest $\dot{M} \sim 5\ 10^{14}\,\mathrm{g\,s^{-1}}$. A similar problem arises in the case of SXTs. It is worth investigating if the Meyer & Meyer-Hofmeister (1994) model could apply also in this case. It is clear that the Livio & Pringle (1992) model cannot work in systems containing black holes.

Finally, both models may suppress "inside-out" outbursts by removing the "inside".

3. The Thermal-Tidal versus the Enhanced Mass Transfer Instability Models

The SU UMa systems are characterised by the appearance of so-called "superoutbursts": stronger outbursts that are longer than "normal" ones. Typically they last more than two weeks. During these events photometric disturbances ("superhumps") are observed in the optical light curve with periods a few percent longer than the orbital one. Studies by Whitehurst (1988), Whitehurst & King (1991), Hirose & Osaki (1990) and Lubow (1991a,b) established that the superhump is due to a tidal distortion of the outer disc resulting from the presence of the 3:1 resonance inside the disc. A short but comprehensive review may be found in King (1994). The presence of a 3:1 resonance inside the disc requires mass ratios (secondary/primary) $q = M_2/M_1 < q_{\mathrm{crit}} \sim 0.25$–$0.33$. This explains why all SU UMa systems are found to have periods $\lesssim 3\,\mathrm{hr}$. Observations of "superhumps" in some black-hole SXTs are a nice confirmation of the model, since one expects those systems to have low mass ratios as well.

The superhump is therefore well explained, but the origin of superoutbursts is subject to debate (e.g., Ichikawa, Hirose & Osaki 1993; Whitehurst 1994).

According to Osaki (1989) the superoutburst is a pure disc phenomenon. In his model, normal outbursts occurring between superoutbursts (the "supercycle") are due to the thermal disc instability. Since, as I have mentioned above, normal outbursts are not efficient in causing accretion on to the central body, both the disc mass and radius grow during the supercycle. When the radius is large enough to contain the 3:1 resonance, the disc becomes eccentric due to a tidal instability. The tidal instability then enhances removal of angular momentum from the outer part of the disc allowing accretion of a large fraction of matter accumulated during the supercycle. This last point, as Whitehurst (1994) pointed out, assumes a very efficient coupling ("viscosity") between the outer and inner disc.

The tidal-thermal instability model makes several, very definite, predictions which can be tested by observation. First, it predicts a secular growth of disc radius (in a 'normal' cycle the radius first increases then decreases). Second, it predicts a rapid decrease of the superhump period during the superoutburst (Whitehurst 1994). Third, it requires the length of the 'normal' cycle to increase during the supercycle. Observations of the disc radius of the system Z Cha show behaviour opposite to the predictions of the tidal-thermal instability model (Smak 1991). There are not enough data to test the second prediction. Finally, observation of VW Hyi shows that generally the cycle length increases during the supercycle but decreases in the final cycles just before the superoutburst. The data in both cases show a large scatter but no tendency to follow the predictions of the Osaki model can be detected. One may conclude, that this model is not confirmed by observations.

On the other hand, as Smak (1991) has pointed out, a slight enhancement of the mass transfer during the last outbursts is capable of making the cycle length decrease. Such an enhancement is indeed observed (Vogt 1983), and as Smak (1991) has stressed, it is seen only in outbursts preceding the superoutbursts – those with decreasing length. Van der Woerd & Van Paradijs (1987) concluded that the recurrence behaviour of VW Hyi suggests that superoutbursts are due to enhanced mass transfer from the secondary.

The (moderate) mass transfer enhancement model for superoutbursts was proposed by Whitehurst (1994). Here the enhanced mass transfer explains the length of the superoutburst. The exact cause of this enhancement is not known, but it is natural to think that it is due to the effects of illumination by the light emitted during normal outbursts. This model has the advantage of being supported by observation. In addition, the mass transfer enhancement episodes are independent of q, so that only for $q \lesssim 0.33$ would the resulting long outburst develop a superhump. U Gem showed a very long (45 day) outburst which was not classified as a superoutburst because a superhump was absent. The mass ratio for this system is ~ 0.46. This type of dwarf nova behaviour cannot be explained by the Osaki model. In any case short-term mass transfer variations in dwarf-novae are not well studied (see Wood *et al.* 1994).

3.1. WZ SGE AND RELATED SYSTEMS

WZ Sge has the particularity of showing only superoutbursts with a recurrence time of ~ 33 years. A few other systems show similar but less extreme behaviour. The properties of WZ Sge are most probably connected to its short period: it is very close to the minimum period of CVs distribution. The mass transfer in this system, as estimated by Smak (1993) is low ($\sim 2\ 10^{15}\,\mathrm{g\,s^{-1}}$) and is close to the one obtained from evolution models (Kolb 1994, private communication). It is no more than 2.5 times less than that of, e.g., Z Cha (Kolb 1994). The disc instability model requires WZ Sge to have an extremely low quiescent viscosity: $\alpha_{\mathrm{cold}} < 5\ 10^{-5}$ (Smak 1993) as compared to the usual value of ~ 0.01. The physical reason for such a small value is not known, so the explanation of the WZ Sge phenomenon (Osaki 1994) by a low viscosity is obviously not satisfactory. The disc instability model for SXTs also requires very low values of viscosity (see below).

4. Models for Transient Outbursts of Low-Mass X-ray Binaries

4.1. DISC INSTABILITY MODEL FOR SOFT X-RAY TRANSIENTS

The dwarf nova–SXT connection was analysed by Van Paradijs & Verbunt (1984). Cannizzo, Wheeler & Ghosh (1985) and Lin & Taam (1984) proposed that the disc instability model used to describe dwarf nova outbursts may explain the SXT phenomenon.

Comparing the "normal" dwarf novae with SXTs one realises immediately that they share only one property: the rise time to outburst is similar in the two classes of events. Other characteristics are totally different: decay times as well as recurrence times are much longer in SXTs. Light curves of SXTs are rather similar to superoutbursts of WZ Sge. Indeed both types of objects shared the erroneous name of *Nova* because of the shape of their optical light curves.

One should, therefore, expect the disc instability model for SXTs to encounter difficulties similar to those appearing in the case of SU UMa systems and WZ Sge.

4.1.1. *The Mineshige & Wheeler model*

Up to now there have been only two detailed calculations modelling SXTs by the disc-instability model: Huang & Wheeler (1989) and Mineshige & Wheeler (1989). In a recent paper (Cannizzo 1994) a new study by Cannizzo, Chen and Livio was announced as being in preparation.

I will discuss here only the Mineshige & Wheeler paper (1989; hereafter MW) because the inner disc boundary in Huang & Wheeler (1989) is assumed to be at 10^8 cm which is certainly at least one order of magnitude too large for a "realistic" model of an SXT. However, I have to comment on a claim by Chen, Livio & Gehrels (1993) and Mineshige *et al.* (1994) that the observed rise of hard X rays preceding the rise in soft X-rays (Ricketts, Pounds & Turner 1975; Lund 1993) in some SXTs is consistent with the "inside-out" character of outbursts predicted by the disc instability model. First, there is no unique model for hard and soft X-ray emission in the inner regions of accretion discs around compact objects, so in itself an "inside-out" character of an outburst proves nothing. Second, whereas it is true that in the case of dwarf novae, models assuming constant values of α in quiescence get "inside-out" outbursts for low accretion rates (Smak 1984), the MW model for SXTs succeeds in igniting the inner disc regions *only* for "outside-in" outbursts. This is clearly explained in MW and will be discussed below. This point is nevertheless extremely confusing because of a sentence in Mineshige *et al.* (1994) which suggests that figure 7 in MW represents a "thermal instability initiated from the inner part", while, in discussing the same figure, MW ask the reader to note that "this is an 'outside-in' burst...". In this particular model (for which the recurrence time is about 7 months, and the peak bolometric luminosity is $< 10^{35}$ erg s^{-1} – parameters not very close to those of SXTs) the inner disc never falls to the 'cold' state because the cooling wave is reflected at $r \sim 20\, r_{\rm in}$ as a outward propagating heating wave. As explained in MW, before this reflected heating wave manages to propagate to more than $r \sim 20\, r_{\rm in}$ the new "outside-in" heating front created at $r \sim 800\, r_{\rm in}$ propagates inward and merges with the hot inner region. The main outburst is therefore of the "outside-in" type.

On the other hand, Huang & Wheeler (1989) get "inside-out" outbursts but their "inside" is at 10^8 cm, obviously not relevant to the problem of the soft/hard X-ray priority.

For a constant α the MW $T_{\rm eff} - \Sigma$ relation at 10^7 cm shows no unstable branch below $\sim 3\, 10^4$ K. To get an "S-curve" they have to tune the viscosity prescription. MW use the $\alpha = \alpha_0 \, (H/r)^n$ viscosity prescription and try to adapt both α_0 and n to obtain light curves corresponding to SXTs. They never succeed in getting recurrence times longer than ~ 16 years. The sequences of outbursts show only a vague resemblance to observations (see their figures 5, 6, 8 and 9). MW use 21 grid points in their calculation. It is therefore interesting to compare their results with those of Cannizzo (1993b) who shows the difference between calculations using 25 and 100 grid points. One has the impression that the MW results suffer seriously from too coarse a resolution.

The best results, in the sense of amplitudes and recurrence times closest to observations, are obtained for $n > 1$ and $\alpha_0 \geq 10$. With the prescription used by MW, the viscous time $\tau_{\rm vis} \sim (H/r)^{-2}(1/\alpha\Omega)$ (see, e.g., Frank, King & Raine 1992) is proportional to $r^{(1-n)/2}$. For $n > 1$, $\tau_{\rm vis}$ is *decreasing* with radius so that even for low accretion rates (long accumulation times) MW get only "outside-in" outbursts.

As for WZ Sge, the α required by the disc instability model is very small in quiescence. MW do not give its value at the disc inner edge (3.16 10^6 cm), but at 10^7 cm it is $8\ 10^{-5}$ for $\alpha_0 = 3.16\ 10^3$, $n = 2$, and $\alpha = 2\ 10^{-4}$ for $\alpha_0 = 10^2$, $n = 1.5$.

From fig. 10 in MW one sees that the accretion rate at the inner disc radius in quiescence is extremely low ($\lesssim 10^6\,\mathrm{g\,s^{-1}}$). One can show that this is a necessary consequence of the disc instability model. Indeed, this model requires that the surface density remain everywhere below the critical surface density $\Sigma_{\max}$ during the quiescence. For black-hole SXTs the recurrence time is longer than 10 years. This implies that the viscosity at the inner edge of the disc satisfies the inequality:

$$\nu \lesssim 10^6 t_8^{-1} r_7^2 \ \mathrm{cm^2\ s^{-1}}. \tag{2}$$

If one uses the formula for $\Sigma_{\max}$ from Smak (1992):

$$\Sigma_{\max} = 16.22\ r_{10}^{1.11} m_1^{-0.37} \alpha^{-0.79} \ \mathrm{g\ cm^{-2}}, \tag{3}$$

one gets an inequality for the value of the accretion rate at the inner boundary of the disc:

$$\dot{M}_{\mathrm{in}} \lesssim \frac{8\pi}{3}\nu\Sigma_{\max} \sim 2.76\ 10^4\ t_8^{-1} r_7^{3.11} m_{10}^{-0.37} \alpha^{-0.79}\ \mathrm{g\ s^{-1}} \tag{4}$$

in very good agreement with MW. Notice that I used a constant α but this does not affect the result.

Several black-hole SXTs have been observed with GINGA (Mineshige et al. 1992) and ROSAT (see the article by Verbunt in these proceedings and Verbunt et al. 1994). Two systems, A 0620–00 and V404 Cyg, have been detected at levels corresponding to accretion rates of at least $\sim 1.4\ 10^{10}\,\mathrm{g\,s^{-1}}$ for A 0620–00 and of $3\ 10^{12}\,\mathrm{g\,s^{-1}}$ for V404 Cyg. Those accretion rates are several orders of magnitude larger than the values predicted by the disc instability model of MW. Detections of X rays from quiescent SXTs contradict the disc instability model, at least in the version proposed by MW (Mineshige et al. 1992).

Other observations with GINGA and ROSAT which provide upper limits on X-ray emission from quiescent SXTs contradict the mass transfer instability model proposed by Hameury, King & Lasota (1986), as I discuss later.

One of the main ingredients of the disc instability model is the presence of an inward propagating "cooling wave" which brings the hot disc back to the (non-equilibrium) cool state through a series of quasi-equilibria. As noticed by Cheng et al. (1992) the HST/FOS spectra of the so called X-ray Nova Muscae 1991 show no evidence for the propagation of a cooling front in the accretion disc of this system. Chen et al. (1993) suggest that this is due to hard X-ray heating of the outer disc regions. Effects of the disc illumination by X rays on SXTs outbursts still await a detailed treatment (see Mineshige, Kim & Wheeler 1990; Kim, Wheeler & Mineshige 1994; Mineshige et al. 1994).

The physics of the hot inner disc around black holes and neutron stars is still poorly known. For recent results on this subject see Abramowicz et al. (1994); Chen & Taam (1994); Milsom, Chen & Taam (1994).

4.2. THE MASS–TRANSFER INSTABILITY MODEL FOR SXTS

A mass-transfer instability model for SXTs was proposed in a series of papers by Hameury, King & Lasota (1986, 1987, 1988, 1990; hereafter HKL). HKL realised that the mass transfer from an X-ray heated secondary star is unstable for a range of accretion rates between two critical values. The lower value corresponds to an illuminating X-ray flux comparable to the intrinsic stellar flux and is typically $\sim 10^{12}$–$10^{14}\,\mathrm{g\,s^{-1}}$ depending on the binary parameters. The upper critical accretion rate above which X-ray heating dominates is $\sim 10^{16.5}\,\mathrm{g\,s^{-1}}$ for $P_{\mathrm{orb}} \sim 3\,\mathrm{h}$ and increases more or less linearly with the orbital period. The two accretion rate limits, expressed in terms of X-ray luminosities, correspond closely to the limits of the "luminosity gap" of steady LMXBs (White, Kaluzienski & Swank 1984; Johnston, these proceedings) which is a very attractive feature of the model.

According to HKL, during the quiescence accretion rates are slightly smaller than the lower critical value. It is assumed that a substantial amount of the accretion luminosity is emitted in hard ($E > 7\,\mathrm{keV}$) X rays which are able to increase the companion's effective temperature. The secondary expands under the effect of illumination and at some moment enters the unstable range of accretion rates. A mass transfer runaway produces the SXT outburst. When the accretion disc thickness is large enough to shield the region around the Lagrange L_1–point, the runaway stops. The secondary contracts and the system goes back to quiescence. During the decay from outburst the L_1 region may become uncovered and produce new outbursts.

This version of the mass transfer instability model suffers from one major difficulty: it requires the secondary star to react to the effects of X-ray illumination on time scales of the order of months. The secondary's heated layer must be able to expand by at least a fraction of the atmospheric scale-height in less 10^6–10^7 s. For example, GINGA observations of GS 2000+25 150 days *before* the outburst have not detected X-rays between 1.2 and 37 keV, implying an upper limit $\sim 6\ 10^{33}\,\mathrm{erg\,s^{-1}}$ (Mineshige et al. 1992). This value is not in contradiction with the HKL model, but let us assume that the actual X-ray luminosity is much lower. In itself this would not invalidate the model: one could imagine that hard X rays appear only three months before the outburst, due for example to an instability in the *inner* disc. The problem is whether the outer layers of the star may expand by a scale-height during three months.

Gontikakis & Hameury (1993) showed (see also Ritter in these proceedings) that for a main-sequence star with a mass less than $\sim 1\,M_\odot$ the expansion time is always longer than $\sim 10^2$ years. For a $0.8\,M_\odot$ main-sequence secondary the characteristic time is $\sim 10^4$ years. Low-mass main-sequence stars possess deep, massive convective zones which react *globally* to surface heating on their thermal time scale. The local thermal time scale introduced by HKL (1986) makes no sense in a convective envelope. Only massive main-sequence stars and evolved low-mass stars (subgiants or stripped giants) can react to X-ray illumination on times scales as short as 10^6 s.

The orbital period of ~ 5.1 hr determined for J0422+32 seems to rule out the mass transfer instability as a cause of its outburst since the companion star, if at the main sequence, should have a mass of $\sim 0.56\,M_\odot$.

One could, of course, try to defend the model by saying that SXTs have the particularity of containing evolved stars at short orbital period, or by saying that the HKL model applies only to systems with $P_{\rm orb} \gtrsim 9$ hr, but unless there is an independent reason to believe that this is true, the HKL mass-transfer instability model cannot describe the SXT phenomenon.

Chen, Livio & Gehrels (1993) and Augusteijn, Kuulkers & Shaham (1993) have proposed models which combine elements of both type of models. Although the proposed versions were not confirmed by observations (Callanan *et al.* 1994; Chevalier & Ilovaisky 1994), this is probably a direction to follow.

Acknowledgements. I am grateful to Joe Smak, Hans Ritter, Darragh O'Donoghue, Uli Kolb, Andrew King, Jean-Marie Hameury and Wolfgang Duschl for useful discussions, advice and comments. I thank Mike Garcia and Janet Wood for sending me their work prior to publication.

References

Abramowicz, M.A. *et al.* 1994, ApJL (in press)
Augusteijn, T., Kuulkers, E. & Shaham, J. 1993, A&A 279, L13
Callanan, P.J. *et al.* 1994, ApJ (in press)
Cannizzo, J.K. 1993a, in *Accretion Disks in Compact Stellar Systems*, J. Craig Wheeler (Ed.), World Scientific Publishing (Singapore), p. 6
Cannizzo, J.K. 1993b, ApJ 419, 318
Cannizzo, J.K. 1994, ApJ 435, 389
Cannizzo, J.K, Wheeler, J.C. & Ghosh, P. 1985, in *Cataclysmic Variables and Low-Mass X-ray Binaries*, J. Patterson & D.Q. Lamb (Eds.), Reidel (Dordrecht), p. 307
Chen, W., Livio, M. & Gehrels, N. 1993, ApJ 408, L5

Chen, X. & Taam, R.E. 1994, ApJ 431, 732
Cheng, F.H. et al. 1992, ApJ 397, 664
Chevalier, C. & Ilovaisky, S. 1994, preprint OHP
Frank, J., King, A.R. & Raine, D.J. 1992, *Accretion Power in Astrophysics*, 2nd Edition, Cambridge Univ. Press
Gontikakis, C. & Hameury, J.M. 1993, A&A 271, 118
Hameury, J.M., King, A.R. & Lasota, J.P. 1986, A&A 161, 71
Hameury, J.M., King, A.R. & Lasota, J.P. 1987, A&A 171, 140
Hameury, J.M., King, A.R. & Lasota, J.P. 1988, A&A 192, 187
Hameury, J.M., King, A.R. & Lasota, J.P. 1990, ApJ 353, 585
Hirose, M. & Osaki, Y. 1990, PASJ 42, 135
Huang, M. & Wheeler, J.C. 1989, ApJ 343, 229
Ichikawa, S., Hirose, M. & Osaki, Y. 1993, PASJ 45, 243
Kim, S.W., Wheeler, J.C. & Mineshige, S. 1994, in *The Evolution of X-ray Binaries*, S.S. Holt and C. Day (Eds.), AIP Conference Proceedings (New York), (in press)
King, A.R. 1994, in *The Evolution of X-ray Binaries*, S.S. Holt and C. Day (Eds.), AIP Conference Proceedings (New York), (in press)
Lin, D.N.C. & Taam, R.E 1984, in *High Energy Transients in Astrophysics*, S.E. Woosley (Ed.), AIP Conf. Proc. No. 115, p. 83
Livio, M. & Pringle, J.E. 1992, MNRAS 259, 23P
Lubow, S.H. 1991a, ApJ 381, 359
Lubow, S.H. 1991b, ApJ 381, 368
Ludwig, K., Meyer-Hofmeister, E. & Ritter, H. 1994, A&A 290, 473
Lund, N. 1993, ApJS 97, 289
Meyer, F. & Meyer-Hofmeister, E. 1984, A&A 132, 184
Meyer, F. & Meyer-Hofmeister, E. 1994, A&A 288, 175
Milsom, J.A., Chen, X. & Taam, R.E. 1994, ApJ 421, 668
Mineshige, S. & Osaki, Y. 1983, PASJ 35, 377
Mineshige, S. & Wheeler, J.C. 1989, ApJ 343, 241
Mineshige, S., Kim, S.W. & Wheeler, J.C. 1990, ApJ 358, L5
Mineshige, S., Yamasaki, T. & Ishizaka, C. 1993, PASJ 45, 707
Mineshige, S. et al. 1992, PASJ 44, 117
Mineshige, S. et al. 1994, ApJ 426, 308
Osaki, Y. 1989, PASJ 41, 1005
Osaki, Y. 1994, in *Theory of Accretion Disks – 2*, W. Duschl, J. Frank, E. Meyer-Hofmeister, F. Meyer & W. Tscharnuter (Eds.), Kluwer (Dordrecht), p. 93
Papaloizou, J., Faulkner, J. & Lin, D.N.C. 1983, MNRAS 205, 487
Ricketts, M.J., Pounds, K.A. & Turner, M.J. 1975, Nat 257, 657
Shakura, N.I. & Sunyaev, R.A. 1973, A&A 24, 337
Smak, J.I. 1984, Acta Astron. 34, 161
Smak, J.I. 1991, Acta Astron. 41, 269
Smak, J.I. 1992, Acta Astron. 42, 323
Smak, J.I. 1993, Acta Astron. 43, 101
Van der Woerd, H. & Van Paradijs, J. 1987, MNRAS 224, 271
Van Paradijs J. & Verbunt, F. 1984, in *High Energy Transients in Astrophysics*, S.E. Woosley (Ed.), AIP Conf. Proc. No. 115, p. 49
Verbunt, F. et al. 1994, A&A 285, 903
Vogt, N. 1983, A&A 118, 95
White, N., Kaluzienski, J.L. & Swank, J.H. 1984, in *High Energy Transients in Astrophysics*, S.E. Woosley (Ed.), AIP Conf. Proc. No. 115, p. 31
Whitehurst, R. 1988, MNRAS 232, 35
Whitehurst, R. 1994, MNRAS 266, 35
Whitehurst, R. & King, A.R. 1991, MNRAS 249, 25
Wood, J., Naylor, T., Hassall, B.J.M. & Ramseyer, T.F. 1994, (preprint)

Discussion

P. Gosh: Attempts to extract values of the viscosity parameter α from the properties of dwarf novae and SXTs have been one of the basic efforts in this field. If the current thinking is that the underlying mechanism is a combination of (1) disk instability and (2) effects of irradiation of the companion, it becomes more difficult to infer α from observations, unless there is a clear way of separating the effects caused by these two phenomena.

J.P. Lasota: I agree

P. Gosh: In answer to Jan van Paradijs' question, directed at the theoreticians: it is true that the (rare) attempts to extract values of α from the Balbus-Hawley instability have yielded numbers ~ 0.1 to 1. However, this kind of mechanism for viscosity has not really been used in the dwarf nova - soft X-ray transient game. For example, the Wheeler and co-workers line of work described by Jean-Pierre uses $\alpha \sim (h/\gamma)^\beta$ (with $\beta \sim 1$), which is a prescription inspired by the mechanisms proposed by Vishniac, Diamond, and co-workers, in which internal (g-) modes of the disk are thought to be the underlying cause of viscosity. So, there has never really been a confrontation between the two approaches, to the best of my knowledge. It is as if one half of the world does not know what the other half is doing!

SPIN EVOLUTION OF THE PROGENITORS OF BINARY AND MILLISECOND PULSARS

PRANAB GHOSH
Tata Institute of Fundamental Research
Bombay 400 005, INDIA

1. Introduction

In this symposium, I have been given the task of summarizing our current understanding of the evolutionary history of spin periods of the neutron stars that we now see as binary and millisecond pulsars, i.e., recycled pulsars. We believe that a newborn, fast-spinning neutron star (with a rather high magnetic field $\sim 10^{11}$–10^{13} G) in a binary system first operates as a spin-powered pulsar, subsequently as an accretion-powered pulsar when accretion begins after the pulsar has been spun down adequately, and finally as a spin-powered pulsar for the second time after having been recycled to become a very fast-rotating neutron star (with a rather low magnetic field $\sim 10^8$–10^{11} G) (see Ghosh 1994a,b, hereafter G94a,b).

A magnetic field–spin period $(B-P)$ diagram of neutron stars, as shown in Fig. 1, is widely used for studying spin evolution (indeed, Dr. Bailes has called it the Hertzsprung-Russell diagram of pulsars!). In addition to the "pulsar island", Fig. 1 shows the recycled pulsars (i.e., those in the second spin-powered phase) at the bottom left of the diagram. Recycled pulsars can also be single, of course, since the binary can be disrupted (by processes that are mentioned in Section 4) before the second spin-powered phase begins. The binary X-ray pulsars (i.e., those in the accretion-powered phase) are at the top right of Fig. 1: only those 10 X-ray pulsars for which *direct measurements of the magnetic field* from a cyclotron line are currently available are shown. Of these, 9 were observed with GINGA (Nagase 1992), and the cyclotron feature in A 0535+26 has been recently observed with OSSE (Grove *et al.* 1994). Fig. 1 also shows the two currently known binary neutron stars which are thought to be in their first spin-powered phase, namely, PSR B1259–63 and PSR J0045–73 (Johnston *et al.* 1994, Johnston

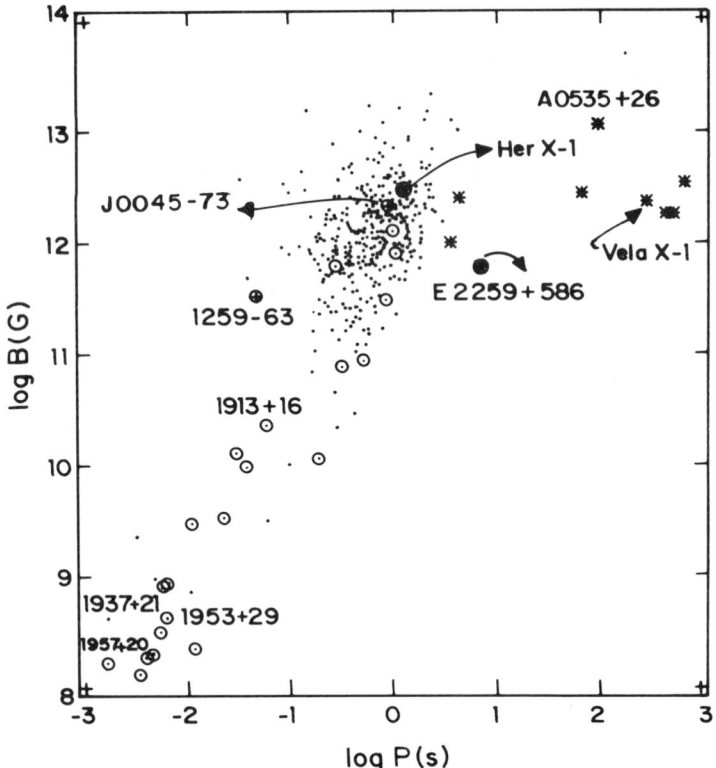

Figure 1. Spin- and accretion-powered pulsars on the $B - P$ diagram. Dots indicate spin-powered pulsars, and asterisks, accretion-powered ones. Spin-powered pulsars in binaries are encircled, and the two recently discovered massive radio pulsar binaries (PSRs B1259–63 and J0045–73) have an additional cross on their symbols. Some well-known pulsars are marked individually: these are (a) several recycled pulsars, (b) PSRs B1259–63 and J0045–73, and, (c) three accretion-powered pulsars, two of them with relatively low-mass companions (Her X-1 and E 2259+586), and one with a massive companion (Vela X-1).

1994, hereafter J94a,b; Kaspi *et al.* 1994): we have heard about these two massive radio binary pulsars in detail from Drs. Johnston and Kaspi in this symposium.

2. Initial Spin Down

During the initial spin down the neutron star moves rightward on the $B - P$ diagram along an approximately horizontal track.

2.1. BRAKING TORQUES

The electromagnetic braking torque, which dominates the spin down at the fastest spin-rates, scales as $N \propto \mu^2(\Omega_s/c)^3$, where μ is the magnetic dipole moment of the neutron star, and Ω_s is its angular velocity. As the star spins down, plasma torques begin to dominate over the electromagnetic torque. This can be readily seen by comparing the electromagnetic torque with the so-called subsonic propeller torque, $N \propto \mu^2 \Omega_s^2 / GM$ (Illarionov & Sunyaev 1975, hereafter IS; Mineshige et al. 1991; see Henrichs 1983 for a review). The latter torque obviously dominates for $P > P_{\rm crit} \propto GM/c^3 \sim 2\,{\rm ms}$. Actually, the subsonic propeller torque is an *upper limit* to the possible plasma torques. In reality, the plasma torque relevant at the highest spin rates is the so-called supersonic propeller torque, which is smaller by a factor $\sim (c_s/\Omega_s r_A)$ than the subsonic torque, so that $P_{\rm crit} \sim 50 - 100\,{\rm ms}$ for typical mass-flow rates from a massive companion to the neutron star. Here, c_s is the sound speed, and r_A is the Alfvén radius. For $c_s \sim v_{\rm ff}$ at r_A, the supersonic propeller torque reduces to the original IS scaling, i.e., $N \propto (\mu^2/r_A^3)(\Omega_K(r_A)/\Omega_s)$. Here, $v_{\rm ff}$ is the free-fall velocity.

2.2. SPIN DOWN OF PSR B1259-63

Studies of the spin down history of PSR B1259–63, a neutron star-Be star (SS 2883) binary with a highly eccentric ($e \simeq 0.87$) 3.4 yr orbit (J94a,b), are now providing us with the first detailed look at the processes of initial spin down. Far away from the periastron, the 47 ms pulsar undergoes spin down with a characteristic age of 3.3 10^5 yr, predominantly by electromagnetic torques, which implies a magnetic field $\sim 3.3\ 10^{11}$ G for the neutron star (J94a). Close to the periastron, propeller-type torques due to the dense plasma in the equatorial disk around the Be star, as well as those due to the more tenuous plasma in the stellar wind, contribute appreciably to spin down, and may well dominate the *instantaneous* spin down rate (G94a,b) as the pulsar passes through the disk. The situation is shown in Fig. 2, which displays the spin down rates ($\dot{P}$) due to the above torques in a polar diagram: to read off $\dot{P}$ due to a particular torque at a given binary phase, one simply draws a line from the origin (the position of SS 2883) towards the position of the pulsar at that phase (the binary orbit is shown in Fig. 2), until the line meets the curve for that particular torque. $\dot{P}$ is just the length of this line. It is clear from Fig. 2 that the values of $\dot{P}$ for the electromagnetic torque and for the propeller torque in the (tenuous) stellar wind are generally comparable, while that for the propeller torque in the (dense) disk of SS 2883 is considerably higher (indeed, parts of the relevant $\dot{P}$ curve are out of scale in Fig. 2).

For the model computations (Ghosh 1994c, hereafter G94c) shown in

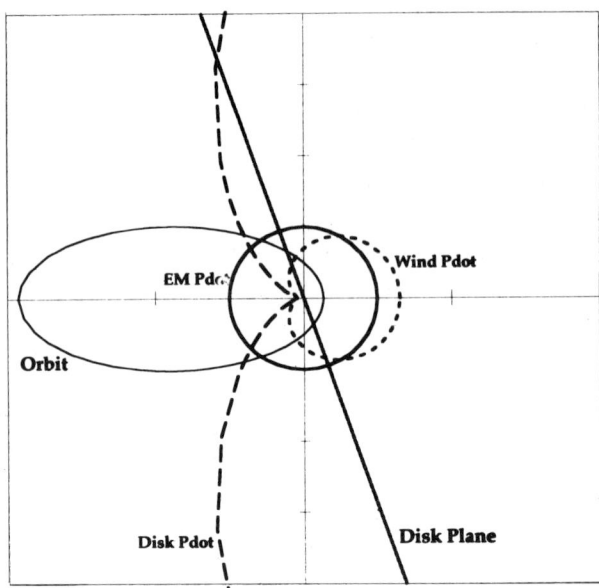

Figure 2. Spin down of PSR B1259–63. Shown as polar plots are the spin down rates for the electromagnetic torque, and the plasma torques due to the disk and the wind of SS 2883 (see text). Also shown are the binary orbit and the line of intersection of the disk plane with the orbital plane. Note that SS 2883 is at the origin, and the angle between the y-axis and the disk plane ($\lesssim 10°$) is shown exaggerated for clarity.

Fig. 2, I have used the disk- and wind-density profiles (ρ) given by both the theoretical calculations of Bjorkman & Cassinelli (1993, hereafter BC) and by the model-fitting of Be-star observations by Waters (1986, hereafter W), i.e., profiles of the type $\rho = \rho_0(r/R_*)^{-n}$, where R_* is the stellar radius. The density scale, ρ_0 is set by the requirements that (a) the density of the disk where it touches the stellar surface is $\sim 10^{-11}$–10^{-12} g cm^{-3} and (b) the wind density at the stellar surface is $\sim 10^{-2}$–10^{-3} of the disk density. These numbers are in accordance with both BC and W. The exponent n is $\simeq 2$ for the wind, and $\simeq 3$ to 4 for the disk, again in accordance with both BC and W. Further, the propeller torque relevant here is the supersonic one in the IS formulation (see above), which yields for the wind spin down rate (G94c) $\dot{P}_{\rm wind}/\dot{P}_{\rm EM} \simeq 0.7(1 + e\cos\theta)$ and the disk spin down rate $\dot{P}_{\rm disk}/\dot{P}_{\rm EM} \simeq 7(1 + e\cos\theta)^2$ relative to the electromagnetic spin down rate $\dot{P}_{\rm EM}$. Here, e is the orbital eccentricity, θ is the true anomaly, and I have taken $n \simeq 4$ for the disk (W, J94b).

While $\dot{P}_{\rm disk}$ is quite large near periastron, the amount of additional spin down produced by this mechanism during periastron passage (which is the observed quantity) depends crucially on the orientation of the disk plane relative to the orbital plane (G94b,c). The two planes tend to become aligned in older systems due to tidal torques. However, this need *not* be the case in a young system like PSR B1259−63/SS 2883, and there are indications already that it is not. There is little perturbation to the Hα emission from the system during periastron passage (Manchester, private communication). This emission is thought to originate in the disk, which would be grossly perturbed by the tidal torques near periastron if the two planes were coaligned (Kochanek 1993), while an appreciable misalignment ($\gtrsim 30°$, say) would be consistent with observation. This picture is also consistent with the dates (J94b) on which the pulsar disappeared ($\sim$ December 20, 1993) before and reappeared ($\sim$ February 2, 1994) after the periastron passage of January 9, 1994, since these dates imply an angular separation $\sim 200°$ between these two points, as would be expected for a misaligned disk with an opening angle $\sim 15°$ (as advocated by W) intersecting the orbit near these points (G94c). For this disk configuration, ΔP is increased by $\sim 50\%$ over the amount expected from the electromagnetic torque (G94c) by the two passages (see Fig. 2) through the disk; this number is consistent with the preliminary results from the peristron observations described above (Johnston, private communication). A much thinner disk (of opening angle $\sim 1°$), as advocated by BC, would give an enhancement which is ~ 10 times as small as the above value, while an *aligned* disk would give an enhancement that is ~ 10 times as big (G94c).

3. Spin Evolution in the Accretion-Powered Phase

Since this is not the main focus of my talk, only the essential points are summarized here; for more details, see G94a,b. Under the action of accretion torques, the pulsar can undergo both spin up and spin down, so that it moves back and forth on approximately horizontal tracks in the region occupied by accretion-powered pulsars in Fig. 1. A crucial feature of the accretion torque is that it changes sign at a critical value of the pulsar's spin rate: the corresponding critical value, ω_c, of the dimensionless fastness parameter (G94a,b) is now becoming constrained by recent studies of the detailed period histories of binary X-ray pulsars: current indications are that $\omega_c \sim 0.4$–0.7.

4. Final Spin Up

During this process, the neutron star moves downward and left in Fig. 1, from the area occupied by accretion-powered pulsars to the (roughly) linear band occupied by recycled pulsars; the upper edge of this band being the so-called spin up line (see Bhattacharya & Van den Heuvel 1991). The essential process at work here is the accretion torque: indeed, the explanation of the periods of the recycled pulsars was a major triumph of the accretion torque theory in the 1980's.

4.1. SPIN EVOLUTION

The final evolution of high-mass X-ray binaries (HMXBs) can go in two ways when the common-envelope (CE) phase begins (Van den Heuvel 1992 and references therein, hereafter vdH92). An initially wide binary (e.g., a Be star system) can eject the entire envelope and produce a neutron star with a helium core companion. The system then evolves either (a) by a supernova explosion of the companion, or, (b) by evolution of the companion into a massive white dwarf. In the former case, the chances are high that the system remains bound, producing a double neutron star system like PSR 1913+16; if it does become unbound, two runaway pulsars are produced, one of which is recycled. In the latter case, a system like PSR 0655+64 is thought to be produced. The spin evolution is qualitatively straightforward in both cases: the neutron star is spun up to short periods ($\sim$50–1000 ms, say) determined by the strength of the full-scale Roche lobe overflow (that initiates the CE phase) and the magnetic field of the neutron star. On the other hand, an initially narrow binary undergoes a complete spiral-in in the CE phase, producing a Thorne-Zytkow object (vdH92) with a disk-accreting neutron star in its core. The end product is a recycled, spun up, single radio pulsar.

Neutron stars with low magnetic fields ($\sim 10^8$–10^9 G) in very bright (near-Eddington) low-mass X-ray binaries (LMXBs) are thought to be spun up to $\sim$millisecond periods by accretion torques, and these systems are then believed to evolve into binary pulsars of the PSR 1953+29 class (vdH92, Webbink 92).

4.2. ACCRETION-INDUCED FIELD DECAY

It is clear from Fig. 1 that magnetic fields of neutron stars are reduced considerably during the passage from the accretion-powered phase to the second spin-powered phase. A connection between this magnetic-field decay and the accretion during the final spin up phase is obviously possible, and this has been a subject of much study recently (see Verbunt 1994). The magnetic-field distribution of recycled pulsars appears bimodal (see Kulkarni 1992; G94b): the low-magnetic-field pulsars are those which descended from LMXBs (i.e., the "1953+29 Class", or LMBPs, see vdH92) with relatively large amounts of mass accreted during the long-lived final spin up phase, and the high-magnetic-field pulsars are those which descended from HMXBs (i.e., the "1913+16 Class", vdH92) with relatively little mass accreted during the short-lived final spin up phase (G94b).

A valuable diagnostic of accretion-induced field decay comes from the correlation between the magnetic fields and orbital periods of LMBPs, first noticed by Van den Heuvel (1994; see also De Kool & Van Paradijs 1987),

SPIN EVOLUTION 63

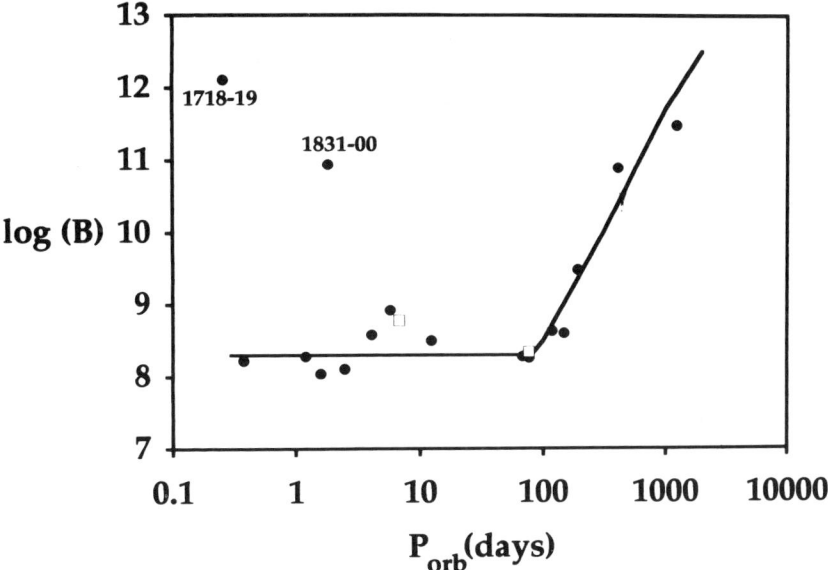

Figure 3. Magnetic field–orbital period correlation for low-mass binary pulsars (LMBPs), after Van den Heuvel (1994). All the 18 LMBPs with measured B as of August 1994 have been included. Squares denote pulsars with only an upper limit on B.

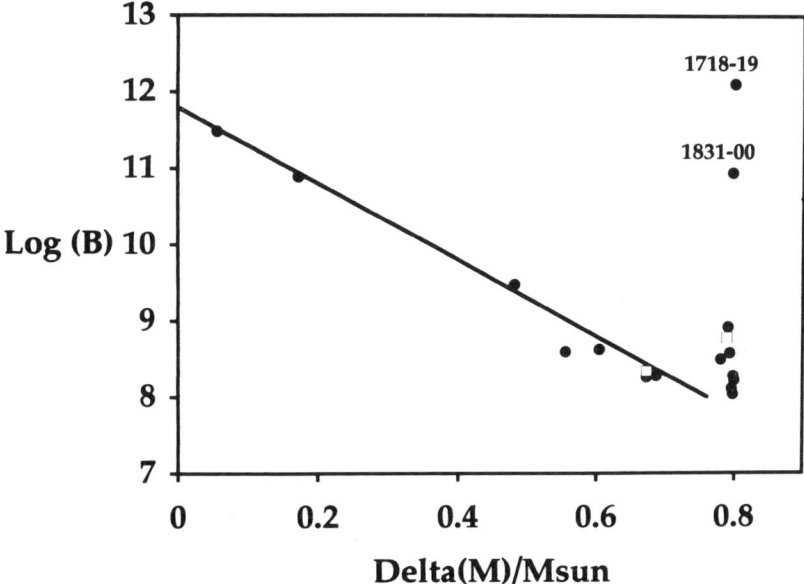

Figure 4. Accretion induced field decay. Shown are the 18 LMBPs from Fig.3.

and shown in Fig. 3. It is well known from the evolutionary calculations (Joss et al. 1987; Pylyser & Savonije 1988) of such binaries that there is a one-to-one relation between the mass transferred (ΔM) during the LMXB evolution and the final orbital period of the system. Using this relation, I get the remarkable description for accretion induced field decay shown in Fig. 4: the relation between B and ΔM can be approximated by $\log B \simeq 11.8 - (\Delta M/0.2 M_\odot)$.

Does Fig. 4 really imply an exponential field deacy, which runs contrary to the prevalent folklore of power-law field decay? On the one hand, we must be careful in interpreting ΔM in Fig. 4: it is the mass lost by the companion, which need not all be accreted. On the other, model calculations of neutron-star conductivity often give $(B - \Delta M)$ relations which are not really power-laws but mimic them over a restricted range. In any case, it is clear that this diagnostic has important implications both for the position of the spin up line (Ghosh & Verbunt, in preparation) and for models of neutron-star conductivity (Ghosh & Pethick, in preparation).

References

Bhattacharya, D. & Van den Heuvel, E.P.J. 1991, Phys. Rep. 203, 1
Bjorkman, J.E. & Cassinelli, J.P. 1993, ApJ 409, 429 (BC)
De Kool, M. & Van Paradijs, J. 1987, A&A 173, 279
Ghosh, P. 1994a, in *The Evolution of X-ray Binaries*, S.S. Holt & C.S. Day (Eds.), AIP Conf. Proc. Vol. 308, p. 439 (G94a)
Ghosh, P. 1994b, in *Pulsars: Festschrift for V. Radhakrishnan*, G. Srinivasan (Ed.), Indian Academy of Sciences (Bangalore), (in press) (G94b)
Ghosh, P. 1994c, ApJ (submitted) (G94c)
Grove, E. et al. 1994, ApJ (submitted)
Henrichs, H. 1983, in *Accretion-Driven Stellar X-ray Sources*, W.H.G. Lewin & E.P.J. van den Heuvel (Eds.), Cambridge Univ. Press, p. 393
Illarionov, A.F. & Sunyaev, R.A. 1975, A&A 39, 185 (IS)
Johnston, S. et al. 1994a, MNRAS 268, 430 (J94a)
Johnston, S. 1994b, these Proceedings (J94b)
Joss, P.C., Rappaport, S. & Lewis, W. 1987, ApJ 319, 180
Kaspi, V.M. et al. 1994, ApJ 423, L43
Kochanek, C. 1993, ApJ 406, 638
Kulkarni, S.R. 1992, Phil. Trans. Roy. Soc. London A341, 77
Mineshige, S., Rees, M.J. & Fabian, A.C. 1991, MNRAS 251, 555
Nagase, F. 1992, in *Proc. Ginga Memorial Symposium*, F. Makino & F. Nagase (Eds.), ISAS, p. 1
Pylyser, E. & Savonije, G.J 1988, A&A 191, 57
Van den Heuvel, E.P.J. 1992, in *X-ray Binaries & Recycled Pulsars*, E.P.J. van den Heuvel & S. Rappaport (Eds.), Kluwer Academic Publishers, p. 233 (vdH92)
Van den Heuvel, E.P.J. 1994, A&A (submitted)
Verbunt, F. 1994, in *The Evolution of X-ray Binaries*, S.S. Holt & C.S. Day (Eds.), AIP Conf. Proc. Vol. 308, p. 351
Waters, L.B.F.M. 1986, A&A 162, 121 (W)
Webbink, R.F. 1992, in *X-ray Binaries & Recycled Pulsars*, E.P.J. van den Heuvel & S. Rappaport (Eds.), Kluwer Academic Publishers, p. 269

THE REACTION OF LOW-MASS STARS TO ANISOTROPIC IRRADIATION AND ITS IMPLICATIONS FOR THE SECULAR EVOLUTION OF CATACLYSMIC BINARIES

H. RITTER, Z. ZHANG AND U. KOLB
Max-Planck-Institut für Astrophysik
Karl-Schwarzschild-Str. 1
Postfach 15 23
85740 Garching, Germany

Abstract. A semi-analytic model for the reaction of a low-mass star to anisotropic irradiation of low incident flux is presented. By applying this model to the donor star of cataclysmic binaries (CBs) it is shown that CBs are likely to be unstable against irradiation-driven runaway mass transfer. The implications of this instability for the long-term evolution of CBs are examined. The possibility is discussed that because of this instability CBs evolve through a limit cycle in which phases of high and low mass transfer rate alternate on a time scale short compared to the evolutionary time scale.

1. Introduction

The possible importance of the reaction of low-mass stars to external irradiation for the long-term evolution of compact binaries has been realized only rather recently in the context of the evolution of low-mass X-ray binaries (LMXBs). So far, most papers have dealt with this problem in spherical symmetry and in the limit of the very high incident fluxes relevant for LMXBs (Podsiadlowski 1991; Harpaz & Rappaport 1991; Frank et al. 1992; Hameury et al. 1993). The limit of low flux, but still in spherical symmetry, has been considered by D'Antona & Ergma (1993). Anisotropic irradiation in the limit of high flux has been studied by Gontikakis & Hameury (1993), Hameury et al. (1993) and by Ritter (1994). To the best of our knowledge, anisotropic irradiation in the limit of low flux has not been considered so far. In this paper we present a first attempt to do so. For this we develop a simple semi-analytical model for the reaction of a low-mass star to anisotropic irradiation and apply this model mainly to CBs.

2. The Reaction of a Low-Mass Star to Anisotropic Irradiation

The following discussion applies only to low-mass stars ($M \lesssim 1\,M_\odot$) on or near to the main sequence. Such stars have either a deep outer convective envelope and a radiative core ($M > M_{\rm conv} \approx 0.35\,M_\odot$) or they are fully convective ($M < M_{\rm conv}$). We model the influence of anisotropic irradiation on the structure of such stars as follows: We assume that a fraction s of the stellar surface is exposed to an incident flux $F_{\rm irr}$ and that the remaining fraction $(1-s)$ of the surface remains in the shadow. Denoting by R_s the radius of the star and by T_1 and T_2 respectively the effective temperatures on the unlit and the irradiated part, the stellar luminosity (energy lost by the star from its interior per unit time) is

$$L = 4\pi R_s^2 \{(1-s)\sigma T_1^4 + s\sigma T_2^4 - sF_{\rm irr}\} \quad . \tag{1}$$

In order to use the modified Stefan-Boltzmann law (1) as an outer boundary condition for stellar evolution calculations one has to know T_2. This, in turn, is determined by the details of how irradiation affects the energy loss through the irradiated surface.

Now it is important to realize that because the stars under investigation have a deep outer adiabatic convective envelope their mechanical and thermal structure remains spherically symmetric to a very high degree of accuracy out to the onset of the superadiabatic convection zone despite anisotropic irradiation. It is only the very thin superadiabatic layer (mass $\sim 10^{-10}\,M_\odot$), where energy is mainly transported via radiation, which is strongly affected by irradiation. So the asphericity of the star is restricted to a very thin surface layer, and it is this property which allows us to determine T_2 in (1) with a simple model. In this model we assume that convection is adiabatic out to a point where the pressure is $P = P_B$ and the temperature $T = T_B$ (see the sketch in Fig. 1).

The point (P_B, T_B) is the base of the superadiabatic convection zone which extends to the photosphere where the pressure is the photospheric pressure $P_{\rm ph}$ and the temperature the effective temperature $T_{\rm eff}$. Because in this zone convection is ineffective, we assume the energy transport to be only radiative, i.e. $\nabla_{\rm rad} = \nabla$ (where $\nabla \equiv \partial \ln T/\partial \ln P$). Because on the irradiated part of the star the effective temperature is $T_2 > T_1$, but T_B is the same for both parts, we see that irradiation, by raising the effective temperature, reduces the temperature gradient ∇ and thus the radiative

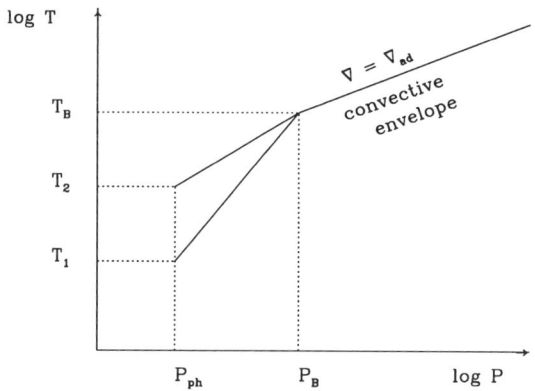

Figure 1. Sketch of the assumed run of temperature versus pressure in the outer layer of our model star.

energy loss through these layers. As is sketched in Fig. 1 we assume for simplicity that $P_{\rm ph}$ is the same on both parts of the star, though this is of course not strictly true. Using now the radiative diffusion approximation

$$F_{\rm rad} = -\frac{ac}{3\kappa\varrho}\frac{dT^4}{dr} \quad , \qquad (2)$$

where ϱ is the density, κ the opacity, and the other symbols have their usual meaning, and a power law approximation for κ

$$\kappa = {\rm const.}\, P^a T^b \quad , \qquad (3)$$

a one zone model for the superadiabatic layers yields (for details see Ritter et al. 1995)

$$\frac{F_{\rm rad,1}}{F_{\rm rad,2}} \equiv \frac{\sigma T_1^4}{\sigma T_2^4 - F_{\rm irr}} = \begin{cases} \frac{T_B^n - T_1^n}{T_B^n - T_2^n} \quad , & n = 5 - b \neq 0 \\[2mm] \frac{\ln T_B - \ln T_1}{\ln T_B - \ln T_2} \quad , & n = 5 - b = 0 \end{cases} \quad . \qquad (4)$$

Eq. (4) can be solved for $T_2(T_1, T_B, F_{\rm irr})$ which together with Eq. (1) can be used as an outer boundary condition for numerical calculations.

3. Stability against Irradiation-Induced Mass Transfer

Let's now examine the situation in which a low-mass star transfers mass to a compact companion and, in turn, is irradiated by the accretion light source. For reasons which will become clear later, we restrict our discussion mainly to CBs, i.e. the compact star is a white dwarf (of mass $M_{\rm WD}$ and radius $R_{\rm WD}$). The average flux with which the donor star is irradiated is

$$\langle F_{\rm irr} \rangle = \frac{\eta}{8\pi} \frac{GM_{\rm WD}(-\dot{M}_{\rm s})}{R_{\rm WD}\, a^2} \quad , \tag{5}$$

where $(-\dot{M}_{\rm s})$ is the mass transfer rate, a the orbital separation and $\eta \lesssim 1$ a dimensionless "efficiency" factor which absorbs such factors as the albedo of the donor and takes into account that the irradiating light source does not necessarily radiate isotropically. Now, from Eqs. (1) and (4) one derives that

$$\frac{\partial L}{\partial F_{\rm irr}} = -4\pi R_{\rm s}^2 s\, g(T_1, T_2, T_{\rm B}) < 0 \quad , \tag{6}$$

where

$$g(T_1, T_2, T_{\rm B}) = \begin{cases} \dfrac{nT_1^4 T_2^{n-1}}{nT_1^4 T_2^{n-1} + 4T_2^3(T_{\rm B}^n - T_1^n)}, & n = 5 - b \neq 0,\ T_{\rm B} > T_2 \\[2mm] \dfrac{T_1^4}{T_1^4 + 4T_2^4(\ln T_{\rm B} - \ln T_1)}, & n = 5 - b = 0,\ T_{\rm B} > T_2 \end{cases} \quad . \tag{7}$$

Thus the main effect of irradiating the star is to reduce its luminosity. This, in turn, means that part of the energy which the star generates in its interior is prevented from leaking through the irradiated surface and, therefore, is stored as internal and gravitational energy with the result that the star swells. In a mass-transferring binary, the swelling of the donor drives additional mass transfer and in this way generates even more irradiating flux. Therefore, we need now to examine under which conditions such a situation is stable against irradiation-induced runaway mass transfer. For this we write the temporal change of the stellar radius as

$$\frac{d\ln R_{\rm s}}{dt} = \zeta_S \frac{\dot{M}_{\rm s}}{M_{\rm s}} + \left(\frac{\partial \ln R_{\rm s}}{\partial t}\right)_{\rm ml} + \left(\frac{\partial \ln R_{\rm s}}{\partial t}\right)_{\rm irr} \quad , \tag{8}$$

where ζ_S is the adiabatic mass radius exponent, $(\partial \ln R_{\rm s}/\partial t)_{\rm ml}$ the thermal relaxation term due to mass loss and $(\partial \ln R_{\rm s}/\partial t)_{\rm irr}$ the one due to irradiation. The condition for (dynamical) stability against mass transfer can be derived in the same way as, e.g., in Ritter (1988). The result is

$$\zeta_S - \zeta_R > \zeta_{\rm irr} \equiv -M_{\rm s} \frac{\partial}{\partial M_{\rm s}} \left(\frac{\partial \ln R_{\rm s}}{\partial t}\right)_{\rm irr}$$

$$= -M_{\rm s} \frac{\partial}{\partial L} \left(\frac{\partial \ln R_{\rm s}}{\partial t}\right)_{\rm irr} \frac{\partial L}{\partial F_{\rm irr}} \frac{\partial F_{\rm irr}}{\partial M_{\rm s}} \quad , \tag{9}$$

where ζ_R is the mass radius exponent of the Roche radius, and $\zeta_{\rm irr}$ is a dimensionless number which measures how sensitively the stellar radius

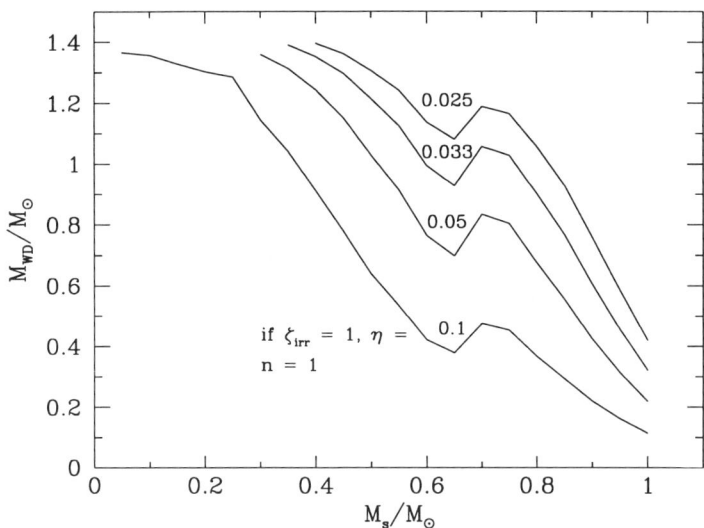

Figure 2. Contour lines for $\eta =$ const. in the $M_s - M_{WD}$ plane along which $\zeta_{irr} = 1$. The donor star is assumed to be on the main sequence with $T_2 = T_1$, $s = 0.5$ and $n = 1$ ($\hat{=} b = 4$, i.e. H$^-$ opacity).

changes in response to irradiation. In order to evaluate $\partial^2 \ln R_s / \partial L \partial t$ we use the bipolytrope model (e.g., Kolb & Ritter 1992) and obtain

$$\frac{\partial}{\partial L}\left(\frac{\partial \ln R_s}{\partial t}\right)_{irr} = \\ -\frac{R_s}{GM_3^2}\left\{\frac{\partial H}{\partial Q}\frac{1}{H_2(Q,n_1)} + \frac{1}{f}\frac{\partial f}{\partial Q}\right\}\left\{\frac{\partial H}{\partial Q} + \frac{H}{f}\frac{\partial f}{\partial Q}\right\}^{-1}, \quad (10)$$

where the quantities in the curly brackets $\{\}\{\}^{-1}$ define a dimensionless number which depends only on the relative size $Q = r_{core}/R_s$ of the radiative core and the polytropic index n_1 in it. In particular, for a single polytrope $n = 3/2$, i.e. $Q = 0$, one has $\{\}\{\}^{-1} = 7/3$. Taking now $F_{irr} = \langle F_{irr}\rangle$, Eqs (5,6,7) yield

$$\zeta_{irr} = \frac{\eta}{2}\frac{M_{WD}}{M_s}\frac{R_s}{R_{WD}}\left(\frac{R_s}{a}\right)^2 s\,\{\}\{\}^{-1}g(T_1,T_2,T_B) \ . \quad (11)$$

Since in a normal CB $\zeta_S - \zeta_R \approx 1$, systems in which $\zeta_{irr} \gtrsim 1$ are unstable against irradiation-induced runaway mass transfer. Under which conditions is $\zeta_{irr} > 1$? Using ZAMS models for the donor (to determine $R_s(M_s)$ and $T_B(M_s)$), Nauenberg's (1972) mass-radius relation for white dwarfs and $T_2 = T_1$ (initial state is an unirradiated star) we find that $\zeta_{irr} > 1$ is reached for surprisingly small values of η (see Fig. 2), i.e. $\eta \gtrsim 0.04\ldots 0.1$ is sufficient for instability. The conclusion to be drawn from this is that at least a large fraction of CBs is likely to be unstable against irradiation-induced runaway mass transfer.

Before we proceed to discuss the long-term evolution of CBs under this instability, we can now make a brief remark regarding the LMXBs. Replacing the white dwarf by a neutron star means that, if everything else remains equal, $\zeta_{\rm irr}$ is larger by a factor $R_{\rm WD}/R_{\rm NS} \gtrsim 10^3$ and that, therefore, $\zeta_{\rm irr} \gtrsim 1$ is reached for much smaller η, i.e. $\eta \gtrsim 10^{-3}$. Thus LMXBs are extremely unstable against irradiation of the donor star.

4. Long-Term Evolution of CBs under the Irradiation Instability

Since there is no observational evidence showing that the known CBs are in a phase of runaway mass transfer and since a number of well-observed properties of CBs such as the period gap can only be explained if their evolution on the long-term average follows closely a standard evolution without taking into account irradiation (e.g. Kolb & Ritter 1992; Kolb 1993), we must conclude that the instability is either not relevant for CBs (because η is too small) or, if it is, that the instability must be quenched such as to prevent the mass transfer from running away too strongly. If the latter is the case, it is then conceivable that CBs go through a limit cycle in which phases of high and low states alternate on a time scale short compared to the evolutionary time scale. The high state would correspond to a phase during which the donor swells in response to irradiation and transfers mass at a rate above the secular mean. Mass transfer above the secular mean, however, cannot be maintained indefinitely and sooner or later the system has to go into a low state during which the mass transfer rate is below the long-term mean and irradiation is correspondingly unimportant. The question, therefore, is whether there are such quenching mechanisms? In fact there are, and in the following we are going to describe two of them.

One quenching mechanism operates only for donor stars with a mass $M_s \gtrsim 0.6 M_\odot$. In this case quenching is possible because with increasing T_2 not only the temperature gradient decreases but also the optical depth in the superadiabatic layer becomes larger (note that the relevant opacity source is H^- and the corresponding value of b is $b \approx 4-5$). As a consequence of this $\partial \zeta_{\rm irr}/\partial T_2 < 0$. Numerical calculations of the long-term evolution of systems with a donor in the mass range $0.6\, M_\odot \lesssim M_s \lesssim 0.9\, M_\odot$ show in fact that the systems can go through a limit cycle, though the oscillations are damped on a time scale of $\sim 10^7$ yr. Details will be presented elsewhere (Ritter et al. 1995).

The other quenching mechanism is more fundamental. Quenching sets in as soon as $T_2 = T_B$. Since for stars with a mass $M \lesssim 0.6\,M_\odot$ already T_1 is only slightly lower than T_B, $T_2 = T_B$ is reached for rather small mass transfer rates. As soon as $T_2 \geq T_B$, the star's luminosity L does no longer depend on $F_{\rm irr}$, i.e. $\partial L/\partial F_{\rm irr} = 0$ and so $\zeta_{\rm irr} = 0$. This is because for $T_2 \geq T_B$ an isothermal layer is formed on the irradiated side of the star which extends from the photosphere ($P = P_{\rm ph}$) down to $P \approx P_B (T_2/T_B)^{1/\nabla_{\rm a}}$, and through which no energy is transported, i.e. $\sigma T_2^4 = F_{\rm irr}$. The mass transfer rate for which quenching sets in, i.e. when $T_2 = T_B$ is

$$\dot{M}_{\rm q} = \frac{1}{\zeta_{\rm irr}} \frac{s}{1-s} \frac{R_{\rm s} L}{GM_{\rm s}^2} \{\}\{\}^{-1} \left(\frac{T_B}{T_1}\right)^4 g(T_1, T_2 = T_B, T_B) \quad (12)$$

$$\lesssim \frac{s}{1-s} \frac{R_{\rm s} L}{GM_{\rm s}^2} \{\}\{\}^{-1} \ .$$

When compared with the mass transfer rates for a standard CB evolution (e.g., Kolb & Ritter 1992) we find that $\dot{M}_{\rm q}$ is less than $\dot{M}_{\rm tr}$, but by not more than a factor of two, if $M_{\rm s} \lesssim 0.5\,M_\odot$. This, in turn, would mean that no limit cycle is possible below $\sim 0.5\,M_\odot$. Rather such systems are stable and the irradiation instability permanently quenched, i.e. $T_2 > T_B$ at all times. However, as far as stars below $0.5\,M_\odot$ are concerned, our conclusions may be premature. This is because application of our model to stars with a mass $M \lesssim 0.6\,M_\odot$ is problematic. The reason is that Eq. (4) was derived in the diffusion approximation which requires that the optical depth in the superadiabatic layer is sufficiently high. This, however, is not the case if T_B is only slightly above T_1 as in main sequence stars with $M \lesssim 0.6\,M_\odot$.

Thus it remains to be demonstrated whether the irradiation instability described here, in combination with a suitable quenching mechanism, allows for a stable and large amplitude limit cycle which, in turn, could account for the observed large scatter of the mass transfer rates of CBs around the long-term mean.

References

D'Antona, F. & Ergma, E. 1993, A&A 269, 219
Frank, J., King, A.R. & Lasota, J.P. 1992, ApJ 385, L45
Gontikakis, C. & Hameury, J.-M. 1993, A&A 271, 118
Harpaz, A. & Rappaport, S. 1991, ApJ 383, 739
Hameury, J.-M., King, A.R., Lasota, J.P. & Raison, F. 1993 A&A 277, 81
Kolb, U. 1993, A&A 271, 149
Kolb, U. & Ritter, H. 1992, A&A 254, 213
Nauenberg, M. 1972, ApJ 175, 417
Podsiadlowski, Ph. 1991, Nat 350, 136
Ritter, H. 1988, A&A 202, 93
Ritter, H. 1994, in *Evolutionary Links in the Zoo of Interacting Binaries*, F. D'Antona (Ed.), Mem. Soc. Astron. It., (in press)
Ritter, H., Zhang, Z. & Kolb, U. 1995, (in preparation)

THE EVOLUTIONARY STATUS OF PSR 1718–19

ENE ERGMA
*Physics Department, Tartu University, Ulikooli 18,
EE2400 Tartu, Estonia.*

AND

MAREK J. SARNA
*N. Copernicus Astronomical Center,
Polish Academy of Sciences, ul. Bartycka 18, 00-716 Warsaw,
Poland.*

Abstract. Possible models for the matter source inside the eclipsing binary system PSR 1718–19, and for the evolution of this system are reviewed, including Zwitter's (1993) stripped main-sequence (MS) turnoff star model. Both the accretion induced collapse (AIC) scenario with a young neutron star, and the capture scenario with an old neutron star are discussed. Although Burderi & King (1994) claim that the size of the Roche lobe ($\sim 0.5\,R_\odot$) unambiguously rules out the AIC formation scenario, we show that in our evolutionary picture an AIC scenario will be possible.

1. Introduction

Lyne *et al.* (1993) discovered the binary radio pulsar PSR 1718–19 in the globular cluster NGC 6342. It is a 6.2 hour eclipsing binary system; the pulsar has a relatively long spin period of 1 s, a very strong magnetic field ($\sim 10^{12}$ G), and a small spin down age (~ 10 Myr). The mass of its companion is only 0.1–0.2$M_\odot$. The discovery of this system increases the number of systems in which the neutron star companion mass is low. Lyne *et al.* pointed out that there is a problem in explaining the source of material in this eclipsing system. They showed that for a MS star with $M_c \sim 0.1$–$0.2\,M_\odot$ the surface of the companion is well inside its Roche lobe. Also the possibility that the eclipsing material results from the ablation of the surface of the companion star by radiation from the pulsar, derived from its spin-down energy, fails. Due to the large spin period of the neutron star the rotational energy flux at the companion is too small ($\sim 2\ 10^9\,\mathrm{erg\,s^{-1}\,cm^{-2}}$) to cause significant evaporative mass loss.

The question arises whether this system is associated with the globular cluster NCG 6342. It lies at about 2.3 arcmin (or 17 core radii) from its center. Lyne et al. (1993) have argued that the position, distance and binary nature suggest that the system is probably associated with the cluster, although there is small (<1%) but not negligible probability that its association with the cluster is a chance superposition (Wijers & Paczynski 1993).

If we accept that the binary is a globular-cluster member then we need an extra mechanism to bring it to the outskirts of the cluster: in the old neutron star capture scenario the binary must have originated in the core of the cluster.

Another important question is whether the pulsar has been recently formed ($\sim 10^7$ yrs ago), or is an old object. Unfortunately, it seems to us that this question can not be easily answered.

So far, several models have been proposed to explain the source of the matter in this eclipsing binary system. We shall give a brief review of these models. Then, using the model of Zwitter (1993), we describe evolutionary scenarios including both the young neutron star version as well as the capture version (old neutron star).

2. How to explain Matter inside the Binary System?

Ergma (1993) invoked X-ray pre-heating of the secondary in AM Her type binaries. She showed that if the hard X-ray luminosity is of the order of $10^{33}\,\mathrm{erg\,s^{-1}}$ then the secondary will be heated and be out of thermal equilibrium. After collapse of the massive white dwarf, and spin down of the neutron star to its current rotational period the pulsar action will switch off. Since the thermal time scale of the companion is much longer ($\sim 100\,\mathrm{Myr}$) than the system time scale ($\sim 10\,\mathrm{Myr}$) the companion has no time to relax to its thermal equilibrium, and its radius will be larger than its equilibrium value. It may nearly fill its Roche lobe and hence be a source of the material inside the system.

Zwitter (1993) has argued that the companion is the stripped turn off star of 0.2–$0.4\,M_\odot$ with an $\sim 0.1\,M_\odot$ helium core, and an ~ 1.8 times bigger radius than a MS star of same mass. Its position in the H-R diagram overlaps that of an $\sim 0.65\,M_\odot$ MS star, and it is then only slightly underfilling its Roche lobe near $P = 6.2\,\mathrm{hrs}$.

Wijers & Paczynski (1993) have discussed the possibility that this system is old, and has been formed via capture of the companion by an old neutron star. To explain the matter inside the system they proposed that the companion is similar to the very active stars found in RS CVn systems. The cause of mass loss in these binaries is thought to be the combination of magnetic activity and the rapid rotation.

To explain the low-mass companion for this system, Verbunt (1994) proposed a new mechanism that leads to such low companion masses – extreme tidal heating as the orbit circularizes when the neutron star is formed, or is exchanged into the binary. The circularization is accompanied by heating of the companion of the neutron star and the amount of energy produced can be comparable to the binding energy of the companion which may cause the strong mass loss. Thus the star will have again a radius larger than a MS star of the same mass.

So three models (Ergma, Zwitter, Verbunt) predict that the secondary is out of thermal equilibrium and hence must be rather bright.

Wijers & Paczynski (1993) have proposed an observational test to clarify the nature of the secondary [in their model neither a normal low-mass star nor a stripped (bloated) star]. If the optical counterpart is a MS star then it should have $m_I > 24.5$ and $R-I > 1.1$. If the pulsar formed via collapse of a white dwarf or has a stripped companion star then $m_I \sim 22.6$ and $R-I \sim 0.5$. To detect the companion a more accurate position determination is needed, which is not yet available (Lorimer, private communication).

3. Scenario with a Young Pulsar

Let us assume an initial system with a massive NeOMg white dwarf and with slightly evolved secondary. For this case initial Roche lobe filling will occur at a relatively large orbital period ($P \sim 19\,\mathrm{hr}$), when the secondary will have a small helium core ($M_{\mathrm{He}} \sim 0.01 M_\odot$). The mass transfer starts and the system evolves, driven by magnetic braking. Near an orbital period of 6 hr the mass of the secondary has decreased to $0.2\,M_\odot$ (Fig. 1). Let us suppose that the accretion induced collapse happened near an orbital period of $\sim 6.1\,\mathrm{hr}$. The maximum value of the mass lost in the form of binding energy is roughly $0.18\,M_\odot$ (Zeldovich & Novikov 1971), but recent computations of accretion induced collapse of a massive NeOMg white dwarf ($M_{\mathrm{NeOMg}} = 1.376\,M_\odot$) by Woosley & Baron (1992) have shown that only about $\sim 0.01\,M_\odot$ has been ejected.

If we accept that there is no other "internal" kick mechanism operating, this mass loss value will lead to the following post-collapse eccentricity, semi-major axis and centre of mass velocity of the system (Dewey & Cordes 1987):

$$e = \frac{\Delta M}{M_1 + M_2 - \Delta M}, \qquad a_{\mathrm{f}} = a_{\mathrm{i}}/(1-e), \qquad v_{\mathrm{s}} = ev_1$$

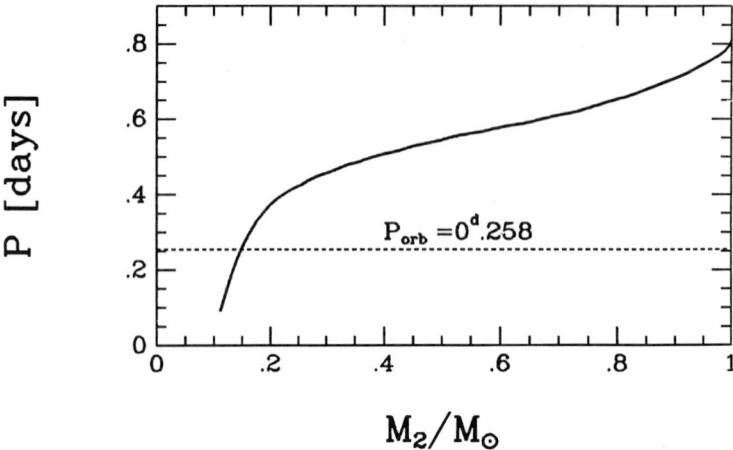

Figure 1. The secondary mass and orbital-period dependence.

For $M_1 = 1.376\,M_\odot$, $M_2 \sim 0.2\,M_\odot$, $\Delta M \sim 0.01\,M_\odot$, we have $e \sim 6\,10^{-3}$. Although the post-collapse orbital separation differs very little from the pre-collapse orbital separation it may be enough to force the secondary to be slightly underfilling its Roche lobe. Since pre- and post-collapse orbital separations are very similar, no violent tidal circularization of the orbit will occur, as it was suggested for large eccentricities by Verbunt (1994), The pre-collapse orbital velocity v_1 (for $P \sim 6.1\,\mathrm{hr}$) is $\sim 50\,\mathrm{km\,s^{-1}}$, hence the post-collapse centre of mass velocity is $\sim 0.3\,\mathrm{km\,s^{-1}}$. This value is less than the central velocity dispersion of NGC 6342 ($v_c = 8\,\mathrm{km\,s^{-1}}$, Pryor & Meylan 1993), so it is insufficient to move the binary outside the center of cluster.

3.1. AIC AND NOVA EXPLOSION

The AIC scenario requires that there must be cataclysmic variables in globular clusters. Up to now, CVs are the rarest type of binaries known in globular clusters (see, e.g., Shara *et al.* 1994). Also, theoretical views on this topic are quite controversial. While Hertz & Grindlay (1983) and Rappaport & Di Stefano (1993) have suggested high numbers of CVs in globular clusters, Verbunt & Meylan (1988) predicted a much lower number on the basis of the lower capture probability of WDs compared to NS.

If we assume a normal CV evolution in globular clusters then we have to face a second problem. It is connected with nova explosions and the subsequent mass loss, which does not allow the increase of the mass of the white dwarf required for AIC. Truran *et al.* (1988) have shown that if the accreted matter is hydrogen poor (the hydrogen content by mass is less

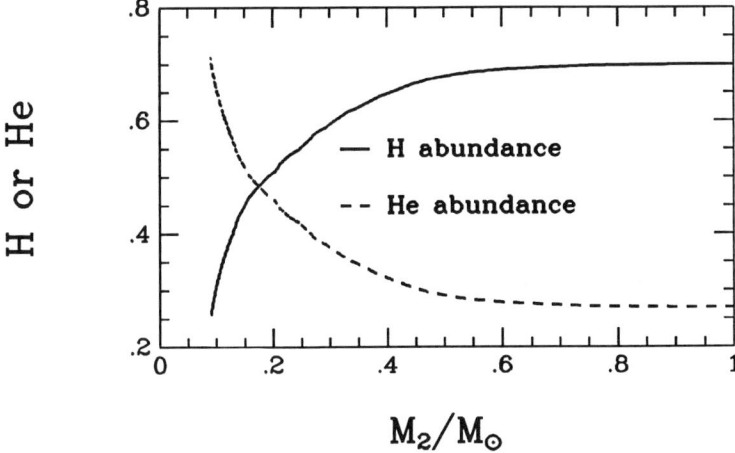

Figure 2. The hydrogen (thick line) and helium content (dashed line) dependence on the secondary mass; it is assumed that $M_1 = 1.4\,M_\odot$.

than 0.6) then significant radius expansion does not occur and no mass is ejected during a thermonuclear flash. According to our calculations (Fig. 2) the surface hydrogen abundance is less than 0.6 when the secondary mass has decreased to $0.36\,M_\odot$. The it is possible to accrete $\sim 0.16\,M_\odot$ without mass ejection, and hence the AIC may occur (see also Livio 1992). We shall present more details in our full paper (Ergma *et al.* 1994).

4. Old Neutron Star Scenario

In this scenario the binary has originated at the center of the cluster by the tidal capture of the secondary by an old neutron star.

For the companion to be captured but to have avoided destruction, its distance of closest approach to the neutron star at the first passage, d, must have been close to $3\,R_c$. If we assume conservation of angular momentum during the circularization the final distance is $a_f = 2\,d = 6\,R_c$ (Wijers & Paczynski 1993). As an estimate we take $a_f = (4-6)\,R_c$. We also propose that the normal companion with mass about $1\,M_\odot$ is a slightly evolved star (as is the case with the young neutron star version). If the $1\,M_\odot$ star has an age of $\sim 10^{10}$ yrs its radius is $1.36\,R_\odot$. After circularization the final orbital separation $a_f = (5.44-8.16)\,R_\odot$ and the orbital period $P_f = 0.95-1.75$ days. Magnetic braking will decrease the orbital separation, and the star with $R_c = 1.36\,R_\odot$ will fill its Roche lobe when $P = 0.57\,\mathrm{d}$. To decrease the period from P_f to $P = 0.57\,\mathrm{d}$ a time interval of $\sim 10^8$ yr is needed. After filling the Roche lobe the evolutionary path for this case is similar to that in the AIC scenario (Fig. 1). Only, we ought to follow also the history of the neutron star and of its field in a manner described by Muslimov & Sarna (1993). Since we have assumed that the captured neutron star was old its initial rotational period (before the capture) may have been of the order of several seconds. After starting the mass transfer the pulsar rotational period decreases. When the orbital period is about 6–7 hours a third body is assumed to arrive near the binary (the pulsar period is then ~ 1 s). The whole system receives some additional kick velocity which may bring the binary to the outskirts of the cluster (Hills 1976). If the binary was ejected from the center of the cluster due to this three-body interaction then its space velocity must be between v_c and $v_{esc} = 2v_c$. We can see that the time required to bring the binary to its current place is $(3-6)\,10^5$ yrs which is much less than the system life time estimated from $\dot{P}$ and P (Lyne et al. 1993).

5. The Fate of the System in both Scenarios

Since the Roche lobe filling star originally was evolved, its future evolution will be as follows. According to Tutukov et al. (1985) (see also Ergma 1991) the star becomes helium rich during the mass loss and it will evolve towards very short orbital periods.

If magnetic-field decay occurs it is possible to predict an evolutionary sequence such as PSR 1718–19 $\Longrightarrow$ PSR 1744–24A $\Longrightarrow$ X 1916–05 $\Longrightarrow$ X 1820–30. If there is no magnetic field decay then we may expect that PSR 1718–19 evolves to a low-mass X-ray binary pulsar with a very short orbital period, like X 1627–67 (Levine et al. 1988).

6. Discussion

In their recent paper Burderi & King (1994) claim that the Roche lobe radius value ($\sim 0.5\,R_\odot$) unambiguously rules out the AIC formation scenario which requires a lobe-filling companion close to the main sequence. In our evolutionary picture we have shown that it is not necessarily the case. Having a slightly evolved star as the secondary at the beginning of the binary evolution, the observed secondary mass as well as the near filling of the Roche lobe are easily achieved. We think that the observational test which has been proposed by Wijers & Paczynski (1993) is very important to restrict the proposed models.

Acknowledgments. This work was supported in part by the Polish National Committee for Scientific Research grant and the Estonian Science Foundation grant N 625. Part of this paper has been prepared during the visit of E^2 to NORDITA, whose financial aid is acknowledged.

References

Burderi, L. & King, A.R. 1994, ApJ 430, L57
Dewey, R.J. & Cordes, J.M. 1987, ApJ 321, 780
Ergma, E. 1991, Comments on Astrophys. 15, 239
Ergma, E. 1993, A&A 273, L38
Ergma, E., Sarna, M.J. & Giersz, M. 1994, (in preparation)
Hertz, P. & Grindlay, J. 1983, ApJ 273, 105
Hills, J. 1976, MNRAS 175, 1P
Levine, A. et al. 1988, ApJ 327, 732
Livio, M. 1992, in *Interacting Binaries*, H. Nussbaumer & A. Orr (Eds.), Springer Verlag (Berlin), p. 250
Lyne, A., Biggs, J., Harrison, P. & Bailes, M. 1993, Nat 361, 47
Muslimov, A.G. & Sarna, M.J. 1993, MNRAS 262, 164
Pryor, C. & Meylan, G. 1993, ESO preprint No. 932
Rappaport, S. & Di Stefano, R. 1993, in *Cataclysmic Variables and Related Physics*, 2nd Technion Haifa Conf., O. Regev & G. Shaviv (Eds.), Israel Phys. Soc. (Jerusalem), p. 48
Shara, M.M., Bergeson, L.E. & Moffat, A.F.J. 1994, ApJ 429, 767
Truran, J.W. et al. 1988, ApJ 324, 345
Tutukov, A.V. et al. 1985, SvAL 11, 52
Verbunt, F. 1994, A&A 285, L21
Verbunt, F. & Meylan, G. 1988, A&A 203, 297
Wijers, R. & Paczynski, B. 1993, ApJ 415, L115
Woosley, S.E. & Baron, E. 1992, ApJ 391, 228
Zeldovich, Ya. & Novikov, I.D. 1971, *Relativistic Astrophysics, Vol. 1, Stars and Relativity*, University of Chicago Press
Zwitter, T. 1993, MNRAS 264, L3

Discussion

S.R. Kulkarni: The location (140 arcsec from the core) of PSR 1718–19 is puzzling. Compare this with the half mass radius, i.e., ∼40 arcsec. This means one has to finetune the systemic velocity in order not to lose the pulsar (to the Galactic field). I wonder how improbable this event is, from a dynamical point of view.

H. Ritter: First, I would like to disagree with the comment made by dr Lipunov, namely that white dwarfs, if sufficiently magnetic, can grow in mass even at low accretion rates. After all, nova Cyg 1975 is an AM Her star, i.e., has a strongly magnetic white dwarf; yet it underwent a nova explosion and ejected mass. Second, as to Nomoto's results about core collapse of an accreting white dwarf at $\dot{M} \approx 10^{-9} M_\odot \, yr^{-1}$, this accretion rate refers to accretion of helium, but not of hydrogen-rich matter.

THE ECOLOGY OF MAGNETIC ROTATORS

V.M. LIPUNOV
Sternberg Astronomical Institute
Universitetskij pr. 13, 119899,
Moscow V-234, Russia

Abstract. A review is given concerning the current state of the theory of evolution of magnetic compact stars. The intrinsic evolution of the magnetized compact star is shown, both theoretically and numerically, to be the decisive factor in explaining observable properties, and in predicting yet unknown properties of high-energy radiation sources in our and other galaxies. The main results are given of recent evolutionary scenario simulations (Scenario Machine) by the Monte-Carlo method.

1. Introduction

It became clear in the early 1980's that the level of comprehension of stellar evolution theory on the one hand, and the observed variety of astrophysical properties of stars gained in numerous space experiments on the other hand, approached a point where construction of particular scenarios for newly discovered exotic sources could be a viable approach. The importance of involving not only a general evolutionary scenario for the joint evolution of normal and compact stars, but also numerical simulations, using the so-called Scenario Machine (Lipunov 1991) became quite obvious (Lipunov 1982a). The first numerical simulation of the evolution of massive binary systems taking into account rotation of neutron stars was carried out by Kornilov & Lipunov (1983a,b; 1984). Subsequently, the ideas and method were extrapolated to low-mass stars (Lipunov & Postnov 1987, 1988). Finally, we now have a general scenario and numerical model for the evolution of binary systems with arbitrary masses (see, e.g., Tatarintzeva et al. 1989; Lipunov et al. 1994a,b).

TABLE 1. Classification of Magnetic Rotators

	Type	Relation between characteristic radii	Accretion rate	Observational appearances
E	ejector	$R_{st} > R_G$ $R_{st} > R_l$	$\dot{M}_c \leq \dot{M}_{cr}$	radio pulsars soft γ-ray repeater Cyg X-3? LS I+61°303?
P	propeller	$R_c < R_{st}$ $R_{st} \leq \max\{R_G, R_l\}$	$\dot{M}_c \leq \dot{M}_{cr}$	X-ray transients? rapid burster? γ-bursters??? magnetic Ap stars
A	accretor	$R_{st} \leq R_G$ $R_{st} \leq R_l$	$\dot{M}_c \leq \dot{M}_{cr}$	X-ray pulsars, bursters, CVs intermediate polars
G	georotator	$R_G < R_{st}$ $R_{st} \leq R_c$	$\dot{M}_c \leq \dot{M}_{cr}$	Earth, Jupiter
M	magnetor	$R_{st} > a$ $R_c > u$???	$\dot{M}_c \leq \dot{M}_{cr}$	AM Her, polars
SE	superejector	$R_{st} > R_l$	$\dot{M}_c > \dot{M}_{cr}$	?
SP	superpropeller	$R_c < R_{st} \leq R_l$ $R_{st} \leq R_l$	$\dot{M}_c > \dot{M}_{cr}$	?
SA	superaccretor	$R_{st} \leq R_c$ $R_{st} \leq R_G$	$\dot{M}_c > \dot{M}_{cr}$	SS 433? T Tau stars? ultrasoft superluminous sources?

2. Classification of Magnetic Rotators and Binary Systems.

The current scenario for the evolution of binaries based upon the original ideas which first appeared in Paczyński (1971), Tutukov & Yungelson (1973), Van den Heuvel & Heise (1972), was joined with the ideas of neutron star evolution [see the pioneering works by Shvartsman (1970, 1971), as well as Illarionov & Sunyaev (1975), Bisnovatyi-Kogan & Komberg (1975), Shakura (1975), Wickramasinghe & Whelan (1975), Lipunov & Shakura (1976), Savonije & Van den Heuvel (1977), and Lipunov (1982a)]. This joint scenario has allowed the construction of a two-dimensional classification of all possible states of binary systems containing neutron stars (Kornilov & Lipunov 1983a).

The full classification of neutron stars and white dwarfs (magnetic rotators) is based upon the idea that the astrophysical manifestations of these stars mainly reflect the character of the interaction of their electromagnetic fields with the surrounding plasma which tends to accrete under the influence of gravitational field.

The complete classification of magnetic compact stars includes eight types (Lipunov 1982a, 1984; see also Lipunov 1992).

2.1. NOMENCLATURE

The interaction of a magnetic rotator with the surrounding plasma to a large extent depends on the relative sizes of four characteristic radii: the stopping radius, R_{st}, the gravitational capture radius, R_G, the light cylinder radius, R_l, and the corotation radius, R_c. The differences between the interaction regimes are so significant, that the magnetic rotators show entirely different behaviour in different regimes. Hence the classification of the interaction regimes may well mean the classification of magnetic rotators. The classification notation and terminology are described below, and summarized in Table 1 (based on Lipunov 1987).

Naturally, not all possible combinations of the characteristic radii can be realized. For example, the inequality $R_l > R_c$ is not possible in principle. Furthermore, some combinations require unrealistically large or small parameters of magnetic rotators. For the same proper and external conditions the same rotator may gradually pass through several interaction regimes. Such a process will be termed the *evolution of a magnetic rotator*.

2.2. CLASSIFICATION OF NORMAL STARS IN BINARY SYSTEMS

The four basic evolutionary stages characterizing a normal star in a binary system can be specified as follows:

I – The size of the star is much smaller than the critical Roche lobe (usually such a star lies on the main sequence). The duration of the first stage is approximately equal to the hydrogen burning time (T_H);
II – The star leaves the main sequence and goes up to the supergiant region (as before, the star does not fill the Roche lobe). The duration of this stage is about one tenth of T_H;
III – The star fills the Roche lobe and begins to flow out to the companion. For large mass ratios, the flow takes place on the thermal time scale of the mass donor;
IV – In the process of mass exchange the envelope of the star flows over and a helium star forms (WR-star). The lifetime of the helium star is determined by the nuclear burning of helium.

In Table 2 we present the two-dimensional classification of massive binary systems with neutron stars and show a matrix of observational candidates.

TABLE 2. Massive Binaries with Neutron Stars; the Matrix of States.

	I main sequence	II supergiant	III Roche lobe overflow	IV Wolf-Rayet star	NS
E	PSR 1259–63[1]	?	?	Cyg X-3(?)[2]	PSR 1913+16
P	?	?	?	?	?
A	A 0535+26	Cen X-3	Her X-1	?	?
G	?	?	—	?	?
M	?	—	—	—	?
SE	?	?	?	?	?
SP	?	?	?	?	?
SA	—	—	SS 433(?)	—	?
BH	?	Cyg X-1	?	?	PSR B0042–73(?)[3,4]
SBH	—	—	SS 433(?)	?	—

[1] Johnston, S. et al. 1992
[2] Van Kerkwijk, M.H. et al. 1992
[3] Kaspi, V.M. et al. 1994
[4] Lipunov, V.M., Postnov, K.A. & Prokhorov, M.E. 1994c

3. Evolution of Neutron Stars

The evolution of neutron stars must be investigated taking into account the evolution of normal stars. This problem was discussed qualitatively by Bisnovatyi-Kogan & Komberg (1974), Van den Heuvel (1977) and Lipunov (1982a). We begin with the qualitative analysis presented in the latter paper.

The most convenient way of analyzing the evolution of a neutron star is with the help of the $P - y$ diagram (Lipunov 1987). In the expression for the stopping radius in the subcritical regime ($\dot{M}_c \geq \dot{M}_{cr}$) one notes that the magnetic dipole moment μ and the accretion rate $\dot{M}_c$ always appear in the same combination,

$$y = \frac{\dot{M}_c}{\mu^2},$$

as was first noticed by Davies & Pringle (1981). The parameter y characterizes the ratio between the gravitational and magnetic "properties" of a star and will, therefore, be called the gravimagnetic parameter. Two magnetic rotators having quite different magnetic fields and external conditions but identical gravimagnetic parameters have similar magnetospheres, as long as the accretion rate is quite low ($\dot{M}_c \leq \dot{M}_{cr}$). Otherwise, the flux of matter

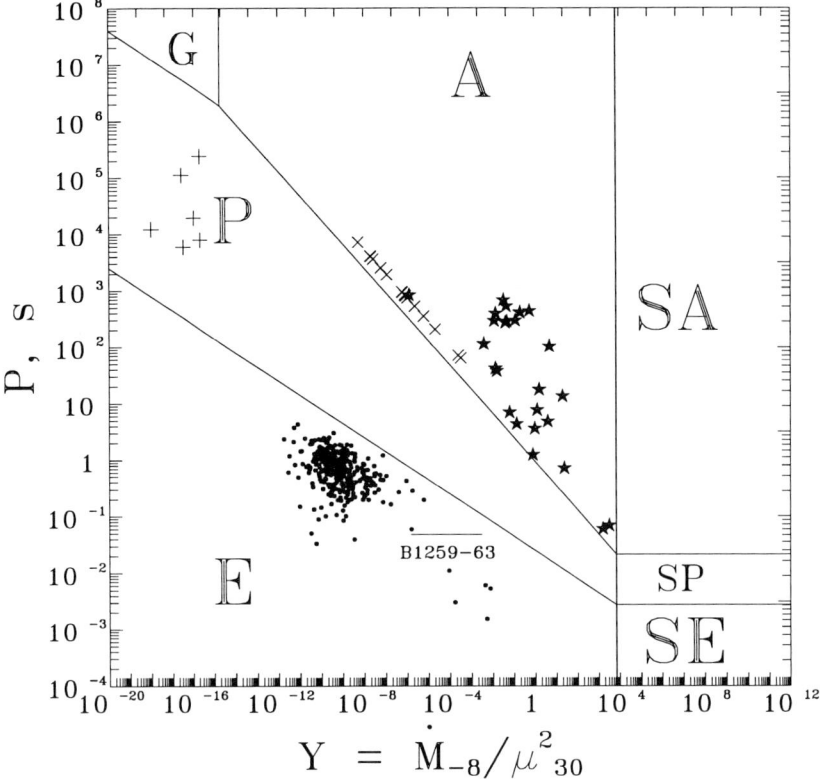

Figure 1. "$P-y$" diagram for neutron stars and white dwarfs. This version of the diagram contains binary WD (intermediate polars - ×), isolated WD (cross), isolated NS (radio pulsars - dots), binary accreting NS (X-ray pulsars – asterisks) (Lipunov, Postnov & Prokhorov 1995).

near the stopping radius no longer depends on the accretion rate at a large distance.

In fact, the number of independent parameters can be further reduced (see Osminkin & Prokhorov 1994) by introducing the parameter

$$y = \frac{\dot{M}_c v_\infty}{\mu^2}.$$

Plotting the rotator's period P versus y we obtain a somewhat less obvious, but more general, classification diagram than the "$P-L$" diagram discussed in Lipunov (1982a) or the "$P-B$" diagram (see, e.g., Ghosh, these proceedings).

In Fig. 2 we show the evolutionary tracks of neutron stars. As a rule, the neutron star is generated at the instant when the companion lies on the main sequence (track b). During the first 10^{5-7} years, the neutron star

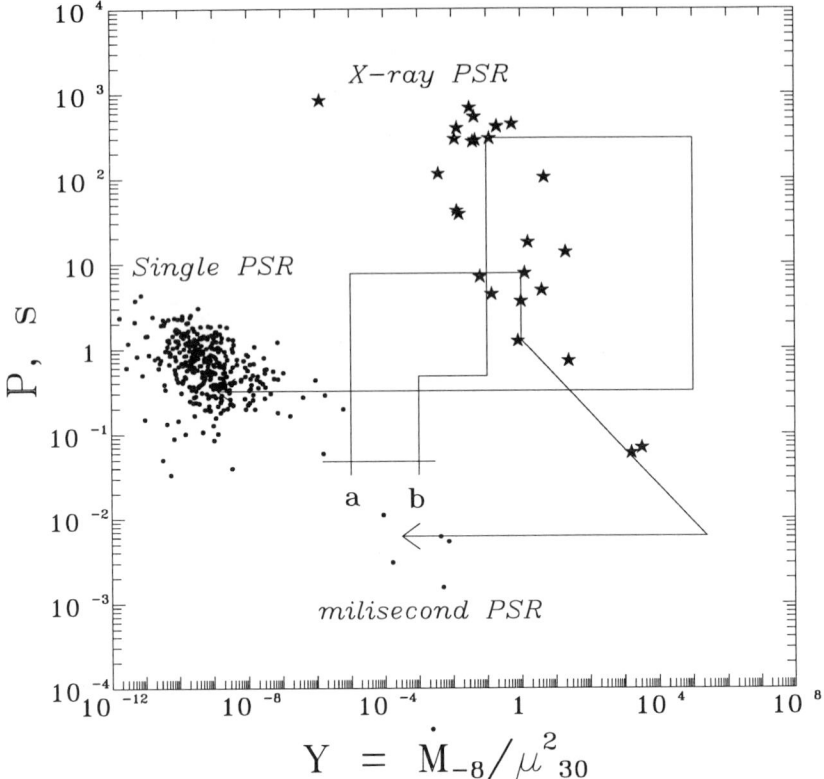

Figure 2. Tracks of neutron stars on a period (P) versus gravimagnetic parameter (y) diagram. (b) – track of a neutron star with standard magnetic field in a massive binary; (a) – track of a neutron star with magnetic-field decay in a binary system with a low-mass companion.

is an ejector; usually it does not manifest itself as a radio pulsar since its radio waves are absorbed in the stellar wind of the normal star. The period of the neutron star increases in accordance with the magnetic dipole losses.

After this, matter penetrates into the light cylinder and the neutron star passes first into the propeller stage, and then into the accretor stage. By this time, the normal star leaves the main sequence and the stellar wind is intensified. This results in the generation of a bright X-ray pulsar. The period of the neutron star stabilizes around its equilibrium value. Finally, the normal star fills its Roche lobe and the accretion rate suddenly increases; the neutron star moves first to the right and then vertically down on the $P-y$ diagram. In other words, the neutron star moves into the supercritical stage **SA** (superaccretor). Its period tends to a new equilibrium value (see Lipunov 1992):

$$P_{eq}(\mathbf{SA}) \simeq (0.17\,s)\, \mu_{30}^{2/3} m_X^{-1/9}.$$

After the mass exchange, only the helium core of the normal star is left (WR star), a separate system is formed, and the neutron star reverts to the propeller or ejector stage. Accretion is hampered by rapid rotation. This is probably the reason behind the absence of X-ray pulsars in pairs with Wolf-Rayet stars (Lipunov 1982b). Since the helium star does not have a long life ($\sim 10^5$ yrs), the neutron star does not have time to spin down considerably: after the explosion of the normal star, the system disintegrates and the neutron star becomes an ejector, that is, a radio pulsar.

The "loop-shaped" track discussed above can be written in the form:

$$\mathbf{I\,E} \to \mathbf{I\,P} \to \mathbf{II\,P} \to \mathbf{II\,A} \to \mathbf{III\,SA} \to$$
$$\mathbf{IV\,P} \to \mathbf{E+E} \text{ (recycled pulsar)} \to \ldots$$

$$\mathbf{I\,E} \to \mathbf{I\,P} \to \mathbf{II\,A} \to \mathbf{III\,SA} \to \mathbf{IV\,E} \text{ (recycled ejector)} \to$$
$$\mathbf{IV\,P} \to \mathbf{E+E} \text{ (recycled pulsar)} \to \ldots$$

The overall lifetime of a neutron star in a binary system depends on the lifetime of the normal star and the parameters of the binary system. However, the rate of transition from one stage to another is proportional to the magnetic field strength of the neutron star.

3.1. EVOLUTION EQUATION

An analysis of the nature of the interaction of a magnetized star with the surrounding plasma allows us to write the approximate evolutionary equation for the angular momentum of a magnetic rotator in the general form (Lipunov 1982a)

$$\frac{dI\omega}{dt} = \dot{M} k_{\rm su} - \kappa_t \frac{\mu^2}{R_t^3},$$

where $k_{\rm su}$ is the specific angular momentum applied by the accretion matter to the rotator. This quantity is given by $k_{\rm su} = (GM_X R_d)^{1/2}$ for Keplerian disk accretion, by $k_{\rm su} = \eta_t \Omega R_G^2$ for wind accretion in a binary, and $k_{\rm su} \sim 0$ for a single magnetic rotator. Here R_d is the radius of the inner disk edge, Ω is the rotational frequency of the binary system, and $\eta_t \sim 1/4$ (Illarionov & Sunyaev 1975). The values of the dimensionless factor κ_t, the characteristic radius R_t and the accretion rate $\dot{M}$ in different regimes are presented in Table 3.

The evolution equation is approximate. In practice, the situation concerning propellers and superpropellers is not yet clear. In Table 3 R_m is the size of the magnetosphere whose value at the propeller stage is not known

TABLE 3. Parameters of the Evolution Equation of a Magnetic Rotator

Parameter	Regime					
	E,SE	P,SP	A	SA	G	M
$\dot{M}$	0	0	$\dot{M}_c$	$\dot{M}_c(R_A/R_s)$	0	$\dot{M}_c$
κ_t	$\sim 2/3$	$\lesssim 1/3$	$\sim 1/3$	$\sim 1/3$	$\sim 1/3$	$\sim 1/3$
R_t	R_l	R_m	R_c	R_c	R_A	a

accurately, and may differ significantly from the standard expressions for the Alfvén radius.

4. Scenario Machine: Computational Method.

To analyze the properties of an ensemble of binaries in the Galaxy, a special numerical program has been designed. In essence, it is a rough model of the real Galaxy. Calculations are carried out using the Monte-Carlo method. A binary system consisting of two normal stars is chosen. The time of its formation is drawn at random, and so are its binary parameters, which are distributed in accordance with currently established empirical laws. The binary system evolves in accordance with the pattern described above. Each star gradually passes through the stages I, II, III, IV. The parameters of the system and the outflow of matter from it are assumed to be constant within each stage. The duration of the stages is calculated from approximation formulae (see, e.g., Lipunov & Postnov 1988). After the appearance of a neutron star or white dwarf in the binary system, the magnetized compact star evolution block is introduced in accordance with the spin evolution equation (see Lipunov 1992). Obviously, the state of the binary system is described by a two-dimensional classification (Kornilov & Lipunov 1983a). For example, state IIA means that we are dealing with a binary system in which the normal star is in stage II, with a supergiant that does not fill its Roche lobe, while the neutron star is in the accretor stage. A typical representative of such systems is the classical massive binary X-ray pulsar Vela X-1. The method of calculating the joint evolution of a NS and the normal companion in the binary system for statistical comparison with observed characteristics (and also for predicting the hitherto undefined stages of evolution of massive binaries) is based on the calculation of the evolution of large number (about 10^6 systems in one experiment) of binary systems with randomly chosen parameters.

5. Monte Carlo Scenario Machine History

The following five main physical processes must be included into the modern scenario of the binary-star evolution:

a Mass exchange;
b Loss of orbital angular momentum due to gravitational waves;
c Loss of orbital angular momentum due to magnetic wind;
d Evaporation of the primary star by a radio pulsar companion;
e Magnetic compact star spin evolution.

History:
$a + e$; Massive binaries – Kornilov & Lipunov (1983a,b,1984)
$a + e + b + c$; Low-mass binaries – Lipunov & Postnov (1987,1988)
a; Massive binaries – Dewey & Cordes (1987)
$a + b + c + d + e$; All masses – Lipunov et al. (1994a,b); see Lipunov (1991)
$a + b + c$; All masses – Tutukov, Yungelson & Iben (1992)
$a + b + c$; Low-mass binaries – Kolb (1993)
$a + b + c$; Low-mass binaries – Rappaport (1994) (see these proceedings)
e; Isolated pulsars – Bailes (1994) (see these proceedings)
$a + b + c$; All masses – Tauris (1994)

6. Scenario Machine: the Matrix of Galactic Binaries

In Table 4, we present the main statistical result of the calculation using the Scenario Machine: the distribution of the number of the different types of binaries (Lipunov et al. 1994a,b). We used 500 000 artificial binaries.

Here α_{CE}, α_q, M_{cr} are the common-envelope parameter, the initial mass ratio exponent, and the minimal presupernova mass which produces a black hole, respectively.

7. Binary Radio Pulsars with Normal Stars

After 25 years of radio pulsar observations, a radio pulsar, PSR 1259–63, was discovered in a binary system containing an optical companion; this radio pulsar is paired with a Be-star in a highly eccentric orbit (Johnston et al. 1992). This discovery is an excellent confirmation of the modern theory of evolution of binary stars including the evolution of magnetized compact stars which was basically elaborated before the beginning of the 1980's (Kornilov & Lipunov 1984).

The natural questions arise – how this eccentricity could be explained in the framework of the theory of binary evolution, and what is the probability to find a radio pulsar on such stretched orbit? To answer these

TABLE 4. SCENARIO MACHINE: The Matrix of Binary Systems States
$\alpha_{CE} = 0.5$, $\alpha_q = 2$, $M_{cr} = 35\,M_\odot$, $k_{BH} = 0.3$, kick = $75\,\text{km s}^{-1}$

		Normal stars				White dwarfs			Neutr. stars		BH
		I	II	III	IV	E	P	A	E	P	
	I	$1\,10^9$	–	–	–	–	–	–	–	–	–
	II	$3\,10^8$	$2\,10^7$	–	–	$7\,10^1$	$3\,10^1$	–	–	–	–
	III	$1\,10^2$	–	–	–	–	–	–	$4\,10^{-1}$	$4\,10^0$	–
	IIIe	$1\,10^7$	–	$1\,10^0$	–	–	–	–	–	–	–
	IV	$2\,10^4$	$4\,10^2$	$2\,10^{-1}$	$9\,10^1$	$3\,10^1$	$6\,10^1$	–	$3\,10^0$	$2\,10^{-1}$	–
	E	$2\,10^5$	$1\,10^5$	–	–	$1\,10^8$	$1\,10^8$	$9\,10^3$	$1\,10^5$	$2\,10^5$	–
	P	$2\,10^2$	$2\,10^4$	$9\,10^4$	$3\,10^2$	$1\,10^8$	$1\,10^8$	$3\,10^4$	$6\,10^4$	$1\,10^5$	–
W	A	$7\,10^3$	$1\,10^6$	$3\,10^5$	$5\,10^1$	$6\,10^5$	$2\,10^5$	$6\,10^3$	–	–	–
D	G	$1\,10^7$	$7\,10^5$	–	$4\,10^1$	$6\,10^3$	$2\,10^3$	–	$1\,10^5$	$2\,10^5$	–
	SA	–	–	$3\,10^2$	–	–	–	–	–	–	–
	M	$5\,10^6$	–	$2\,10^4$	–	–	–	–	$1\,10^3$	–	–
	E	$5\,10^3$	$8\,10^2$	$5\,10^1$	$1\,10^1$	$4\,10^4$	$9\,10^3$	–	$7\,10^4$	$6\,10^4$	$3\,10^4$
	P	$4\,10^2$	$1\,10^2$	$2\,10^3$	$1\,10^1$	$8\,10^4$	$5\,10^4$	–	$3\,10^4$	$2\,10^5$	$6\,10^4$
N	A	$1\,10^3$	$7\,10^1$	$8\,10^1$	$1\,10^{-1}$	–	–	–	$5\,10^3$	$4\,10^4$	–
S	G	$1\,10^3$	$5\,10^1$	–	–	$8\,10^{-1}$	–	–	–	–	–
	SE	–	–	–	–	–	–	–	–	–	–
	SP	–	–	–	–	–	–	–	–	–	–
	SA	–	–	$1\,10^1$	–	–	–	–	–	–	–
	BH	$4\,10^1$	$2\,10^1$	$8\,10^3$	$2\,10^0$	$8\,10^3$	$2\,10^5$	–	$2\,10^4$	$1\,10^5$	$2\,10^6$
	SBH	–	–	$8\,10^1$	–	–	–	–	–	–	–

questions Lipunov et al. (1994b) have numerically simulated the evolution of an ensemble of binaries with the object of calculating the eccentricity distribution of ejectors in binaries.

We have calculated the distribution of eccentricities of binary radio pulsars with normal stars. The calculation has been done for all pulsars in pairs with normal stars, and separately for visible pulsars (i.e., the optical depth τ_{ff} for free-free absorption in the stellar wind is less than 1). The distributions of systems with "visible" pulsars have only one maximum at $e \sim 1$ because the pulsars with high eccentricity are in systems with relatively large semi-major axes (so that $\tau_{ff} < 1$). Moreover, these pulsars spend almost all their time at the distant parts of the orbit, which increases the probability of their detection. The probability to find a pulsar in an orbit with the eccentricity $e \geq 0.8$ is at least ~ 0.90.

Radio pulsars in binary systems with optical companions were considered in detail by Lipunov & Prokhorov (1984, 1987). The factors of radio wave absorption, of radiation delay due to dispersion measure, and of Faraday rotation in a magnetic field of a stellar wind of an optical companion have been estimated. As a matter of fact, the discovery of such objects gives us a unique possibility for investigating the stellar wind flowing from a massive star, a sort of radio probe launched by nature itself. All three effects have recently been discovered (see Johnston 1994).

8. Binary Radio Pulsars with Black Holes

The discovery of binary radio pulsars with massive unseen companions (> 3 to $5\,M_\odot$) would be of great importance for fundamental physics and the modern theory of stellar evolution, giving compelling evidence for blackhole existence in nature. The formation of binaries consisting of a black hole (BH) and a radio pulsar (PSR) has been previously discussed by Bisnovatyi-Kogan & Komberg (1974) and Narayan et al. (1991). As is well known, observations of a radio pulsar in a binary system (Manchester & Taylor 1977; Brumberg et al. 1975; Blandford & Teukolsky 1975) provide the most accurate information about physical parameters of the binary companion. This concerns not only the mass of the companion, which until now has been the main signature of BH. The pulsar radio emission can be used as a probe of plasma emitted by the secondary star (Lipunov & Prokhorov 1984; Lipunov et al. 1994b) and, consequently, by giving a picture of the physical properties of the adjacent medium, can prove the BH nature of the companion. Some relativistic effects specific to BH can be observed in these systems, such as propagation of the radiation through the BH ergosphere.

The calculations with the Scenario Machine predict a subclass of binary radio pulsars with black holes to exist, their galactic number being large enough (Lipunov et al. 1994a) for them to be discovered in the near future. The pulsars themselves must be similar to standard isolated pulsars.

Recently, the 1-s PSR B0042-73 in a highly eccentric ($e = 0.8$) 51-day orbit around a massive companion was discovered in the Small Magellanic Cloud (Kaspi et al. 1994). This may be the first such pulsar with a BH (see, however, the contribution by Kaspi to these proceedings). The fact that the first pulsar discovered in the SMC proved to be a binary seems natural in the framework of a recent burst of star formation in the SMC. Lipunov et al. (1994c) confirm this fact by a numerical calculation of the evolution of radio pulsars after a star formation burst.

References

Bailes, M. 1994, these Proceedings
Bisnovatyi-Kogan, G.S. & Komberg, B.V. 1974, AZh 51, 373
Bisnovatyi-Kogan, G.S. & Komberg, B.V. 1975, AZh 52, 457
Blandford, R.D. & Teukolsky, S.A. 1975, ApJ 198, L27
Brumberg, V.A. et al. 1975, SvAL 1, 5

Davies, R.E. & Pringle, J.E. 1981, MNRAS 196, 209
Dewey, R.J. & Cordes, J.M. 1987, ApJ 321, 780
Johnston, S. et al. 1992, ApJ 387, L37
Johnston, S. 1994, these Proceedings
Illarionov, A.F. & Sunyaev, R.A. 1975, A&A 39, 185
Kaspi, V.M. et al. 1994, ApJ 423, L43.
Kolb, U. 1993, in *Proceedings International Workshop in Memory of Livio Gratton* (Monte Porzio, 21 June 1993), (in press)
Kornilov, V.G. & Lipunov, V.M. 1983a, AZh 60, 284
Kornilov, V.G. & Lipunov, V.M. 1983b, AZh 60, 574
Kornilov, V.G. & Lipunov, V.M. 1984, AZh 61, 686
Lipunov, V.M. 1982a, Ap&SS 85, 451
Lipunov, V.M. 1982b, Pis'ma AZh 8, 358
Lipunov, V.M. 1984, Adv. Space Res. 3, 323
Lipunov, V.M. 1987, Ap&SS 132, 1
Lipunov, V.M. 1991, in *Frontier Objects in Astrophysics and Particle Physics*, Vulcano Workshop 1990, F. Giovannelli and G. Mannocchi (Eds.), Italian Physical Society, p. 29
Lipunov, V.M. 1992, *Astrophysics of Neutron Stars*, Springer-Verlag
Lipunov, V.M. & Postnov, K.A. 1987, Astr. Zh. 64, 773
Lipunov, V.M. & Postnov, K.A. 1988, Astr. Sp. Sc. 145, 1
Lipunov, V.M. & Prokhorov, M.E. 1984, Ap&SS 98, 221
Lipunov, V.M. & Prokhorov, M.E. 1987, AZh 64, 1189
Lipunov, V.M. & Shakura, N.I. 1976, Pis'ma AZh 2, 343
Lipunov, V.M. et al. 1994a, ApJ 423, L121
Lipunov, V.M. et al. 1994b, A&A 282, 61
Lipunov, V.M., Postnov, K.A. & Prokhorov, M.E. 1994c, ApJ (in press)
Lipunov, V.M., Postnov, K.A. & Prokhorov, M.E. 1995, A&A (in press)
Manchester, R.N. & Taylor, J.H. 1977, *Pulsars*, Freeman
Narayan, R., Piran, T. & Shemi, A. 1991, ApJ 379, L17
Osminkin, E.Yu. & Prokhorov, M.E. 1995, SvAL (in press)
Paczyński, B. 1971, ARA&A, 9, 183
Rappaport, S. 1994, these Proceedings
Savonije, G.J. & Van den Heuvel, E.P.J. 1977, ApJ 214, L19
Shakura, N.I. 1975, Pis'ma AZh 1, 23
Shvartsman, V.F. 1970, Radiofizika 13, 1852
Shvartsman, V.F. 1971, AZh 48, 438
Tatarintzeva, V.S. et al. 1989, in *Proc. 23rd ESLAB Symposium on Two Topics in X-ray Astronomy*, ESA SP-296, p. 653
Tauris, T.M. 1994, in *XXIInd IAU General Assembly Posters*, H. van Woerden (Ed.), p. 110
Tutukov, A.V. & Yungelson, L.R. 1973, Astrofizika 8, 381
Tutukov, A.V., Yungelson, L.R., Iben Jr, I. 1992, ApJ 386, 197
Van Kerkwijk, M.H. et al. 1992, Nat 355, 703
Van den Heuvel, E.P.J. 1977, Ann. N.Y. Acad. Sci. 302, 13
Van den Heuvel, E.P.J. & Heise, J. 1972, Nat. Phys. Sci. 239, 67
Wickramasinghe, D.T. & Whelan, J.A.J. 1975, Nat 258, 502

THE FORMATION AND EVOLUTION OF BLACK-HOLE BINARIES

ROGER W. ROMANI
Department of Physics
Stanford University
Stanford, CA 94305-4060, U.S.A.

1. Introduction

The presence of accreting black holes (BH) among the X-ray binaries has been recognized for many years. Traditionally, Cyg X-1 and the handful of other candidates have been thought of as cousins of the HMXB neutron star systems. Recent studies of the soft X-ray transients such as A 0620–00 have, however, shown that the dynamical evidence makes these low-mass systems very strong black-hole candidates. Further, analysis of the eventual end-states of various high-mass X-ray binaries suggest that some could end as observable BH-pulsar binaries, although the first such system is yet to be discovered.

The study of black-hole binaries (BHB) is thus becoming a rich field observationally. Still, our theoretical understanding of the evolutionary tracks leading to such systems is somewhat rudimentary. Pioneering studies (e.g., Van den Heuvel & Habets 1984; De Kool *et al.* 1987) have outlined some basic paths that could generate BHB, in fairly strict analogy with the neutron star (NS) X-ray binary evolutions. However, a number of other scenarios for specific systems have been presented in the literature; we describe some of these in Section 2 below. A second important development has been the discovery of a large and diverse family of radio pulsar binaries offering a rich variety of evolutionary endpoints for models to match. This, coupled with observations of the pulsar population showing that the core collapse process is much more violent than previously believed, has substantially modified our picture of X-ray binary evolution. While similar observational constraints on black-hole formation processes are still far in the future, we can invoke some theoretical prejudice to contrast the genesis of NS-containing XRB with the equivalent black-hole systems (Section 3).

The differences in the evolution through core collapse and in the X-ray phase suggest that a number of peculiar types of black-hole binaries might have been formed from channels that are not available to the neutron star population. We describe a few of these briefly in Section 4. Finally, with the availability of simple analytic approximations for many of the important evolutionary processes, access to substantial computer time and a bit of hubris, it has become easy and popular to simulate the eventual fate of an entire population of primordial binaries. With such population sums one can estimate the probability of evolution into particular end states. Combined with estimates of lifetimes and visibility, these give 'predictions' that may be useful in guiding the interpretation of present surveys, as described in Section 5. Here we note that the broad conclusions of these models provide useful guidance, but that we will have to put the refining fire of observational tests to the scenarios before detailed prediction will merit much credence.

2. Formation Models

Several scenarios for the formation of black holes in binary systems have been proposed. These are similar to the evolutionary paths described for neutron star XRB by Webbink (1992). The basic difference is that here the primary has a mass greater than some critical value M_{BH}, the mass beyond which the core cannot propagate a successful supernova explosion. The value of this mass is quite uncertain. The presence of NS in certain high-mass XRB argues that in some cases primaries with initial masses as high as $\sim 40\,M_\odot$ can eject their outer cores and envelopes in a successful supernova explosion, leaving $\sim 1.4\,M_\odot$ neutron stars (Van den Heuvel & Habets 1984). Other authors, however, favor lower masses $M_{BH} \sim 25\,M_\odot$ (Wheeler & Shields 1976).

In fact it seems likely that M_{BH} may not be uniform in the Galaxy. In particular, at early times, at large galactocentric radii and in environments such as the Magellanic clouds, lower metalicity in regions of massive star formation result in lower opacities in the stellar envelopes. The result is a suppression of main-sequence winds and mass loss, which may favor the growth of a heavier core for a star of a given ZAMS mass. Conversely, in close interacting binaries, early transfer of envelope material and reduced escape velocity leading to enhanced stellar wind might increase mass loss and raise the required initial mass for black-hole formation. The picture adopted here is that mass loss is always increasingly important for stars approaching $M_{BH} \sim 40\,M_\odot$. Lower-mass stars will also experience significant mass loss well before core H exhaustion, and for the heavier stars even the outer layers of the core are shed before core collapse. A rough formula for the final pre-collapse core mass suggests that ~ 7–$15\,M_\odot$ typically remains of the massive progenitor (Romani 1994). We also picture the post-bounce shock wave stalling as it propagates through this massive core, leaving the outer layers of the core in dynamic collapse and trapping the neutrinos emitted from the hot central regions. Thus here the collapse is 'silent', resulting in no neutrino burst, little or no envelope ejection and little optical display. For very strong stellar winds or very close binaries it is possible that the heaviest stars' cores lose enough mass to avoid this silent collapse, restricting the BH formation range to intermediate masses (Woosley et al. 1993). Unless this effect dominates down to nearly $M_{\rm BH}$ this channel does not dominate the fate of the most massive stars and we ignore it in the discussion below. An attractive feature of our picture is that the resulting BH are in the ~ 7–$15\,M_\odot$ range, similar to the values observed for the present strong candidates.

We next summarize the scenarios that lead to binary black-hole systems. We shall see that even for steep IMFs ensuring that BH forming progenitors are rarer than the primaries that can form NS by a factor of ~ 30 (cf., Van den Heuvel 1993), the severe disruption associated with the supernova explosion greatly favors the survival of the BH systems. It is this effect which leads to a significant galactic population of certain types of BH binaries.

2.1. "HE STAR CORE COLLAPSE"

We distinguish three varieties of the model in which the black hole is formed from the He core remnant of a star with $M > M_{\rm BH}$ in a close binary with a companion that will eventually become the mass donor. In the 'direct' scenario which forms HMXB, two high-mass stars in a close binary evolve via strong stellar winds. After the primary reaches core collapse, a continued strong mass loss from the secondary can result in a bright wind fed system like Cyg X-1. This requires a secondary mass greater than ~ 15–$20\,M_\odot$. With a silent core collapse leading to a ~ 7–$15\,M_\odot$ BH, it seems unlikely that there will be many 'Be' systems accreting from the equatorial disk of a rapidly rotating secondary in an eccentric orbit. Instead, with a typical hole mass of ~ 7–$10\,M_\odot$, secondaries with mass ~ 8–$15\,M_\odot$ will have relatively weak stellar winds and will only initiate a bright X-ray phase during a brief period of radius expansion resulting in 'incipient Roche lobe overflow' (Savonije 1983). This will likely be followed by a common envelope spiral in, which may leave either a close binary or a merged system. For the secondaries that do survive, core collapse leads to a neutron star (pulsar) + black hole system, which can remain bound after the supernova because of the relatively high mass of the black-hole companion. The pulsar will be a young object with a high field, spinning down and disappearing in $\sim 10^7$ y.

The second variety of He star core collapse can be labeled the Reverse Mass Transfer case (Narayan, Piran & Shemi 1991). Here we start from a close ZAMS binary with $M_1 \sim M_2 \lesssim M_{BH}$. The primary evolves, transferring mass to its companion via a strong stellar wind. When the primary undergoes a supernova explosion the system will remain bound in a modest fraction of the cases because of the large secondary mass, now $M_2 > M_{BH}$. As the secondary evolves, it transfers matter onto its companion in a classic HMXB with a blue supergiant donor. The mass transfer drives the initial neutron star to lower fields and shorter periods, keeping it above the pulsar death line. After the second core collapse one now has a binary with a BH in orbit around a mildly recycled pulsar. This pulsar will have a relatively long active phase – such systems are attractive targets for radio searches.

Finally, if the secondary is of low mass we have black-hole systems evolving in analogy to the He star model of LMXB formation (De Kool et al. 1987), for which $M_2 \lesssim 1\,M_\odot$. Here, as the primary evolves through a giant phase, it comes into Roche lobe (or tidal) contact with the low-mass secondary, which becomes engulfed in a common envelope spiral-in. With adequate common envelope ejection efficiency (Taam, these proceedings) the system can emerge from this phase as a detached binary consisting of a massive He+Z core and a low-mass star (Romani 1994). The silent core collapse poses no problem here and after evolution via angular momentum loss or radius expansion through core hydrogen exhaustion of the low mass companion, the resulting binary can have a long phase of mass transfer at a moderate rate. For secondary masses $\lesssim 8\,M_\odot$ this transfer will be stable.

2.2. AIC OF A NEUTRON STAR

The viability of NS formation via Accretion-Induced Collapse of a massive white dwarf (e.g., Nomoto & Kondo 1991) is still the subject of some debate, but AIC seems an attractive model for the formation of certain LMXB and wide radio pulsar binary systems. There has been discussion of accretion driving a neutron star above its maximum stable mass (Chevalier 1989), which might be as low as $\sim 2\,M_\odot$ for some soft equations of state (Brown et al. 1992).

One arena where this might happen is in a two-stage collapse of a massive core (Chevalier 1989). After core collapse and the formation of a proto-neutron star, a significant phase of late infall driven by the reverse shock in the progenitor envelope might push the neutron star above the maximum mass. There are some problems with this being the primary formation channel for the binary BH systems that we do see. For one, the resulting holes would be light, typically $\sim 3\,M_\odot$, in contrast to the $\sim 10\,M_\odot$ estimated for the strong BH candidates. In addition the neutrino burst, velocity kick and mass ejection associated with the original supernova event make it very difficult for the system to survive as a detached binary, as seen below. So it seems unlikely that this mechanism can contribute many interacting BH binaries.

Alternatively, steady Roche lobe over-flow in an LMXB might be sufficient to effect the NS collapse. However, there does not seem to be much parameter space accessible to this channel if $\sim 1\,M_\odot$ of accreted mass is needed. If the secondary is any lighter than $\sim 1.5\,M_\odot$ it will not be able to donate the needed mass, while if it is heavier then $\sim 1.8\,M_\odot$ transfer will tend to runaway, leading to high $\dot{M}$ and little mass acceptance.

However, if the secondary is substantially more massive than the neutron star, the runaway to CE may push the effective $\dot{M}$ on the neutron star to very high values. AIC under this 'hypercritical accretion', where the accretion luminosity is carried off by neutrinos, may provide an interesting alternative channel (Chevalier 1993). More work on the flow around the embedded NS is however needed to see if the transition to hypercritical accretion actually takes place. Certainly in the case of the NS-NS binaries and the NS massive WD binaries such as PSR 0655+54, only very modest mass accretion can have occurred (from the presence of recycled pulsars and from component mass estimates) so in many cases the transition to high value of $\dot{M}$ will not be made. Chevalier, in fact, shows that the transition to the hypercritical solution may most easily occur for CE in relatively dense envelopes. For these CE phases, initiating at small radius, it seems likely that complete spiral-in will result. This leads to the final AIC channel, described by Podsiadlowski (these proceedings): collapse in the core of a TŻO. Again viability is not clear, but the model has some attractive features, if the remnants of the disrupted TŻO envelope at the end of the accretion phase can form a low mass companion star. If such a star is formed and its nuclear evolution or the system angular momentum loss initiates transfer onto the BH, a low-mass black-hole binary may be formed.

2.3. TRIPLE SCENARIOS FOR LMBHB PRODUCTION

A close relative of this last model is the scenario created for the low-mass black-hole binary A 0620–00 by Eggleton & Verbunt (1986). In this 'Triple System' picture, a high-mass initial binary in a hierarchical triple with a distant low-mass companion evolves into a HMXB containing a BH, while retaining the distant $\sim 1\,M_\odot$ star. After the MXRB secondary evolves, a CE forms, and the resulting spiral-in and final nuclear evolution causes the merged product with a compact core to expand to very large radii. The result is that the distant low-mass companion also enters the common envelope. However, with its large orbital energy, it may survive spiral-in to create a system like A 0620–00, a BH in a close orbit with a low mass, main sequence companion. If one adds the possibility of AIC in the TŻO product of the initial spiral-in, then a classic NS-containing MXRB might suffice for the inner binary system.

3. Contrast with NS Systems

Even though the basic evolutionary paths outlined above have parallels in NS-XRB evolution, the evolution of BHB should differ from that of the NS-systems in a number of important ways. First, the higher-mass primaries will have especially high mass loss rates $\dot{M}$ prior to core collapse. Indeed, estimates of single-star mass loss (Woosley et al. 1993) suggest that wind loss can lead to bare He cores at collapse for $M > 25 \, M_\odot$, providing a source of SN Ibc, although these may also be largely formed via binary-induced wind loss (Podsiadlowski et al. 1992). In any case the high-mass BH-producing primaries in close binaries will lose substantial mass on the main sequence. This should assist the survival of low-mass companions in the CE phase, since the high gravity of the massive core and the relatively low envelope mass will require lower CE efficiencies and smaller initial orbits to survive the CE phase. Indeed, MS mass loss is essential in allowing CE formation of a close binary at high initial mass ratios. A second interesting possibility is that for sufficiently strong winds, the primary may never become a red giant, instead evolving directly to a WR-star phase and then making a silent collapse to a BH. Romani (1992) argued that this may occur in high-metallicity regions such as the galactic center and may give observable X-ray binaries after the secondary evolves to contact (WR-evolution).

The second important difference lies in the core collapse itself. New estimates of initial pulsar velocities indicate that SN events give new-born neutron stars a typical kick velocity of $v_{rms} \sim 450 \, \mathrm{km \, s^{-1}}$, albeit with a wide distribution around this value (Lyne & Lorimer 1994). At a minimum, neutron star forming cores must eject ~ 1–$5.5 \, M_\odot$ to leave a $\sim 1.4 \, M_\odot$ compact object. The attrition of binary systems with low-mass companions due to the Blaauw mechanism will be severe; in fact, Yamaoka et al. (1993) compute that kick velocities of order v_{rms} above are need to simply keep some of the observed NS-NS systems bound. The BH systems, we have argued, have a silent collapse with no kick or mass ejection. They should therefore have a much better chance of retaining a low-mass companion after collapse. Even though the corresponding initial binaries with very high mass ratio may be formed rarely, after passing through the CE and core collapse phases, they will gain dramatically in the galactic population compared to their NS cousins. The test of this picture lies in the space velocities (and galactic scale heights) of the systems; NS-LMXB should have typical velocities $\sim 200 \,\mathrm{km\, s^{-1}}$, while the BH X-ray transients should have $v < 50\,\mathrm{km\, s^{-1}}$ and low $\langle z \rangle$. Differences should exist for the high-mass systems, as well. Even here, the large NS kicks cause disruption and/or merger for $\sim 2/3$ of the HMXBs; the rarer BH systems should thereby increase in representation.

We note that the presence of kicks does confer one advantage on the NS systems. When the secondary has a mass $8\,\mathrm{M_\odot} < M_2 < 15\,\mathrm{M_\odot}$, the SN mass loss and core kick tends to leave the resulting system in eccentric orbits; with kicks opposite the orbital velocity the semi-major axis will in fact shrink. NS with such close periastron approaches and these lower-mass companions will be able to spin up the secondary via tidal effects, generate an equatorial disk and thereby achieve a substantial phase of episodic accretion in a Be/X-ray binary. Black-hole collapse with a secondary in the same mass range will not only not generate an eccentric orbit, but will likely suffer runaway mass transfer when accretion starts since $M_2 \gtrsim M_{\mathrm{BH}}$. This gives only a short X-ray bright phase as noted above.

The next important difference lies in the fate after the initial compact object is formed. When the secondary can itself form a neutron star, the attrition during core collapse is again severe for the neutron star systems: depending on the post-CE period distribution, roughly $1/30$ of these survive to become NS-NS binaries detectable by their recycled radio pulsars. Black-hole systems, having the SN occur with a ~ 7–$10\,\mathrm{M_\odot}$ companion, will survive $\sim 1/3$ of the time, thereby enhancing the fraction of BH-NS binaries.

When the secondary has $2\,\mathrm{M_\odot} \lesssim M_2 \lesssim 10\,\mathrm{M_\odot}$, transfer onto a neutron star primary will be unstable, leading to CE. In some cases, these systems may survive as detached binaries to form the PSR+massive WD systems (Bhattacharya, these proceedings). In any case, though, these will have a short bright X-ray phase. However, when the accretor is a BH the mass ratio is favorable and stable accretion can proceed until the secondary finishes its evolution. X-ray sources with companion masses in the 3–5 $\mathrm{M_\odot}$ range are likely to harbour BH, as in LMC X-3 (Cowley et al. 1983). Thus black-hole X-ray binaries will not have the secondary mass gap of the NS XRB, but will instead have the 'gap' described above for $M_2 \sim 8$–$15\,\mathrm{M_\odot}$, and many fewer spun up Be-star companions.

Finally, for low-mass secondaries the LMXB statistics and luminosities indicate that some sort of accelerated evolution reduces the lifetime below the time scale for angular-momentum loss. Comparison with survival rates through SN in population computations suggests that $\tau_{\mathrm{ev}} \sim 3\ 10^7\,\mathrm{y}\ (\langle v_{\mathrm{rms}}\rangle/100\,\mathrm{km\, s^{-1}})$ to match the observed LMXB population, i.e., lifetimes of $\sim 10^8\,\mathrm{y}$. In contrast, for the BH LMXB quiescence accretion rates estimated from hot-spot fluxes seem more in line with rates predicted from $\dot{J}$ evolution. Thus these systems will have long $\gtrsim 10^9$ y mass transfer lifetimes. However, the low secondary mass ratios $q \lesssim 0.1$, possibly abetted by the lack of soft flux from the NS surface (Romani 1992), ensure that the BH LMXB systems will be disk instability soft X-ray transients (cf., Whitehurst 1988; Mineshige et al. 1990; see also the contribution of Lasota to these proceedings).

4. Odd Endpoints

To summarize the consequences of the evolutionary differences in Section 3, we note that certain unusual classes of X-ray binaries may be found containing BH. First when the primary mass loss is sufficiently severe to allow direct evolution to a WR phase and silent collapse, a distant lower-mass secondary can remain bound without ever having experienced significant mass transfer from the primary. When this star in turn evolves, it can in fact initiate transfer to the primary after it running some distance up the giant branch. A few of these giant-fed, steady bright BH systems might be found in the inner galaxy (Romani 1992).

The second interesting product is the result of the 'Reverse Mass Transfer' scenario above, leading to a mildly recycled pulsar, with a BH from the *lower*-mass secondary star. The BH X-ray binaries should also have a significant number of 'Intermediate-Mass' donors. For NS there should be only a handful of such systems, like Her X-1, in the Galaxy. Presumably, these will be disk transients in many cases. Finally, there should be a surprisingly large number of low-mass X-ray binaries with black holes among the soft X-ray disk transients. The present identifications of A 0620–00, V404 Cyg and Nova Muscae 1991 represent only a small fraction of a large population undergoing steady slow mass transfer and outburst phase accretion.

5. Population Estimates and Conclusions

What exactly should we see and how many of the unusual systems above can be hidden in the present XRB and Nova catalogs? The solution is to model the processes above for a full range of progenitor systems. Comparison with observed samples will refine our understanding of the evolutionary processes. Computations of this sort have been made by Tutukov & Yungel'son (1993), Narayan et al. (1991), Romani (1992, 1994) and Lipunov et al. (1994). Largely these sums are directed at matching the NS X-ray binary and radio pulsar binary populations, but their extrapolation to the BH-producing regime provides some insights. Other population synthesis were presented at a poster by Kalogera & Tauris at this Symposium, but again the results concentrate on the NS systems. In general, the conclu-

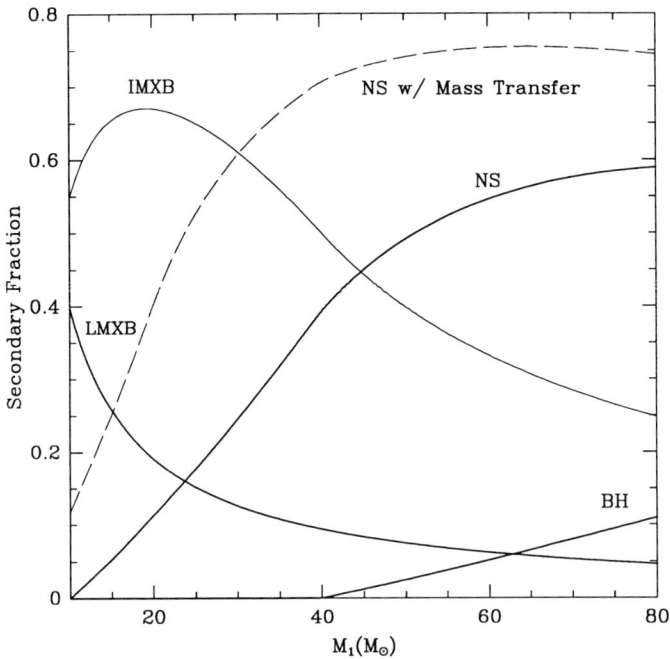

Figure 1. Fraction of secondaries by type, for IMF secondary distribution $q > 0.25$, flat in $\log(q)$ below. Primary mass transfer may result in more NS-producing secondaries (dashed line).

sions of these studies are similar, although a more modern treatment of the decimation at supernova events is needed in some cases.

Rough results representative of these calculations can be summarized as follows. The normalization of the star formation rate and the initial IMF produces $\sim 10^8$ NS and $\sim 3\ 10^6$ BH during the lifetime of the Galaxy; roughly $3\ 10^5$ of the former are active radio pulsars, of which about 500 have been detected. The distribution of the mass ratio $q = M_2/M_1$ in the primordial binary population is presently uncertain: observational evidence for secondaries distributed like the IMF exists for modest q and primaries of moderate mass. For smaller q the number of secondaries probably flattens while for very massive primaries, more equal mass ratios may be preferred. Taking, for example, an IMF secondary distribution to $q = 0.25$ and flat in $\log(q)$ thereafter, gives the secondary breakdown shown in Fig. 1; this distribution is particularly uncertain for large M_1.

After evolving an initial population, typical galactic numbers found for binaries transferring mass in the X-ray producing phase are listed in Table 1. Here $\dot{J}$ indicates systems driven by angular-momentum losses, while $\dot{N}$ indicates mass transfer driven by nuclear evolution at longer $P_{\rm b}$. The

TABLE 1. Galactic Numbers of NS and Black-Hole Binary Systems

X-Ray Binary Phase	Active End Phase	N_{Gal}
NS – LMXB		100
– IMXB		1–3
– MXRB		30
	NS – NS	10^4
	PSR – NS (detected)	5
BH – LMXB		10^3 ($\sim$500 $\dot{J}$/500 $\dot{N}$)
– LMXB(WR)		10
– IMXB		300
– HMXB		3
	BH–NS	$3\;10^5$
	PSR–BH (detected)	$\sim 1/2$

low-mass and intermediate-mass systems should in most cases be transients (cf., papers above for details). For the high-mass systems, it is also useful to estimate the number of active pulsar binaries produced. Normalizing to the observed pulsar sample, these computations give roughly 10^4 NS-PSR binaries produced by the HMXB, of which $\sim$5 should be present in today's pulsar sample. The high-mass BH systems will produce $\sim 3\;10^5$ BH-NS binaries over the lifetime of the Galaxy, of which $\sim 10^3$ are active pulsars. Narayan et al. (1991) and Lipunov et al. (1994) estimate that $\lesssim 1$ of these objects should be present in today's pulsar sample, so future prospects look bright.

The identification of black-hole candidates in the soft X-ray transients has spurred a re-appraisal of the evolutionary channels that can generate black-hole binaries. Interestingly, sums that duplicate the NS XRB population fairly well predict some surprises in the BH sample. Given the uncertain evolutionary processes and the substantial extrapolation of the presumed binary progenitors from the observed galactic binaries, these results are still quite uncertain. When more BH candidates are carefully characterized and matched by detailed evolutions, our evolutionary assumptions can merit greater trust. A match to the population statistics and the properties of the candidate systems is also absolutely essential for any satisfactory description. An increase in the present sample by about a factor of 3, which should occur over the next several years in the case of low-mass black-hole transients, will allow significant progress. If in the process we discover a few of the other unusual end-states predicted by the population synthesis models, then these evolutionary modeling exercises will have been a success.

Acknowledgements. This work was supported in part by NASA grant NAGW-2963 and the Alfred P. Sloan Foundation.

References

Brown, G.E., Bruenn, S.W. & Wheeler, J.C. 1992, Com. Astrophys. 13, 153
Chevalier, R.A. 1989, ApJ 346, 847
Chevalier, R.A. 1993, ApJ 411, L33
Cowley, A.P., Crampton, D. & Hutchings, J.B. 1983, ApJ 272, 118
De Kool, M., Van den Heuvel, E.P.J. & Pylyser, E. 1987, A&A 183, 47
Eggleton, P.P. & Verbunt, F. 1986, MNRAS 220, 13
Lipunov, V.M. et al. 1994, ApJ 423, L121
Lyne, A.G. & Lorimer, D.R. 1994, Nat 369, 127
Mineshige, S., Kim, S. & Wheeler, J.C. 1990, ApJ 358, L5
Nomoto, K. & Kondo, Y. 1991, ApJ 367, L19.
Narayan, R., Piran, T. & Shemi, A. 1991, ApJ 379, L17
Podsiadlowski, P., Joss, P.C. & Hsu, J.J.L. 1992, ApJ 391, 246
Romani, R.W. 1992, ApJ 399, 621
Romani, R.W. 1994, in *Interacting Binary Stars*, A.E. Shafter (Ed.), ASP Conf. Proc. Vol. 56, p. 196
Savonije, G.J. 1983, in *Accretion Driven Stellar X-ray Sources*, W.H.G. Lewin & E.P.J. van den Heuvel (Eds.), Cambridge Univ. Press, p. 434
Tutukov, A.V. & Yungel'son, L.R. 1993, Ast. Rep. 37, 411.
Van den Heuvel, E.P.J. 1993, in *Space Sciences with particular emphasis on High Energy Astrophysics*, (ESA: ESTEC Noordwijk)
Van den Heuvel, E.P.J. & Habets, G.M.H.J. 1984, Nat 309, 689.
Webbink, R.F. 1992, in *X-ray Binaries and Recycled Pulsars*, E.P.J. van den Heuvel & S.A. Rappaport (Eds.), Kluwer Academic Publishers, p. 269
Wheeler, J.C. & Shields, G.A. 1976, Nat 259, 642.
Whitehurst, R. 1988, MNRAS 232, 35
Woosley, S.E., Langer, N. & Weaver, T.A. 1993, ApJ 411, 823
Yamaoka, H., Shigeyama, T. & Nomoto, K. 1993, A&A 267, 433.

2

Supernovae in Binaries

TYPE IB/C SUPERNOVAE AND THEIR RELATION TO BINARY STARS

BRUNO LEIBUNDGUT
European Southern Observatory
Karl-Schwarzschild-Strasse 2
D-85478 Garching
Germany

Abstract. The present understanding of type Ib/c supernovae and their connection to interacting binaries is reviewed. The problems of the classification and the lack of well-observed events exclude direct inference of progenitor characteristics. The absence of hydrogen lines in the observed spectrum, nevertheless, requires restricted evolutionary schemes to produce suitable progenitor stars for core collapse explosions with no hydrogen envelope. New relative statistics among the supernova types are presented which indicate that SN Ib/c are on average brighter than SN II, and with the dense sampling of supernova searches in nearby galaxies, a small intrinsic incidence of SN Ib/c is determined. The small rates might be in conflict with the observed ratio of massive stars in binaries in the Galaxy.

1. Introduction

Of all supernova classes the type Ib/c supernovae (SN Ib/c) are the most mysterious. Their apparent similarity with SN Ia in light curves and early spectral evolution, while a core collapse probably initiates the explosion, makes them the case of "cross dressing" supernovae. They were discovered as a separate subclass only about a decade ago and the difficulty of distinguishing them clearly from the other classes hamper meaningful statistics.

The spectrum near maximum light lacks any obvious signs of hydrogen lines and displays remarkably weak lines of Si II ($\lambda 6355$ Å) which made the classification (Harkness & Wheeler 1990) basically one by exclusion of the other supernova types. Confusion arose from the resemblance of the maximum light spectrum of SN Ib/c with the one of SN Ia about four weeks past peak, which led to the expression that SN Ib/c are "born old" (Panagia 1984). The nebular spectrum is distinguished by the strong emission lines of forbidden oxygen and calcium. Again no hydrogen is observed. The optical light curves are almost indistinguishable from the ones of SN Ia (Vacca & Leibundgut 1995), while the near infrared brightness evolution is characteristically different (Elias et al. 1985).

The discovery of SN 1993J in M81, a type II supernova, has strengthened the connection of SN Ib/c with core collapse events. SN 1993J displayed many signatures of SN Ib/c (Filippenko et al. 1994). The optical light curves are very similar to SN Ib/c (Leibundgut 1994) and the spectrum did develop very strong oxygen and calcium lines in the nebular phase (e.g., Lewis et al. 1994). It has to be emphasized that, despite early expectations, SN 1993J always displayed Hα emission (up to 500 days; RGO data archive, Filippenko et al. 1994; Patat et al. 1995).

The progenitors of SN Ib/c have to date been fairly elusive. Several studies have tried to connect SN Ib/c with star formation regions, but so far without conclusive results (Panagia & Laidler 1991; Van Dyk 1992). Recent theoretical work has concentrated on models with stars of small mass which lost all their hydrogen (Nomoto et al. 1990; Woosley et al. 1993, 1994b; Wheeler et al. 1994). To achieve this with current stellar evolution models mass exchange in close binaries is invoked (Nomoto et al. 1994).

Attempts to determine the rates of supernovae have been fairly restricted in the case of SN Ib/c due to the difficulties in separating them from the other classes and the limited sample size (only 32 SN Ib/c have been identified to date). The latter may be caused by the apparent faintness (relative to the other supernovae) and, possibly, large extinction from the environment of the explosions, or an intrinsic rareness of the phenomenon. An attempt to resolve this question will be made.

In the following we will present the current classification scheme and discuss its status within the framework of our understanding of supernovae (Section 2). Especially the sub-classifications into SN Ib and SN Ic will be reviewed critically. Some very simple supernova statistics are presented in Section 3 to investigate the nature of SN Ib/c relative to other supernova types. The discussion (Section 4) describes the minimal knowledge achieved so far and some cautious conclusions.

2. The nature of SN Ib/c

Several observables have been exploited to determine the nature of SN Ib/c. Among the techniques are association with star formation regions (Panagia & Laidler 1991; Van Dyk 1992; Van Dyk et al. in these proceedings) starting from the notion that no SN Ib/c has ever been observed in an elliptical galaxy (e.g., van den Bergh & Tammann 1991), spectral modeling (Harkness et al. 1987; Swartz et al. 1993; Jeffery et al. 1991; Wheeler et al. 1994), and radio observations indicating dense circumstellar material around the explosion (Van Dyk et al. 1993). A cursory survey of the published data, discovery announcements, and the CfA data archive provided some coarse statistics on observed Hα emission within the slit width (typically 1 to 2 arcseconds) of SN Ib/c. For the 25 supernovae since 1983 we find 14 events with narrow Hα emission superposed on the supernova spectrum reported. Considering the inhomogeneity with which these data have been assembled (some from long exposures at late phases others from short integrations of bright supernovae near maximum) the agreement with other determinations (Van Dyk 1992; Van Dyk et al. in these proceedings) is remarkable.

The absence of hydrogen has triggered the association of SN Ib/c with Wolf-Rayet stars (e.g., Wheeler & Levreault 1985) and very massive progenitors as well as stars stripped of the hydrogen envelope induced by binary interaction (Nomoto et al. 1994; Van den Heuvel 1994). Progress has been slow due to the rareness of well-observed SN Ib/c. The supernovae leading to the introduction of the new subclass still represent the prime examples. The observations of SN 1985F (Filippenko & Sargent 1986), SN 1983N (Panagia 1985), and SN 1984L (Wheeler & Levreault 1985; Uomoto & Kirshner 1985; Schlegel & Kirshner 1989) have been supplemented only with SN 1987M (Filippenko et al. 1990). The bright SN 1994I in M51 will change this situation profoundly (Wheeler et al. 1994). Thus, most knowledge has been gathered from statistical evidence and interference with other supernova types. The *hybrid* supernovae SN 1987K (Filippenko 1988) and SN 1993J (Lewis et al. 1994) displayed many characteristics of SN Ib/c at late phases and implied a close relation between SN II and SN Ib/c. Nevertheless, SN 1993J exhibited sustained hydrogen emission and never fully changed its appearance to a SN Ib/c (Lewis et al. 1994; Filippenko et al. 1994).

The separation of the subclasses SN Ib, i.e., helium-rich, and SN Ic, helium-poor, has proven to be rather difficult to implement. The spectroscopic differences are subtle and it has never been shown that there could not exist a continuum rather than a dichotomy in the distribution. Harkness et al. (1987) argued on the basis of the strength of the He I lines in the optical spectrum for a separation, but recently Wheeler et al. (1994) prefer a distinction on the basis of the neutral oxygen line at $\lambda 7774$Å. This absorption is strong in SN Ic while fairly weak in SN Ib. An open issue is also the appearance of the He I 10830 Å line. Its strength is largely undetermined for any supernova. SN 1990W has been reclassified to a SN Ib due to the strong emission in this line (Wheeler et al. 1994), although it did not clearly display strong He I lines in the optical. The reclassification is based on spectral synthesis calculations which indicate enhanced He in the optical spectrum.

Additional distinguishing characteristics between the two subtypes are also rather sparse. The declines of the late-time light curves are suspected to be steeper for SN Ic while the smaller rates in SN Ib have been interpreted as due to larger envelope masses in the explosions (Swartz & Wheeler 1991). The sampling of light curves of SN Ib/c, however, is at best still marginal (Vacca & Leibundgut 1995) and firm conclusions are not possible yet.

Hydrogen in SN Ib/c is a controversial issue. The identification of $H\alpha$ in the early time spectra of SN Ic (Jeffery et al. 1991; Filippenko 1992) has not been generally accepted (Swartz et al. 1993). The main worry is the expansion velocity measured for hydrogen which appears smaller than for calcium and oxygen contrary to any reasonable explosion model. Swartz et al. (1993) experimented with models including a small amount of hydrogen and found strong inconsistencies with the observed spectra of SN 1987M. SN 1993J with its close resemblance in certain aspects with SN Ib/c but its obvious hydrogen emission has further complicated the interpretation.

Another possible distinction pattern in the peak light spectra has now been proposed by Wheeler et al. (1994). The lines of C II ($\lambda 6580$ Å) and Si II ($\lambda 6355$ Å) appear in SN Ic like SN 1994I and SN 1987M, while an absorption line observed at almost the same position in the spectrum of SN Ib is interpreted to arise from $H\alpha$, e.g., SN 1983N (Wheeler et al. 1994). If these identifications are correct, the evolution of these lines is expected to be different and might provide a rather stringent test on the nature of the two subclasses.

Classification of supernovae is a difficult issue. For analyses connecting supernova explosions with global parameters, like star formation rates or chemical enrichment, a simple scheme which can quickly and easily separate clearly distinct objects is needed. For a detailed understanding of the physics of the individual objects a more specific description is called for. The difficulty in separating type Ib objects from type Ic probably outlines this border. Spectral synthesis calculations are needed to find the subtle distinctions between the two classes. This precludes classifications with spectra of newly discovered supernovae at the telescope, as it has been done in most cases in the past. Statistics based on supernova catalogs become difficult with such disparate classification systems.

3. Statistics of SN Ib/c

Two main problems plague supernova statistics. First, the numbers are rather small and meaningful statistics are hard to find. This is especially so

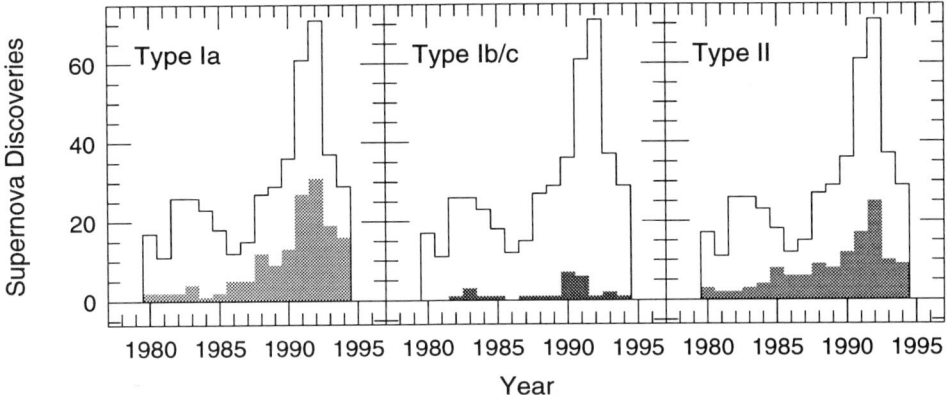

Figure 1. Supernova discoveries between 1980 and 1994. The data and classifications are from the Asiago Supernova Catalog (Barbon *et al.* 1989) for all supernovae before 1989 and according to IAU Circulars and the literature thereafter. The line histogram shows the total number of supernovae discovered per calendar year and the shaded areas the supernovae of the particular type.

when further subdivision of the samples is attempted, a procedure obviously necessary to investigate the different underlying physics of the events. Second, strong selection biases are inherent in the detection mechanisms and the classification procedures as described for the case of SN Ib/c above. The basic ingredients for the statistics are (see Strom 1994 for an excellent review): the observed supernova rate, i.e., the number of supernovae per unit time interval, and the corrections depending on the form of the luminosity profiles and distribution of the individual objects. The tricky part is how to correct to find the true number of explosions in a galaxy (e.g., Cappellaro *et al.* 1993; van den Bergh & McClure 1994).

With the limited number of SN Ib/c known to date (32 objects in total) it is difficult to derive significant results from statistics. Since in the past the lack of the fine tool of spectral synthesis prevented classification of the objects into SN Ib and SN Ic, we will apply statistics only to the overall sample of known SN Ib/c. An important deficit of this study will also be objects misclassified as SN Ia, due to their similar spectra and light curves throughout the peak phase; furthermore, the separation from SN II has become difficult as there are cases where a clear association to either class is not obvious. The examples of SN 1987K and SN 1993J should serve as warnings. Unless a sample can be constructed which has a uniform classification applied to it, we will have to deal with a rather inhomogeneous data set. We have chosen to use the classification near maximum light reported in the literature for the selection of our samples.

An overall picture of the supernova discovery rates and the number of the individual classes for the last 15 years is presented in Fig. 1. It is evident how limited the number of SN Ib/c is relative to the other two classes. There is also an interesting trend discernible. While SN Ia and SN II were found roughly at a constant fraction of the total number of discoveries, the number of SN Ib/c appears anti-cyclic at the beginning of this decade. The distributions in Fig. 1 are mainly the signatures of two complementary supernova searches. The Calan-Tololo supernova search (Hamuy et al. 1993) was targeted at rather faint, distant supernovae while the Berkeley supernova search (Perlmutter et al. 1990) was aimed to find supernovae in relatively nearby galaxies. The former was running during the years 1990–1993 (Hamuy et al. 1995) and has been a major contributor to the pronounced peak of discoveries. The latter, employing an unfiltered CCD, started seriously in 1988 and ended in 1991. The Berkeley search is also responsible for the large number of SN Ib/c at the turn of the decade (Muller et al. 1992). The red sensitivity of CCDs has been proposed as the cause for the increased *relative* SN Ib/c rate found by the Berkeley group (Cappellaro et al. 1993), but only six out of the 15 SN Ib/c between 1988 and 1991 are from this search. No real explanation is offered here for the strange distribution of SN Ib/c discoveries.

Relative changes among the various supernova classes can also be garnered from Fig. 1. The increased rate of SN Ia is most likely due to their intrinsic brightness and the fainter detection limits of recent searches. The increase of SN II discoveries is more moderate while the SN Ib/c have fluctuated substantially. The lower discovery rate of SN Ib/c compared with SN II are the signature of the combination of two effects. For illustration consider two extreme cases. Let's first assume SN Ib/c have the same luminosity as SN II which means that the former must be rarer as the detection probability is roughly the same. On the other hand, if both types have about equal rates, Fig. 1 implies SN Ib/c to be fainter (either because they are less luminous or suffer from significant extinction). Complications arise from the differently shaped light curves. If SN II remain brighter for an extended period, which is certainly the case for the plateau-type light curves, their discovery chance is increased proportionally.

To investigate the relative differences among supernova classes further we employ a very simple statistical scheme. The only inputs required are a classification of the supernova and the recession velocity of the parent galaxy. The distribution for each type Is presented in Fig. 2. We have chosen to display all samples on the same scale to emphasize the variation in the *total* number of objects known in each class. The small sample of known SN Ib/c is once more striking. This is also true for the more recent samples starting in 1989 (the shaded areas). In passing we note that the last five years contain 44 percent of all classified supernovae which makes this subsample statistically useful, especially when considerations like the

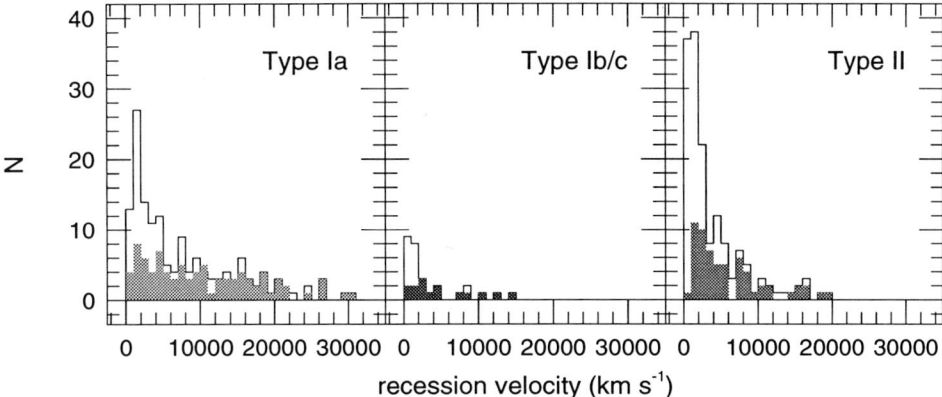

Figure 2. Histograms of the distribution of supernovae in distance bins. The shaded areas are statistics starting only 1989. The bin width is $1000\,\mathrm{km\,s^{-1}}$. Thus only the two closest bins are likely to be sensitive to peculiar galaxy motions.

restricted number of classifying people are taken into account. Not shown are the few objects with redshifts substantially larger than 0.1. There are several identifications of objects at higher redshift with rather uncertain classifications. They include a few SN Ia (SN 1988U, and SN 1994F; possibly SN 1992bi, SN 1994G and SN 1994H), one SN Ib/c (SN 1992ar), and one probable SN II (SN 1988T).

Not surprisingly all distributions in Fig. 2 are peaked towards small distances. In particular the overall samples offer very little room for distinction among the distributions. The situation is slightly different if one considers only the more recent supernovae. With the increase of the limiting magnitudes of the searches over the past few years, the distribution of SN Ia is much flatter than before. In other words, most of the more distant supernovae have been found in recent years. Although this equally applies to all classes the effect is less pronounced for SN II, which still are discovered preferentially within a distance of $cz \leq 10000\,\mathrm{km\,s^{-1}}$. The higher peak luminosity of SN Ia is easily inferred from this diagram. The histogram for SN Ib/c is clearly less peaked than the one of SN II, although the small numbers make this comparison rather shaky. What is further complicating the picture is the discovery of SN 1992ar (Hamuy *et al.* 1992). This SN Ib/c was located in a galaxy with a redshift of 0.145 and appears to have been similar in brightness to SN Ia. The picture presented in Fig. 2 is then not so simple anymore. But the detection of a SN Ib/c with such a high luminosity hints to the possibility that these objects are in their mean not fainter than SN II.

A high *relative* rate for SN Ib/c was found in data extracted from the Berkeley Supernova Search (Muller *et al.* 1992). Although plagued with very

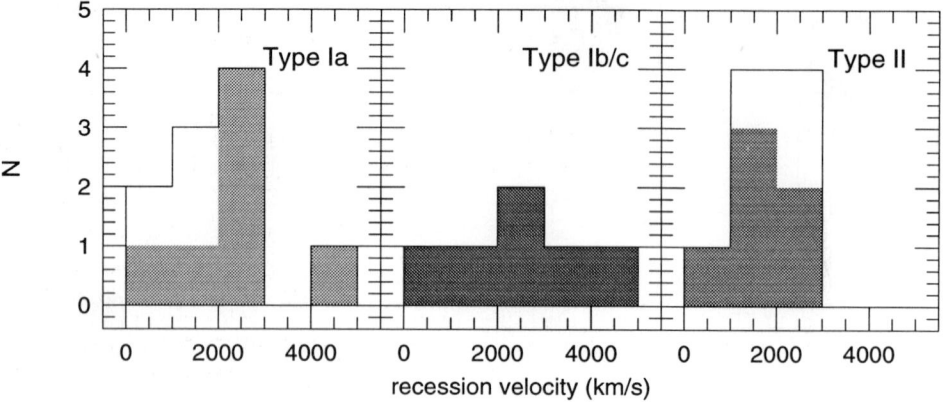

Figure 3. Same as Fig. 2 for the supernovae of Muller *et al.* (1992), shaded area, and the sample including the supernovae discovered during the first year of LOSS (Treffers *et al.* 1994).

small numbers (only a total of 12 objects were included in their analysis), the highest total number of any subclass was found for SN Ib/c. While there is a fair agreement for the rates of SN Ia and SN II Muller *et al.* find an enhancement in their SN Ib/c rate by a factor of three over other determinations (van den Bergh & Tammann 1991; Cappellaro *et al.* 1993; van den Bergh & McClure 1994). We tried to assess this result with the statistics presented above. Fig. 3 displays the number of supernovae in the different distance bins. For these diagrams we have combined the samples from Muller *et al.* and the supernovae discovered during the first year of the Leuschner Observatory Supernova Search (LOSS; Treffers *et al.* 1994). The two searches have employed the same equipment but slightly different detection algorithms. LOSS is not fully automatic but rather provides a list of candidates every morning which are then inspected by an observer. While the original Berkeley search was performed with the unfiltered CCD, LOSS is using an R filter. The limiting magnitudes of the two searches are probably comparable at $R = 17$. An important aspect of these searches is the high frequency with which the galaxies are observed (typically once a week) which for a distant limited sample of galaxies (like the one employed in LOSS) reduces the dependence on light curve shapes considerably.

While the numbers are still very small we find that searches targeted at nearby galaxies do indeed find a fair fraction of SN Ib/c in the total sample. The distribution in Fig. 3 is 10:6:9 (SN Ia:SN Ib/c:SN II). Also the clustering towards small distances is obvious for SN Ia and SN II. Not so clear is the distribution for SN Ib/c which appears rather flat. Incidentally, LOSS has only added SN Ia and SN II to the total sample during the first year, but no SN Ib/c. Most likely this is due to small number statistics.

TABLE 1. Galaxies with multiple supernova events

SN type	N_{SNe}	percentage	SN type	N_{SNe}	percentage
I	6	17	Ia	16	20
II	20	57	Ib/c	3	4
unclassified	8	23	I	7	9
peculiar	1	3	II	21	26
			unclassified	35	41

Note: 2 galaxies with 6 supernovae
2 galaxies with 4 supernovae
5 galaxies with 3 supernovae

Note: 41 galaxies with 2 supernovae

The distance distribution clearly indicates that SN Ib/c certainly are *not* fainter than SN II. The argument that the rate of SN Ib/c should be higher than the rates for the other supernova classes can not be supported. Fig. 3 clearly demonstrates that the argument of Muller et al. (1992) does not hold for their own data. We then conclude that the rate of SN Ib/c is certainly smaller than for SN II and also that they have a comparable detection chance to SN II.

A last test to compare the relative frequencies of supernovae is the comparison of rates in galaxies with multiple supernova events (Richter & Rosa 1988). This assumes that the survey time for all parts of a galaxy are equal and that global effects triggering supernova events, e.g., starbursts, are not influencing the relative rates between the different kind of core collapse supernovae. There have now been a fair number of galaxies with multiple supernova events over the last century. The distributions by type are listed in Table 1. For the 9 galaxies with more than two supernovae (a total of 35 events) SN II are three times as frequent as SN I. We have not distinguished among SN Ia and SN Ib/c as most events are of historical character. Unless a large fraction of SN II have been misclassified SN I, there is a strong preponderance of hydrogen-displaying supernovae. Inclusion of the 41 galaxies with 2 events increases the fraction of unclassified supernovae, but still shows that SN II have been discovered at least twice more often than SN Ib/c.

Overall, it appears clear that SN Ib/c are observed less frequently than SN II. This is implied by the smaller total number of objects while the mean distance in the samples appears to be larger (Figs. 2 and 3). With similar detection probabilities (especially in densely sampled searches like the two Leuschner projects) we have to conclude that SN Ib/c are intrinsically rarer than SN II (and SN Ia).

4. Discussion

To determine the stellar progenitors of supernovae has proven a very difficult enterprise. The explosion destroys the progenitor and the chemistry changes erase many traces of the composition of the progenitor star. This is much more the case for explosions inside small envelopes, typically assumed for SN Ia and SN Ib/c. Thus, it is very difficult to estimate the characteristics of the star before it hit catastrophe. Observing supernova progenitors before the explosion is restricted to the nearest galaxies. The progenitor of SN 1987A is the only clear identification to date, while a progenitor of the type Ic SN 1994I in M51 could not be identified (Kirshner et al. 1994). A possible identification has been reported for the progenitor of SN 1993J (Aldering et al. 1994). Photometry of the stellar object coincident with the position of SN 1993J indicates an excess of blue light compared with the progenitor structure inferred from models of the supernova (e.g., Wheeler & Filippenko 1994; Woosley et al. 1994a). The blue excess could be due to a nearby cluster of blue stars or a massive, hot companion. Although single-star models can be devised to retain only a small hydrogen envelope such an evolution occurs more naturally in a binary-star scenario (Woosley et al. 1994b; Nomoto et al. 1993, 1994). The observed non-spherical distribution of matter in the ejecta of SN 1993J (Fransson et al. 1994) is also indicative of a binary system. This is to illustrate that the borders of supernova classes not necessarily outline different evolutionary channels. Nevertheless, we might derive very general conclusions, if we assume that SN Ib/c are most likely originating in binary systems (Nomoto et al. 1994; Wheeler et al. 1994; Woosley et al. 1994a,b). Paradoxically, this statement very strongly relies on the identification of one SN II (SN 1993J) with a binary progenitor.

The current study restricted itself to find the relative differences in the occurrence of core collapse supernovae. Taking the results of Section 3, we have to conclude that SN Ib/c are on average brighter at maximum than SN II as they are observed at larger mean distances. We cannot *directly* infer that SN Ib/c are rarer than SN II due to the distinct light curve shapes of the two classes. Since SN Ib/c drop about two magnitudes in 20 days in blue light whereas many SN II remain on a high-luminosity plateau for up to 2 months (Patat et al. 1994) the chance of discovery is strongly enhanced for SN II. At redder wavelengths, as is the case of the Berkeley searches, this effect is less pronounced due to the redder colors of SN Ib/c at maximum. Thus, we believe that with SN Ib/c on average brighter than SN II the number statistics reflect a truly smaller incidence of SN Ib/c compared to SN II. Such a small frequency has been found in several other studies before (van den Bergh & McClure 1994; Cappellaro et al. 1993) contrasting with the rates derived by Muller *et al.* (1992). An interesting unresolved question remains why SN Ib/c discoveries peaked in a different year than the discoveries of all other types (Fig. 1). It might be that the absolute magnitudes at peak and the rareness of the events resulted in a strong bias against these objects in the nearby and distant searches conducted at the beginning of the decade.

The rareness of SN Ib/c and the tentative identification of these events with progenitors in interacting binary systems means that this evolutionary channel is constrained fairly strongly and very specific scenarios are needed (e.g., Van den Heuvel 1994). Most models favor evolutions involving common-envelope phases to strip stars from their H and He envelopes. With low rates for SN Ib/c an apparent conflict is further obtained in the estimate of the progenitor systems and the observed ratio of SN Ib/c to SN II. Using the formalism described in Van den Heuvel (1994) we find that only roughly a quarter of all stars with $M > 10\,M_\odot$ are in interacting binary systems. Some explosions in close binary systems might, however, retain enough hydrogen to disguise as SN II, as observed in SN 1993J.

Studying SN rates to learn about local properties, like the fraction of close binaries, assumes that what we find in a morphological mixture of external galaxies still is applicable to the Galaxy. SN rates are derived from an ensemble of galaxies with very distinct properties and it is by no means obvious that an average over all possible solutions provides something close to the situation in the Galaxy. Lacking other more direct observational routes to determine the fate of stellar evolution we have to make a big swing through the nearby universe and stretch our imagination to obtain any results.

References

Aldering, G., Humphreys, R.M. & Richmond, M. 1994, AJ 107, 662
Barbon, R., Cappellaro, E. & Turatto, M. 1989, A&AS 81, 421
Cappellaro. E. *et al.* 1993, A&A 273, 383
Elias, J.H. *et al.* 1985, ApJ 196, 379
Filippenko, A.V. 1988, AJ 96, 1941
Filippenko, A.V. 1992, ApJ 384, L37
Filippenko, A.V. & Sargent, W.L.W. 1986 AJ, 691
Filippenko, A.V., Matheson, T. & Barth, A.J. 1994 AJ, 108, 2220
Filippenko, A.V., Porter, A.C. & Sargent, W.L.W. 1990, AJ 100, 1575
Fransson, C., Lundqvist, P. & Chevalier, R.A. 1994, ApJ (in press)
Hamuy, M. *et al.* 1993, AJ 106, 2392
Hamuy, M. *et al.* 1992, IAU Circ. 5574
Hamuy, M. *et al.* 1995, AJ (January 1995 issue)
Harkness, R. P. & Wheeler, J.C. 1990, in *Supernovae*, A.G. Petschek (Ed.), Springer (New York), p. 1
Harkness, R.P. *et al.* 1987, ApJ 317, 355
Jeffery, D.J. *et al.* 1991, ApJ 377, L89

Kirshner, R. P. et al. 1994, IAU Circ. 5981
Leibundgut, B. 1994, in *The Lives of Neutron Stars*, M.A. Alpar, Ü. Kızıloğlu & J. van Paradijs (Eds.), Kluwer Academic Publishers, (in press)
Lewis, J.R. et al. 1994, MNRAS 266, L27
Muller, R.A. et al. 1992, ApJ 384, L9
Nomoto, K., Filippenko, A.V. & Shigeyama, T. 1990 A&A, 240, L1
Nomoto, K. et al. 1993, Nat 364, 507
Nomoto, K. et al. 1994, Nat 371, 227
Patat, F. et al. 1994, A&A 282, 731
Patat, F., Chugai, N. & Mazzali, P.A. 1995, A&A (submitted)
Panagia, N. 1984, Proc. 4th European IUE Conference, ESA SP-218
Panagia, N. 1985, in *Supernovae as Distance Indicators*, N. Bartel (Ed.), Springer (Berlin), p. 14
Panagia, N. & Laidler, V.G. 1991, in *Supernovae*, S.E. Woosley (Ed.), Springer (New York), p. 559
Perlmutter, S. et al. 1989, in *Particle Astrophysics: Forefront Experimental Issues*, E.B. Norman (Ed.), World Scientific (Singapore), p. 196
Richter, O.-G. & Rosa, M. 1988, A&A 206, 219
Schlegel, E.M. & Kirshner, R.P. 1989, AJ 98, 577
Strom, R. 1994, in *The Lives of the Neutron Stars*, M.A. Alpar, Ü. Kızıloğlu & J. van Paradijs (Eds.), Kluwer Academic Publishers (in press)
Swartz, D.A. & Wheeler, J.C. 1991, ApJ 379, L13
Swartz, D.A. et al. 1993, ApJ 411, 313
Treffers, R.R. et al. 1994, in *Robotic Telescopes*, G. Henry (Ed.), Astron. Soc. Pacific (in press)
Uomoto, A. & Kirshner, R.P. 1985, A&A 149, L7
Vacca, W. D. & Leibundgut, B. 1995, (in preparation)
van den Bergh, S. & McClure, R.D. 1994, ApJ 425, 205
van den Bergh, S. & Tammann, G.A. 1991, ARA&A 29, 36
Van den Heuvel, E.P.J. 1994, in *Interacting Binaries*, H. Nussbaumer & A. Orr (Eds.), Springer (Berlin), p. 263
Van Dyk, S.D. 1992, AJ 103, 1788
Van Dyk, S.D. et al. 1993, ApJ 419, L69
Wheeler, J.C. & Filippenko, A.V. 1994, in *Supernovae and Supernova Remnants*, R. McCray (Ed.), Cambridge University Press (in press)
Wheeler, J.C. & Levreault, R., 1985, ApJ 294, L17
Wheeler, J.C. et al. 1994, ApJ (in press)
Woosley, S.E., Langer, N. & Weaver, T.A. 1993, ApJ 411, 823
Woosley, S.E. et al. 1994a, ApJ 429, 300
Woosley, S.E., Langer, N. & Weaver, T.A. 1994b, ApJ (in press)

THE ORIGIN OF TYPE IB-IC-IIB-IIL SUPERNOVAE AND BINARY STAR EVOLUTION

K. NOMOTO, K. IWAMOTO AND T. SUZUKI
Department of Astronomy, University of Tokyo

O.R. POLS
Institute of Astronomy, University of Cambridge

H. YAMAOKA AND M. HASHIMOTO
Department of Physics, Kyushu University

P. HÖFLICH
Department of Astronomy, Harvard University

AND

E.P.J. VAN DEN HEUVEL
Astronomical Institute, University of Amsterdam

Abstract. Supernovae are classified as type I and type II and further subdivided into Ia, Ib, Ic, II-P, II-L, and IIb. The origin of this observational diversity has not been well understood. The recent nearby supernovae SN 1993J and SN 1994I have provided particularly useful material to clarify the supernova - progenitor connection. For a progenitor of type IIb supernova 1993J, we propose that merging of two stars in a close binary is responsible for the formation of a thin H-rich envelope. As a progenitor of type Ic supernova 1994I, we propose a bare C+O star that has lost both its H and He envelope after a common-envelope phase. By generalizing these scenarios, we show that common-envelope evolution in massive close binary stars leads to various degrees of stripping off of the envelope of a massive star. This naturally leads to an explanation of the origin of type II-L, IIn, IIb, Ib, and Ic in a unified manner. The binary hypothesis to explain the diversity of supernovae can be substantiated with new information on SN IIb 1993J and SN Ic 1994I. Model light curves are compared with observations. Since extensive mass loss is essential for the binary scenario, circumstellar interactions are examined for comparison with X-ray observations.

1. Introduction

Supernovae are classified as type I and type II according to the absence or presence of hydrogen lines in their spectra, and further subdivided into Ia, Ib, Ic, II-P, II-L, IIb, and IIn (e.g., Branch et al. 1991; Filippenko 1991). The presence of strong Si lines and He lines defines type Ia and type Ib, respectively, while type Ic is characterized by the lack or weakness of these lines. The light curves of type II-P supernovae (SNe II-P) have a plateau, while those of type II-L (SNe II-L) show a linear decline.

It has been suggested that the diversity of SNe II originates from the progenitor's different main-sequence mass ranges, i.e., SNe II-L from 7–$10\,M_\odot$ (Swartz et al. 1991) and above $10\,M_\odot$ SNe II-P, and that type Ib/Ic supernovae (SNe Ib/Ic) originate from He stars of different mass range in binary systems (Nomoto et al. 1990). However, the exact supernova - progenitor connection for these types has been a controversial issue.

The recent nearby supernovae, SN 1993J in M81 and SN 1994I in M51, have been identified as type IIb (SN IIb) (e.g., Schmidt et al. 1993) and type Ic (e.g., Wheeler et al. 1994), respectively, adding more diversity of supernovae (i.e., IIb) but shedding new light on the classification and their evolutionary origins. Here we show that common-envelope evolution in massive close binary stars can provide a plausible explanation of the observational diversity.

For SN Ic 1994I, we will discuss a C+O star progenitor model, with three possible evolutionary paths to form it (Section 2). Light curves are particularly useful probes to discriminate the progenitor's mass (Section 3). For SN IIb 1993J, we propose that merging of two stars in a close binary is responsible for the formation of a thin H-rich envelope of the progenitor, as opposed to the conservative mass transfer scenario (Section 4). We then generalize the binary scenario to show that common-envelope evolution in massive close binary stars leads to various degree of stripping off of the envelope mass of massive star. This naturally explains the origin of supernova types, namely, II-L, IIb, Ib, and Ic, in a unified manner, depending on the mass ratio q of component stars and the initial separation R_0 (Section 4).

Another clue to the understanding of the nature of these nearby supernovae in relation to the binary hypothesis is the extensive mass loss from progenitors to form dense circumstellar matter. Circumstellar interactions are studied using a realistic ejecta model of SN 1993J to compare with X-ray observations (Section 5).

2. Progenitors of type Ic Supernovae

At the time of explosion, progenitors of SNe Ib/Ic have lost their hydrogen-rich envelope, and most of the helium envelope as well for SNe Ic. Two cases are possible (Wheeler & Harkness 1990): (1) stellar wind in Wolf-Rayet stars and (2) Roche lobe overflow in binary stars. Both these cases may actually occur as SNe Ib since SNe Ib light curves show a significant diversity from slow to fast decline.

For SNe Ic, however, earlier models have some difficulties to account for the observations. (1) WC/WO Wolf-Rayet stars are so massive that the light curve declines too slowly (Woosley et al. 1993). (2) Low-mass helium star models in binaries (Shigeyama et al. 1990; Nomoto et al. 1990; Woosley et al. 1995) have too much helium to be consistent with the lack of He features in the spectra of type Ic SN 1987M (Filippenko et al. 1990; Lucy 1991; Swartz et al. 1993).

These difficulties have led to the suggestion that C+O stars which have lost even their He envelope are the progenitors of SNe Ic (Yamaoka et al. 1993; Swartz et al. 1993). In a massive binary, a bare C+O star can form after two stages of mass transfer (Bhattacharya & Van den Heuvel 1991). The first mass transfer occurs when the more massive star has formed a helium core, and its envelope expands to fill the Roche lobe. The H-rich envelope is lost and a helium star is produced. After core He burning, the helium star expands and may again fill its Roche lobe, depending on its mass M_α. This second mass transfer is more likely to occur for lower mass helium stars because they attain larger radii (Habets 1986); for example, the maximum radii of helium stars with $M_\alpha = 3.3, 4.0, 6.0,$ and $8.0\,M_\odot$ are 3.7, 3.0, 1.9, and $1.3\,R_\odot$, respectively (Nomoto & Hashimoto 1988), while helium stars less massive than $2.7\,M_\odot$ even expand to red giant dimensions (Habets 1986).

In case the second mass transfer occurs to a more massive companion, mass transfer will be conservative (i.e., most of the transferred mass is accreted by the companion), and the He star probably retains part of its envelope. On the other hand, if the companion is less massive, the helium star may lose its entire envelope and produce a bare C+O star. The following three different evolutionary paths (A, B, C) are possible for the formation of C+O stars as illustrated in Fig. 1.

Path A: Initially, the binary system consists of star 1 and star 2 whose main-sequence masses are $M_1 \approx 11$–$16\,M_\odot$ and $M_2 \approx 1$–$4\,M_\odot$, respectively, i.e., their mass ratio q is between ~ 0.1 and ~ 0.25. Because of the extreme mass ratio, the first mass transfer is highly non-conservative. This almost inevitably leads to the formation of a common envelope (Van den Heuvel 1994) and the subsequent spiral-in (Pols et al. 1991) of star 2 and the core of star 1. In many cases, the spiral-in causes the stars to merge into a single star, but if the initial orbital separation is large enough, the binary system probably survives the spiral-in. It then consists of helium star 1 of 2.2–$4\,M_\odot$ and main-sequence star 2

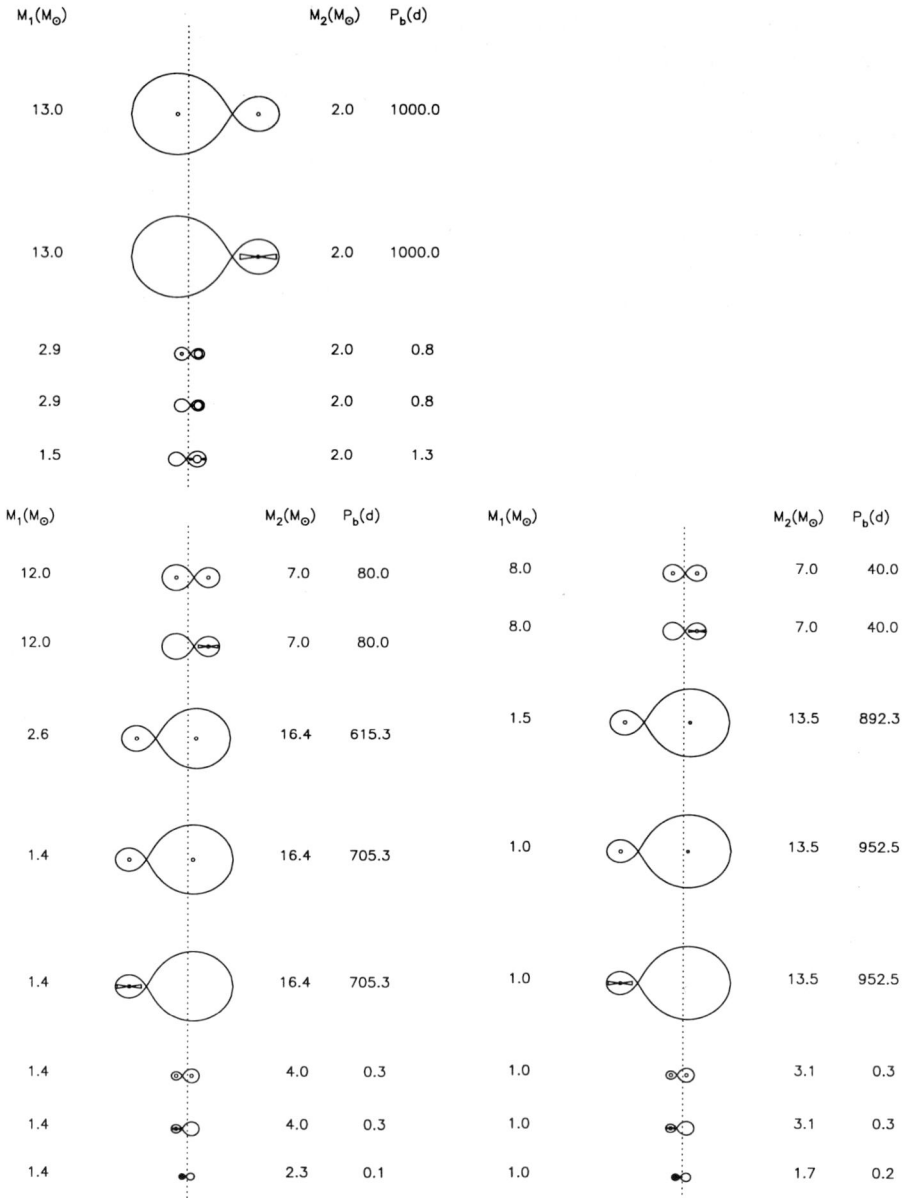

Figure 1. Schematic overview of evolutionary paths leading to the formation of bare C+O stars (see text): Path A with a 2.0 M$_\odot$ main-sequence companion (upper), Path B with a 1.4 M$_\odot$ neutron star comparnion (lower left), and Path C with a 1.0 M$_\odot$ white dwarf companion (lower right). Evolutionary changes in the masses and orbital period are shown.

of 1–4 $M_\odot$, in a much closer orbit. Helium star 1 subsequently expands and undergoes a second, non-conservative mass transfer onto star 2, losing its He envelope. Star 1 then becomes an almost bare C+O star with a low-mass main-sequence companion.

Path B: Initially star 1 is massive enough to evolve through core collapse ($M_1 \gtrsim 11\,M_\odot$), leaving a neutron star. The initial mass ratio is not too small, $q \gtrsim 0.4$, so that mass transfer from star 1 to star 2 is quasi-conservative and star 2 becomes more massive than $11\,M_\odot$. Star 2 subsequently evolves to fill its Roche lobe. Since star 1 is a compact star and is much less massive than star 2, it will inevitably spiral into the envelope of star 2. Like in Path A, the binary may either merge, or it may survive the spiral-in. The resulting binary system then consists of helium star 2 of 2.2–4 $M_\odot$ and compact star 1 of 1.4 $M_\odot$ in a close orbit. Helium star 2 expands to fill its Roche lobe and undergoes another non-conservative mass transfer to star 1. This leads to the formation of C+O star 2 and a neutron star companion.

Path C: Initially star 1 has a mass in the range $\sim$6–11 $M_\odot$ and becomes a C+O or ONeMg white dwarf. The initial mass ratio is close to unity, so that the mass transfer from star 1 to star 2 is almost conservative, and star 2 becomes more massive than $11\,M_\odot$. Further evolution is similar to Path B since star 1 is a compact white dwarf. Star 2 inevitably undergoes non-conservative mass transfer to star 1 twice and becomes a bare C+O star with a white-dwarf companion.

Fig. 2 shows the second mass transfer in Path B, i.e., evolutionary tracks in the H-R diagram of $4\,M_\odot$ helium star binaries with a $1.4\,M_\odot$ neutron star companion (Pols et al. 1994). Here the solid, dashed, and dotted lines show the cases for a single He star, $P_{orb} = 0.3\,\mathrm{d}$, and $P_{orb} = 0.1\,\mathrm{d}$, respectively. The tracks have been computed with non-conservative mass transfer; the ejected mass is assumed to leave the system via a jet or symmetric wind from the neutron star. The calculations terminate at central carbon ignition. Mass transfer in the $P_{orb} = 0.3\,\mathrm{d}$ system is stable (on a thermal time scale), but becomes unstable in the 0.1 day system after transfer of a few tenths of a solar mass. The mass transfer rate in the 0.3 day system is $\sim 10^{-5}\,M_\odot\,\mathrm{yr}^{-1}$. The final mass of the He star in the 0.3 day system is $2.7\,M_\odot$ with the CO core mass being $2.24\,M_\odot$. It seems likely that part or all of the remaining $0.46\,M_\odot$ of He envelope can be transferred during carbon burning, leaving an almost bare C+O star remnant.

The C+O star thus formed through these paths explodes as a SN Ic. If a substantial He envelope is retained in the second mass transfer, the explosion would be a SN Ib. For the latter case of SNe Ib, the ejected mass is significantly smaller than that for the $>5\,M_\odot$ He star model, thereby forming a light curve with steeper decline. In this scenario, the spectroscopic

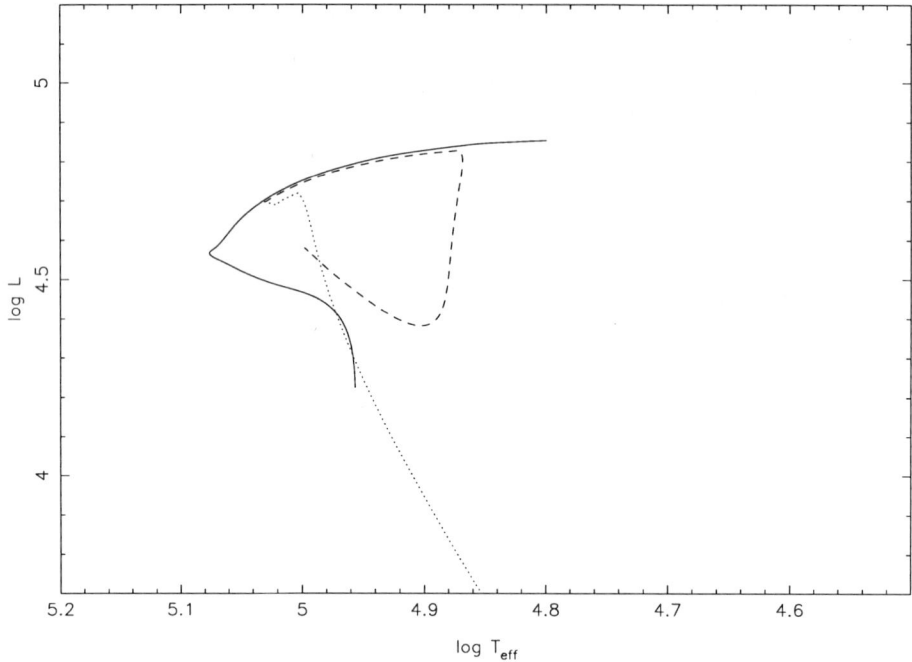

Figure 2. The second mass transfer in Path B, i.e., evolutionary tracks in the H-R diagram of $4\,M_\odot$ helium star binaries with a $1.4\,M_\odot$ neutron star companion (Pols et al. 1994). Shown are the cases for a single He star (solid line), $P_{orb} = 0.3\,d$ (dashed line), and $P_{orb} = 0.1\,d$ (dotted).

and photometric features of SNe Ib and SNe Ic are predicted to form a continuous sequence rather than distinct phenomena.

The explosion of a C+O star is likely to leave a neutron star behind. Unless a kick velocity is too high, the binary system would survive because of the small ejecta mass. A companion of the neutron star would be a low mass main-sequence star, a C+O white dwarf, or a neutron star in a short and eccentric orbit.

The formation rate of C+O stars and the resultant SN Ic rate from the above paths may be estimated as follows. We use an initial mass function $\Psi(m) \propto m^{-2.7}$ and a q-distribution $\Phi(q) = 2\,(1+q)^{-2}$. We estimate the probability S that the binary survives after the spiral-in without merging to be $S \sim 0.25$ (path A) and 0.5 (paths B and C) from the relative range of initial separations. The fraction of binaries that evolve conservatively as a function of primary mass is taken from Pols et al. (1991). We then obtain the relative rates of type Ic with respect to type II supernovae as ~ 0.015 (path A), 0.025 (B), and 0.065 (C) for a close binary fraction of 0.5. Note that Path C is the most important channel, mainly due to the shape of the initial mass function. In total, the expected type Ic frequency is ~0.1 times that of SNe II. This is probably consistent with the observed relative ratio between SNe Ic and SNe II, if SN Ib and SN Ic rates are comparable (van den Bergh 1991; see, however, Muller et al. 1992).

3. Light Curve Models for SN Ic 1994I

We apply the above scenario to SN Ic 1994I. The simplest evolutionary path is as follows. The progenitor was a 13–18 $M_\odot$ star on the main sequence. Through Roche lobe overflow the 13, 15, and 18 $M_\odot$ stars became He stars with M_α = 3.3, 4.0, and 5.0 $M_\odot$ and then lost their helium envelopes to become C+O stars with M_{C+O} = 1.8, 2.1, and 2.9 $M_\odot$. Hereafter these models are called CO18, CO21, and CO29, respectively (Nomoto et al. 1994).

The mass cut is chosen to produce 0.07 $M_\odot$ of ^{56}Ni. The deposited energy is set to produce the kinetic energy of explosion $E = 10^{51}$ erg s^{-1} for CO18, CO21, and CO29, and $E = 6\ 10^{51}$ erg s^{-1} for the lower explosion energy model CO21L. It is noticeable that the ejecta masses of $M_{\rm ej}$ = 0.5 and 0.9 $M_\odot$ for CO18 and CO21 are significantly smaller than the Chandrasekhar mass. Even for CO29, $M_{\rm ej}$ is still as small as 1.5 $M_\odot$ (Hashimoto et al. 1993).

Since the C+O star is compact with a radius of ~0.2 $R_\odot$, the light curve is not due to shock heating but is powered by the radioactive decay chain ^{56}Ni → ^{56}Co → ^{56}Fe. For a detailed comparison with observations, monochromatic light curves for the C+O star models have been calculated and are compared with observations (Schmidt & Kirshner 1994) in Fig. 3. The slopes of the calculated light curves are found to be sensitive to the models as follows:

CO18: Both the rise time and the decline are far too short compared with the observations. The main cause is that the diffusion time scales in the envelope are too short, suggesting the need for a more massive model to increase the diffusion time scale and, consequently, to produce a broader and flatter maximum.

CO21 : This model gives almost perfect agreement between the slopes of the theoretical and observed shapes of the light curves for the monochromatic B, V, R and I band and the bolometric light curve as well (Iwamoto et al. 1994). This means that the diffusion time scales, the energy input and the temperature structure are about correct. From the light curve fits, a distance modulus of 29.2 ± 0.3 mag is derived for M51. Also a high interstellar reddening, $E(B - V) = 0.45$

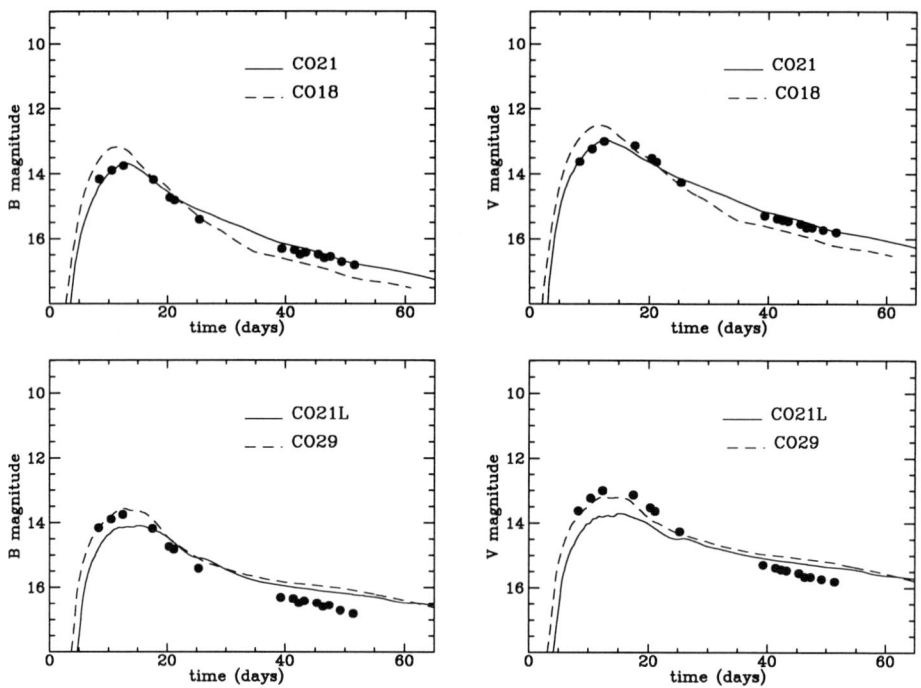

Figure 3. Theoretical monochromatic light curves for models CO18, CO21, CO29, and CO21L (Iwamoto *et al.* 1994), in comparison with observations of SN 1994I (Schmidt & Kirshner 1994).

($A_V \sim 1.4$ mag) is required. The ejected ^{56}Ni mass is found to be $0.07^{+0.035}_{-0.025}\,M_\odot$.

CO29 : The photosphere recedes nicely at maximum light. At later times, however, the light curve declines too slowly because the escape probability changes only little with time due to the lower expansion velocities compared with CO21.

CO21L : Due to the small expansion rate, the escape probability for γ rays is significantly higher than for CO21. This keeps the photosphere hot, i.e., the opacity hardly drops at maximum light and, consequently, the maximum is not well pronounced compared to CO21. At later times, the decline of the light curve is too slow for the same reasons as for CO29.

Wolf-Rayet star models are clearly too massive to be consistent with the light curves of SN 1994I, because even the smallest WC star model has $2.7\,M_\odot$ of ejecta (Woosley *et al.* 1993), which is significantly more massive than the $1.5\,M_\odot$ for CO29.

However, the late time oxygen line emissions of SN1994I suggest that models with somewhat lower expansion rates than CO21 are favored (Fransson 1994). Such slower models CO29 and CO21L are not consistent with the light curve, however. Mixing of ^{56}Ni would lead to a faster decline of the light curves for these models. Large scale mixing due to Rayleigh-Taylor instabilities may not be expected for these C+O star models because of the lack of both H and He layers and the associated density jumps (Hachisu et al. 1991). If the expansion should certainly be slower than for CO21, a new mechanism of mixing is required unless the star has a rather massive He envelope.

4. Progenitor of SN 1993J

SN 1993J has revealed important new features of supernovae [see Wheeler & Filippenko (1994) for a review and references]. It has been identified as a type II supernova (SN II) from hydrogen features. The light curve of SN 1993J is distinctly different from those of previously known SNe II. It was obvious that this peculiar light curve of SN 1993J cannot be accounted for by an explosion of an ordinary red supergiant (RSG) with a massive hydrogen-rich envelope, which produces a light curve of a SN II-P (Nomoto et al. 1993).

As will be described in Section 3, the light curve of SN 1993J can be understood as the explosion of an RSG whose hydrogen-rich envelope is as small as $\lesssim 1\,M_\odot$ (Nomoto et al. 1993; Podsiadlowski et al. 1993). The thin-envelope model has been confirmed by the spectral changes which shows growing features of helium and oxygen, so that SN 1993J can be classified as a SN IIb (Woosley et al. 1988; Filippenko et al. 1988).

The progenitor of SN 1993J is likely to have lost most of its H-rich envelope due to the interaction with its companion star in a binary system. The binary scenario raised the following questions: (1) what controls the mass of the remaining H-rich envelope of the progenitor, and (2) what is the relation of SN 1993J with other types of supernovae, such as SNe IIn, II-L and Ib/Ic.

4.1. CONSERVATIVE MASS TRANSFER SCENARIO

As a possible evolutionary scenario leading to the progenitor of SN 1993J, a *Case C* binary evolution has been proposed (Podsiadlowski et al. 1993; Woosley et al. 1994; Ray et al. 1993). This scenario postulates that (1) the initial separation between the two component stars is so large that the mass transfer from the progenitor started only after helium was exhausted in the core, and (2) mass ratio $q = M_1/M_2$ between the progenitor 1 and the companion star 2 is close to unity, so that the mass transfer from the progenitor is more or less conservative.

In Case C mass transfer, however, the primary star has a convective envelope, thereby transferring mass to the companion star on a dynamical time scale, at least in its early phase. Such an evolution is described by assuming the non-dimensional specific angular-momentum loss from the system α and the ratio β between the mass accreted by star 2 and the mass lost by star 1 as arbitrary parameters (Podsiadlowski *et al.* 1992; Rathnasree & Ray 1992). It is seen that, even for $q < 1$ and $\beta \sim 1$, the radius of the mass donor exceeds the Roche lobe. Nevertheless the response of the companion star or the formation of the common envelope was not calculated, i.e., there has been no consistent set of calculations of both the dynamical mass loss from the progenitor and the response of mass receiving companion. Thus it is possible that even if q is initially smaller than 1 due to earlier wind mass loss from star 1, Case C mass transfer leads quickly to the formation of a common envelope. Thus, although the Case C scenario may be possible for a narrow parameter space, an alternative scenario is worth exploring.

4.2. NON-CONSERVATIVE MASS TRANSFER SCENARIO

Evolutionary paths of close binaries depend significantly on the mass ratio q of component stars and the initial separation R_0. Here we consider binary systems consisting of star 1 and star 2 whose main-sequence mass ratio q is significantly smaller than unity. Star 1 evolves to form a He core and its H-rich envelope expands to fill its Roche lobe. Because of the extreme mass ratio, the mass transfer is highly non-conservative. This almost inevitably leads to the formation of a common envelope and the subsequent spiral-in of star 2 and the core of star 1.

The spiral-in deposits the orbital energy in the envelope due to viscous evolution. Subsequent-common envelope evolution is so complicated that we take a simplified approach. From energy considerations, we assume that spiral-in yields the following outcome.

1. If the deposited orbital energy, E_g, is larger than the binding energy of the common envelope, E_b, almost all envelope material is ejected before star 2 is dissolved. In other words, the binary system survives the spiral-in and then consists of helium star 1 and main-sequence star 2 in a much closer orbit. Since E_g is larger for larger R_0, this case occurs when R_0 is larger than a certain limit. (Here the time scale is so short that radiation loss is negligible.)
2. If $E_g < E_b$, on the contrary, the two stars merge into a single star, i.e., star 2 is completely dissolved in the common envelope before all

the envelope mass is ejected. The resulting single star 1 retains some envelope material. The envelope mass M_{env} after merging depends on the deposited energy E_g relative to E_b. Larger E_g/E_b induces larger amount of mass loss and forms a lower-mass envelope when the merging is completed. In terms of R_0, such a merging occurs when R_0 is smaller than a certain limit. The remaining envelope mass is smaller for larger R_0. Afterwards star 1 would expand to become an RSG but its M_{env} could be significantly smaller than, say, $5\,M_\odot$.

The non-conservative scenario predicts the formation of a single neutron star, in contrast to the binary neutron star (in an eccentric orbit) predicted by the conservative mass transfer scenario.

5. Supernova Types and Merging

Based on the above spiral-in scenarios 1 and 2, we can present a new interpretation of the origin of various types of supernovae, in particular IIb, II-L, and IIn.

1. **SN Ib:** In this case, the spiral-in forms a pair of helium star 1 and main-sequence star 2. If the helium star mass exceeds $\sim 2.5\,M_\odot$, it evolves through Fe core collapse and explodes as a SN Ib because of the presence of helium (e.g., Ensman & Woosley 1988; Shigeyama et al. 1990).
   SN Ic: If the mass of helium star 1 is smaller than $\sim 5\,M_\odot$, its helium envelope expands possibly to exceed the Roche lobe (Habets 1986; Nomoto & Hashimoto 1988). After the loss of the helium envelope, star 1 becomes a C+O star. The explosion of the C+O star is triggered by Fe core collapse and must be observed as a SN Ic.
2. In this case, the two stars merge to form a single core and lose a significant fraction of their common envelope due to frictional heating. Unless the envelope mass becomes as small as $10^{-2}\,M_\odot$, the envelope of the merged star should expand to an RSG size. If the helium core mass exceeds $\sim 2.5\,M_\odot$, a supernova explosion is triggered by Fe core collapse and observed as a type II. We propose that the progenitors of IIb, II-L, and possibly IIn are these merged stars and the difference in the types originates from the difference in the mass of the H-rich envelope M_{env} as follows.
   SN IIb: If $M_{env} \lesssim 1\,M_\odot$, the pre-supernova configuration would be similar to that of SN 1993J. Their light curves around the second peak and tails must be similar to SNe Ib (Section 5). SN 1993J continues to exhibit strong Hα emissions, while SN 1987K (also IIb, see Filippenko 1988) did not show H-emission lines at late phase. Late time Hα emission in SN 1993J is powered by X-rays from circumstellar interaction

(Section 6, Clocchiatti & Wheeler 1994). Further study is needed to understand whether such a difference stems from the difference in $M_{\rm env}$ (and thus density structure) or in the circumstellar matter density (or some other parameters).

SN II-L: If $M_{\rm env} \sim 2\text{--}3\,M_\odot$, the radius of the RSG is as large as that of an ordinary RSG. However, because $M_{\rm env}$ is small, the expansion velocity at the bottom of the H-rich envelope is higher. This leads to a shorter duration of the plateau, i.e., SN II-L (e.g., Shigeyama & Nomoto 1990; Swartz et al. 1991; Blinnikov & Bartunov 1993).

SN IIn: If the envelope mass remains as large as $M_{\rm env} \gtrsim 5\,M_\odot$ possibly due to the large initial mass of star 1, the star becomes an ordinary RSG after merging. It eventually explodes as a SN II-P. However, its CSM would consist of the material ejected during spiral-in and the RSG wind material. The structure of CSM originating from spiral-in is likely to be asymmetric; the mass ejection may form a bipolar jet or disk like material. The mass loss rate during the RSG phase could be larger than for an ordinary RSG because of the larger mass and extra heating due to merging. This case might correspond to SN IIn, like SN 1988Z (e.g., Filippenko 1991).

6. Circumstellar Interaction in SN 1993J

The merging model for the progenitor of SN 1993J predicts the presence of (probably) asymmetric outer circumstellar matter (CSM) formed through a merging process and more symmetric inner CSM formed by the wind from the red supergiant.

Suzuki et al. (1995) have constructed a hydrodynamical model of interaction between the ejecta of SN 1993J and CSM to account for the basic features of X-ray emissions from SN 1993J as observed with OSSE, ASCA, and ROSAT for the first 570 days (also Fransson et al. 1994). This model consists of a realistic ejecta model and clumpy CSM.

The collision between the ejecta and CSM creates a reverse shock which is radiative, to form a cooling dense shell in the ejecta. X rays emitted from the reverse shock are mostly absorbed by this shell. Early hard X rays are well modeled as thermal emissions from shocked CSM. The CSM density inferred from X-ray observations is as high as $\dot{M}/v_w = (3\text{--}4)\,10^{-5}\,M_\odot\,{\rm yr}^{-1}/(10\,{\rm km\,s^{-1}})$ which is rather high compared with the mass loss rate estimated for the 13–15 $M_\odot$ model. In the merging progenitor model, this problem could be resolved because mass loss rate can be enhanced due to extra heating in the envelope produced by merging.

The above model indicates that CSM has a spatially variable density gradient. In the inner layer, the gradient has to be shallower than that of

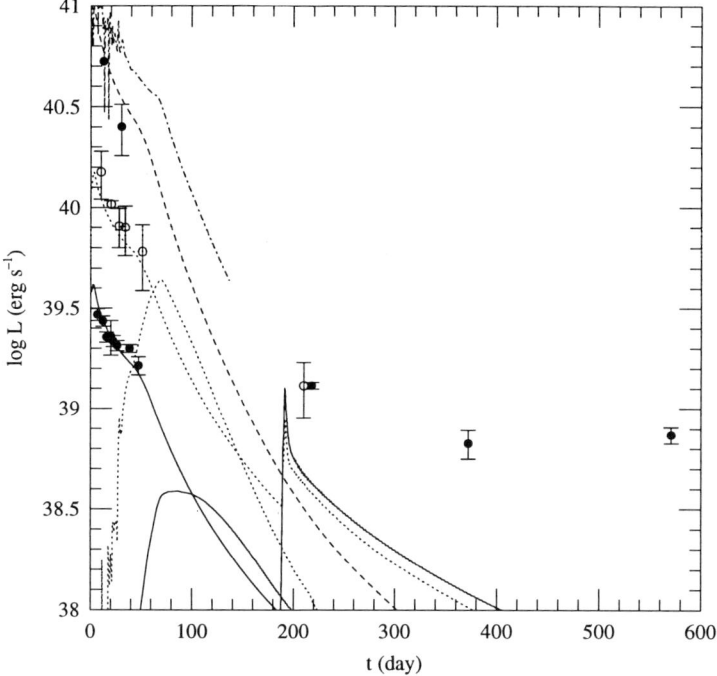

Figure 4. Calculated X-ray light curves in the 0.1–2.4 keV (solid), 1–10 keV (dotted), and 50–150 keV (dashed) bands, which are compared with observations with ROSAT, ASCA (open circles), and OSSE, respectively. The upper three lines which decline monotonically after day 6 show the homogeneous CSM components, while the lower two lines (solid and dotted) show the ejecta components that escape from being absorbed by the dense shell. The total luminosity of X-rays emitted from the ejecta is shown by the dash-dotted curve through day 140. Faster decline from day 50 is due to the steeper density gradient of homogeneous CSM. The sudden increase around day 190 is due to the collision with a spherical thin shell clump.

a steady wind to be consistent with the slow decline of the 0.1–2.4 keV light curve. In the outer layers, CSM is highly clumpy so that the density gradient of inter-clump matter is steeper. Soft X-rays at late times are mostly emitted from the shocked high density clumps as seen in Fig. 4 (see legend).

The clumpy circumstellar structure can account for other important features of SN 1993J as well (Van Dyk *et al.* 1994). In particular, the steep density gradient of inter-clump matter leads to weak deceleration of the ejecta and shocked CSM, i.e., the expansion velocities of the shocked ejecta and CSM decrease only slowly. This is consistent with (i) a roughly constant maximum velocity of hydrogen ($\sim 10\,000\,\mathrm{km\,s^{-1}}$) as observed in H$\alpha$ features (Patat et al. 1995; Fransson 1994), and (ii) the average expansion velocity of the radio shell over 240 days, which is as high as $15\,000\pm2\,000\,\mathrm{km\,s^{-1}}$ for a distance of $3.63\pm0.34\,\mathrm{Mpc}$ (Marcaide et al. 1995).

In the non-conservative binary evolution scenario for the progenitor of SN 1993J (Nomoto et al. 1994, 1995) clumpy CSM could form as follows. Firstly the spiral-in of a companion star into the progenitor leads to the ejection of a common envelope. Later stellar-wind material from the red supergiant progenitor would collide with previously ejected common-envelope matter. As a result of the deceleration of wind matter, the wind velocity would be lower at a larger distance, i.e., the density gradient would be shallower than for a steady wind, and CSM would be more clumpy as it is closer to the common-envelope matter. This model predicts the collision of the blast shock wave with former common-envelope matter, which would lead to some enhancement of soft X-ray fluxes and relatively low-velocity Hα emissions; this might correspond to the latest ROSAT observations showing recent leveling off of the X-ray flux (Zimmermann et al. 1994) and the HST observations of Hα features suggesting a collision with ring-like matter (Kirshner 1994).

Another future event would be an enhancement of X-ray fluxes when the reverse shock reaches the H/He interface where the density sharply increases (Shigeyama et al. 1994; Woosley et al. 1994). If the enhancement is observed, its date will provide information on the thickness of the H-rich envelope and propagation speed of the reverse shock.

In this way, X-ray emissions from circumstellar interactions provide important information on the internal structure of the supernova ejecta (such as the density gradient), as well as mass loss from the progenitor. Detailed study of circumstellar interaction would thus provide an important clue to the progenitor's evolution, especially its binary nature, for SNe IIb, II-L, and IIn.

7. Concluding Remarks

We have proposed that except for *classical* SNe I and SNe II (SNe Ia and II-P), all other supernova subtypes (SNe Ib, Ic, IIb, II-L and IIn) can be explained as Fe core collapse of massive ($\gtrsim 10\,M_\odot$) stars in close binary systems. In other words, the main-sequence mass range of the progenitor of SNe Ib, Ic, IIb, and II-L and thus their nucleosynthesis may be similar to those of SNe II-P, except for the SNe II-P from the 8–10 $M_\odot$ AGB stars (Hashimoto et al. 1993).

If the above proposal is correct, light curves and spectra of SNe Ib, Ic, IIb, II-L, especially those of recent nearby events, are particularly useful to obtain the ^{56}Ni and O masses as a function of the main-sequence mass. The ^{56}Ni mass estimates are closely related with the neutron star mass estimate (Thielemann et al. 1995). From the light curve shape, we may infer the ejecta mass (i.e., the progenitor's mass) and the ^{56}Ni mass for the calculated extent of mixing. The produced ^{56}Ni mass (and the progenitor's main-sequence mass) is estimated as $0.09\pm0.02\,M_\odot$ (13–$14\,M_\odot$) for SN 1993J ($A_V = 0.4$ mag) and $0.07^{+0.035}_{-0.025}\,M_\odot$ (14–$15\,M_\odot$) for SN 1994I. Combining with $0.075\pm0.01\,M_\odot$ ^{56}Ni for SN 1987A ($\sim 20\,M_\odot$, Arnett et al. 1989; Nomoto et al. 1994), the ejected ^{56}Ni mass seems to be not too sensitive to the progenitor's mass. [Earlier estimates of the $\sim 0.15\,M_\odot$ ^{56}Ni for SNe Ib/Ic being $\sim 1/4$ of those for SNe Ia (Panagia 1987) are higher than the above values, which probably suggests the necessity of reexamination of the distances and/or extinction for SNe Ib/Ic.]

The light curve tails of SNe II-L would provide a useful information on whether SNe II-L have massive progenitors. Recent observations of SN II-L 1990K (Cappellaro et al. 1994) have shown that its late light curve is similar to that of SN 1987A, suggesting the progenitors of *bright* II-L may be similarly massive. This is consistent with the binary merger model but not with the AGB model (Swartz et al. 1991). For normal II-L, more observations are needed.

To conclude, our hypothesis implies that the observational diversity of supernovae can be the result of a single mechanism of explosion, i.e., a gravitational collapse of an Fe core, combined with merging of stars in close binary systems.

Acknowledgements. This research has been supported in part by the Grants-in-Aid for Scientific Research of the Ministry of Education, Science, and Culture in Japan (05242102, 06233101) and the Fellowships of the Japan Society for the Promotion of Science for Japanese Junior Scientists (2539, 4227).

References

Arnett, W.D. et al. 1989, ARA&A 27, 629
Bhattacharya, D. & Van den Heuvel, E.P.J. 1991, Phys. Rep. 203, 1
Blinnikov, S.I. & Bartunov, O.S. 1993, A&A 273, 106
Branch, D., Nomoto, K. & Filippenko, A.V. 1991, Comm. Astrophysics 15, 221
Cappellaro, E. et al. 1994, A&A (in press)
Clocchiatti, A. & Wheeler, J.C. 1994, IAU Circ. 6005
Ensman, L.M. & Woosley, S.E. 1988, ApJ 333, 754
Filippenko, A.V. 1988, AJ 96, 1941
Filippenko, A.V. 1991, in *Supernovae and Stellar Evolution*, A. Ray & T. Velusamy (Eds.), World Scientific (Singapore), p. 58
Filippenko, A.V., Porter, A.C. & Sargent, W.L.W. 1990, AJ 100, 1575

Fransson, C. 1994, talk at the 17th Texas Symposium on Relativistic Astrophysics
Fransson, C., Lundqvist, P. & Chevalier, R.A. 1994, ApJ (submitted)
Habets, G.M.H.J. 1986, A&A 187, 209
Hachisu, I. et al. 1991, ApJ 368, L27
Hashimoto, M., Iwamoto, K. & Nomoto, K. 1993, ApJ 414, L105
Hashimoto, M. et al. 1993, in *Nuclei in the Cosmos*, F. Käppeler & K. Wisshak (Eds.), IOP Pub. (Bristol), p. 587
Iwamoto, K. et al. 1994, ApJ 437, L115
Kirshner, R. 1994, talk at the 17th Texas Symposium on Relativistic Astrophysics
Lucy, L.B. 1991, ApJ 383, 308
Marcaide, J. M. et al. 1995, Nat 373, 44
Muller, R.A., et al. 1992, ApJ 384, L9
Nomoto, K. & Hashimoto, M. 1988, Phys. Rep. 163, 13
Nomoto, K., Filippenko, A.V. & Shigeyama, T. 1990, A&A 240, L1
Nomoto, K. et al. 1993, Nat 364, 507
Nomoto, K. et al. 1994, Nat 371, 227
Nomoto, K. et al. 1994, in *New Horizon of X-Ray Astronomy*, F. Makino & T. Ohashi (Eds.), Universal Academy Press (Tokyo), p. 139
Nomoto, K. et al. 1994, in *Supernovae*, S.A. Bludman et al. (Ed.), Elsevier (New York), p. 489
Nomoto, K., Iwamoto, K. & Suzuki, T. 1995, Phys. Rep. (in press)
Panagia, N. 1987, in *High Energy Phenomena Around Collapsed Stars*, F. Pacini (Ed.), Reidel (Dordrecht), p. 33
Patat, F., Chugai, N.N. & Mazzali, P.A. 1995, A&A (in press)
Podsiadlowski, Ph., Joss, P.C. & Hsu, J.J.L. 1992, ApJ 391, 246
Podsiadlowski, Ph. et al. 1993, Nat 364, 509
Pols, O.R. et al. 1991, A&A 241, 419
Pols, O.R. et al. 1994, Poster presented at IAU Symposium 165
Rathnasree, N. & Ray, A. 1992, J. Astrophys. Astron. 13, 3
Ray, A., Singh, P. & Sutaria, F.K. 1993, J. Astrophys. Astron. 14, 53
Schmidt, B.P. et al. 1993, Nat 364, 600
Schmidt, B. & Kirshner, R. 1994, E-mail circulation
Shigeyama, T. & Nomoto, K. 1990, ApJ 360, 242
Shigeyama, T. et al. 1990, ApJ 361, L23
Shigeyama, T. et al. 1994, ApJ 420, 341
Swartz, D.A. & Wheeler, J.C. 1991, ApJ 379, L13
Swartz, D.A., Wheeler, J.C. & Harkness, R.P. 1991, ApJ 374, 266
Swartz, D.A. et al. 1993, ApJ 411, 313
Thielemann, F.-K., Nomoto, K. & Hashimoto, M. 1995, ApJ (in press)
van den Bergh, S.A. 1991, ARA&A 29, 363
Van den Heuvel, E.P.J. 1994, in *Interacting Binaries*, H. Nussbaumer & A. Orr (Eds.), Springer Verlag (Berlin), p. 263
Van Dyk, S.D. et al. 1994, ApJ 432, L115
Wheeler, J.C. & Filippenko, A.V. 1994, in *Supernovae and Supernova Remnants*, IAU Coll. 145, R. McCray (Ed.), Cambridge University Press (Cambridge), (in press)
Wheeler, J.C. & Harkness, R.P. 1990, Rep. Progr. Phys. 53, 1467
Wheeler, J.C. et al. 1994, ApJ 436, L135
Woosley, S.E., Langer, N. & Weaver, T.A. 1993, ApJ 411, 823
Woosley, S.E., Langer, N. & Weaver, T.A. 1995, ApJ (submitted)
Woosley, S.E., Pinto, A. & Ensman, L.M. 1988, ApJ 324, 466
Woosley, S.E. et al. 1994, ApJ 429, 300
Yamaoka, H., Shigeyama, T. & Nomoto, K. 1993, A&A 267, 433
Zimmermann, H.-U. et al. 1994, IAU Circ. 6120

THE ENVIRONMENTS OF TYPE IB/C SUPERNOVAE

S. D. VAN DYK, A. J. BARTH AND A. V. FILIPPENKO
Astronomy Department, University of California,
Berkeley CA 94720-3411 USA

1. Introduction

Up to about 1985, supernovae (SNe) generally were placed into the two Minkowski classes, type I and type II, defined by the absence or presence, respectively, of hydrogen in their optical spectra. Around that time it was acknowledged that several type I SNe were systematically peculiar, both spectroscopically and photometrically (Elias *et al.* 1985; Wheeler & Levreault 1985; Uomoto & Kirshner 1985; Branch 1986; Filippenko 1986), by missing the characteristic Si II spectral feature near 6150 Å, having distinct infrared light curves, being optically redder and subluminous, and showing radio emission (Sramek *et al.* 1984). These SNe were designated as type Ib (Elias *et al.* 1985; Branch 1986) to distinguish them from the classical type Ia. Harkness *et al.* (1987) identified He I lines in spectra of the SN Ib 1984L, but some subsequent examples showed no He in their spectra and were further subclassified as Type Ic (Wheeler & Harkness 1990). The two subtypes, however, are nearly indistinguishable at late times. In this Symposium the entire class has been referred to as type Ib/c SNe. A recent bright example is SN 1994I in M51 (Filippenko *et al.* 1994).

It has been realized that what makes SNe Ib/c different must also extend to the nature of their progenitor stars (e.g., Wheeler & Levreault 1985). Constraints on the progenitor can be obtained not only through observations of each event (see Leibundgut, these proceedings) and through theoretical modelling (see Nomoto, these proceedings), but also by examining the relationship of these SNe with their environment. Here we look at this environment on progressively smaller scales. Since only 32 *bona fide* members of this subclass exist, it must be kept in mind that small-number statistics are clearly a problem. Yet, even with this small sample, we are developing a clearer picture of the progenitor population for SNe Ib/c.

2. Global Environment

SNe Ib/c are found only in late-type galaxies (Porter & Filippenko 1987; van den Bergh & Tammann 1991). This implies that these SNe likely arise only from young massive stars, as do type II SNe. However, unlike SNe II, somehow during the lifetime of the progenitor star, the hydrogen envelope must have been entirely stripped away. Two hypotheses have been offered to accomplish this: (1) a Wolf-Rayet (W-R) star progenitor, or, (2) a progenitor in a massive interacting binary system. A previous discussion of W-R stars as SNe Ib/c progenitors can be found in Filippenko (1991). From the global standpoint, the rate of SNe Ib/c in these galaxies (van den Bergh & Tammann 1991) is in conflict with hypothesis (1), since W-R stars, which evolve from stars with $M_{ZAMS} \gtrsim 25\text{--}40\,M_\odot$ (Maeder & Conti 1994), cannot provide enough progenitors.

3. Local Interstellar Environment

Anecdotally, it is known that SNe Ib/c tend to occur in or near H II regions (e.g., Porter & Filippenko 1987). Kennicutt *et al.* (1989) showed that giant H II regions in late-type galaxies are the sites of the majority of massive star formation. Thus, if one finds SNe in or near H II regions, it is likely that their progenitors were massive stars. Van Dyk (1992) examined the association of SNe of all types, including 9 SNe Ib/c, with H II regions in late-type galaxies. Recently, Van Dyk & Hamuy (1993) extended this survey, considering 31 SNe II and 15 SNe Ib/c. Both studies found little difference between the association of SNe Ib/c and of SNe II with H II regions. Here, we continue to explore this with a sample of 16 SNe Ib/c.

H II regions are mapped through narrow-band Hα imaging. Association of SNe with H II regions is established by comparing the projected angular separation (Δ) of each SN from the center of its nearest H II region with the maximum angular radial extent (r_{max}) of the H II region measured toward the SN position. Continuum-subtracted, flux-calibrated CCD data on nearly face-on galaxies have been acquired at KPNO, CTIO, Lowell Observatory, and Lick Observatory, as well as from other investigators. The H II region detection limit is $\gtrsim (2\text{--}3)\,10^{37}\,\mathrm{erg\,s^{-1}}$. We have measured r_{max} at a surface brightness level $\gtrsim 3\,10^{-18}\,\mathrm{erg\,cm^{-2}\,s^{-1}\,arcsec^{-2}}$. As in Van Dyk & Hamuy (1993), we use absolute positions or precise nuclear offsets when available. Astrometry for the CCD images with $\lesssim 1''$ accuracy is achieved using star positions on the images measured with the STScI GASP[1] system.

---

[1] GASP is the Guide Star Astrometric Support Program available at the Space Telescope Science Institute (STScI).

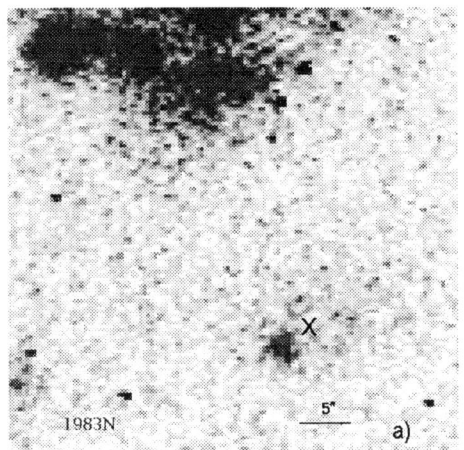 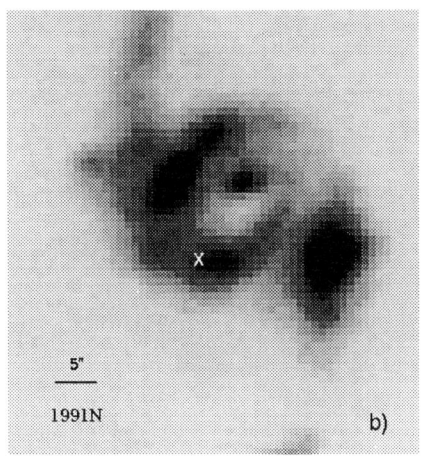

Figure 1. Hα images of the environments of (a) the type Ib SN 1983N in M83 and (b) the Type Ic SN 1991N in NGC 3310, showing the differing relationships of these two SNe with nearby H II regions. North is up, and East is to the left.

In Fig. 1 we show the local interstellar environments of the two SNe Ib/c 1983N and 1991N.

For each SN Ib/c we determine the ratio $R \equiv \Delta/r_{\max}$. We then consider the proportion of SNe Ib/c with $R \leq 1$. Unfortunately, not all SNe in our sample can be treated equally. For lack of a better weighting scheme when calculating this proportion, those SNe having absolute positions or precise nuclear offsets are arbitrarily given 3 times the weight of those SNe before about 1980 with poorer nuclear offsets and 1.5 times the weight of those SNe after about 1980 with poorer offsets. (What are needed, of course, are accurate positions for all SNe.) We find this proportion to be 0.57±0.12 (statistical error) for SNe Ib/c. Comparing this to 0.59±0.09 of 32 SNe II with $R \leq 1$, measured and weighted in the same manner, we find no significant difference. This suggests that the progenitors for both types may be very similar in ZAMS mass range, probably $8 \lesssim M(M_\odot) \lesssim 25$–40. W-R star progenitors are less likely, since a closer association of SNe Ib/c with H II regions might be expected for consistency with this hypothesis, although not all W-R stars in galaxies will necessarily be associated with bright starburst H II regions.

In addition to the lower-resolution ground-based data, we have also examined two SNe Ib/c environments using pre-repair *Hubble Space Telescope* (HST) archive data. These are broad-band, rather than narrow-band, images. In Fig. 2a we show a 200-s FOC UV image through the F220W filter for the site of SN 1991N. The site is midway between two probable large clusters of O-type stars. Photometry and improvement of the astrometry

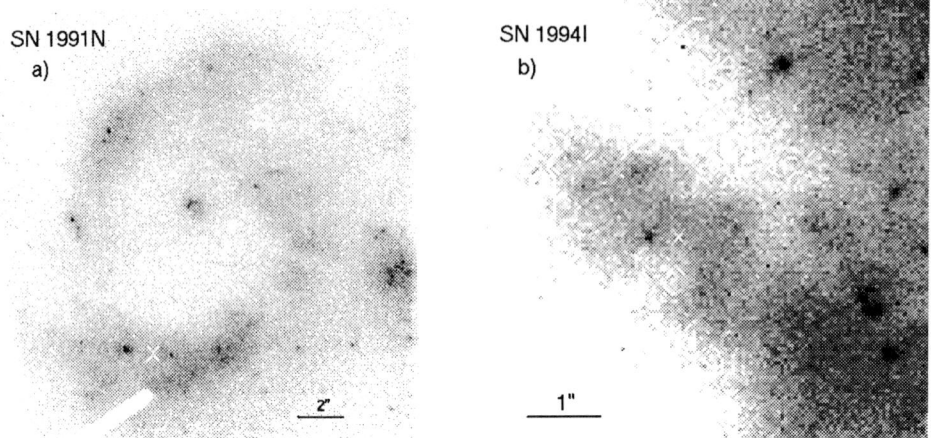

Figure 2. HST images of the environments of the type Ic supernovae (a) 1991N in NGC 3310 and (b) 1994I in M51. North is up, and East is to the left.

is being done for this image.

In Fig. 2b we show a 1100-s PC image made through the F555W filter for the site of SN 1994I. We find a diffuse background of unresolved stars at the site. We estimate an upper limit $(3\,\sigma)$ for the progenitor of $V \gtrsim 23.5$ mag. Assuming a distance of 6.8–9.7 Mpc and $A_V \simeq 1.5$–2.0 mag (Ho & Filippenko 1995), the progenitor had $M_V \gtrsim -7.7$ to -8.4 mag. Unfortunately, this upper limit to the luminosity is not very restrictive; it is brighter than supergiants and many W-R stars (e.g., Torres-Dodgen & Massey 1988).

Based on just these two HST images we can neither eliminate the W-R hypothesis nor support the competing massive binary hypothesis. Additional archive images, especially the more valuable post-repair ones, will soon become available. Inevitably, HST imaging will be the best way to detail the interstellar and stellar environments for SNe and begin to place rigorous mass constraints on the progenitors. In the cases of SNe 1983N and 1994I it should also eventually be possible to locate, through HST imaging, any possible surviving binary companion to the progenitor.

4. Circumstellar Environment

SNe Ib/c, like SNe II, are radio emitters (Weiler et al. 1986). This was first discovered for SN 1983N by Sramek et al. (1984). The other cases include SNe 1984L, 1990B, and 1994I. Most radio supernovae (RSNe) share the common properties of non-thermal emission with high brightness temperature, light curve "turn-on" at shorter wavelengths first and longer wavelengths later, a rapid increase in flux density with time at each wavelength, with a power-law decline after maximum, and a decreasing spectral index

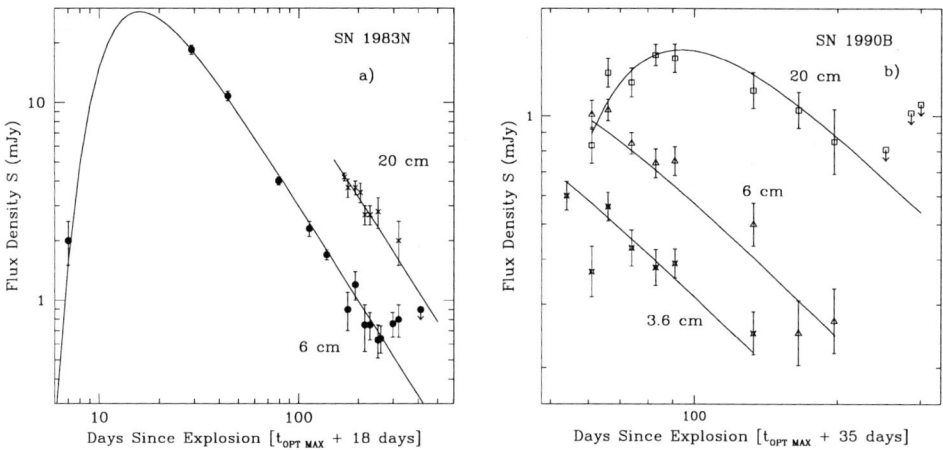

Figure 3. Radio light curves for (a) the type Ib SN 1983N (Weiler et al. 1986) and (b) the type Ic SN 1990B in NGC 4568 (Van Dyk et al. 1993).

between two wavelengths, with the spectral index α inevitably approaching an optically thin, non-thermal, constant negative value (Weiler et al. 1986). RSNe Ib/c show a steep spectral index, rapid "turn-on" at 6 cm before optical maximum, rapid decline after maximum, and homogeneity in spectral luminosity at 6 cm ($\sim 10^{27}$ erg s^{-1} Hz^{-1}). This behavior is in contrast to the slower, more gradual, more heterogeneous behavior for RSNe II.

In Fig. 3 we show radio light curves for the two well-studied RSNe Ib/c 1983N and 1990B.

The radio emission from SNe has been successfully interpreted using the Chevalier (1984) "mini-shell" model, where non-thermal synchrotron emission is produced via interaction of the SN shock with the high-density circumstellar matter produced by a red supergiant progenitor star through mass-loss in the form of a stellar wind prior to explosion. The synchrotron emission is assumed to be free-free absorbed by the fully ionized wind matter. Both the synchrotron luminosity and the absorption depend on the circumstellar density, which is proportional to the ratio of the mass-loss rate $\dot{M}$ to the wind speed w (Weiler et al. 1986) for a spherically symmetric wind. For RSNe Ib/c, $\dot{M}/w \approx 10^{-6}$ M$_\odot$ yr^{-1}/(km s^{-1}). For a W-R star with a fast ($w \sim 10^3$ km s^{-1}), low-density wind, the required $\dot{M}$ ($\sim 10^{-3}$ M$_\odot$ yr^{-1}) is much larger than typically observed for these stars (e.g., Conti 1988), and would result in a low-density bubble around the star prior to explosion, rather than a high-density shell necessary to produce the rapidly evolving, luminous radio emission. We can then reasonably exclude the W-R model. The homogeneity in radio properties for SNe Ib/c implies a homogeneity in circumstellar environments, which must be different from SNe II circumstellar environments. Interaction in a massive binary system could provide this difference.

Acknowledgements. This work was partially based on observations made with the NASA/ESA HST, obtained from the data archive at STScI, which is operated by the Association of Universities for Research in Astronomy, Inc., under NASA contract NAS 5-26555. Financial support was provided by NASA through grant number AR-4302.01-92A from STScI.

References

Branch, D. 1986, ApJ 300, L51
Chevalier, R.A. 1984, ApJ 285, L63
Conti, P.S. 1988, in *O Stars & Wolf-Rayet Stars*, P.S. Conti & A.B. Underhill (Eds.), p. 81
Elias, J.H., Matthews, K., Neugebauer, G. & Persson, S.E. 1985, ApJ 296, 379
Filippenko, A.V. 1986, in *Highlights of Astronomy*, J.-P. Swings (Ed.), Reidel (Dordrecht), p. 589
Filippenko, A.V. 1991, in *Wolf-Rayet Stars and Interrelations with Other Massive Stars in Galaxies*, K.A. van der Hucht & B. Hidayat (Eds.), Kluwer (Dordrecht), p. 529
Filippenko, A.V., Matheson, T. & Barth, A.J. 1994, IAU Circ. 5964
Harkness, R.P. et al. 1987, ApJ 317, 355
Ho, L.C. & Filippenko, A.V. 1995, ApJ (in press)
Kennicutt, R.C., Jr., Edgar, B.K. & Hodge, P.W. 1989, ApJ 337, 761
Maeder, A. & Conti, P.S. 1994, ARA&A 32, 227
Porter, A.C. & Filippenko, A.V. 1986, AJ 93, 1372
Sramek, R.A., Panagia, N. & Weiler, K.W. 1984, ApJ 285, L59
Torres-Dodgen, A.V. & Massey, P. 1988, AJ 96, 1076
Uomoto, A. & Kirshner, R.P. 1985, A&A 149, L7
van den Bergh, S. & Tammann, G.A. 1991, ARA&A 29, 363
Van Dyk, S.D. 1992, AJ 103, 1788
Van Dyk, S.D., & Hamuy, M. 1993, in *Massive Stars: Their Lives in the Interstellar Medium*, J.P. Cassinelli & E.B. Churchwell (Eds.), ASP Conf. Ser. Vol. 35, p. 440
Van Dyk, S.D., Sramek, R.A., Weiler, K.W. & Panagia, N. 1993, ApJ 409, 162
Weiler, K.W., Sramek, R. A., Panagia, N., van der Hulst, J.M. & Salvati, M. 1986, ApJ 301, 790
Wheeler, J.C. & Harkness, R. P. 1990, Rept. Progr. Phys. 53, 1467
Wheeler, J.C. & Levreault, R. 1985, ApJ 294, L17

TYPE II SUPERNOVAE IN BINARY SYSTEMS

P.C. JOSS
Massachusetts Institute of Technology
Room 6-203, Cambridge MA 02139, U.S.A.

Abstract. The presence of a close binary companion can affect the evolution of a massive star through one or more episodes of mass transfer, or by merger in a common-envelope phase. Monte Carlo calculations indicate that $\sim$20–35% of all massive supernovae are affected by such processes, and that a substantial fraction of these events will be supernovae of type II. The properties of the progenitor star, the distribution of circumstellar material, the peak supernova luminosity, the shape of the supernova light curve, and other observable features of the supernova event can be affected by prior binary membership. Binary interactions may be the cause of much of the variability among type II supernova light curves. In particular, many of the peculiarities of SN 1987A and SN 1993J may well have resulted from the prior duplicity of the progenitors.

1. Introduction

A large fraction of all stars are members of binary systems. It is therefore reasonable to consider the possibility that the properties of many massive supernovae (i.e., supernovae whose progenitors had initial main-sequence masses greater than $\sim 8\,M_\odot$) are influenced by prior interactions of the progenitor with a binary companion star. This possibility was brought into focus in recent years by the nearby type II supernovae SN 1987A and SN 1993J, many of whose properties differed markedly from theoretical expectations. As a consequence, several studies have been undertaken to estimate the frequency of massive supernovae in binaries and to investigate the unique properties of the progenitors and supernova events that result from the evolution of a massive star in a close binary system (Podsiadlowski, Joss & Hsu 1992, and references therein; Tutukov, Yungelson & Iben 1992;

Hsu et al. 1995). We here describe the main results of recent theoretical work on type II supernovae in binaries and discuss the application of this work to SN 1987A and SN 1993J. Work on the possible origin of at least some type Ib and Ic supernovae in binaries is reviewed elsewhere in these proceedings by Leibundgut (1995) and Nomoto (1995).

2. Evolution of Type II Supernova Progenitors in Binaries

The principal effects of evolution in a close binary system on the progenitor of a type II supernova can be broadly divided into three categories (Podsiadlowski, Joss & Hsu 1992; hereafter PJH): (1) loss of a portion of the stellar envelope to the companion star, (2) accretion of matter from the companion star, or (3) merger of the two stars in a common-envelope phase. In addition, a star in a close but detached binary may lose a large fraction of its envelope in an enhanced stellar wind (Vanbeveren 1987; Tout & Eggleton 1988).

On the basis of Monte Carlo calculations, PJH concluded that $\sim$20–35% of all massive stars experience binary interactions before undergoing a supernova explosion (which may be of type Ib, Ic, or II). This is consistent with the findings of Tutukov et al. (1992), who concluded, by somewhat different means, that $\sim$25–45% of all supernovae (including those involving low-mass progenitors) originate in initially close binaries.

In the following paragraphs, we describe the salient features of each of the three modes of pre-supernova binary evolution listed above.

2.1. MASS LOSS SCENARIOS

If the supernova progenitor was originally the more massive of the binary components, it can lose mass to its companion via Roche-lobe overflow. This scenario has been considered in detail by Joss et al. (1988) and PJH. If the star first fills its Roche lobe while it is still on the main sequence, a contact system and eventual merger of the binary components is likely to result; the merged star should then have the properties of a rejuvenated main-sequence star. Of greater interest is the possibility that the primary first fills its Roche lobe during the course of its post-main-sequence evolution. In this case, the masses of the binary components will not be very different, and the resultant mass transfer can take place on a sufficiently long time scale that a common envelope does not form. In instances where the entire hydrogen-rich envelope is lost, the progenitor will become a helium star and will likely end its life as a type Ib/Ic supernova (see PJH; Nomoto 1995; and references in these works). If, however, the mass transfer process terminates when the progenitor still retains at least a few tenths of a solar mass of its envelope, the final pre-supernova radius and effective temperature will be nearly the same as those the star would have had in the absence of mass loss, and the resultant explosion will be of type II. Monte Carlo calculations (PJH) indicate that a few percent of all massive stars (perhaps up to $\sim 5\%$ if systems with binary-enhanced winds are included) become type II supernovae of this class.

2.2. ACCRETION SCENARIOS

The original secondary in a close binary system with a Roche-lobe filling primary should accrete a substantial fraction of the mass lost by the primary. If the mass transfer commences before the original secondary has completed core hydrogen burning, the subsequent evolution of the secondary should mimic that of a more massive main-sequence star (Hellings 1983; PJH). Analogously to the case of mass-loss models, of greater interest is the situation where mass transfer commences only after the original secondary has left the main sequence (Podsiadlowski & Joss 1989; De Loore & Vanbeveren 1992; PJH). Due to the accreted mass, the original secondary becomes the more massive of the two stars, and its concomitantly accelerated evolution may cause it to reach the supernova stage prior to the original primary. If the original secondary is the first star to become a supernova, it will have a normal post-main-sequence companion at the time of the explosion; if, instead, the original primary reaches the supernova stage first, it should leave a neutron star or black-hole remnant that will remain gravitationally bound to the original secondary until the latter, too, becomes a supernova. In either case, however, the explosion of the original secondary will eject more than half of the residual systemic mass, generally causing the system to become unbound. Nevertheless, the companion object may become detectable after the supernova photosphere has receded sufficiently. Another diagnostic of supernova events of this class is the color of the immediate progenitor; if the original secondary accretes a sufficient amount of mass, it will end its life as a blue supergiant rather than a red supergiant, which is the generally expected precursor for a type II supernova that has evolved in isolation (see, e.g., Falk & Arnett 1977; Woosley & Weaver 1985).

2.3. MERGER SCENARIOS

If the initial masses of the two stars are sufficiently different, the time scale for mass transfer, once it commences, will be much less than the Kelvin time of the secondary. As a result, the mass transfer will be unstable, and the system will develop a common envelope (see, e.g., Paczyński 1976; Kippenhahn & Meyer-Hofmeister 1977; Podsiadlowski, Joss & Rappaport 1990; PJH); the primary should lose its entire hydrogen-rich envelope to the common envelope. Thereafter, dynamical friction between the secondary and the common envelope will cause the secondary to spiral in toward the system center-of-mass. It is uncertain whether or when the common envelope will subsequently be ejected (see Hsu et al. 1995 for a discussion). If the envelope is ejected before either the secondary is dissolved or the binary components merge to form a single star, and if the core mass of the primary is greater than $\sim 1.4\,M_\odot$ at the time of the ejection, a type Ib/Ic event may result; however, if the explosion strips off a significant amount of the hydrogen-rich envelope of the secondary, the supernova may have the appearance of a type II event. In cases where mass transfer commences when the primary is still on the first red giant branch (case B transfer) and the common envelope is not subsequently ejected, the spiral-in time scale should be much shorter than the remaining evolutionary time for the primary; the binary components should therefore merge before a supernova event occurs. If, instead, mass transfer does not commence until the primary has reached the asymptotic giant branch (case C transfer) and the common envelope is not ejected, it is uncertain whether merger will occur before the the primary becomes a supernova. When the binary components merge before the occurrence of the supernova event, the net effect is very similar to that of the accretion scenario described in Section 2.2, and the merged star may well end its life as a blue supergiant. If the merger is not yet complete by the time of the supernova event, the immediate progenitor (i.e., the common envelope itself) may have the appearance of either a red or a blue supergiant, depending on the values of various parameters for the initial binary system and the details of the common-envelope evolution. Monte Carlo calculations (PJH) indicate that ~ 3–6% of all massive stars end their lives as blue supergiants, due to accretion from or merger with a binary companion, before exploding as type II supernovae.

3. Hydrodynamics of Massive Supernovae in Binaries

We have recently completed a series of hydrodynamic calculations to explore the consequences of mass-loss and accretion/merger scenarios for the observational properties of the resultant type II supernova events (Hsu et al. 1995). We here briefly summarize the results of these calculations.

To investigate the effects of mass loss from a type II supernova progenitor, either through mass transfer to a close-binary stellar companion (see Section 2.1) or via a strong intrinsic stellar wind, we followed the explosion of a star with an initial main-sequence mass, M_{ms}, of $12\,M_\odot$. In the absence of mass loss, such a star would have a hydrogen-rich envelope of mass $M_{env} \simeq 8.7\,M_\odot$; we considered cases with residual envelope masses of 0.4, 1.9, and $4.9\,M_\odot$, as well as a case with no mass loss.

The visual light curves of our four mass-loss models are shown in the

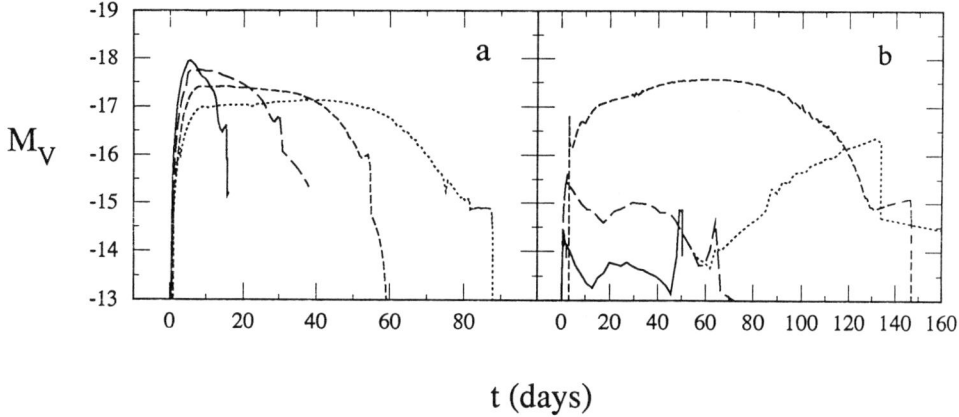

Figure 1. (a) Light curves (absolute visual magnitude, M_V, as a function of elapsed time, t, since core collapse) for four supernova models whose progenitors had initial main-sequence masses, M_{ms}, of $12\,M_\odot$ but had lost their hydrogen-rich envelopes to varying degrees during the course of their post-main-sequence evolution. All four hydrodynamic calculations assumed an explosion energy, E, of 1 foe and no energy input from the decay of radioactive material. *Solid curve*, $M_{env} = 0.4\,M_\odot$. *Long-dashed curve*, $M_{env} = 1.9\,M_\odot$. *Short-dashed curve*, $M_{env} = 4.9\,M_\odot$. *Dotted curve*, $M_{env} = 8.7\,M_\odot$ (corresponding to no mass loss). (b) Same as (a), but for three models whose progenitors underwent accretion from, or merger with, a binary stellar companion during the course of their post-main-sequence evolution; the final pre-supernova mass was $20\,M_\odot$ in all cases. All calculations again assumed $E = 1$ foe and no energy input from radioactive decay, except where otherwise noted. *Solid curve*, $M_{ms} = 15\,M_\odot$. *Long-dashed curve*, $M_{ms} = 17\,M_\odot$. *Dotted curve*, $M_{ms} = 17\,M_\odot$, with additional energy from the radioactive decay of $0.071\,M_\odot$ of Ni^{56} and its decay product, Co^{56}, deposited in the innermost layers of the ejecta; this light curve comes closest to matching the general properties of the light curve of SN 1987A, although it does not fit the observed light curve in detail (see Hsu et al. 1995 for a discussion). *Short-dashed curve*, $M_{ms} = 20\,M_\odot$, with no mass gained or lost by the progenitor during the course of its pre-supernova evolution, shown for comparison. (From Joss et al. 1994; adapted from Hsu et al. 1995.)

left-hand panel of Fig. 1. The most dramatic effects of a reduced envelope mass are (1) a much more rapid rate of decline of the light curve, (2) a higher peak luminosity (by as much as a magnitude in the V band), and (3) peak photospheric velocities that are higher by as much as a factor of ~ 2 ($\sim 2\,10^4\,\mathrm{km\,s^{-1}}$ for the model with the smallest residual envelope, compared to $\sim 1\,10^4\,\mathrm{km\,s^{-1}}$ for the model with no mass loss).

In order to explore the effects of an increase in the envelope mass of the progenitor via accretion from or merger with a binary companion during the course of its post-main-sequence evolution, we calculated the explosion of three stars, each with a final pre-supernova mass of $20\,M_\odot$. The first star had an initial main-sequence mass of $20\,M_\odot$ and underwent no mass loss or gain during the course of its evolution; the other two stars had $M_{ms} = 17\,M_\odot$ and $15\,M_\odot$ and suffered a net gain of 3 and $5\,M_\odot$, respectively, during their post-main-sequence evolution. Both models that gained mass were blue supergiants, rather than red ones, at the time of the supernova event (see Sections 2.2 and 2.3, Podsiadlowski & Joss 1989, and PJH).

The visual light curves for these three models are shown in the right-hand panel of Fig. 1. The principal effects of the addition of mass are (1) a peak luminosity that is fainter by as much as 3.5 magnitudes in the V band (if we exclude the initial flash near $t = 0$, which is not modeled very accurately in our calculations), (2) a more rapid decline of the light curve, and (3) a reduction of the peak photospheric velocity by as much as a factor of ~ 2.5, from $\sim 1.6 \; 10^4 \, \mathrm{km\,s^{-1}}$ for the constant-mass model to only $\sim 6 \; 10^3 \, \mathrm{km\,s^{-1}}$ for the model that has gained $5 \, \mathrm{M}_\odot$ during the course of its evolution.

4. Application to Recent Supernovae

4.1. SN 1987A

A number of authors (Fabian & Rees 1988; Joss et al. 1988; Barkat & Wheeler 1989; Hillebrandt & Meyer 1989; Podsiadlowski & Joss 1989; Podsiadlowski, Joss & Rappaport 1990; De Loore & Vanbeveren 1992; PJH; Rathnasree 1993) have explored the possibility that Sk $-69°202$, the progenitor of SN 1987A, had been a member of a binary system prior to the supernova event. Among the various binary scenarios that have been proposed, the most promising appears to be one in which Sk $-69°202$ underwent merger with a binary stellar companion in a common-envelope phase (Hillebrandt & Meyer 1989; Podsiadlowski, Joss & Rappaport 1990; PJH). (The plausibility of accretion scenarios has been somewhat diminished by the lack of evidence for either a prior supernova event or a normal or neutron-star companion following the recession of the photosphere of SN 1987A.) The major lines of evidence in support of a merger scenario include (1) the blue color of Sk $-69°202$, which was in contrast to most prior theoretical expectations (see Sections 2.2 and 2.3, Podsiadlowski & Joss 1989, and PJH), (2) chemical peculiarities in the progenitor and in the supernova ejecta, which may have resulted from the dredge-up of nuclear-processed material during the merger process (see Hillebrandt & Meyer 1989; PJH; and references in these works), (3) the low peak luminosity of SN 1987A and the exceptionally strong effect of energy input from radioactive decay upon its light curve, which is in accord with the results of hydrodynamic calculations by Hsu et al. (1995) for accretion/merger models of type II supernovae (see Section 3 and Fig. 1b), and (4) the approximate axial symmetry of the circumstellar material, which consists of a central ring (Crotts & Heathcote 1991), two outer rings that are displaced approximately symmetrically on either side of the central ring and are approximately concentric with it (Lemonick 1994), and an outer, diffuse nebulosity known as "Napoleon's Hat" (Wampler et al. 1990). In regard to this last point, Cumming & Podsiadlowski (1994) have recently presented a model for the triple-ring structure based on the hypothesis that Sk $-69°202$ underwent merger with a binary companion in the not-too-distant past; this model can successfully explain not only the physical appearance of the rings but their measured velocity structure as well.

4.2. SN 1993J

Supernova 1993J, which exploded in the nearby spiral galaxy M81, was the brightest supernova since SN 1987A and, like the latter, it was a type II event whose properties were peculiar in some key respects. A candidate progenitor, whose position was consistent with that of the supernova, displayed a blended spectrum that can be best fit by a late-B to early-A supergiant (or an OB association) and a G to early-K supergiant (Aldering, Humphreys & Richmond 1994; see also Podsiadlowski et al. 1993); as in the case of SN 1987A, these spectral types are inconsistent with most theoretical expectations concerning the evolutionary state of the progenitor of a type II supernova that has evolved in isolation. (The actual progenitor may, however, have been a fainter star in the same field, especially if it had been the mass-losing component of an interacting binary; see Podsiadlowski et al. 1993.) Moreover, the initial peak of the supernova light curve was very sharp (qualitatively similar to the solid curve in Fig. 1a). This latter feature, in particular, led almost immediately to the suggestion by several authors (Nomoto et al. 1993; Podsiadlowski et al. 1993, 1994; Ray, Singh & Sutaria 1993; Woosley et al. 1994) that the progenitor had lost most of its hydrogen-rich envelope by transfer to a close binary companion, in the manner discussed in Section 2.1 (see Fig. 2). This suggestion was confirmed by the subsequent evolution of the spectrum of the supernova from type II to type I (Filippenko, Matheson & Ho 1993), indicating that the hydrogen-rich envelope was of anomalously low mass at the time of the explosion. (The alternative hypothesis that the progenitor had lost most of its hydrogen-rich envelope in a strong intrinsic stellar wind requires fine tuning, in order for the star to retain a small but non-vanishing amount of hydrogen at the time of the supernova event.)

We observe that there may be an interesting evolutionary link between SN 1993J and SN 1987A. If the companion of SN 1993J accreted several solar masses of material during the mass-transfer process, at a time when it had already evolved off the main sequence, it should end its life ($\sim 10^5$–10^6 years hence) as a blue supergiant, and this second supernova event should resemble SN 1987A.

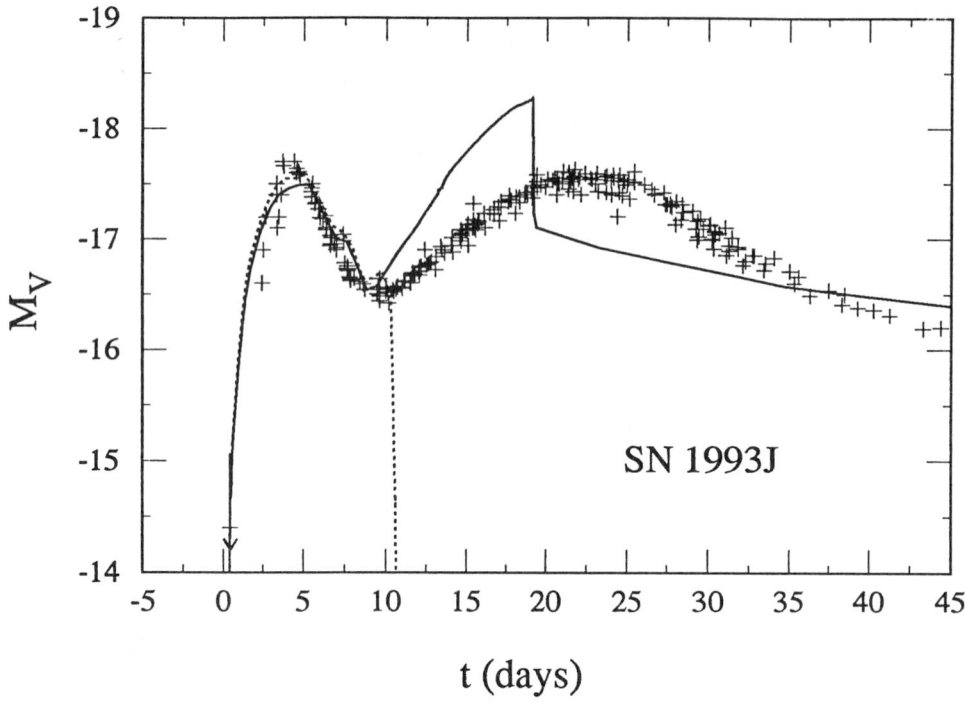

Figure 2. Comparison of theoretical visual light curves (M_V as a function of t) with observations of SN 1993J. The time of core collapse ($t = 0$) corresponds to March 26.0 UT, and the shock breaks out at $t = 8.8$ hr. The crosses represent observed V magnitudes, as compiled by T. Kato and communicated by R.J. Cumming and A.V. Filippenko, which have been converted to absolute magnitudes by use of an assumed distance modulus to M81 of 27.6 mag and a visual extinction of 0.8 mag; the cross marked with an arrow near $t = 0.5$ days is a lower limit on M_V. For both theoretical curves, the progenitor was assumed to be a K supergiant which had an initial main-sequence mass of 15 $M_\odot$ but lost all but 0.2 $M_\odot$ of its hydrogen-rich envelope during the course of its post-main-sequence evolution. *Dotted curve*, model with $E = 1$ foe and no input of energy from radioactive decay. *Solid curve*, model with $E = 0.9$ foe plus energy from the radioactive decay of 0.15 $M_\odot$ of Ni^{56}/Co^{56} deposited in the innermost layers of the ejecta. Several effects that were not taken into account in the theoretical calculations, including Rayleigh-Taylor instabilities (which may have produced clumping of the ejecta and enhanced opacities within the inner ejecta due to the admixture of a small amount of hydrogen), would tend to smooth the light curve for $t > 10$ days and thereby produce a better fit between the latter theoretical light curve and the observed light curve. (From Hsu et al. 1995; see also Podsiadlowski et al. 1993.)

It is remarkable that both SN 1987A and SN 1993J, the two nearest known supernovae of the past century, have both displayed substantial evidence for origin in massive close-binary systems. Of course, evidence of prior duplicity becomes easier to obtain with increasing proximity of the supernova event. It is distinctly possible that, for supernovae of all types, prior duplicity will turn out to be the rule rather than the exception.

Acknowledgements. It is a pleasure to acknowledge my collaborators J. Hsu, Ph. Podsiadlowski, S. Rappaport, and R. Ross, whose efforts were essential to the completion of all aspects of the work described in this article. This work was supported in part by the U.S. National Aeronautics and Space Administration under grant NAGW-1545.

References

Aldering, G., Humphreys, R.M. & Richmond, M. 1994, AJ 107, 662
Barkat, Z. & Wheeler, J.C. 1989, ApJ 342, 940
Crotts, A.P.S. & Heathcote, S.R. 1991, Nat 350, 683
Cumming, R.J. & Podsiadlowski, Ph. 1994, (preprint)
De Loore, C. & Vanbeveren, D. 1992, A&A 260, 273
Fabian, A.C. & Rees, M.J. 1988, Nat 335, 50
Filippenko, A.V., Matheson, T. & Ho, L.C. 1993, ApJ 415, L103
Falk, S.W. & Arnett, W.D. 1977, ApJS 33, 515
Hellings, P. 1983, Ap&SS 96, 37
Hillebrandt, W. & Meyer, F. 1989, A&A 219, L3
Hsu, J.J.L., Joss, P.C., Ross, R.R. & Podsiadlowski, Ph. 1995, ApJ (in press)
Joss, P.C., Hsu, J.J.L., Podsiadlowski, Ph. & Ross, R.R. 1994, in *Proc. 34th Herstmonceux Conf., Circumstellar Media in the Late Stages of Stellar Evolution*, R. Clegg et al. (Eds.), Cambridge Univ. Press (Cambridge, England), (in press)
Joss, P.C., Podsiadlowski, Ph., Hsu, J.J.L. & Rappaport, S. 1988, Nat 331, 237
Kippenhahn, R. & Meyer-Hofmeister, E. 1977, A&A 54, 539
Leibundgut, B. 1995, these Proceedings
Lemonick, M.D. 1994, Time, Vol. 143, No. 22, 51
Nomoto, K. 1995, these Proceedings
Nomoto, K. et al. 1993, Nat 364, 507
Paczyński, B. 1976, in *IAU Symp. No. 73, Structure and Evolution of Close Binary Systems*, P.P. Eggleton et al. (Eds.), Reidel (Dordrecht), p. 75
Podsiadlowski, Ph. & Joss, P.C. 1989, Nat 338, 401
Podsiadlowski, Ph., Hsu, J.J.L., Joss, P.C. & Ross, R.R. 1993, Nat 364, 509
Podsiadlowski, Ph., Hsu, J.J.L., Joss, P.C. & Ross, R.R. 1994, in *Proc. 34th Herstmonceux Conf., Circumstellar Media in the Late Stages of Stellar Evolution*, R. Clegg et al. (Eds.), Cambridge Univ. Press (Cambridge, England), (in press)
Podsiadlowski, Ph., Joss, P.C. & Hsu, J.J.L. 1992, ApJ 391, 246 (PJH)
Podsiadlowski, Ph., Joss, P.C. & Rappaport, S. 1990, A&A 227, L9
Rathnasree, N. 1993, ApJ 411, 848
Ray, A., Singh, K.P. & Sutaria, F.K. 1993, J. Astrophys. Astr. 14, 53
Tout, C.A. & Eggleton, P.P. 1988, ApJ 334, 357
Tutukov, A.V., Yungelson, L.R. & Iben, I. 1992, ApJ 386, 197
Vanbeveren, D. 1987, A&A 182, 207
Wampler, E.J. et al. 1990, ApJ 362, L13
Woosley, S.E. & Weaver, T.A. 1985, in *Nucleosynthesis and Its Implications On Nuclear and Particle Physics, Proc. 5th Moriand Astrophys. Conf.*, J. Audouze & T. van Thuan (Eds.), Reidel (Dordrecht), p. 145
Woosley, S.E., Eastman, R.G., Weaver, T.A. & Pinto, P.A. 1994, ApJ 429, 300

3

Gravitational Waves from Binaries

GRAVITATIONAL WAVES FROM COMPACT BODIES

KIP S. THORNE
California Institute of Technology
Pasadena, California USA

1. Introduction

According to general relativity theory, compact concentrations of energy (e.g., neutron stars and black holes) should warp spacetime strongly, and whenever such an energy concentration changes shape, it should create a dynamically changing spacetime warpage that propagates out through the Universe at the speed of light. This propagating warpage is called *gravitational radiation* — a name that arises from general relativity's description of gravity as a consequence of spacetime warpage.

There are a number of efforts, worldwide, to detect gravitational radiation. These efforts are driven in part by the desire to "see gravitational waves in the flesh", but more importantly by the goal of using the waves as a probe of the Universe and of the nature of gravity. They should be a powerful probe, since they carry very detailed information about gravity and their sources.

In this lecture I shall describe the prospects to study gravitational waves from astronomical systems that exist in our Universe today. Such systems are expected to radiate in the "high-frequency" (1–10^4 Hz) and "low-frequency" (10^{-4}–1 Hz) gravitational-wave bands. By contrast, gravitational waves from the early universe (which I shall not discuss here) should populate a far wider band of frequencies, $\sim 10^{-18}$ Hz–10^{+8} Hz.

The high-frequency band is the domain of Earth-based gravitational-wave detectors: resonant-mass antennas, and most especially laser interferometers. A number of interesting compact sources fall in this band: the stellar collapse to a neutron star or black hole in our Galaxy and distant galaxies, which triggers supernovae; the rotation and vibration of neutron stars in our Galaxy; and the coalescence of neutron star and stellar-mass black-hole binaries ($M \lesssim 1000\,M_\odot$) in distant galaxies.

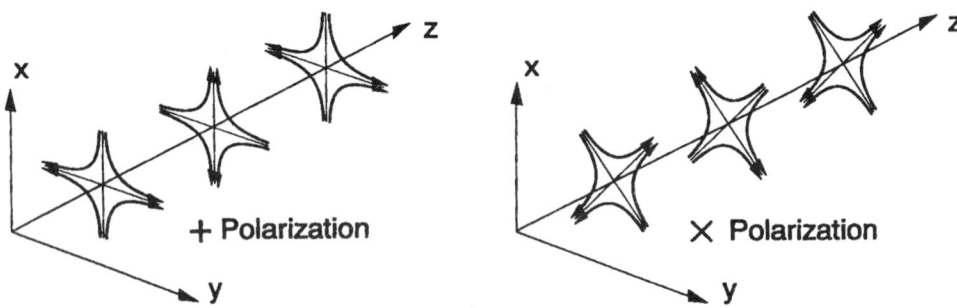

Figure 1.   The lines of force associated with the two polarizations of a gravitational wave. (From Abramovici *et al.* 1992.)

The low-frequency band is the domain of detectors flown in space (in Earth orbit or in interplanetary orbit). The most important of these are the Doppler tracking of spacecraft via microwave signals sent from Earth to the spacecraft and there transponded back to Earth (a technique that NASA has pursued since the early 1970's), and optical tracking of spacecraft by each other (laser interferometry in space, a technique now under development for possible flight in ~2014). The low-frequency band should be populated by waves from short-period binary stars in our own Galaxy (main-sequence binaries, cataclysmic variables, white-dwarf binaries, neutron star binaries, ...); by waves from white dwarfs, neutron stars, and small black holes spiraling into supermassive black holes ($M \sim 10^3$ to $10^8 M_\odot$) in distant galaxies; and by waves from the inspiral and coalescence of supermassive black-hole binaries in distant galaxies.

The body of this lecture consists of four principal sections. Section 2 describes a network of high-frequency, ground-based laser interferometers that is currently under construction, and then Section 3 describes the high-frequency sources the network will search for, and the information that we hope to glean from their waves. Section 4 describes a set of low-frequency, space-based interferometers planned for launch roughly 15 years from now, and Section 5 describes the low-frequency sources they will search for and the information carried by their waves.

2. Ground-Based Laser Interferometers

2.1. WAVE POLARIZATIONS, WAVEFORMS, AND HOW AN INTERFEROMETER WORKS

According to general relativity theory, a gravitational wave has two linear polarizations, conventionally called + (plus) and × (cross). Associated

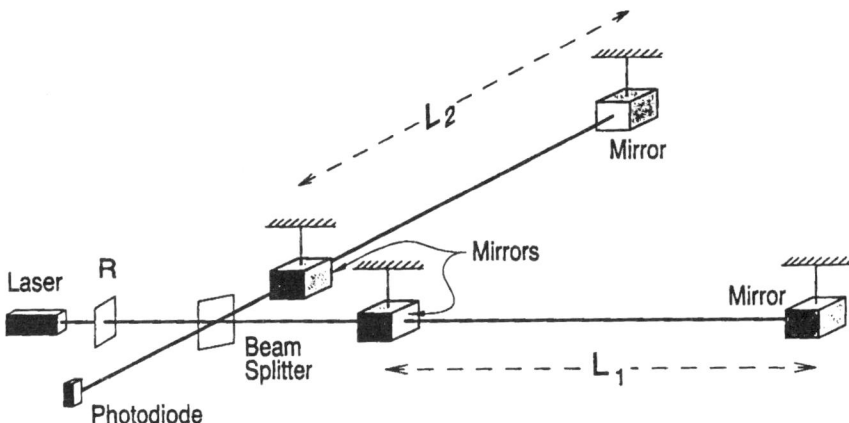

Figure 2. Schematic diagram of a laser interferometer gravitational wave detector. (From Abramovici *et al.* 1992.)

with each polarization there is a gravitational-wave field, h_+ or $h_\times$, which oscillates in time and propagates with the speed of light. Each wave field produces tidal forces (stretching and squeezing forces) on any object or detector through which it passes. If the object is small compared to the waves' wavelength (as is the case for ground-based interferometers and resonant mass antennas), then relative to the object's center, the forces have the quadrupolar patterns shown in Fig. 1. The names "plus" and "cross" are derived from the orientations of the axes that characterize the force patterns (Thorne 1987).

A laser interferometer gravitational-wave detector ("interferometer" for short) consists of four masses that hang from vibration-isolated supports as shown in Fig. 2, and the indicated optical system for monitoring the separations between the masses (Thorne 1987; Abramovici *et al.* 1992). Two masses are near each other, at the corner of an "L", and one mass is at the end of each of the L's long arms. The arm lengths are nearly equal, $L_1 \simeq L_2 = L$. When a gravitational wave, with frequencies high compared to the masses' ~1 Hz pendulum frequency, passes through the detector, it pushes the masses back and forth relative to each other as though they were free from their suspension wires, thereby changing the arm-length difference, $\Delta L \equiv L_1 - L_2$. That change is monitored by laser interferometry in such a way that the output of the photodiode (the interferometer's output) is directly proportional to $\Delta L(t)$.

If the gravitational waves are coming from overhead or underfoot and the axes of the + polarization coincide with the arms' directions, then it is the waves' + polarization that drives the masses, and $\Delta L(t)/L = h_+(t)$.

More generally, the interferometer's output is a linear combination of the two wave fields:

$$\frac{\Delta L(t)}{L} = F_+ h_+(t) + F_\times h_\times(t) \equiv h(t) . \tag{1}$$

The coefficients F_+ and $F_\times$ are of order unity and depend in a quadrupolar manner on the direction to the source and the orientation of the detector (Thorne 1987). The combination $h(t)$ of the two waveforms is called the *gravitational-wave strain* that acts on the detector; and the time evolutions of $h(t)$, $h_+(t)$, and $h_\times(t)$ are sometimes called *waveforms*.

2.2. WAVE STRENGTHS AND INTERFEROMETER ARM LENGTHS

The strengths of the waves from a gravitational-wave source can be estimated using the "Newtonian/quadrupole" approximation to the Einstein field equations. This approximation says that $h \simeq (G/c^4)\ddot{Q}/r$, where $\ddot{Q}$ is the second time derivative of the source's quadrupole moment, r is the distance of the source from Earth, and G and c are Newton's gravitation constant and the speed of light. The strongest sources will be highly non-spherical and thus will have $Q \simeq ML^2$, where M is their mass and L their size, and correspondingly will have $\ddot{Q} \simeq 2Mv^2 \simeq 4E_{\text{kin}}^{\text{ns}}$, where v is the their internal velocity and $E_{\text{kin}}^{\text{ns}}$ is the non-spherical part of their internal kinetic energy. This provides us with the estimate

$$h \sim \frac{1}{c^2} \frac{4G(E_{\text{kin}}^{\text{ns}}/c^2)}{r} ; \tag{2}$$

i.e., h is about 4 times the gravitational potential produced at Earth by the mass-equivalent of the source's non-spherical, internal kinetic energy — made dimensionless by dividing by c^2. Thus, in order to radiate strongly, the source must have a very large non-spherical, internal kinetic energy.

The best known way to achieve a huge internal kinetic energy is via gravity; and by energy conservation (or the virial theorem), any gravitationally-induced kinetic energy must be of order the source's gravitational potential energy. A huge potential energy, in turn, requires that the source be very compact, not much larger than its own gravitational radius. Thus, the strongest gravity-wave sources must be highly compact, dynamical concentrations of large amounts of mass (e.g., colliding and coalescing black holes and neutron stars).

Such sources cannot remain highly dynamical for long; their motions will be stopped by energy loss to gravitational waves and/or the formation of an all-encompassing black hole. Thus, the strongest sources should be transient. Moreover, they should be very rare — so rare that to see a reasonable event rate will require reaching out through a substantial fraction of the Universe. Thus, just as the strongest radio waves arriving at Earth tend to be extragalactic, so also the strongest gravitational waves are likely to be extragalactic.

For highly compact, dynamical objects that radiate in the high-frequency band, e.g. colliding and coalescing neutron stars and stellar-mass black holes, the internal, non-spherical kinetic energy E_{kin}/c^2 is of order the mass of the Sun; and, correspondingly, Eq. (2) gives $h \sim 10^{-22}$ for such sources at the Hubble distance; $h \sim 10^{-21}$ at 200 Mpc (a best-guess distance for several neutron-star coalescences per year; see below); and $h \sim 10^{-20}$ at the Virgo cluster (15 Mpc). These numbers set the scales of sensitivities that ground-based interferometers seek to achieve: $h \lesssim 10^{-21}$ to 10^{-22}.

When one examines the technology of laser interferometry, one sees good prospects to achieve measurement accuracies $\Delta L \sim 10^{-16}$ cm (1/1000 the diameter of the nucleus of an atom). With such an accuracy, an interferometer must have an arm length $L = \Delta L/h \sim 1$ to 10 km, in order to achieve the desired wave sensitivities, 10^{-21} to 10^{-22}. This sets the scale of the interferometers that are now under construction.

2.3. LIGO, VIRGO, AND THE INTERNATIONAL INTERFEROMETRIC NETWORK

Interferometers are plagued by non-Gaussian noise, e.g. due to sudden strain releases in the wires that suspend the masses. This noise prevents a single interferometer, by itself, from detecting with confidence short-duration gravitational-wave bursts (though it might be possible for a single interferometer to search for the periodic waves from known pulsars). The non-Gaussian noise can be removed by cross correlating two, or preferably three or more, interferometers that are networked together at widely separated sites.

The technology and techniques for such interferometers have been under development for nearly 25 years, and plans for km-scale interferometers have been developed over the past 13 years. An international network consisting of three km-scale interferometers, at three widely separated sites, is now in the early stages of construction. It includes two sites of the American LIGO Project ("Laser Interferometer Gravitational Wave Observatory") (Abramovici et al. 1992), and one site of the French/Italian VIRGO Project, named after the Virgo cluster of galaxies (Bradaschia et al. 1990).

LIGO will consist of two vacuum facilities with 4-kilometer-long arms, one in Hanford, Washington (in the northwestern United States) and the other in Livingston, Louisiana (in the southeastern United States). These facilities are designed to house many successive generations of interferometers without the necessity of any major facilities upgrade; and after a planned future expansion, they will be able to house several interferometers at once, each with a different optical configuration optimized for a different type of wave (e.g., broad-band burst, or narrow-band periodic wave, or stochastic wave). The LIGO facilities and their first interferometers are being constructed by a team of about 80 physicists and engineers at Caltech and MIT, led by Barry Barish (the PI) and Gary Sanders (the Project Manager). Substantial contributions are also being made by scientists at other institutions.

The VIRGO Project is building one vacuum facility in Pisa, Italy, with 3-kilometer-long arms. This facility and its first interferometers are a collaboration of more than a hundred physicists and engineers at the INFN (Frascati, Napoli, Perugia, Pisa), LAL (Orsay), LAPP (Annecy), LOA (Palaiseau), IPN (Lyon), ESPCI (Paris), and the University of Illinois (Urbana), under the leadership of Alain Brillet and Adalberto Giazotto.

Both LIGO and VIRGO are scheduled for completion in the late 1990s, and their first gravitational-wave searches are likely to be performed in 2000 or 2001.

LIGO alone, with its two sites which have parallel arms, will be able to detect an incoming gravitational wave, measure one of its two waveforms, and (from the time delay between the two sites) locate its source to within a $\sim 1°$ wide annulus on the sky. LIGO and VIRGO together, operating as a *coordinated international network*, will be able to locate the source (via time delays plus the interferometers' beam patterns) to within a 2-dimensional error box with size between several tens of arcminutes and several degrees, depending on the source direction and on the amount of high-frequency structure in the waveforms. They will be able to monitor both waveforms $h_+(t)$ and $h_\times(t)$ (except for frequency components above about 1 kHz and below about 10 Hz, where the interferometers' noise becomes severe).

The accuracies of the direction measurements and the ability to monitor more than one waveform will be severely compromised when the source lies anywhere near the plane formed by the three LIGO/VIRGO interferometer locations. To get good all-sky coverage will require a fourth interferometer at a site far out of that plane; Japan and Australia would be excellent locations, and research groups there are carrying out research and development on interferometric detectors, aimed at such a possibility. A 300 meter prototype interferometer called TAMA is under construction in Tokyo, and a 400 meter prototype called AIGO400 has been proposed for construction north of Perth.

Two other groups are major players in this field, one in Britain led by James Hough, the other in Germany, led by Karsten Danzmann. These groups each have two decades of experience with prototype interferometers (comparable experience to the LIGO team and far more than anyone else) and great expertise. Frustrated by inadequate financing for a kilometer-scale interferometer, they are constructing, instead, a 600 meter system called GEO600 near Hanford, Germany. Their goal is to develop, from the outset, an interferometer with the sort of advanced design that LIGO and VIRGO will attempt only as a "second-generation" instrument, and thereby achieve sufficient sensitivity to be full partners in the international network's first gravitational-wave searches; they then would offer a variant of their interferometer as a candidate for second-generation operation in the much longer arms of LIGO and/or VIRGO. It is a seemingly audacious plan, but with their extensive experience and expertise, the British/German collaboration might pull it off successfully.

3. High-Frequency Gravitational-Wave Sources

3.1. COALESCING COMPACT BINARIES

The best understood of all gravitational-wave sources are coalescing, compact binaries composed of neutron stars (NS) and black holes (BH). These NS/NS, NS/BH, and BH/BH binaries may well become the "bread and butter" of the LIGO/VIRGO diet.

The Hulse-Taylor binary pulsar (Hulse & Taylor 1975; Taylor 1994) PSR 1913+16, is an example of a NS/NS binary whose waves could be measured by LIGO/VIRGO, if we were to wait long enough. At present PSR 1913+16 has an orbital frequency of about 1/(8 hours) and emits its waves predominantly at twice this frequency, roughly 10^{-4} Hz, which is in the low-frequency band—far too low to be detected by LIGO/VIRGO. However, as a result of their loss of orbital energy to gravitational waves, the PSR 1913+16 NS's are gradually spiraling inward. If we wait roughly 10^8 years, this inspiral will bring the waves into the LIGO/VIRGO high-frequency band. As the NS's continue their inspiral, the waves will then sweep upward in frequency, over a time of about 15 minutes, from 10 Hz to $\sim 10^3$ Hz, at which point the NS's will collide and coalesce. It is this last 15 minutes of inspiral, with $\sim$16,000 cycles of waveform oscillation, and the final coalescence, that LIGO/VIRGO seeks to monitor.

3.1.1. *Wave Strengths Compared to LIGO Sensitivities*

Fig. 3 compares the projected sensitivities of interferometers in LIGO with the wave strengths from the last few minutes of inspiral of BH/BH, NS/BH, and NS/NS binaries at various distances from Earth. The two solid curves at the bottoms of the stippled regions (labeled $h_{\rm rms}$) are the r.m.s. noise levels for broad-band waves that have optimal direction and polarization. The tops of the stippled regions (labeled $h_{\rm SB}$ for "sensitivity to bursts") are the sensitivities for highly confident detection of randomly polarized, broad-band waves from random directions (i.e., the sensitivities for high

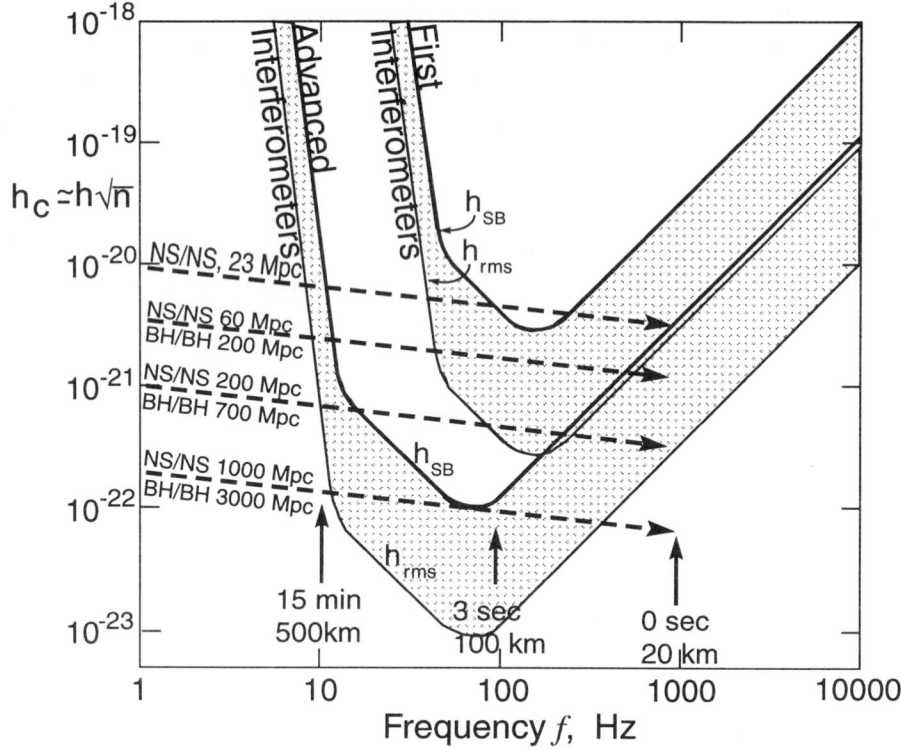

Figure 3. LIGO's projected interferometer sensitivities (Abramovici et al. 1992) compared with the strengths of the waves from the last few minutes of inspiral of compact binaries. The signal to noise ratios are $\sqrt{2}$ higher than in Abramovici et al. because of a factor 2 error in equation (29) of Thorne (1987).

confidence that any such observed signal is not a false alarm due to Gaussian noise). The upper stippled region and its bounding curves are the expected performances of the first interferometers in LIGO; the lower stippled region and curves are performances of more advanced LIGO interferometers.

As the NS's and/or BH's spiral inward, their waves sweep upward in frequency (left to right in the diagram). The dashed lines show their "characteristic" signal strength h_c (approximately the amplitude h of the waves' oscillations multiplied by the square root of the number of cycles spent near a given frequency, $\sqrt{n}$); the signal-to-noise ratio is this h_c divided by the detector's $\sqrt{5}h_{\rm rms}$, where the $\sqrt{5}$ converts $h_{\rm rms}$ from "optimal direction and polarization" to "random direction and polarization" (Thorne 1987; Abramovici et al. 1992). The arrows along the NS/NS inspiral track indicate the time until final coalescence and the separation between the NS centers of mass. Each NS is assumed to have a mass of 1.4 suns and a radius ~ 10 km; each BH, 10 suns and ~ 20 km.

Notice that the signal strengths in Fig. 3 are in good accord with rough estimates based on Eq. (2); at the endpoint (right end) of each inspiral, the number of cycles n spent near that frequency is of order unity, so the quantity plotted, $h_c \simeq h\sqrt{n}$, is about equal to h — and at distance 200 Mpc is about 10^{-21}, as we estimated.

3.1.2. *Coalescence Rates*

Such final coalescences are few and far between in our own galaxy: about one every 100,000 years, according to 1991 estimates by Phinney (1991) and by Narayan, Piran & Shemi (1991), based on the statistics of binary pulsar searches in our galaxy which found three that will coalesce in less than 10^{10} years. Extrapolating out through the Universe on the basis of the density of production of blue light (the color produced predominantly by massive stars), Phinney (1991) and Narayan et al. (1991) infer that to see several NS/NS coalescences per year, LIGO/VIRGO will have to look out to a distance of about 200 Mpc (give or take a factor ~ 2); cf. the "NS/NS inspiral, 200 Mpc" line in Fig. 3. Since these estimates were made, the binary pulsar searches have been extended through a significantly larger volume of the galaxy than before, and no new ones with coalescence times $\lesssim 10^{10}$ years have been found; as a result, the binary-pulsar-search-based best estimate of the coalescence rate should be revised downward (Bailes 1995), perhaps to as little as one every million years in our galaxy, corresponding to a distance 400 Mpc for several per year (Bailes 1995).

A rate of one every million years in our galaxy is ~ 1000 times smaller than the birth rate of the NS/NS binaries' progenitors: massive, compact, main-sequence binaries (Phinney 1991; Narayan et al. 1991). Therefore, either 99.9 per cent of progenitors fail to make it to the NS/NS state [e.g., because of binary disruption during a supernova or forming Thorne-Żytkow objects (TŻO's; Thorne & Zytkow 1975, see also the contribution by Podsiadlowski to these Proceedings)], or else they do make it, but they wind up as a class of NS/NS binaries that has not yet been discovered in any of the pulsar searches. Several experts on binary evolution have argued for the latter (Tutukov & Yungelson 1993; Yamaoka et al. 1993; Van den Heuvel 1994; Lipunov et al. 1994): most NS/NS binaries, they suggest, may form with such short orbital periods that their lifetimes to coalescence are significantly shorter than normal pulsar lifetimes ($\sim 10^7$ years); and with such short lifetimes, they have been missed in pulsar searches. By modeling the evolution of the galaxy's binary star population, the binary experts arrive at best estimates as high as $3 \ 10^{-4}$ coalescences per year in our galaxy, corresponding to about 1 per year out to 60 Mpc distance (Tutukov & Yungelson 1993). Phinney (1991) describes other plausible populations of NS/NS binaries that could increase the event rate, and he argues for "ultraconservative" lower and upper limits of 23 Mpc and 1000 Mpc for how far one must look to see several coalescences per year.

By comparing these rate estimates with the signal strengths in Fig. 3, we see that (i) the first interferometers in LIGO/VIRGO have a moderate but not high probability of seeing NS/NS coalescences; (ii) advanced interferometers are almost certain of seeing them (the requirement that this be so was one factor that forced the LIGO/VIRGO arm lengths to be so long, several kilometers); and (iii) they are most likely to be discovered roughly half-way between the first and advanced interferometers — which means by an improved variant of the first interferometers several years after LIGO operations begin.

We have no good observational handle on the coalescence rate of NS/BH or BH/BH binaries. However, theory suggests that their progenitors might not disrupt during the stellar collapses that produce the NS's and BH's, so their coalescence rate could be about the same as the birth rate for their progenitors, i.e., $\sim 1/100,000$ years in our galaxy. This suggests that within 200 Mpc distance there might be several NS/BH or BH/BH coalescences per year (Phinney 1991; Narayan et al. 1991; Tutukov & Yungelson 1993; Lipunov et al. 1994). This estimate should be regarded as a plausible upper limit on the event rate and lower limit on the distance to look (Phinney 1991; Narayan et al. 1991).

If this estimate is correct, then NS/BH and BH/BH binaries will be seen before NS/NS, and might be seen with the first LIGO/VIRGO interferometers or soon thereafter; cf. Fig. 3. However, this estimate is far less certain than the (rather uncertain) NS/NS estimates!

Once coalescence waves have been discovered, each further improvement of sensitivity by a factor 2 will increase the event rate by $2^3 \simeq 10$. Assuming a rate of several NS/NS per year at 200 Mpc, the advanced interferometers of Fig. 3 should see ~ 100 per year.

3.1.3. *Inspiral Waveforms and the Information They Can Bring*

Neutron stars and black holes have such intense self gravity that it is exceedingly difficult to deform them. Correspondingly, as they spiral inward in a compact binary, they do not gravitationally deform each other significantly until several orbits before their final coalescence (Kochanek 1992; Bildsten & Cutler 1992). This means that the inspiral waveforms are determined to high accuracy by only a few, clean parameters: the masses and spin angular momenta of the bodies, and their initial orbital elements (i.e. the elements when the waves enter the LIGO/VIRGO band).

Though tidal deformations are negligible during inspiral, relativistic effects can be very important. If, for the moment, we ignore the relativistic effects — i.e., if we approximate gravity as Newtonian and the wave gener-

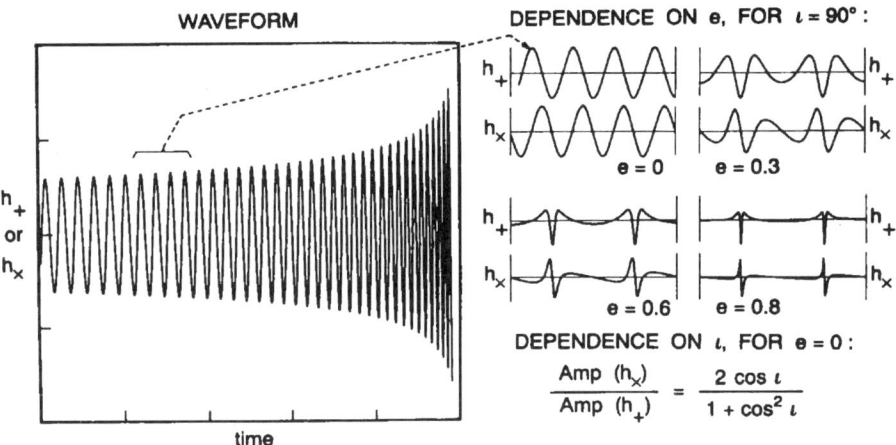

Figure 4. Waveforms from the inspiral of a compact binary, computed using Newtonian gravity for the orbital evolution and the quadrupole moment approximation for the wave generation. (From Abramovici et al. 1992.)

ation as due to the binary's oscillating quadrupole moment (Thorne 1987), then the shapes of the inspiral waveforms $h_+(t)$ and $h_\times(t)$ are as shown in Fig. 4.

The left-hand graph in Fig. 4 shows the waveform increasing in amplitude and sweeping upward in frequency (i.e., undergoing a "chirp") as the binary's bodies spiral closer and closer together. The ratio of the amplitudes of the two polarizations is determined by the inclination ι of the orbit to our line of sight (lower right in Fig. 4). The shapes of the individual waves, i.e. the waves' harmonic content, are determined by the orbital eccentricity (upper right). [Binaries produced by normal stellar evolution should be highly circular due to past radiation reaction forces, but compact binaries that form by capture events, in dense star clusters that might reside in galactic nuclei (Quinlan & Shapiro 1987), could be quite eccentric.] If, for simplicity, the orbit is circular, then the rate at which the frequency sweeps or "chirps", df/dt [or equivalently the number of cycles spent near a given frequency, $n = f^2(df/dt)^{-1}$] is determined solely, in the Newtonian/quadrupole approximation, by the binary's so-called *chirp mass*, $M_c \equiv (M_1 M_2)^{3/5}/(M_1 + M_2)^{1/5}$ (where M_1 and M_2 are the two bodies' masses). The amplitudes of the two waveforms are determined by the chirp mass, the distance to the source, and the orbital inclination. Thus (in the Newtonian/quadrupole approximation), by measuring the two amplitudes, the frequency sweep, and the harmonic content of the inspiral waves, one can determine as direct, resulting observables, the source's dis-

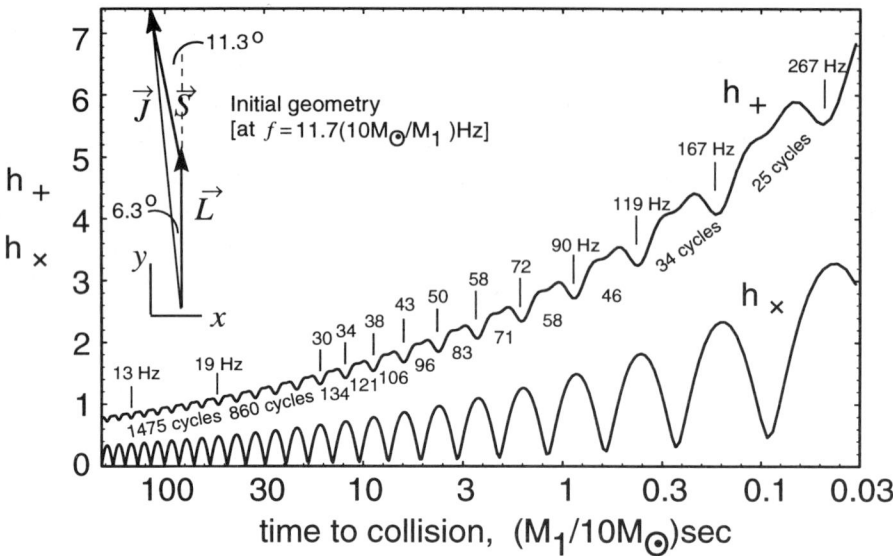

Figure 5. Modulational envelope for the waveform from a $1\,M_\odot$ non-spinning NS spiraling into a $10\,M_\odot$, rapidly spinning Kerr black hole (spin parameter $a = 1$). The orbital angular momentum **L** is inclined by $\alpha = 11.3$ degrees to the hole's spin angular momentum **S**, and the two precess around $\mathbf{J} = \mathbf{L} + \mathbf{S}$, whose direction remains fixed in space as $L = |\mathbf{L}|$ shrinks and $S = |\mathbf{S}| = M_{\rm BH}a$ remains constant. The precession modulates the waves by an amount that depends on (i) the direction to Earth (here along the initial $\mathbf{L} \times \mathbf{S}$, i.e. out of the paper) and (ii) the orientation of the detector's arms (here parallel to the figure's initial **L** and to $\mathbf{L}\times$(direction to Earth) for h_+, and rotated 45 degrees for $h_\times$). The figure shows the waveforms' modulational envelopes (in arbitrary units, the same for h_+ and $h_\times$), parametrized by the wave frequency f and the number of cycles of *oscillation* between the indicated f's. The total number of *precessions* from f to coalescence is $N_{\rm prec} \simeq (5/64\pi)(Ma/\mu)(\pi Mf)^{-2/3} \simeq 20(f/10\,{\rm Hz})^{-2/3}$. (From Cutler et al. 1993; Apostolatos et al. 1994.)

tance, chirp mass, inclination, and eccentricity (Schutz 1986, 1989).

As in binary pulsar observations (Taylor 1994), so also here, relativistic effects add further information: they influence the rate of frequency sweep and produce waveform modulations in ways that depend on the binary's dimensionless ratio $\eta = \mu/M$ of reduced mass $\mu = M_1M_2/(M_1+M_2)$ to total mass $M = M_1 + M_2$ (Lincoln & Will 1990) and on the spins of the binary's two bodies (Kidder, Will & Wiseman 1993); These relativistic effects are reviewed and discussed at length by Cutler et al. (1993) and Will (1994). Two deserve special mention: (i) As the waves emerge from the binary, some of them get backscattered one or more times off the binary's spacetime curvature, producing wave *tails*. These tails act back on the binary, modifying its inspiral rate in a measurable way. (ii) If the orbital plane is inclined to one or both of the binary's spins, then the spins drag inertial frames in the binary's vicinity (the "Lense-Thirring effect"), this frame dragging causes the orbit to precess, and the precession modulates the waveforms (Cutler et al. 1993; Apostolatos et al. 1994; Kidder 1995). Fig. 5 shows the resulting modulation for a $1\,M_\odot$ NS spiraling into a rapidly spinning, $10\,M_\odot$ BH.

Remarkably, the relativistic corrections to the frequency sweep can be measurable with very high accuracy, even though they are typically $\lesssim 10$ per cent of the Newtonian contribution, and even though the typical signal to noise ratio will be only ~ 9 even after optimal signal processing. The reason is as follows (Cutler & Flanagan 1994; Finn & Chernoff 1993; Cutler et al. 1993):

The frequency sweep will be monitored by the method of "matched filters"; in other words, the incoming, noisy signal will be cross correlated with theoretical templates. If the signal and the templates gradually get out of phase with each other by more than $\sim 1/10$ cycle as the waves sweep through the LIGO/VIRGO band, their cross correlation will be significantly reduced. Since the total number of cycles spent in the LIGO/VIRGO band will be $\sim 16{,}000$ for a NS/NS binary, ~ 3500 for NS/BH, and ~ 600 for BH/BH, this means that LIGO/VIRGO should be able to measure the frequency sweep to a fractional precision $\lesssim 10^{-4}$, compared to which the relativistic effects are very large. [This is essentially the same method as Joseph Taylor and colleagues use for high-accuracy radio-wave measurements of relativistic effects in binary pulsars (Taylor 1994).]

Preliminary analyses, using the theory of optimal signal processing, predict the following typical accuracies for LIGO/VIRGO measurements based solely on the frequency sweep, i.e., ignoring modulational information (Poisson & Will 1995; Cutler & Flanagan 1994; Finn & Chernoff 1993; Jaronowski & Krolak 1994; Cutler et al. 1993): (i) The chirp mass M_c will typically be measured, from the Newtonian part of the frequency sweep, to $\sim 0.04\%$ for a NS/NS binary and $\sim 0.3\%$ for a system containing at least one BH. (ii) If we are confident (e.g., on a statistical basis from measurements of many previous binaries) that the spins are a few percent or less of the maximum physically allowed, then the reduced mass μ will be measured to $\sim 1\%$ for NS/NS and NS/BH binaries, and $\sim 3\%$ for BH/BH binaries. (Here and below NS means a $\sim 1.4\,M_\odot$ neutron star and BH means a $\sim 10\,M_\odot$ black hole.) (iii) Because the frequency dependences of the (relativistic) μ effects and spin effects are not sufficiently different to give a clean separation between μ and the spins, if we have no prior knowledge of the spins, then the spin/μ correlation will worsen the typical accuracy of η by a large factor, to $\sim 20\%$ for NS/NS, $\sim 30\%$ for NS/BH, and $\sim 60\%$ for BH/BH (Poisson & Will 1995; Cutler & Flanagan 1994). These worsened accuracies might be improved somewhat by waveform modulations caused by the spin-induced precession of the orbit (Apostolatos et al. 1994; Kidder

1995), and even without modulational information, a certain combination of μ and the spins will be determined to a few per cent. Much additional theoretical work is needed to firm up the measurement accuracies.

To take full advantage of all the information in the inspiral waveforms will require theoretical templates that are accurate, for given masses and spins, to a fraction of a cycle during the entire sweep through the LIGO/VIRGO band. Such templates are being computed by an international consortium of relativity theorists, Blanchet and Damour in France, Iyer in India, Will and Wiseman in the U.S., and others, using post-Newtonian expansions of the Einstein field equations (Will 1994; Blanchet *et al.* 1995). This enterprise is rather like computing the Lamb shift to high order in powers of the fine structure constant, for comparison with experiment. The terms of leading order in the mass ratio $\eta = \mu/M$ are being checked by a Japanese-American consortium (Poisson, Nakamura, Sasaki, Tagoshi, Tanaka) using the Teukolsky formalism for weak perturbations of black holes (Poisson & Sasaki 1995; Shibata *et al.* 1995). These small-η calculations have been carried to very high post-Newtonian order for circular orbits and no spins (Tagoshi & Nakamura 1994; Tagoshi & Sasaki 1994), and from those results Cutler & Flanagan (1995) have estimated the order to which the full, finite-η computations must be carried in order that systematic errors in the theoretical templates will not significantly impact the information extracted from the LIGO/VIRGO observational data. The answer appears daunting: radiation-reaction effects must be computed to three full post-Newtonian orders [six orders in v/c =(orbital velocity)/(speed of light)] beyond the leading-order radiation reaction, which itself is 5 orders in v/c beyond the Newtonian theory of gravity.

It is only about a ten years since controversies over the leading-order radiation reaction (Ashtekar 1983) were resolved by a combination of theoretical techniques and binary pulsar observations. Nobody dreamed then that LIGO/VIRGO observations will require pushing post-Newtonian computations onward from $O[(v/c)^5]$ to $O[(v/c)^{11}]$. This requirement epitomizes a major change in the field of relativity research: At last, 80 years after Einstein formulated general relativity, experiment has become a major driver for theoretical analyses.

Remarkably, the goal of $O[(v/c)^{11}]$ is achievable. The most difficult part of the computation, the radiation reaction, has been evaluated to $O[(v/c)^9]$ beyond Newton by the French/Indian/American consortium (Blanchet *et al.* 1995) and as of this writing, rumors have it that $O[(v/c)^{10}]$ is coming under control.

These high-accuracy waveforms are needed only for extracting information from the inspiral waves, after the waves have been discovered; they are not needed for the discovery itself. The discovery is best achieved using a different family of theoretical waveform templates, one that covers the space of potential waveforms in a manner that minimizes computation time instead of a manner that ties quantitatively into general relativity theory (Cutler *et al.* 1993). Such templates are in the early stage of development (Krolak *et al.* 1995; Apostolatos 1995; Sathyaprakash 1994).

LIGO/VIRGO observations of compact binary inspiral have the potential to bring us far more information than just binary masses and spins:

- They can be used for high-precision tests of general relativity. In scalar-tensor theories [some of which are highly attractive alternatives to general relativity (Damour & Nordveth 1993)], radiation reaction due to emission of scalar waves places a unique signature on those waves that LIGO/VIRGO would detect — a signature that can be searched for with high precision (Will 1994).
- They can be used to measure the Hubble constant, deceleration parameter, and cosmological constant (Schutz 1986, 1989; Markovic 1993; Chernoff & Finn 1992). The keys to such measurements are that (i) advanced interferometers in LIGO/VIRGO will be able to see NS/NS out to cosmological redshifts $z \sim 0.3$, and NS/BH out to $z \sim 2$. (ii) The direct observables that can be extracted from the observed waves include the source's luminosity distance r_L (measured to accuracy ~ 10 per cent in a large fraction of cases), and its direction on the sky (to accuracy ~ 1 square degree)—accuracies good enough that only one or a few electromagnetically-observed clusters of galaxies should fall within the 3-dimensional gravitational error boxes, thereby giving promise to joint gravitational/electromagnetic statistical studies. (iii) Another direct gravitational observable is $(1+z)M$ where z is redshift and M is any mass in the system (measured to the accuracies quoted above). Since the masses of NS's in binaries seem to cluster around $1.4 \, M_\odot$, measurements of $(1+z)M$ can provide a handle on the redshift, even in the absence of electromagnetic aid.
- For a NS or small BH spiraling into a massive ~ 50 to $500 \, M_\odot$ BH, the inspiral waves will carry a "map" of the spacetime geometry around the big hole — a map that can be used, e.g., to test the theorem that "a black hole has no hair" (Ryan, Finn & Thorne 1995); cf. Section 5.3 below.

3.1.4. *Coalescence Waveforms and their Information*

The waves from the binary's final coalescence can bring us new types of information.

BH/BH Coalescence: In the case of a BH/BH binary, the coalescence will excite large-amplitude, highly non-linear vibrations of spacetime curvature near the coalescing black-hole horizons — a phenomenon of which we have very little theoretical understanding today. Especially fascinating will be the case of two spinning black holes whose spins are not aligned with each other or with the orbital angular momentum. Each of the three angular momentum vectors (two spins, one orbital) will drag space in its vicinity into a tornado-like swirling motion — the general relativistic "dragging of inertial frames", so the binary is rather like two tornados with orientations skewed to each other, embedded inside a third, larger tornado with a third orientation. The dynamical evolution of such a complex configuration of spacetime warpage (as revealed by its emitted waves) may well bring us surprising new insights into relativistic gravity. Moreover, if the sum of the BH masses is fairly large, $\sim$40 to 200 $M_\odot$, then the waves should come off in a frequency range $f \sim$ 40 to 200 Hz where the LIGO/VIRGO broad-band detectors have their best sensitivity and can best extract the information the waves carry.

To get full value out of such wave observations will require having theoretical computations with which to compare them (Flanagan & Hughes 1995). There is no hope to perform such computations analytically; they can only be done as supercomputer simulations. The development of such simulations is a major effort within the world's relativity community.

NS/NS Coalescence: The final coalescence of NS/NS binaries should produce waves that are sensitive to the equation of state of nuclear matter, so such coalescences have the potential to teach us about the nuclear equation of state (Cutler *et al.* 1993). In essence, we will be studying nuclear physics via the collisions of atomic nuclei that have nucleon numbers $A \sim 10^{57}$ — somewhat larger than we are normally accustomed to. The accelerator used to drive these nuclei up to the speed of light is the binary's self gravity, and the radiation by which the details of the collisions are probed is gravitational.

A number of research groups (Shibata *et al.* 1992, 1993; Rasio & Shapiro 1992; Nakamura 1994; Kochanek 1992; Bildsten & Cutler 1992; Zhuge *et al.* 1994; Davies *et al.* 1994) are engaged in numerical astrophysics simulations of NS/NS coalescence, with the goal not only to predict the emitted gravitational waveforms and their dependence on equation of state, but also (more immediately) to learn whether such coalescences might power the γ-ray bursts that have been a major astronomical puzzle since their discovery in the early 1970s. If advanced LIGO interferometers were now in operation, they could report definitively whether or not the γ-bursts are produced by NS/NS binaries; and if the answer were yes, then the combination of γ-burst data and gravitational-wave data could bring valuable information that neither could bring by itself. For example, we could determine when, to within a few msec, the γ-burst is emitted relative to the moment the NS's first begin to touch; and by comparing the γ and gravitational times of arrival, we might test whether gravitational waves propagate with the speed of light to a fractional precision of $\sim 0.01 \sec/(3 \; 10^9 \, \mathrm{lyr}) = 10^{-19}$.

Unfortunately, the final NS/NS coalescence will emit its gravitational waves in the kHz frequency band ($800 \, \mathrm{Hz} \lesssim f \lesssim 2500 \, \mathrm{Hz}$) where photon shot noise will prevent them from being studied by the standard, "workhorse," broad-band interferometers of Fig. 3. However, a specially configured ("dual-recycled") interferometer invented by Brian Meers (1988), which could have enhanced sensitivity in the kHz region at the price of reduced sensitivity elsewhere, may well be able to measure the waves and extract their equation of state information, as might massive, spherical bar detectors (Cutler et al. 1993; Kennefick et al. 1995). Such measurements will be very difficult and are likely only when the LIGO/VIRGO network has reached a mature stage.

NS/BH Coalescence: A NS spiraling into a BH of mass $M \gtrsim 10 \, \mathrm{M}_\odot$ should be swallowed more or less whole. However, if the BH is less massive than roughly $10 \, \mathrm{M}_\odot$, and especially if it is rapidly rotating, then the NS will tidally disrupt before being swallowed. Little is known about the disruption and accompanying waveforms. To model them with any reliability will likely require full numerical relativity, since the circumferences of the BH and NS will be comparable and their physical separation at the moment of disruption will be of order their circumferences. As with NS/NS, the coalescence waves should carry equation of state information and will come out in the kHz band, where their detection will require advanced, specialty detectors.

Christodoulou Memory: As the coalescence waves depart from their source, their energy creates (via the nonlinearity of Einstein's field equations) a secondary wave called the "Christodoulou memory" (Christodoulou 1991; Thorne 1992; Wiseman & Will 1991). Whereas the primary waves may have frequencies in the kHz band, the memory builds up on the timescale of the primary energy emission profile, which is likely to be of order 0.01 sec, corresponding to a memory frequency in the optimal band for the LIGO/VIRGO workhorse interferometers, $\sim 100 \, \mathrm{Hz}$. Unfortunately, the memory is so weak that only very advanced interferometers have much chance of detecting and studying it — and then, perhaps only for BH/BH coalescences and not for NS/NS or NS/BH (Kennefick 1994).

3.2. STELLAR CORE COLLAPSE AND SUPERNOVAE

Several features of the stellar core collapse, which triggers supernovae, can produce significant gravitational radiation in the high-frequency band. We shall consider these features in turn, the most weakly radiating first.

3.2.1. *Boiling of the Newborn Neutron Star*

Even if the collapse is spherical, so it cannot radiate any gravitational waves at all, it should produce a convectively unstable neutron star that "boils" vigorously (and non-spherically) for the first ~ 0.1 second of its life (Bethe 1990). The boiling dredges up high-temperature nuclear matter ($T \sim 10^{12}$ K) from the neutron star's central regions, bringing it to the surface (to the "neutrino-sphere"), where it cools by neutrino emission before being swept back downward and reheated. Burrows (1995) and Burrows *et al.* (1995) estimate that the boiling should generate $n \sim 10$ cycles of gravitational waves with frequency $f \sim 100$ Hz and amplitude $h \sim 3 \ 10^{-22}(30 \, \text{kpc}/r)$ (where r is the distance to the source), corresponding to a characteristic amplitude $h_c \simeq h\sqrt{n} \sim 10^{-21}(30 \, \text{kpc}/r)$; cf. Fig. 6. LIGO/VIRGO will be able to detect such waves only in the local group of galaxies, where the supernova rate is probably no larger than ~ 1 each 10 years. However, neutrino detectors have a similar range, and there could be a high scientific payoff from correlated observations of the gravitational waves emitted by the boiling's mass motions and neutrinos emitted from the boiling neutrino-sphere.

3.2.2. *Axisymmetric Collapse, Bounce, and Oscillations*

Rotation will centrifugally flatten the collapsing core, enabling it to radiate as it implodes. If the core's angular momentum is small enough that centrifugal forces do not halt or significantly slow the collapse before it reaches nuclear densities, then the core's collapse, bounce, and subsequent oscillations are likely to be axially symmetric. Numerical simulations (Finn 1991; Mönchmeyer *et al.* 1991) show that in this case the waves from collapse, bounce, and oscillation will be quite weak: the total energy radiated as gravitational waves is not likely to exceed $\sim 10^{-7}$ solar masses (about 1 part in a million of the collapse energy) and might often be much less than this; and correspondingly, the waves' characteristic amplitude will be $h_c \lesssim 3 \ 10^{-21}(30 \, \text{Mpc}/r)$. These collapse-and-bounce waves will come off at frequencies ~ 200 Hz to ~ 1000 Hz, and will precede the boiling waves by a fraction of a second. Like the boiling waves, they probably cannot be seen by LIGO/VIRGO beyond the local group of galaxies and thus will be a very rare occurrence.

3.2.3. *Rotation-Induced Bars and Break-Up*

If the core's rotation is large enough to centrifugally flatten the core significantly, before or as it reaches nuclear density, then a dynamical and/or secular instability is likely to break the core's axisymmetry. The core will be transformed into a bar-like configuration that spins end-over-end like an American football, and that might even break up into two or more massive

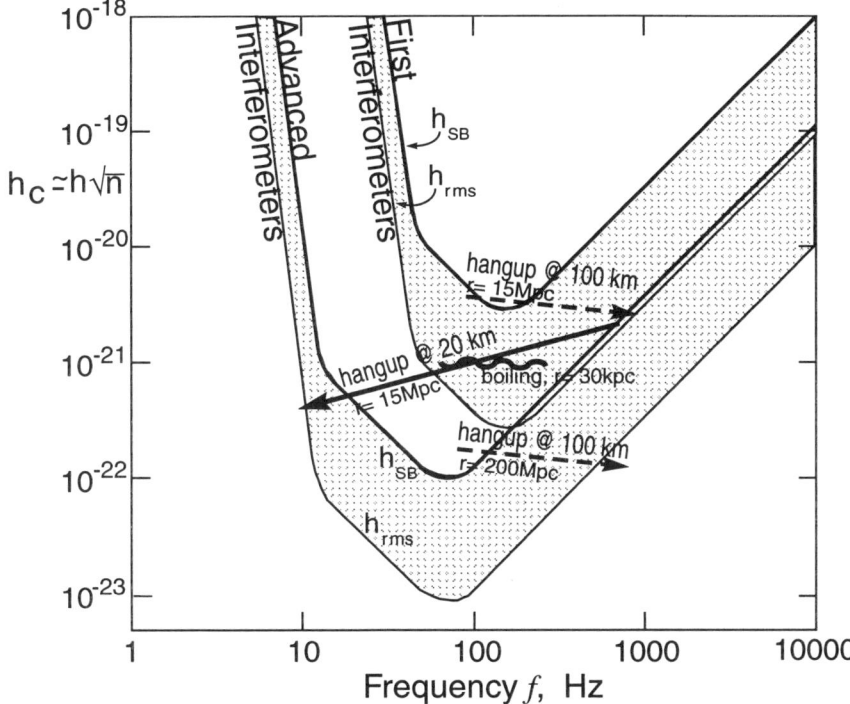

Figure 6. Characteristic amplitudes of the gravitational waves from various processes accompanying stellar core collapse and supernovae, compared with projected sensitivities of LIGO's interferometers.

pieces. In this case, the radiation from the spinning bar or orbiting pieces *could* be almost as strong as that from a coalescing neutron-star binary, and thus could be seen by the LIGO/VIRGO first interferometers out to the distance of the Virgo cluster (where the supernova rate is several per year) and by advanced interferometers out to several hundred Mpc (supernova rate $\sim 10^4$ per year); cf. Fig. 6. It is far from clear what fraction of collapsing cores will have enough angular momentum to break their axisymmetry, and what fraction of those will actually radiate at this high rate; but even if only $\sim 1/1000$ or $1/10^4$ do so, this could ultimately be a very interesting source for LIGO/VIRGO.

Several specific scenarios for such non-axisymmetry have been identified:

Centrifugal hangup at ~ 100 km radius: If the pre-collapse core is rapidly spinning (e.g., if it is a white dwarf that has been spun up by accretion from a companion), then the collapse may produce a highly flattened, centrifugally supported disk with most of its mass at radii $R \sim 100$ km, which then (via instability) may transform itself into a bar or may bifurcate. The bar or bifurcated lumps will radiate gravitational waves at twice their rotation frequency, $f \sim 100\,\mathrm{Hz}$ — the optimal frequency for LIGO/VIRGO interferometers. To shrink on down to $\sim 10\,\mathrm{km}$ size, this configuration must shed most of its angular momentum. If a substantial fraction of the angular momentum goes into gravitational waves, then independently of the strength of the bar, the waves will be nearly as strong as those from a coalescing binary. The reason is this: The waves' amplitude h is proportional to the bar's ellipticity e, the number of cycles n of wave emission is proportional to $1/e^2$, and the characteristic amplitude $h_c = h\sqrt{n}$ is thus independent of the ellipticity and is about the same whether the configuration is a bar or is two lumps (Schutz 1989). The resulting waves will thus have h_c roughly half as large, at $f \sim 100\,\mathrm{Hz}$, as those from a NS/NS binary (half as large because each lump might be half as massive as a NS), and they will chirp upward in frequency in a manner similar to those from a binary.

It is rather likely, however, that the excess angular momentum does *not* go into gravitational waves, but instead goes largely into hydrodynamic waves as the bar or lumps, acting like a propeller, stir up the surrounding stellar mantle. In this case, the radiation will be correspondingly weaker.

Centrifugal hangup at $R \sim 20\,\mathrm{km}$: Lai & Shapiro (1995) have explored the case of centrifugal hangup at radii not much larger than the final neutron star, say $R \sim 20\,\mathrm{km}$. Using compressible ellipsoidal models, they have deduced that, after a brief period of dynamical bar-mode instability with wave emission at $f \sim 1000\,\mathrm{Hz}$ (explored by Houser, Centrella & Smith 1994), the star switches to a secular instability in which the bar's angular velocity gradually slows while the material of which it is made retains its high rotation speed and circulates through the slowing bar. The slowing bar emits waves that sweep *downward* in frequency through the LIGO/VIRGO optimal band $f \sim 100\,\mathrm{Hz}$, toward $\sim 10\,\mathrm{Hz}$. The characteristic amplitude (Fig. 6) is only modestly smaller than for the upward-sweeping waves from hangup at $R \sim 100\,\mathrm{km}$, and thus such waves should be detectable near the Virgo Cluster by the first LIGO/VIRGO interferometers, and at distances of a few $100\,\mathrm{Mpc}$ by advanced interferometers.

Successive fragmentations of an accreting, newborn neutron star: Bonnell & Pringle (1995) have focused on the evolution of the rapidly spinning, newborn neutron star as it quickly accretes more and more mass from the pre-supernova star's inner mantle. If the accreting material carries high angular momentum, it may trigger a renewed bar formation, lump formation, wave emission, and coalescence, followed by more accretion, bar and lump formation, wave emission, and coalescence. Bonnell & Pringle speculate that hydrodynamics, not wave emission, will drive this evolution, but that the total energy going into gravitational waves might be as large as $\sim 10^{-3}\,M_\odot$. This corresponds to $h_c \sim 10^{-21}(10\,\mathrm{Mpc}/r)$.

3.3. SPINNING NEUTRON STARS; PULSARS

As the neutron star settles down into its final state, its crust begins to solidify (crystallize). The solid crust will assume nearly the axisymmetric shape, with poloidal ellipticity $\epsilon_p \propto$ (angular velocity of rotation)^2, that centrifugal forces are trying to maintain. However, the principal axis of the star's moment of inertia tensor may deviate from its spin axis by some small "wobble angle" θ_w, and the star may deviate slightly from axisymmetry about its principal axis; i.e., it may have a slight ellipticity ϵ_e in its equatorial plane.

As this slightly imperfect crust spins, it will radiate gravitational waves (Zimmermann & Szedenits 1979): ϵ_e radiates at twice the rotation frequency, $f = 2f_{\rm rot}$ with $h \propto \epsilon_e$, and the wobble angle couples to ϵ_p to produce waves at $f = f_{\rm rot} + f_{\rm prec}$ (the precessional sideband of the rotation frequency) with amplitude $h \propto \theta_w \epsilon_p$. For typical neutron star masses and moments of inertia, the wave amplitudes are

$$h \sim 6\ 10^{-25} \left(\frac{f_{\rm rot}}{500\,{\rm Hz}}\right)^2 \left(\frac{1\,{\rm kpc}}{r}\right) \left(\frac{\epsilon_e \text{ or } \theta_w \epsilon_p}{10^{-6}}\right). \quad (3)$$

The neutron star gradually spins down, due in part to gravitational-wave emission but perhaps more strongly due to electromagnetic torques associated with its spinning magnetic field and pulsar emission. This spin down reduces the strength of centrifugal forces, and thereby causes the star's poloidal ellipticity ϵ_p to decrease, with an accompanying breakage and resolidification of its crust's crystal structure (a "starquake", Shapiro & Teukolsky 1983). In each starquake, θ_w, ϵ_e, and ϵ_p will all change suddenly, thereby changing the amplitudes of the star's two gravitational "spectral lines" $f = 2f_{\rm rot}$ and $f = f_{\rm rot} + f_{\rm prec}$. After each quake, there should be a healing period in which the star's fluid core and solid crust, now rotating at different speeds, gradually regain synchronism. By monitoring the amplitudes, frequencies, and phases of the two gravitational-wave spectral lines, and by comparing with timing of the electromagnetic pulsar emission, one might learn much about the physics of the neutron star interior.

How large will the quantities ϵ_e and $\theta_w \epsilon_p$ be? Rough estimates of the crustal shear moduli and breaking strengths suggest an upper limit in the range $\epsilon_{\rm max} \sim 10^{-4}$ to 10^{-6}, and it might be that typical values are far below this. We are extremely ignorant, and correspondingly there is much to be learned from searches for gravitational waves from spinning neutron stars.

One can estimate the sensitivity of LIGO/VIRGO (or any other broadband detector) to the periodic waves from such a source by multiplying the waves' amplitude h by the square root of the number of cycles over which one might integrate to find the signal, $n = f\hat{\tau}$ where $\hat{\tau}$ is the integration time. The resulting effective signal strength, $h\sqrt{n}$, is larger than h by

$$\sqrt{n} = \sqrt{f\hat{\tau}} = 10^5 \left(\frac{f}{1000\,\text{Hz}}\right)^{1/2} \left(\frac{\hat{\tau}}{4\,\text{months}}\right)^{1/2}. \qquad (4)$$

This $h\sqrt{n}$ should be compared to the detector's r.m.s. broad-band noise level h_{rms} to deduce a signal-to-noise ratio, or to h_{SB} to deduce a sensitivity for high-confidence detection when one does not know the waves' frequency in advance (Thorne 1987). Such a comparison suggests that the first interferometers in LIGO/VIRGO might possibly see waves from nearby spinning neutron stars, but the odds of success are very unclear.

The deepest searches for these nearly periodic waves will be performed by narrow-band detectors, whose sensitivities are enhanced near some chosen frequency at the price of sensitivity loss elsewhere — e.g., "dual recycled" interferometers or resonant bars. With "advanced-detector technology," dual-recycled interferometers might be able to detect with confidence all spinning neutron stars that have (Thorne 1987)

$$(\epsilon_e \text{ or } \theta_w \epsilon_p) \gtrsim 3 \; 10^{-10} \left(\frac{500\,\text{Hz}}{f_{\text{rot}}}\right)^2 \left(\frac{r}{1000\,\text{pc}}\right)^2. \qquad (5)$$

There may well be a large number of such neutron stars in our galaxy; but it is also conceivable that there are none. We are extremely ignorant.

Some cause for optimism arises from several physical mechanisms that might generate radiating ellipticities large compared to $3 \; 10^{-10}$:

- It may be that, inside the superconducting cores of many neutron stars, there are trapped magnetic fields with mean strength $B_{\text{core}} \sim 10^{13}$ G or even 10^{15} G. Because such a field is actually concentrated in flux tubes with $B = B_{\text{crit}} \sim 6 \; 10^{14}$ G surrounded by field-free superconductor, its mean pressure is $p_B = B_{\text{core}} B_{\text{crit}}/8\pi$. This pressure could produce a radiating ellipticity $\epsilon_e \sim \theta_w \epsilon_p \sim p_B/p \sim 10^{-8} B_{\text{core}}/10^{13}$ G (where p is the core's material pressure).
- Accretion onto a spinning neutron star can drive precession (keeping θ_w substantially non-zero), and thereby might produce measurably strong waves (Schutz 1995).
- If a neutron star is born very rapidly rotating, then it may experience a gravitational-radiation-reaction-driven instability. In this "CFS" (Chandrasekhar 1970; Friedman & Schutz 1978) instability, density waves propagate around the star in the opposite direction to its rotation, but are dragged forward by the rotation. These density waves produce gravitational waves that carry positive energy as seen by observers far from the star, but negative energy from the star's viewpoint; and because the star thinks it is losing negative energy, its density waves get

amplified. This intriguing mechanism is similar to that by which spiral density waves are produced in galaxies. Although the CFS instability was once thought ubiquitous for spinning stars, we now know that neutron star viscosity will kill it, stabilizing the star and turning off the waves, when the star's temperature is below some limit $\sim 10^{10}$ K (Lindblom 1995) and above some limit $\sim 10^9$ K (Lindblom & Mendell 1995); and correspondingly, it should operate only during the first few years of a neutron star's life.

4. LISA: The Laser Interferometer Space Antenna

Turn, now, from the high-frequency band, 1–10^4 Hz, to the low-frequency band, 10^{-4}–1 Hz. At present, the most sensitive gravitational-wave searches at low frequencies are those carried out by researchers at NASA's Jet Propulsion Laboratory, using microwave-frequency Doppler tracking of interplanetary spacecraft. These searches are done at rather low cost, piggyback on missions designed for other purposes. Although they have a definite possibility of success, the odds are against them. Their best past sensitivities to bursts, for example, have been $h_{\text{SB}} \sim 10^{-14}$, and prospects are good for reaching $\sim 10^{-15}$–10^{-16} in the next 5 to 10 years. However, the strongest low-frequency bursts arriving several times per year might be no larger than $\sim 10^{-18}$; and the domain of an assured plethora of signals is more like $h_{\text{SB}} \sim 10^{-19}$–$10^{-20}$.

In the 2014 time frame, the European Space Agency (ESA) and/or NASA is likely to fly a *Laser Interferometer Space Antenna* (LISA) which will achieve $h_{\text{SB}} \lesssim 10^{-20}$ over the frequency band $3 \; 10^{-4}$ Hz $\lesssim f \lesssim 3 \; 10^{-2}$ Hz.

4.1. MISSION STATUS

LISA is largely an outgrowth of 15 years of studies by Peter Bender and colleagues at the University of Colorado. Unfortunately, the prospects for NASA to fly such a mission have not looked good in the early 1990s. By contrast, prospects in Europe have looked much better, so a largely European consortium was put together in 1993 with Bender's participation but under the leadership of Karsten Danzmann (Hannover) and James Hough (Glasgow), to propose LISA to the European Space Agency. The proposal has met with considerable success; LISA might well achieve approval to fly as an ESA Cornerstone Mission around 2014 (Bender et al. 1994). Members of the American gravitation community hope that NASA will join together with ESA in this endeavor, and that working jointly, ESA and NASA will be able to fly LISA considerably sooner than 2014.

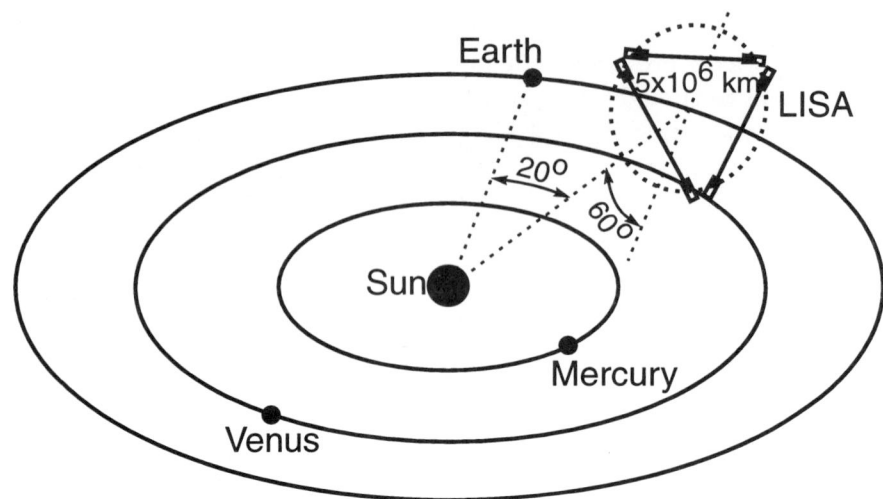

Figure 7. LISA's orbital configuration.

4.2. MISSION CONFIGURATION

As presently conceived, LISA will consist of six compact, drag-free spacecraft (i.e. spacecraft that are shielded from buffeting by solar wind and radiation pressure, and that thus move very nearly on geodesics of spacetime). All six spacecraft would be launched simultaneously by a single Ariane rocket. They would be placed into the same heliocentric orbit as the Earth occupies, but would follow 20° behind the Earth; cf. Fig. 7. The spacecraft would fly in pairs, with each pair at the vertex of an equilateral triangle that is inclined at an angle of 60° to the Earth's orbital plane. The triangle's arm length would be 5 million km (10^6 times larger than LIGO's arms!). The six spacecraft would track each other optically, using one-Watt YAG laser beams. Because of diffraction losses over the 5 10^6 km arm length, it is not feasible to reflect the beams back and forth between mirrors as is done with LIGO. Instead, each spacecraft will have its own laser; and the lasers will be phase locked to each other, thereby achieving the same kind of phase-coherent out-and-back light travel as LIGO achieves with mirrors. The six-laser, six-spacecraft configuration thereby functions as three, partially independent but partially redundant, gravitational-wave interferometers.

4.3. NOISE AND SENSITIVITY

Fig. 8 depicts the expected noise and sensitivity of LISA in the same language as we have used for LIGO (Fig. 3). The curve at the bottom of the

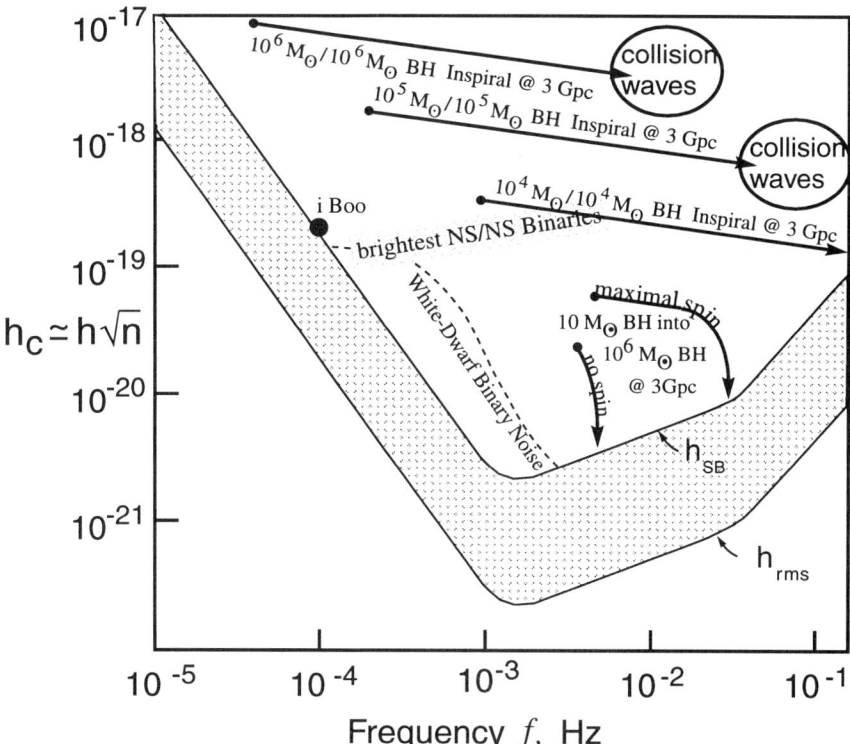

Figure 8. LISA's projected broad-band noise $h_{\rm rms}$ and sensitivity to bursts $h_{\rm SB}$, compared with the strengths of the waves from several low-frequency sources. [*Note:* When members of the LISA team plot curves analogous to this, they show the sensitivity curve (top of stippled region) in units of the amplitude of a periodic signal that can be detected with $S/N = 5$ in one year of integration; that sensitivity to periodic sources is related to the $h_{\rm SB}$ used here by $h_{\rm SP} = h_{\rm SB}/\sqrt{f \cdot 3\ 10^7\ {\rm sec}}$.]

stippled region is $h_{\rm rms}$, the r.m.s. noise, in a bandwidth equal to frequency, for waves with optimum direction and polarization. The top of the stippled region is $h_{\rm SB} = 5\sqrt{5} h_{\rm rms}$, the sensitivity for high-confidence detection ($S/N = 5$) of a broad-band burst coming from a random direction, assuming Gaussian noise.

At frequencies $f \gtrsim 10^{-3}$ Hz, LISA's noise is due to photon counting statistics (shot noise). The noise curve steepens at $f \sim 3\ 10^{-2}$ Hz because at larger f than that, the waves' period is shorter than the round-trip light travel time in one of LISA's arms. Below 10^{-3} Hz, the noise is due to buffeting-induced random motions of the spacecraft that are not being properly removed by the drag-compensation system. Notice that, in terms of dimensionless amplitude, LISA's sensitivity is roughly the same as that of

LIGO's first interferometers (Fig. 3, but at 100,000 times lower frequency. Since the waves' energy flux scales as $f^2 h^2$, this corresponds to 10^{10} better energy sensitivity than LIGO.

4.4. OBSERVATIONAL STRATEGY

LISA can detect and study, simultaneously, a wide variety of different sources scattered over all directions on the sky. The key to distinguishing the different sources is the different time evolution of their waveforms. The key to determining each source's direction, and confirming that it is real and not just noise, is the manner in which its waves' amplitude and frequency are modulated by LISA's complicated orbital motion — a motion in which the detector triangle rotates around its center once per year, and the detector plane rotates around the normal to the Earth's orbit once per year. Most sources will be observed for a year or longer, thereby making full use of these modulations.

5. Low-Frequency Gravitational-Wave Sources

5.1. WAVES FROM BINARY STARS

LISA has a large class of "guaranteed" sources: short-period binary stars in our own galaxy. A specific example is the classic binary 44 i Boo (HD 133640), a $1.35\,M_\odot/0.68\,M_\odot$ system just 12 parsecs from Earth, whose wave frequency f and characteristic amplitude $h_c = h\sqrt{n}$ are depicted in Fig. 8. (Here h is the waves' actual amplitude and $n = f\hat{\tau}$ is the number of wave cycles during $\hat{\tau}=1$ year of signal integration.) Since 44 i Boo lies right on the $h_{\rm SB}$ curve, the signal to noise ratio in one year of integration should be $S/N = 5$.

To have an especially short period, a binary must be made of especially compact bodies — white dwarfs (WD), neutron stars (NS), and/or black holes (BH). WD/WD binaries are thought to be so numerous that they might produce a stochastic background of gravitational waves, at the level shown in Fig. 8, that will hide some other interesting waves from view (Hills, Bender & Webbink 1990). Since WD/WD binaries are very dim optically, their actual numbers are not known for sure; Fig. 8 might be an overestimate.

Assuming a NS/NS coalescence rate of 1 each 10^5 years in our galaxy (Phinney 1991; Narayan et al. 1991), the shortest period NS/NS binary should have a remaining life of about $5\ 10^4$ years, corresponding to a gravitational-wave frequency today of $f \simeq 5\ 10^{-3}\,\mathrm{Hz}$, an amplitude (at about 10 kpc distance) $h \simeq 4\ 10^{-22}$, and a characteristic amplitude (with one year of integration time) $h_c \simeq 2\ 10^{-19}$. This is depicted in Fig. 8 at the right edge of the region marked "brightest NS/NS binaries". These brightest NS/NS binaries can be studied by LISA with the impressive signal to noise ratios $S/N \sim 50$ to 500.

5.2. WAVES FROM THE COALESCENCE OF MASSIVE BLACK HOLES IN DISTANT GALAXIES

LISA would be a powerful instrument for studying massive black holes in distant galaxies. Fig. 8 shows, as examples, the waves from several massive black hole binaries at 3 Gpc distance. The waves sweep upward in frequency (rightward in the diagram) as the holes spiral together. The black dots show the waves' frequency one year before the holes' final collision and coalescence, and the arrowed lines show the sweep of frequency and characteristic amplitude $h_c = h\sqrt{n}$ during that last year. For simplicity, the figure is restricted to binaries with equal-mass black holes: $10^4 \, M_\odot / 10^4 \, M_\odot$, $10^5 \, M_\odot / 10^5 \, M_\odot$, and $10^6 \, M_\odot / 10^6 \, M_\odot$.

By extrapolation from these three examples, we see that LISA can study much of the last year of inspiral, and the waves from the final collision and coalescence, whenever the holes' masses are in the range $3 \; 10^4 \, M_\odot \lesssim M \lesssim 10^8 \, M_\odot$. Moreover, LISA can study the final coalescences with remarkable signal to noise ratios: $S/N \gtrsim 1000$. Since these are much larger S/N's than LIGO/VIRGO is likely to achieve, we can expect LISA to refine the experimental understanding of black-hole physics, and of highly non-linear vibrations of warped spacetime, which LIGO/VIRGO initiates — *provided* the rate of massive black-hole coalescences is of order one per year in the Universe or higher. The rate might well be that high, but it also might be much lower.

By extrapolating Fig. 8 to lower BH/BH masses, we see that LISA can observe the last few years of inspiral, but not the final collisions, of binary black holes in the range $100 \, M_\odot \lesssim M \lesssim 10^4 \, M_\odot$, out to cosmological distances.

Extrapolating the BH/BH curves to lower frequencies using the formula (time to final coalescence) $\propto f^{-8/3}$, we see that equal-mass BH/BH binaries enter LISA's frequency band roughly 1000 years before their final coalescences, more or less independently of their masses, for the range $100 \, M_\odot \lesssim M \lesssim 10^6 \, M_\odot$. Thus, if the coalescence rate were to turn out to be one per year, LISA would see roughly 1000 additional massive binaries that are slowly spiraling inward, with inspiral rates df/dt readily measurable. From the inspiral rates, the amplitudes of the two polarizations, and the waves' harmonic content, LISA can determine each such binary's luminosity distance, redshifted chirp mass $(1+z)M_c$, orbital inclination, and eccentricity; and from the waves' modulation by LISA's orbital motion,

5.3. WAVES FROM COMPACT BODIES SPIRALING INTO MASSIVE BLACK HOLES IN DISTANT GALAXIES

When a compact body with mass μ spirals into a much more massive black hole with mass M, the body's orbital energy E at fixed frequency f (and correspondingly at fixed orbital radius a) scales as $E \propto \mu$, the gravitational-wave luminosity $\dot E$ scales as $\dot E \propto \mu^2$, and the time to final coalescence thus scales as $t \sim E/\dot E \propto 1/\mu$. This means that the smaller is μ/M, the more orbits are spent in the hole's strong-gravity region, $a \lesssim 10\,GM/c^2$, and thus the more detailed and accurate will be the map of the hole's spacetime geometry, which is encoded in the emitted waves.

For holes observed by LIGO/VIRGO, the most extreme mass ratio that we can hope for is $\mu/M \sim 1\,M_\odot/300\,M_\odot$, since for $M > 300\,M_\odot$ the inspiral waves are pushed to frequencies below the LIGO/VIRGO band. This limit on μ/M seriously constrains the accuracy with which LIGO/VIRGO can hope to map out the spacetime geometries of black holes and test the black-hole no-hair theorem (Section 3.1). By contrast, LISA can observe the final inspiral waves from objects of any mass $M \gtrsim 0.5\,M_\odot$ spiraling into holes of mass $3\;10^5\,M_\odot \lesssim M \lesssim 3\;10^7\,M_\odot$.

Fig. 8 shows the example of a $10\,M_\odot$ black hole spiraling into a $10^6\,M_\odot$ hole at 3 Gpc distance. The inspiral orbit and waves are strongly influenced by the hole's spin. Two cases are shown (Finn & Thorne 1995): an inspiraling circular orbit around a non-spinning hole, and a prograde, circular, equatorial orbit around a maximally spinning hole. In each case the dot at the upper left end of the arrowed curve is the frequency and characteristic amplitude one year before the final coalescence. In the non-spinning case, the small hole spends its last year spiraling inward from $r \simeq 7.4\,GM/c^2$ (3.2 Schwarzschild radii) to its last stable circular orbit at $r = 6\,GM/c^2$ (3 Schwarzschild radii). In the maximal spin case, the last year is spent traveling from $r = 6\,GM/c^2$ (3 Schwarzschild radii) to the last stable orbit at $r = GM/c^2$ (half a Schwarzschild radius). The $\sim 10^5$ cycles of waves during this last year should carry, encoded in themselves, rather accurate values for the massive hole's lowest few multipole moments (Ryan 1995). If the measured moments satisfy the "no-hair" theorem (i.e., if they are all determined uniquely by the measured mass and spin in the manner of the Kerr metric), then we can be sure the central body is a black hole. If they violate the no-hair theorem, then (assuming general relativity is correct), either the central body was not a black hole, or an accretion disk or other material was perturbing its orbit (Molteni et al. 1994). From the evolution of the waves one can hope to determine which is the case, and to explore the properties of the central body and its environment.

Models of galactic nuclei, where massive holes reside, suggest that inspiraling stars and small holes typically will be in rather eccentric orbits (Hils & Bender 1995). This is because they get injected into such orbits via gravitational deflections off other stars, and by the time gravitational radiation reaction becomes the dominant orbital driving force, there is not enough inspiral left to fully circularize their orbits. Such orbital eccentricity will complicate the waveforms and complicate the extraction of information from them. Efforts to understand the emitted waveforms are just now getting underway.

The event rates for inspiral into massive black holes are not at all well understood. However, since a significant fraction of all galactic nuclei are thought to contain massive holes, and since white dwarfs and neutron stars, as well as small black holes, can withstand tidal disruption as they plunge toward the massive hole's horizon, and since LISA can see inspiraling bodies as small as $\sim 0.5\,M_\odot$ out to 3 Gpc distance, the event rate is likely to be interestingly large.

6. Conclusion

It is now 35 years since Joseph Weber initiated his pioneering development of gravitational-wave detectors (Weber 1960) and 25 years since Forward and Weiss initiated work on interferometric detectors. Since then, hundreds of talented experimental physicists have struggled to improve the sensitivities of these instruments. At last, success is in sight. If the source estimates described in this lecture are approximately correct, then the planned interferometers should detect the first waves in 2001 or several years thereafter, thereby opening up this rich new window onto the Universe. One payoff should be deep new insights into compact astrophysical bodies and their roles in binary systems.

References

Abramovici, A. et al. 1992, Science 256, 325
Apostolatos, T.A. 1995, Phys. Rev. D (in press)
Apostolatos, T.A. et al. 1994, Phys. Rev. D 49, 6274
Ashtekar, A. 1983, in *Gravitational Radiation*, N. Deruelle & T. Piran (Eds.), North Holland, p. 421
Bailes, M. 1995, these proceedings
Bender, P. et al. 1994, in *LISA, Laser interferometer space antenna for gravitational wave measurements: ESA Assessment Study Report*, R. Reinhard (Ed.), ESTEC
Bethe, H.A. 1990, Rev. Mod. Phys. 62, 801
Bildsten, L. & Cutler, C. 1992, ApJ 400, 175
Blanchet, L. et al. 1995, Phys. Rev. Lett. (submitted)

Bonnell, I.A. & Pringle, J.E. 1995, MNRAS 273, L12
Bradaschia, C. et al. 1990, Nucl. Instrum. & Methods A289, 518
Burrows, A., Hayes, J. & Fryxell, B.A. 1995, ApJ (in press)
Burrows, A. 1995, (private communication)
Chandrasekhar, S. 1970, Phys. Rev. Lett. 24, 611
Chernoff, D.F. & Finn, L.S. 1993, ApJ 411, L5
Christodoulou, D. 1991, Phys. Rev. Lett. 67, 1486
Cutler, C. & Flanagan, E.E. 1994, Phys. Rev. D 49, 2658
Cutler, C. & Flanagan, E.E. 1995, Phys. Rev. D, (in preparation)
Cutler, C. et al. 1993, Phys. Rev. Lett. 70, 1984
Damour, T. & Nordtvedt, K. 1993, Phys. Rev. D 48, 3436
Davies, M.B. et al. 1994, ApJ 431, 742
Finn, L.S. 1991, Ann. N.Y. Acad. Sci. 631, 156
Finn, L.S. & Chernoff, D.F. 1993, Phys. Rev. D 47, 2198
Finn, L.S. & Thorne, K.S. 1995, Phys. Rev. D, (in preparation)
Flanagan, E.E. & Hughes, S.A. 1995, Phys. Rev. D, (in preparation)
Friedman, J.L. & Schutz, B.F. 1978, ApJ 222, 281
Hils, D. & Bender, P. 1995, (preprint)
Hils, D., Bender, P. & Webbink, R.F. 1990, ApJ 360, 75
Houser, J.L., Centrella, J.M. & Smith, S.C. 1994, Phys. Rev. Lett. 72, 1314
Hulse, R. & Taylor, J.H. 1975, ApJ 324, 355
Jaranowski, P. & Krolak, A. 1994, Phys. Rev. D 49, 1723
Kennefick, D. 1994. Phys. Rev. D 50, 3587
Kennefick, D., Laurence, D. & Thorne, K.S. 1995, Phys. Rev. D., (in preparation)
Kidder, L.E. 1995, Phys. Rev. D (in press)
Kidder, L.E., Will, C.M. & Wiseman, A.G. 1993, Phys. Rev. D 47, 3281
Kochanek, C. 1992, ApJ 398, 234
Królak, A., Kokkotas, K.D. & Schäfer, G. 1995, in *Proceedings of the 17th Texas Symposium on Relativistic Astrophysics*, Ann. N.Y. Acad. Sci., (in press)
Lai, D. & Shapiro, S.L. 1995, ApJ 442, 259
Lincoln, C.W. & Will, C.M. 1990, Phys. Rev. D 42, 1123
Lindblom, L. 1995, ApJ 438, 265
Lindblom, L. & Mendell, G. 1995, ApJ 444, 804
Lipunov, V.M., Postnov, K.A. & Prokhorov, M.E. 1994, ApJ 423, L121, and related, unpublished work
Markovic, D. 1993, Phys. Rev. D 48, 4738
Meers, B.J. 1988, Phys. Rev. D 38, 2317
Molteni, D., Gerardi, G. & Chakrabarti, S.K. 1994, ApJ 436, 249
Mönchmeyer, R. et al. 1991, A&A 246, 417
Nakamura, T. 1994, in *Relativistic Cosmology*, M. Sasaki (Ed.), Universal Academy Press, p. 155
Narayan, R., Piran, T. & Shemi, A. 1991, ApJ 379, L17
Phinney, E.S. 1991, ApJ 380, L17
Poisson, E. & Sasaki, M. 1995, Phys. Rev. D (in press)
Poisson, E. & Will, C.M. 1995, Phys. Rev. D (submitted)
Quinlan, G. & Shapiro, S.L. 1987, ApJ 321, 199
Rasio, F.A. & Shapiro, S.L. 1992, ApJ 401, 226
Ryan, F. 1995, Phys. Rev. D, (in preparation)
Ryan, F., Finn, L.S. & Thorne, K.S. 1995, Phys. Rev. Lett., (in preparation)
Sathyaprakash, B.S. 1994, Phys. Rev. D 50, R7111
Schutz, B.F. 1986, Nat 323, 310
Schutz, B.F. 1989, Class. Quant. Grav. 6, 1761
Schutz, B.F. 1995, (private communication)
Shapiro, S.L. & Teukolsky, S.A. 1983, in *Black Holes, White Dwarfs and Neutron Stars: The Physics of Compact Objects*, Wiley: Interscience, Section 10.10 and references

cited therein
Shibata, M., Nakamura, T. & Oohara, K. 1992, Prog. Theor. Phys. 88, 1079
Shibata, M., Nakamura, T. & Oohara, K. 1993, Prog. Theor. Phys. 89, 809
Shibata, M. et al. 1995, Phys. Rev. D (in press)
Tagoshi, H. & Nakamura, T. 1994, Phys. Rev. D 49, 4016
Tagoshi, H. & Sasaki, M. 1994, Prog. Theor. Phys. 92, 745
Taylor, J.H. 1994. Rev. Mod. Phys. 66, 711
Thorne, K.S. 1987, in *Three Hundred Years of Gravitation*, S.W. Hawking & W. Israel (Eds.), Cambridge University Press, p. 330
Thorne, K.S. 1992. Phys. Rev. D 45, 520
Thorne, K.S. & Zytkow, A.N. 1975, ApJ 199, L19
Tutukov, A.V. & Yungelson, L.R. 1993, MNRAS 260, 675
Van den Heuvel, E.P.J. 1994, (preprint)
Weber, J. 1960, Phys. Rev. 117, 306
Will, C.M. 1994, Phys. Rev. D 50, 6058
Will, C.M. 1994, in *Relativistic Cosmology*, M. Sasaki (Ed.), Universal Academy Press, p. 83
Wiseman, A.G. & Will, C.M. 1991, Phys. Rev. D 44, R2945
Yamaoka, H., Shigeyama, T. & Nomoto, K. 1993, A&A 267, 433
Zhuge, X., Centrella, J.M. & McMillan, S.L.W. 1994, Phys. Rev. D 50, 6247
Zimmermann, M. & Szedenits, E. 1979, Phys. Rev. D 20, 351

4

Radio Pulsars

PLANETS AROUND PULSARS

A. WOLSZCZAN

Penn State University, Dept. of Astronomy and Astrophysics
525 Davey Laboratory, University Park, PA 16802, U.S.A.

Abstract. A discovery of three terrestrial-mass planets orbiting the millisecond pulsar PSR B1257+12 and a subsequent detection of the predicted effect of gravitational interaction between the two more massive planets confirms that the first extrasolar planetary system has been identified and that pulsars can be successfully used as probes of planetary dynamics. In the absence of detections of planet-sized objects around Sun-like stars, planets orbiting a precise pulsar clock represent a unique source of information concerning the origin and evolution of planetary systems.

1. Introduction

A detection of planetary companions to stars other than the Sun has been one of the most challenging tasks of modern observational astrophysics. Either a positive or negative outcome of searches for extrasolar planets would directly address fundamental problems associated with the origin of the Solar System and it would be instrumental in the process of understanding the relation of Earth and terrestrial life to the rest of the universe. So far, no direct detection of planets orbiting a Sun-like star has been reported, in spite of a growing number of tantalizing hints (Sargent & Beckwith 1993). Instead, a discovery of planets around a rapidly rotating, old neutron star, the 6.2-millisecond radio pulsar PSR B1257+12, has been announced in early 1992 (Wolszczan & Frail 1992).

This paper describes the detection of the PSR B1257+12 planets, their confirmation through a successful measurement of the planetary perturbations and the most important consequences of this discovery.

2. Planets Around PSR B1257+12

A 6.2-millisecond pulsar, PSR B1257+12, was discovered in 1990 during a pulsar search conducted with the 305-m Arecibo radio telescope (Wolszczan 1991). The analysis of the follow-up timing observations has revealed large, quasi-periodic deviations of the times-of-arrival (TOAs) of pulses predicted by the standard timing model from the actually observed TOAs. Further examination of the post-fit residuals has shown that they can be decomposed into two steady, almost sinusoidal oscillations with the periods of 66.6 and 98.2 days (Figs. 1b,c).

A number of possible sources of this uncommon timing behavior of PSR B1257+12 have been considered. These included a timing noise caused by the seismic activity of a rapidly rotating neutron star, propagation effects in the circumpulsar medium, neutron star precession, errors in timing analysis and a variety of possibilities of instrumental origin. All these alternatives have been successively eliminated leaving a Keplerian orbital motion of two planet-mass objects around the pulsar as the most plausible explanation of the periodicities in the pulsar's timing residuals. More detailed analysis of the TOA measurements has led to a detection of an additional, much lower amplitude periodicity which was interpreted as a signature of the presence of yet another, very low-mass object, in a 25.3-day orbit around the pulsar (Fig. 1a). The resulting 3-planet timing model and the basic characteristics of the PSR B1257+12 planetary system are shown in Table 1 with the planets labeled A, B and C in order of increasing distance from the pulsar (see Wolszczan (1994) and references therein for a detailed description of the development of the timing model for PSR B1257+12).

3. Detection of Planetary Perturbations

The question of detectability of gravitational perturbations of the orbits of the pulsar planets has been raised soon after the announcement of the PSR B1257+12 system (Rasio et al. 1992; Malhotra et al. 1992). It has been pointed out that an approximately 3:2 ratio of the orbital periods of the two larger planets creates a near-resonance condition which leads to accurately predictable and possibly measurable periodic perturbations of the two orbits. A detection of planetary perturbations has been commonly regarded as the most unambiguous and perhaps the only way to produce a final proof that the timing behavior of PSR B1257+12 is due to orbital dynamics involving at least two planet-sized bodies.

Detailed analyses of the dynamics of the PSR B1257+12 planetary system and their effects on pulsar timing have been carried out by Malhotra (1993), Rasio et al. (1993) and Peale (1993). The observable consequences of a mutual gravitational interaction between the two planets include near-

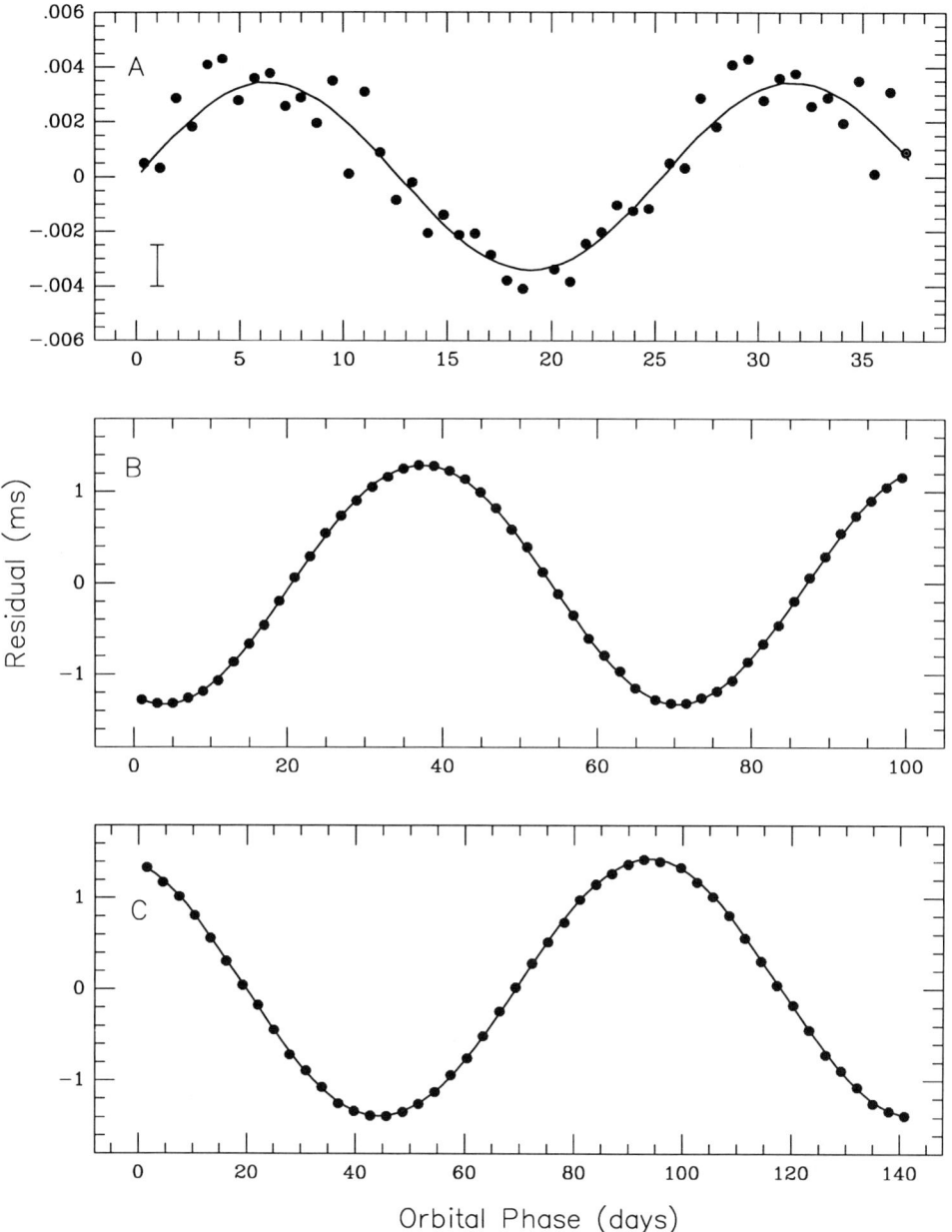

Figure 1. The post-fit residuals of pulse arrival times from PSR B1257+12 folded modulo the orbital periods of 25.34 days (planet A), 66.54 days (planet B) and 98.22 days (planet C), over a 3-year span of timing observations. In each case, the arrival time variations due to the other two planets have been fitted out. For planet A, a 2σ uncertainty in residuals is indicated by the error bar. For planets B and C, the uncertainties are too small to be shown.

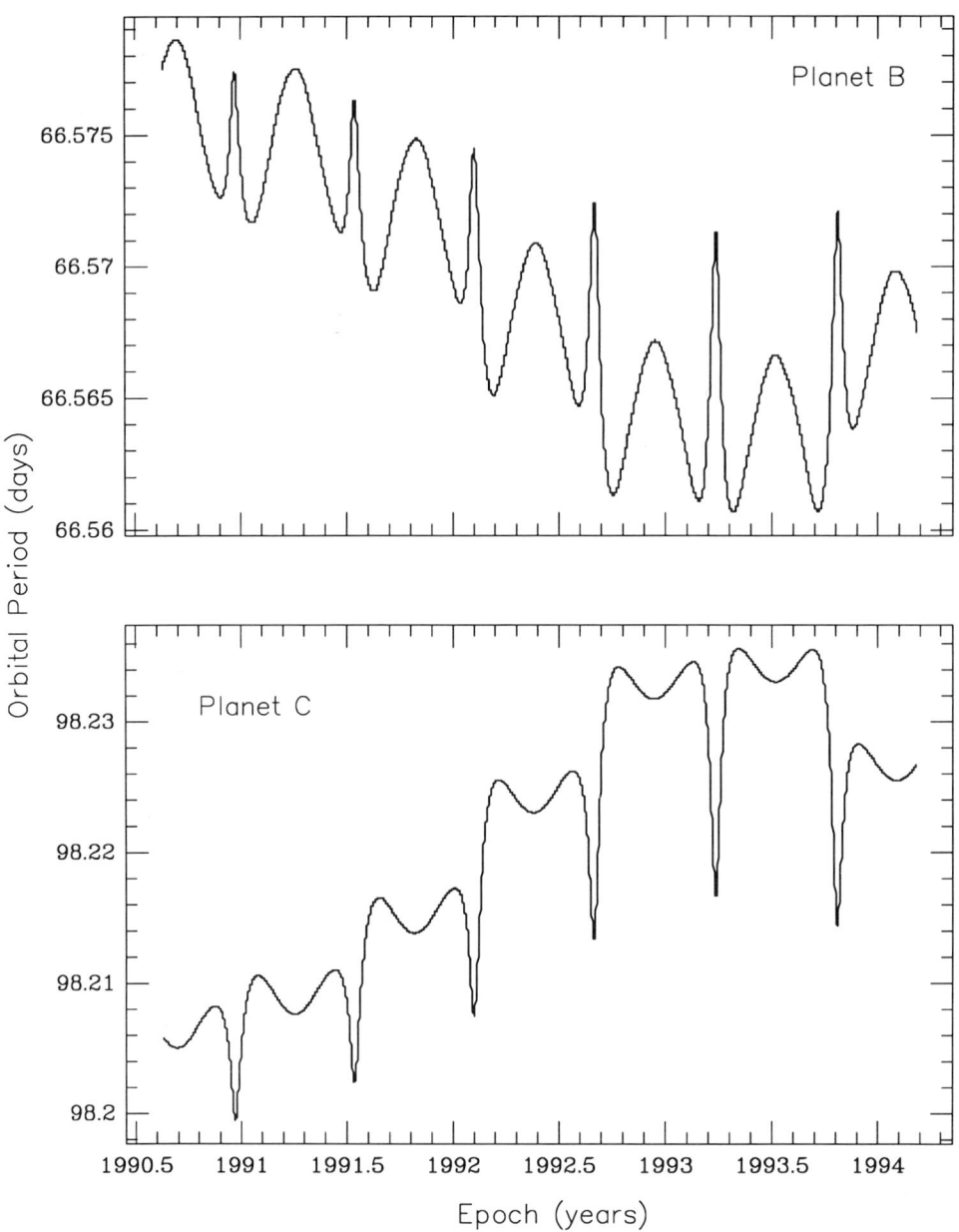

Figure 2. The predicted perturbations of orbital periods derived from numerical integration of equations of motion of planets *B* and *C*.

TABLE 1. Parameters of the PSR B1257+12 planetary system

Pulsar parameters

Rotational period	0.0062185319388187(2) s
Period derivative	$1.14334(6)\,10^{-19}$ s s^{-1}
Second period derivative	$4.5(9)\,10^{-30}$ s^{-1}
Right ascension, α_{1950}	$12^h\,57^m\,33^s.12730(3)$
Declination, δ_{1950}	$12°\,57'\,06''.406(1)$
Proper motion in α	46.4(6) mas yr^{-1}
Proper motion in δ	$-82.9(9)$ mas yr^{-1}
Epoch	JD 2448788.9
Dispersion measure	10.186(1) pc cm^{-3}

Keplerian orbital parameters

	A	B	C
Semi-major axis (light ms)	0.0035(6)	1.3106(6)	1.4121(6)
Eccentricity	0.0	0.0182(9)	0.0264(9)
Epoch of periastron (JD)	2448754.3(7)	2448770.3(6)	2448784.4(6)
Orbital period (s)	2189645(4000)	5748713(90)	8486447(180)
Longitude of periastron (deg)	0.0	249(3)	106(2)

Parameters of the planetary system

Planet mass ($M_\oplus$)	$0.015/\sin i_1$	$3.4/\sin i_2$	$2.8/\sin i_3$
Distance from the pulsar (AU)	0.19	0.36	0.47
Orbital period (days)	25.34	66.54	98.22

resonant, periodic variations of the elements of their orbits and the superimposed short-term, non-resonant fluctuations (Fig. 2). If the planetary masses are not too big, the relevant period is simply given in terms of the mean angular velocities of the planets, n_1 and n_2, as $2\pi/(2n_1 - 3n_2) = 5.56$ years and the predicted maximum amplitude of the corresponding timing residuals (after fitting out the two non-interacting orbits) is a function of planetary masses and the total time span of observations (Peale 1993).

In timing observations, the perturbations should manifest themselves in the form of oscillating residuals, if the least-squares fit of a timing model to data assumes fixed-parameter, non-interacting orbits. These oscillations are characterized by time scales of the order of orbital periods of the planets and by amplitudes which depend on the planet-to-pulsar mass ratios. Using the timing model of Table 1, and with the planetary masses, $m_{1,2}$ expressed in terms of the Earth mass and $M_{\rm psr}$ in units of $1.4\,M_\odot$, these mass ratios are given by:

$$\frac{m_1}{M_{\rm psr}} = \frac{3.4}{\sin i_1\, M_{\rm psr}^{1/3}} \qquad (1)$$

$$\frac{m_2}{M_{\text{psr}}} = \frac{2.8}{\sin i_2 \, M_{\text{psr}}^{1/3}}, \qquad (2)$$

for planets B, C and orbital inclinations i_1, i_2, respectively. Since the effect of a mutual inclination of the two orbital planes would have to be quite large to become detectable (Malhotra 1993), it is reasonable to assume coplanar orbits. Consequently, with $i_1 = i_2 = i$, the perturbation amplitude becomes a function of one variable, $(\sin i)^{-1} M_{\text{psr}}^{-1/3}$, which is the only parameter governing the effect of planetary perturbations on pulse arrival times.

The result of a search for the best-fit perturbation model using the downhill simplex algorithm (Nelder & Mead 1965) is shown in Fig. 3a. Compared to a timing model without planetary perturbations (Table 1), inclusion of this effect in the modelling process reduces the value of χ^2 for the global fit by nearly 3%, which is about 50 times the formal accuracy of the chosen minimization procedure. The presence of a well-defined minimum in the $\chi^2 = \chi^2[(\sin i)^{-1} M_{\text{psr}}^{-1/3}]$ curve confirms that planetary perturbations influence the observed pulse arrival times in the manner predicted by theory. The detection of this effect represents a proof that the pulse arrival time variations observed in PSR B1257+12 are due to orbital motion of planet-mass bodies with the dynamical characteristics that are not unlike those of the inner planets of the Solar System. This discovery is the first of a planetary system around a star other than the Sun. A more detailed description of the above analysis can be found in Wolszczan (1994).

Although the planetary perturbation modelling does not allow determination of all system parameters in a unique way, useful constraints on the masses of the pulsar and its planetary companions can be established in a straightforward manner. With the aid of Eqs. (1) and (2), the result of the perturbation model fitting (Fig. 3a) can be presented in the form of a pulsar mass-normalized planetary mass diagram (Fig. 3b). The uncertainty of the determination of a χ^2 minimum in Fig. 3a defines a range of allowable masses in this diagram. It is further constrained by the highest possible orbital inclination ($i = 90°$) and by the maximum neutron star mass (taken to be $2\,M_\odot$). Within these limits, the minimum pulsar mass is $\sim 1.2\,M_\odot$ and the range of possible pulsar masses for orbital inclinations close to $i = 90°$ includes both the typical theoretical value ($\sim 1.3\,M_\odot$) (Woosley 1987) and the observed average neutron star mass ($1.35 \pm 0.27\,M_\odot$) (Thorsett et al. 1993). Furthermore, the masses of planets B and C must be similar to their respective "canonical", Earth-like values of $3.4\,M_\oplus$ and $2.8\,M_\oplus$ (Table 1 and Eqs. 1, 2) and the orbital inclinations are unlikely to be less than 60° for any reasonable choice of a neutron star mass. No additional information concerning the orbit of planet A can be extracted from this analysis, because its effect on the mutual perturbations of planets B and C is entirely negligible. However, if the orbits of all three planets are approximately coplanar, planet A must be a very low-mass, Moon-like object (Table 1).

4. Pulsar Planets and Other Planetary Systems

The observed characteristics of the planets and PSR B1257+12 itself, when confronted with current ideas concerning planetary formation (Levy 1993, Ruden 1993) and the origin and evolution of millisecond pulsars (Bhattacharya & Van den Heuvel 1991), indicate that the planets are likely to have evolved in a circumpulsar disk of matter created from the remains of the pulsar's binary stellar companion. Therefore, most of the scenarios of the millisecond pulsar planet formation concern themselves with possible ways to transform a fraction of the companion's mass into a protoplanetary disk, implying that the planets would subsequently form in a manner similar to that envisioned for the origin of the Solar System (Podsiadlowski 1993; Phinney & Hansen 1993). This indicates an interesting possibility that the planets around solar-type stars and pulsars may differ in their physical and chemical characteristics, but the fundamental features of the dynamics of their parent planetary systems should be comparable.

Pulsars accompanied by planets can be used as highly accurate probes of planetary dynamics. In the case of PSR B1257+12, a positive identification of planetary perturbations and the detection of a Moon-mass planet A involved measurements and analysis of pulse arrival times at a microsecond precision level, which is equivalent to radial-velocity resolution of the order of $1\,\mathrm{mm\,s^{-1}}$ (a factor of 10^4 better than the typical $10\,\mathrm{m\,s^{-1}}$ resolution achieved in modern single-line Doppler spectroscopy). These two techniques are compared in Fig. 4, in which the minimum detectable amplitudes of timing residuals, δt and radial velocities, δV_r are expressed in terms of masses and orbital radii as:

$$\delta t = c^{-1}\frac{m}{M}a \qquad (3)$$

$$\delta V_r = G^{1/2}\frac{m}{(Ma)^{1/2}}, \qquad (4)$$

where m is the planetary mass, M is the mass of the central body, a is the semi-major axis of the orbit and circular, "edge-on" orbits are assumed. Evidently, detections and dynamical studies of terrestrial-mass planets will remain beyond the reach of optical astrometry and Doppler spectroscopy in the foreseeable future, but they will be quite feasible with the pulse timing method.

The detection of pulsar planets dramatically emphasizes the value of enriching the strategies of planetary searches with non-standard approaches. At present, among more than twenty known millisecond pulsars, there are

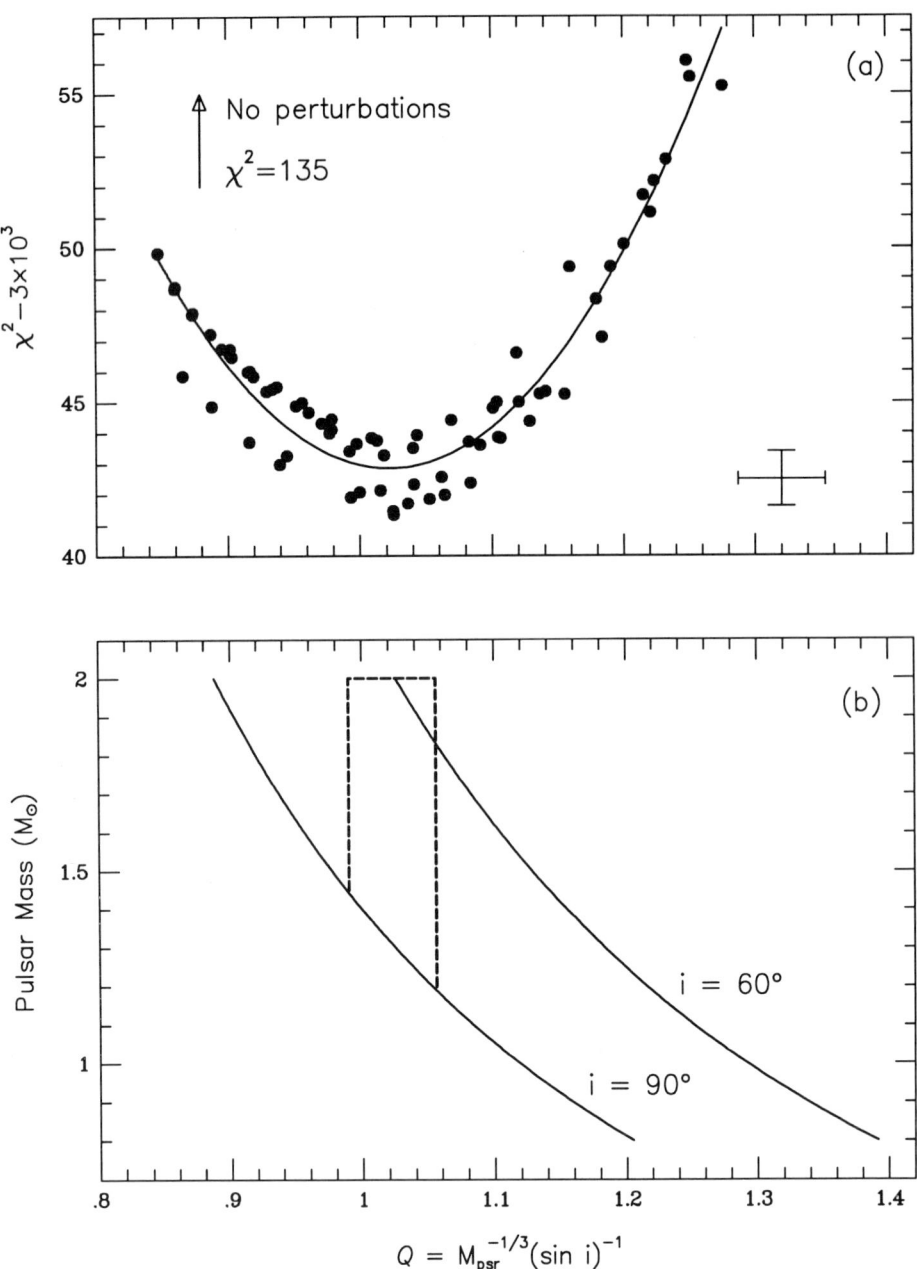

Figure 3. (a) A fit of the numerical models of the gravitational perturbations of planets B and C (filled circles) to pulse arrival times. The solid line represents the resultant best parabolic fit to the χ^2 data. A 2σ uncertainty of the minimum χ^2 obtained from this fit is shown by the horizontal error bar. The vertical bar denotes the accuracy of the downhill simplex minimization procedure. (b) Constraints on the pulsar mass, the masses of planets B and C and the common orbital inclination. The dashed lines delimit the area containing the most likely combinations of these parameters.

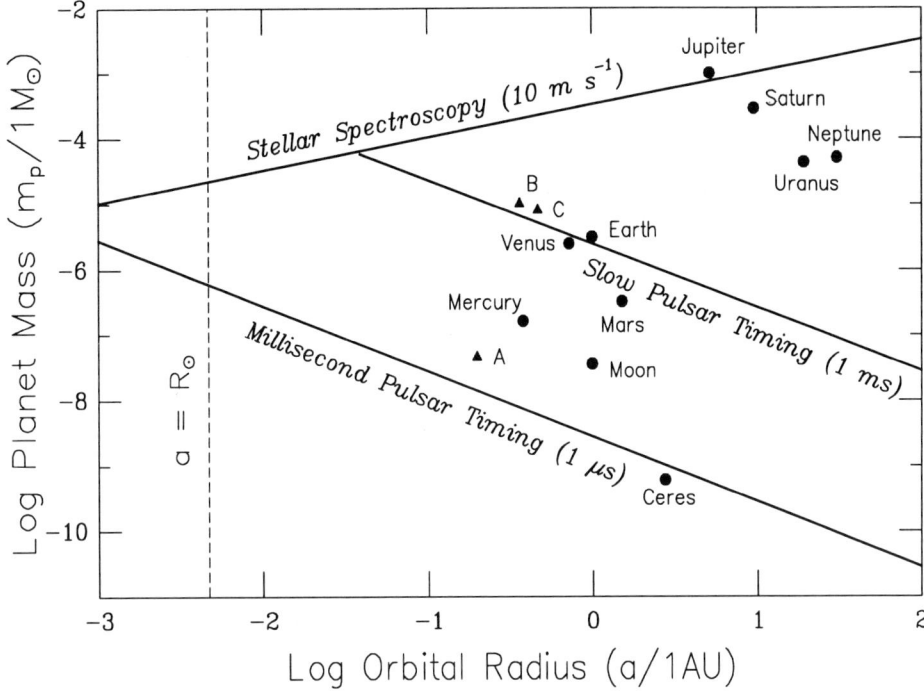

Figure 4. A comparison of the planet detection capabilities of the techniques of stellar spectroscopy and pulse timing. The assumed stellar and pulsar masses are $1\,M_\odot$ and $1.4\,M_\odot$, respectively. The filled circles and triangles mark the respective positions of the Solar System planets and the planets around PSR B1257+12 in the discovery space. The vertical dashed line sets the lower limit to orbital size defined by the solar radius.

five solitary objects that have apparently managed to dispose of their binary stellar companions. If a missing stellar companion to the pulsar indicates a possibility of "leftover" planets around it, PSR B1257+12 is the only confirmed case. Clearly, the pulsar planet formation is not 100% efficient, but the available statistics are too small to reliably constrain this efficiency. In addition, there is a large body of almost 600 younger, "slow" pulsars, some of which may have retained planets of their parent stars (Thorsett & Dewey 1993). In fact, two of such pulsars have been reported to exhibit TOA variations that could be explained in terms of planetary dynamics (Demiański & Prószyński 1979; Bailes et al. 1993; Lyne 1994). Extensive timing measurements and more direct observations of as many of these objects as possible will establish the significance of pulsar planetary systems as a class of astrophysical objects and their relationship to still hypothetical planets around Sun-like stars.

Acknowledgements. I thank F. Camilo, R. Foster and A. Vázquez for help with observations, R. Malhotra and F. Rasio for discussions and S. Peale and J. Taylor for their contributions to the results presented in this paper. This research was supported by NASA under grant NAGW-3405 and by the NSF under grant AST-9317757. Arecibo Observatory is part of the National Astronomy and Ionosphere Center, which is operated by Cornell University under contract with the NSF.

References

Bailes, M., Lyne, A.G. & Shemar, S.L. 1993, in *Planets around Pulsars*, J.A. Phillips, S.E. Thorsett & S.R. Kulkarni (Eds.), ASP Conf. Ser. 36, 19
Bhattacharya, D. & Van den Heuvel, E.P.J. 1991, Phys. Rep. 203, 1
Demiański, M. & Prószyński, M. 1979, Nat 282, 383
Levy, E.H. 1993, in *Planets around Pulsars*, J.A. Phillips, S.E. Thorsett & S.R. Kulkarni (Eds.), ASP Conf. Ser. 36, 181
Lyne, A.G. 1994, in *Millisecond Pulsars: A Decade of Surprise*, A.S. Fruchter, M. Tavani & D.C. Backer (Eds.), ASP Conf. Ser. (in press)
Malhotra, R. 1993, ApJ 407, 266
Malhotra, R., Black, D., Eck, A. & Jackson, A. 1992, Nat 356, 583
Nelder, J.A. & Mead, R. 1965, Computer J. 7, 308
Peale, S.J. 1993, AJ 105, 1562
Phinney, E.S. & Hansen, B.M.S. 1993, in *Planets around Pulsars*, J.A. Phillips, S.E. Thorsett & S.R. Kulkarni (Eds.), ASP Conf. Ser. 36, 371
Podsiadlowski, P. 1993, in *Planets around Pulsars*, J.A. Phillips, S.E. Thorsett & S.R. Kulkarni (Eds.), ASP Conf. Ser. 36, 149
Rasio, F.A. et al. 1992, Nat 355, 325
Rasio, F.A. et al. 1993, in *Planets around Pulsars*, J.A. Phillips, S.E. Thorsett & S.R. Kulkarni (Eds.), ASP Conf. Ser. 36, 107
Ruden, S.P. 1993, in *Planets around Pulsars*, J.A. Phillips, S.E. Thorsett & S.R. Kulkarni (Eds.), ASP Conf. Ser. 36, 197
Sargent, A.I. & Beckwith, S.V.W. 1993, Phys. Today 46, 22
Thorsett, S.E. & Dewey, R.J. 1993, ApJ 419, L65
Thorsett, S.E. et al. 1993, ApJ 405, L29
Wolszczan, A. 1991, Nat 350, 688
Wolszczan, A. 1994, Science 264, 538
Wolszczan, A. & Frail, D.A. 1992, Nat 355, 145
Woosley, S.E. 1987, in *The Origin and Evolution of Neutron Stars*, IAU Symp. 125, D.J. Helfand & J.H. Huang (Eds.), Reidel, p. 255

TIMING OF MILLISECOND PULSARS

D. C. BACKER
Astronomy Department & Radio Astronomy Laboratory
University of California
Berkeley, CA 94720, U.S.A.

Abstract.
Arrival times of some millisecond pulsars can be measured with submicrosecond precision. This allows unprecedented measurements of the rotation of these stars, astrometric parameters including proper motion and parallax, motion in binary and complex dynamical systems, perturbations resulting from propagation through small scale turbulence in the interstellar plasma and other effects. In addition to the independent parameters which are required to model each pulsar, there is a set of global parameters that affect the array of pulsars in a correlated manner. These parameters involve the international atomic time scale, the ephemeris of the Earth's orbit and the gravitational wave background and have monopole, dipole and quadrupole and higher order angular signatures, respectively.

1. INTRODUCTION

Shortly after the discovery of the first millisecond pulsar, B1937+21, analysis of the arrival times of its pulses indicated that its rotation was extremely stable, $\frac{\Delta\Omega}{\Omega} < 10^{-13}$. The initial observations could be modeled by three rotational parameters – phase, rotation rate, and spindown rate – and two astrometric parameters – right ascension and declination. The measurement precision was unprecedented for pulsars – a few microseconds of time. This precision was the result of the high flux density, the short period and concomitant pulse width, and the unusually sharp structure of the average pulse profile. Pulse widths scale with the inverse square root of the period in simple polar cap emission models. The extreme stability of the millisecond pulsars allows their use for a variety of fundamental physical and astrophysical investigations, and arises from the extremely large mechanical Q of these objects, $Q = \dot{P} \simeq 10^{-19}$ to 10^{-21}.

In this review I discuss first the *independent* parameters of each pulsar that are required to model pulse arrival time observations with the focus on millisecond pulsars and pulsars with companions. Next, I describe how these parameters provide information about the neutron star and, in some cases, the system in which it lives. For some pulsars, other areas of astrophysics can be explored with pulsar timing observations. In my final section I will discuss a number of applications that require a set of *global* parameters for a spatial array of millisecond pulsars. In brief these are: to stabilize atomic time; to tie ecliptic and equatorial reference frames together; to determine the masses of the outer planets; and to probe the cosmic background of gravitational radiation. Kaspi (1994) has reported on progress in millisecond pulsar timing, and Phinney & Kulkarni (1994) have recently reviewed other topics pertinent to millisecond and binary pulsars.

2. INDEPENDENT TIMING OF PULSARS

2.1. FORMULATION AND RECENT RESULTS FOR SINGLE PULSARS

Pulsar signals are faint. Large radio telescopes, wide bandwidths, and long integration times are required for precise pulsar timing. Use of large bandwidths is limited by the need to accurately remove the severe dispersion incurred as the signal traverses the interstellar plasma. A variety of solutions for dispersion removal and subsequent signal processing have been developed in recent years. The resulting data product is an average pulse profile whose first sample is time tagged using the observatory's atomic time scale. The epoch of arrival of a fiducial point on the average pulse is determined by a variety of software techniques. Pulsar timing precision is potentially limited either by the temporal stability of the intrinsic average pulse profile (Helfand *et al.* 1975; Weisberg *et al.* 1989; Cordes *et al.* 1990; Blaskiewicz 1993; Suleymanova & Shitov 1994) or by our ability to measure accurately the profile given calibration errors and strong polarization. The typical precision of these measurements is 0.01 to 0.001 periods. The topocentric pulse arrival time – a space-time event – is referred to an international atomic time scale (see below) using GPS time transfer techniques at most observatories. Pulsar timing proceeds with a number of arrival times per session on a given day, and a number of sessions per year. The analysis task outlined below is to generate a model for these many arrival times with no ambiguity of rotation number. Recent references on pulsar timing techniques include: Backer & Hellings (1986), Backer (1993), and Taylor (1992).

The observatory is a constantly accelerating laboratory. A relativistic space-time transformation of the observatory space-time reception event to the emission event in the pulsar's frame must be made to explore the link between the observed pulse arrival times and the rotational properties of the star. Pulsar emission is modeled as spherical waves emanating from a point source. The transformation can be broken into a relativistic clock term and spatial terms such as from the observatory to the Earth center and from the Earth center to the Solar System barycenter. Transformation from the Solar System barycenter to the pulsar cannot be done with any level of accuracy and therefore it is ignored. The scale of the different terms may be large or small with respect to the pulse period. For example, there are 640 982 pulses from B1937+21 stacked up across the Earth's orbit at any instant.

The pulsar's equatorial coordinates are required to remove these effects. Iteration is required in the process of determining pulsar rotation and astrometric parameters. The phase residuals are used to solve for new parameter estimates. Typically one solves for phase, frequency, frequency first derivative, right ascension and declination. With long data sets and high precision timing one can solve for proper motion components (e.g., Nice & Taylor 1994) and even parallax (Ryba & Taylor 1991; Camilo et al. 1994). Proper motions and parallaxes provide fundamental data for discussions of the origin and evolution of the millisecond pulsar population. The proper motions of pulsars in globular clusters, which are just now being determined, can be combined with their Doppler velocities to discuss cluster evolution and the mass distribution in the galaxy.

While one cannot model the free space propagation of the pulsar signal, one must model the dispersive delay through the cold, interstellar plasma. This is reasonably straightforward owing to the simple dependence of group velocity on frequency. Measurements at two radio frequencies spaced by an octave provide differential arrival times that yield the required column density of electrons needed to extrapolate the arrival time to infinite frequency. Often alignment of the fiducial point between frequencies is confounded by spectral index variations across the pulse profile and by instrumental effects (e.g., Foster et al. 1991). Measurements at three or more radio frequencies are recommended.

The interstellar plasma is turbulent on a wide range of length scales. For the highest timing precision the electron column density must be monitored (Phillips & Wolszczan 1992; Backer et al. 1993a; Kaspi et al. 1994). The uneven electron density distribution leads to refractive and diffractive effects that further complicate the removal of plasma propagation effects (Foster & Cordes 1990; Hu et al. 1991). Fig. 1 illustrates a 4.5 year record of refractive flux density variations and dispersion measure changes for B1937+21.

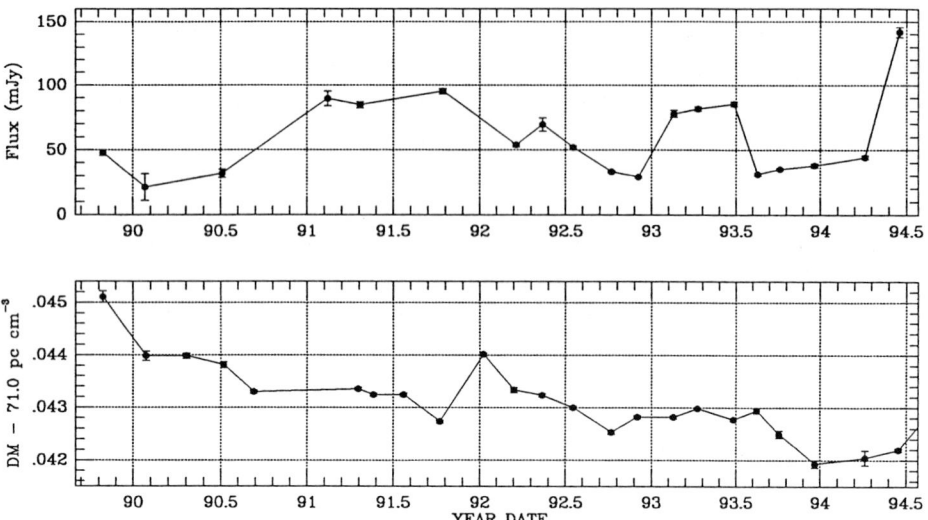

Figure 1. Flux density at 800 MHz (a) and dispersion measure (b) of B1937+21 as a function of time from the pulsar timing array experiment at NRAO Green Bank. For experimental details see Backer *et al.* (1993a). On these long time scales there is not a strong correlation between variations of the two quantities.

Cordes *et al.* (1986), Blandford *et al.* (1985) and others discussed the expected relationships between scintillation amplitude, dispersion measure variations and pulse arrival times for turbulence with various logarithmic slopes. Lestrade (1994) has demonstrated a correlation between scintillation amplitude and pulse arrival time for PSR 1937+21 on short time scales. There is now evidence for propagation events that are attributed to traversal of a caustic caused by discrete interstellar plasma lenses (Cognard *et al.* 1993). Precise timing through these events may be possible with dense sampling in time at a number of frequencies although the simplest approach would be to observe at sufficiently high frequency where the amplitude of such effects is reduced below measurement error. While these effects are a nuisance to precision pulsar clock experiments, they do provide crucial constraints on the nature and distribution of this turbulence.

A second, frequency independent, 'propagation' effect occurs as the signals pass through the distorted space-time metric in the vicinity of a massive body (Shapiro 1964). This delay is the *longitudinal* counterpart of the more familiar bending of light which is a *transverse* effect. The time variable Shapiro delay through the Solar System is removed by simple calculation using a PPN GR parameter that is adequately determined by radar and VLBI experiments. Delays from the passage by stars along the line of sight are large, but only slowly varying.

The primary rotation parameters – phase, frequency, and frequency derivative – are sufficient to model most millisecond pulsars. The ratio of frequency and frequency derivative provide a time scale for the evolution of the pulsar's rotational energy. The age of the pulsar can be estimated from this ratio, if one assumes both that the slowing down follows the simple vacuum dipole radiation formula and that the initial rotation frequency is much larger than the present one. These assumptions may be questioned for the millisecond pulsars. In particular, the initial period of the shortest period objects may be comparable to the present period. An additional uncertainty arises in that galactic motion – transverse velocity and acceleration – give rise to significant contributions to the frequency derivative (Camilo et al. 1994). Pulsars in globular clusters will have an additional source of acceleration owing to the combined effects of nearest neighbor stars and the smooth, deep central potential (Anderson et al. 1990).

In the past few years the millisecond pulsars which have the longest record of timing have shown signs of *instability* at the level of a few microseconds for B1937+21 (Kaspi et al. 1994) and B1821-24 (Cognard et al. 1994) over time scales of several years. Fig. 2 displays our data which also shows the instability of B1821-24 arrival times along with the record of dispersion measure variations. The contribution of atomic time scale and ephemeris uncertainties to this are discussed further below. The conclusion is that these instabilities are internal to the pulsars. While solid state physicists have proposed detailed models of sudden glitches in the otherwise smooth rotation of pulsars, there is not a good understanding of the origins of the $1/f$-like spectrum of noise. All forms of internal timing noise are driven by figure readjustment forces that increase monotonically with $\dot{\Omega}$. The noise for millisecond pulsars with low $\dot{\Omega}$'s is expected to be small, but not zero. The dependence of timing noise on $\dot{\Omega}$ reaching to millisecond periods has been discussed by Dewey & Cordes (1989) and more recently by Arzoumanian et al. (1994a).

2.2. PULSARS WITH COMPANIONS

Pulsar companions add complexity for timing measurements in that an additional transformation must be done from the system barycenter to the pulsar. Initial analysis using the apparent rotation frequency to follow the Doppler curve as in a single-line spectroscopic binary is often required for millisecond pulsars whose period may be much smaller than the orbit size. Five parameters are required to model the general Keplerian binary system. The orbit size projected along the line of sight, the orbital period, and an epoch when the star passes through the plane of the sky are needed for all systems. If the orbit is eccentric, then the eccentricity and the orientation

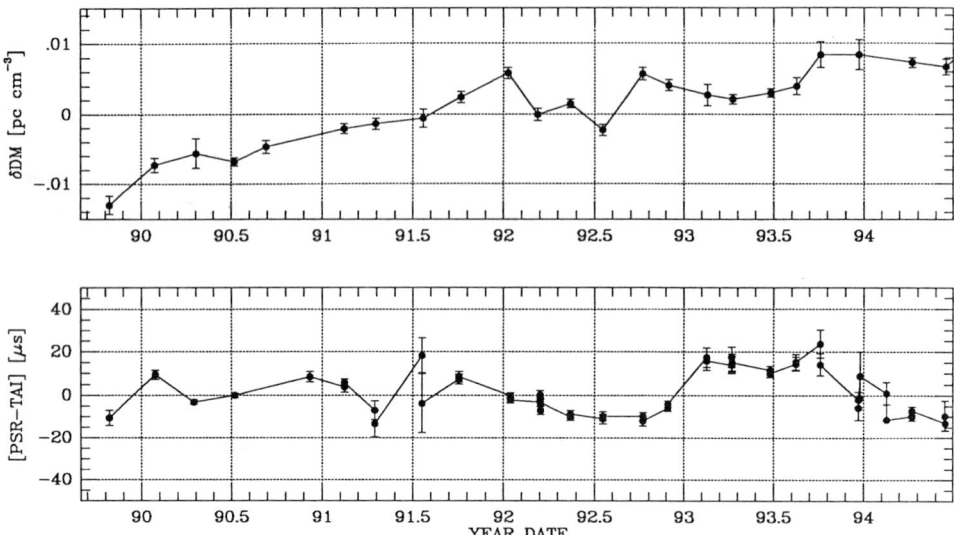

Figure 2. Temporal variations of the dispersion measure (a) and the arrival time (b) of B1821−24. The timing model includes pulse phase, frequency, frequency first derivative, right ascension, declination and proper motion in right ascension. The timing of B1821−24 is not sensitive to declination proper motion. A constant dispersion measure of 119.8 pc cm^{-3} has been removed.

of the periastron point with respect to the line of nodes are required.

B1620−26 is an 11-ms period pulsar with a 0.3 M$_\odot$ companion in a 191-d orbit that is located in the nearby globular cluster M4 (Lyne et al. 1988). Timing observations following its discovery were not stable until the presence of a large frequency second derivative was added to the model. The large value of $\ddot\Omega$ is best explained by the secular influence of a second bound companion with an orbital period of 100's of years (e.g., Backer et al. 1993b; Thorsett et al. 1993a). The principal observed effect in our seven-year data set is then an acceleration derivative of the center of mass frame of the 191-d binary system which produces the apparent frequency second derivative. Fig. 3 displays residuals from a recent fit to the data that shows the effects of a third derivative of the rotation frequency as a quartic timing residual. This provides a measure of the component of the acceleration second derivative which is estimated as 3.3 10^{-24} cm s^{-2}. The formation and stability of this system are under considerable debate (Sigurdsson 1993; Rasio 1994). Like other pulsars in globular clusters B1620−26 is providing a new window into dynamical processes that contribute to the stability of clusters against gravitational collapse (Phinney 1992). Confirmation of this hypothesis for B1620−26 will eventually come from long duration timing observations. A tentative identification of the optical counterpart has been

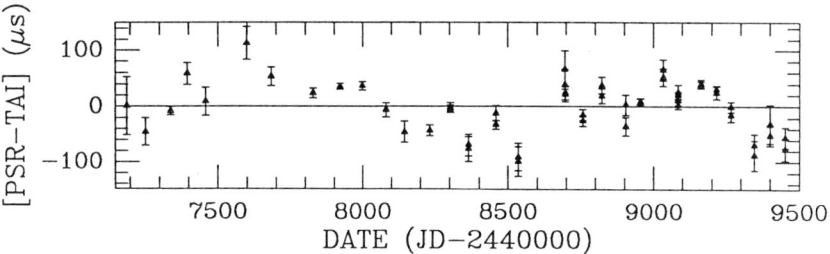

Figure 3. Timing residuals for pulsar B1620−26 which has two companions in hierarchical orbits and resides in the globular cluster M4. The model removed has parameters of phase, frequency, frequency first derivative, frequency second derivative, right ascension, declination and right ascension proper motion.

made (Bailyn *et al.* 1994), and followup HST observations have been proposed. If the orbital planes of the two companions are not collinear, the projected orbit size will change secularly. Current data analysis is beginning to show this effect, but the significance is not high (Backer & Thorsett 1994).

An entirely different, multiple companion system is found in B1257+12 as reported by Wolszczan (1994; these proceedings). This pulsar with planets is unique amongst the well studied millisecond pulsars. Yet another subclass of millisecond pulsars are objects like B1957+20 whose companion produces an eclipse during inferior conjunction (Fruchter *et al.* 1990). The companions have masses intermediate between those in low mass binary systems like B1620−26 and the planets in B1257+12. High precision timing of B1957+20 has revealed secular variations of the orbital period (Arzoumanian *et al.* 1994b) that provides important constraints on the physical state of the companion star.

A number of post Keplerian parameters are required to explain the relativistic effects of gravity. These are observable when the effects are large owing to the combined effects of a high eccentricity, a short orbital period, a high mass companion, a strong pulsar, and/or a large telescope with sensitive electronics. Briefly the effects are: precession of periastron as in Einstein's test for Mercury's orbit; relativistic clock correction which results from transverse Doppler effect and gravitational redshift; the Shapiro delay; and the decay of the orbit owing to the back reaction from radiation of gravitational waves. Some of the latest results including the first tests of strong field effects in General Relativity are described by Taylor *et al.* (1992). Apart from the fundamental test of physical theory provided by these observations, one also obtains accurate values for neutron star masses (Thorsett *et al.* 1993b).

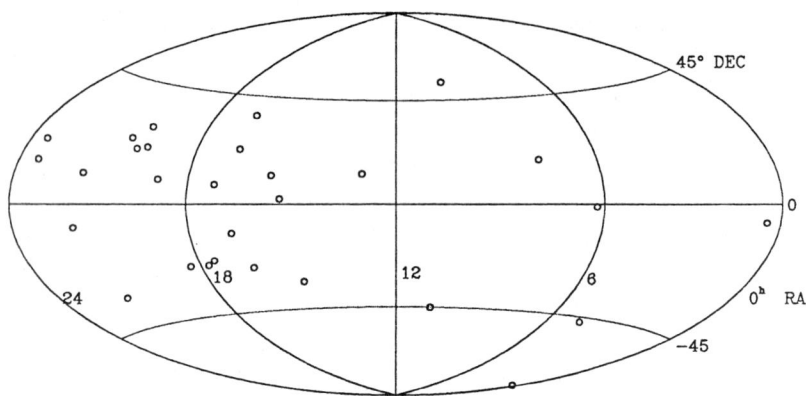

Figure 4. Millisecond pulsar timing array plotted in equatorial coordinates. Pulsars included have periods less than 15 ms. The non-uniformity in distribution reflects a combination of search completeness and true galactic distribution. Owing to large differences in flux density and pulse profile morphology different pulsars can be timed to different levels of precision.

3. GLOBAL TIMING OF PULSARS

In the previous section the timing of isolated and binary pulsars was discussed. The data for each pulsar were treated independently using a model of the Earth's space-time coordinate as a function of epoch that is assumed to be more accurate than the precision of the pulsar data. In this section the limits of this model are discussed, and the influence of a stochastic background of gravitational radiation on pulsar arrival times is presented. The section concludes with a description of the use of an array of precisely timed pulsars to solve for *global* parameters that affect all objects in a correlated manner. Fig. 4 displays the current array of millisecond pulsars.

3.1. TIME AND TIME TRANSFER

The definition of the second of Terrestrial Time (TT) is 9 192 631 770 cycles of the frequency corresponding to a fine structure transition in the ground state of the Cesium 133 atom at sea level on the geoid. An initial epoch is required to define a time scale. TT replaces International Atomic Time, TAI. The atomic standards laboratories – in Washington, Boulder, Paris, Braunschweig, Tokyo – realize this scale. The time scale required for pulsar timing is that of a clock at rest in the solar system barycenter inertial reference frame: Barycentric Coordinate Time (TCB). These definitions and their impact on pulsar timing are discussed by Guinot & Petit (1991),

Seidelmann & Fukishima (1992), and Fukishima (1994). The epoch when atomic time reads some value, such as zero or 2440000.0 days 0 seconds, is not of particular concern here. Accurate realization of this definition is made difficult by the fact that Cesium clocks are affected by temperature, magnetic fields, vibrations of the room, interactions with container walls, and so on. Many of these effects can shift the frequency and will vary with time.

Estimates of the stability of TT on yearly time scales are difficult to obtain owing to its inclusion of all of the world's standards in its definition. Estimates by Guinot and Petit suggest that the stability is just under 10^{-14}. On longer time scales there is even less understanding of the stability other than the statements that the standards have been improving on a time scale shorter than a decade and that the stability slowly degrades with time interval from its minimum value on time intervals of about six months. The most precise pulsar timing is now sensitive to choice of atomic time scale (Kaspi et al. 1994).

Timing of an array of millisecond pulsars can provide independent information to help stabilize TT on intervals of years to decades. Time scale errors will have a monopole signature as one moves around the sky – all pulsars will arrive early or late in a correlated manner. Creation of a pulsar time scale then requires defining, in the simplest case, their initial epochs, frequencies, frequency derivatives, and equatorial positions which is just five parameters. As already discussed, further parameters are required and, in addition, the most closely studied millisecond pulsars show signs of instability in the rotation rates. The prospects of creating a pulsar time scale are not clear.

Pulsar timing observations are conducted at remote observatories. While today transfer of a time scale from the standards laboratories to the observatories is straightforward using GPS, care needs to be taken to ensure that the remarkable *short* term time transfer stability inherent in the GPS technique is maintained to ensure *long* term stability. Changes in the measurement technique within the observatory and the location of GPS receivers need careful calibration to ensure stability over decades.

3.2. EPHEMERIS AND FRAME TIE

The transformation from the observatory to the solar system barycenter inertial reference frame requires a precalculated ephemeris of the earth's motion. The ephemeris is constructed from a detailed model of solar system dynamics that includes the Sun, all planetary systems, minor planets, and a band of matter representing the asteroid belt, and uses an appropriate level of relativistic effects (Standish 1990a). Table 1 gives the current estimates of the errors in planetary masses, and the corresponding offset between the estimated barycenter location and the "true" location. These mass errors are large with respect to asteroid and cometary masses, but comparable to masses of the largest minor planets and larger moons. The barycenter will also be shifted by errors in the celestial coordinates of the planets.

TABLE 1. Planetary Mass Errors

System	Δm_p (10^{22} gm)	P_p (y)	r_p (AU)	$\Delta r/c$ (ns)	Reference
Mercury	1.4	0.25	0.39	1	Mariner
Venus	1.5	0.62	0.72	3	Mariner
Mars	0.2	1.88	1.52	1	Mariner
Jupiter	150	11.9	5.20	1950	Voyager
Saturn	300	29.5	9.52	7140	Voyager
Uranus	15	84.0	19.2	720	Voyager
Neptune	30	165	30.0	2250	Voyager

The arrival time transformation is effected by a scalar product between the observatory vector, which is of order 500 s – the AU – in length, and a unit vector in the direction of the source. This means that the planetary timing error for pulsar j in direction $\hat{n}_j$ will be

$$c\Delta t_{pj}(t) = \Sigma_p [\left(\frac{\Delta m_p}{M_\odot}\right) \mathbf{r_p}(t) + \left(\frac{m_p}{M_\odot}\right) \Delta \mathbf{r_p}(t)] \cdot \hat{n}_j,$$

where Δm_p, $\mathbf{r_p}$, and $\Delta \mathbf{r_p}$ are the mass error (Standish 1990b), position, and position error (Standish 1990c), respectively, of planet p.

At any instant the residuals for different pulsars will differ as a result of the dot product in the above expression. There will be a slow variation around the sky with a dipole signature in the residuals. The temporal signature will be a cubic phase residual on short time scales, and periodic on time scales exceeding those of the periods of the outer planets. Kaspi et al. (1994) have demonstrated by timing two nearby pulsars over an 8-year span that ephemeris errors are not important yet at the several microsecond level.

3.3. GRAVITATIONAL WAVE BACKGROUND

Precise pulsar timing may allow direct detection of long-wavelength gravitational radiation. Gravitational radiation is a perturbation of the spatial part of the space-time metric. One can view this perturbation as a change in the index of refraction from unity for propagation of an electromagnetic wave in simple three dimensional, Cartesian space. The transmission time from emission to reception then varies according to the integration of the metric perturbations along the path. The perturbation of the transmission time from the simple geometric path L/c varies with time as the radiation traverses the line of sight. The integration produces contributions at both the Solar System and pulsar ends from incomplete cycles of each spectral component of the gravitational radiation. The local contribution will lead to a correlation between timing residuals around the pulsar timing array with quadrupole and higher order angular signatures. The pulsar end contributions will have equal amplitudes and will be uncorrelated. For a general background of gravitational radiation the correlated component can be represented by independent waves along orthogonal axes with five independent parameters (Hellings 1990). The correlations can be displayed as apparent Doppler shift patterns using the formulation developed by Detweiler (1979). These patterns may be used as a basis of functions for analysis of pulsar timing array data (Backer 1993). Doppler shifts are derived from the temporal derivative of timing residuals. Fig. 5 displays one of the five Doppler patterns and demonstrates the quadrupole and higher order angular structure that allows gravitational radiation effects to be detected in the presence of time (monopole) and ephemeris (dipole) perturbations. The Doppler shifts themselves are not measurable owing to the period fitting done for each pulsar. Detection requires temporal modulations of each Doppler pattern with a spectrum of fluctuations. The amplitude and shape of these spectra will be the primary information to relate to source models.

A number of sources of long-wavelength gravitational background radiation have been proposed in the past decade. Defects in the space-time continuum called cosmic strings were once proposed as the gravitational centers that produced the large-scale structure we see today in the distribution of galaxies (Vilenkin 1981). The principal channel for decay of these strings is gravitational radiation with a spectrum extending from inverse months to inverse centuries (Romani 1988; Bennett & Bouchet 1991; Caldwell & Allen 1992). The millisecond pulsar timing observations by Kaspi et al. (1994) place strong constraints on this theory. Strings could exist and decay at a level lower than that required for structure formation. Detweiler (1979) proposed that precision pulsar observations might detect the slow decay of massive binary black holes in distant quasars and AGNs. Ra-

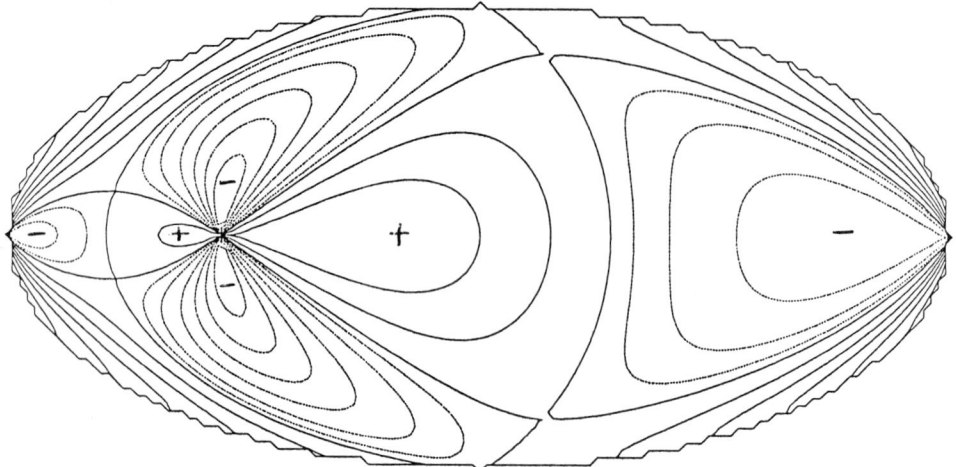

Figure 5. Angular Doppler pattern on the celestial sphere for one independent component of a background of gravitational radiation. Timing residuals of the array of pulsars shown in Fig. 4 would show correlations with these Doppler patterns modulated by a temporal fluctuation spectrum.

jagopal & Romani (1994) have taken a new look at this with estimates of the level of the spectrum over a range of interesting frequencies. Attempts at direct detection of the gravitational radiation background spectrum are important even without a proposed source based solely on serendipity.

3.4. PULSAR TIMING ARRAY EXPERIMENTS

There are a number of pulsar timing array programs underway at major radio observatories around the globe. Any regular monitoring of the strong, millisecond pulsars will contribute to the general data base for global analyses discussed in the previous section. The programs include those at Jodrell Bank in England, Nançay in France, Effelsberg in Germany, Parkes in Australia, Usada in Japan, Green Bank, WV in USA and Arecibo in USA. At a number of these observatories more than one group of investigators are conducting regular timing.

At the NRAO Green Bank site our group initiated a pulsar timing array experiment with the 42m telescope in 1987. Our first results are described in Foster & Backer (1990). The observations are conducted every two months at 800 and 1400 MHz. The two frequencies allow removal of time variable dispersion (Backer *et al.* 1993a). This year we introduced new hardware that will allow higher precision measurements, began sampling a significantly larger set of objects (Table 2), and initiated observations with a 26m 'pulsar monitoring telescope'. We look forward to the commissioning

TABLE 2. Pulsar Timing Array – NRAO Green Bank

Pulsar	Period (ms)	1987	1988	1989	1990	1991	1992	1993	1994
B1821-24	3.1 ms			———	———	———	———	———	———
B1937+21	1.6			———	———	———	———	———	———
B1620-26	11.1				———	———	———	———	———
B1855+09	5.6					———	———	———	———
B1257+12	6.2						———	———	———
J1713+0747	4.6							———	———
J2317+1439	3.4							———	———
J2145-0750	16.1							———	———
J0034-0534	1.9							———	———
J0613-0200	3.1								——
J1730-2304	8.1								——
J0437-4715	5.8								—

of the 100m Green Bank Telescope (GBT) in 1996. The goal is to have timing residuals at or below one microsecond for a modest number of pulsars distributed across the full sky accessible to Green Bank.

4. OUTLOOK

The precise timing of millisecond pulsars has a bright future. New surveys continually find new objects that are independently interesting and augment the pulsar timing array discussed above. New instrumentation is being developed both to search for pulsars with even shorter rotation periods and to investigate known objects with higher precision. Telescopes such as those at Arecibo, Nançay and Westerbork are being refurbished to provide new capabilities for pulsar research. Telescopes such as the GMRT and the GBT which are under construction will be important for pulsar timing and related measurements. A number of propagation phenomena affect pulsar arrival times on short times scales, days to weeks. The time is ripe for a dedicated pulsar monitoring telescope, and some efforts toward this with a 26m antenna are underway. A larger aperture is, of course, important given the weakness of most pulsar signals. Finally radio astronomers are considering the idea of a radio telescope with a square km of collecting area. This would provide spectacular sensitivity for pulsar research at microwave frequencies.

The next decade will then bring: new insights into the origin and evolution of millisecond pulsars; an assessment of the maximum rotation rate of neutron stars; further tests of relativistic gravity including further new constraints on strong field effects; a clearer picture of the distribution and spectrum of interstellar plasma microturbulence; an understanding of the discrete structures in the interstellar plasma; a programmatic link between Terrestrial Time scale and that established by the pulsar timing array; improved masses of the outer planets; an improved tie between the celestial and ecliptic reference frames; and a detection of, or improved limits on, the background of gravitational radiation.

Acknowledgements. This summary is based in part on many excellent presentations at the 1994 Aspen Winter School of Physics – *Millisecond Pulsars : A Decade of Surprise*. I thank my colleagues involved in the NRAO Green Bank pulsar timing array experiment and the Observatory staff. Pulsar research at UC Berkeley is partially supported by NSF grant 90003404 and the NSF Center for Particle Astrophysics.

References

Anderson, S.B. *et al.* 1990, Nat 346, 42
Arzoumanian, Z. *et al.* 1994a, ApJ 422, 671
Arzoumanian, Z., Fruchter, A.S. & Taylor, J.H. 1994b, ApJ 426, L85
Backer, D.C. & Hellings, R.W. 1986, ARA&A 24, 537
Backer, D.C. 1993, in *Galactic High Energy Astrophysics - High Accuracy Timing and Positional Astronomy*, J. van Paradijs & H.M. Maitzen (Eds.), Springer Lecture Notes in Physics, Vol. 418, p. 193.
Backer, D.C. & Thorsett, S.E. 1994, in *Millisecond Pulsars – A Decade of Surprise*, A.S. Fruchter, M. Tavani & D.C. Backer (Eds.), ASP Conf. Proc. (in press)
Backer, D.C., Foster, R.S. & Sallmen, S. 1993a, Nat 365, 817
Backer, D.C. *et al.* 1993b, ApJ 404, 636
Bailyn, C.D. *et al.* 1994, ApJ (submitted)
Bennett, D.P. & Bouchet, F.R. 1991, Phys. Rev. D 43, 2733
Blandford, R.D. & Narayan, R. 1985, MNRAS 213, 591
Blaskiewicz, M. 1993, in *Planets around Pulsars*, J.A. Phillips, S.E. Thorsett & S.R. Kulkarni (Eds.), ASP Conf. Proc. Vol. 36, 43
Caldwell, R.R. & Allen, B. 1992, Phys. Rev. D 45, 3447
Camilo, F., Foster, R.S. & Wolszczan, A. 1994, ApJ (in press)
Camilo, F., Thorsett, S.E. & Kulkarni, S.R. 1994, ApJ 421, L15
Cognard, I. *et al.* 1993, Nat 366, 320

Cognard, I. et al. 1994, (preprint)
Cordes, J.M., Pidwerbetsky, A. & Lovelace, R.V.E. 1986, ApJ 310, 737
Cordes, J.M., Wasserman, I. & Blaskiewicz, M. 1990, ApJ 349, 546
Detweiler, S. 1979, ApJ 234, 1100
Dewey, R.J. & Cordes, J.M. 1989, in *Timing Neutron Stars*, H. Ögelman & E.P.J. van den Heuvel (Eds.), NATO ASI C262, p. 119
Foster, R.S. & Backer, D.C. 1990, ApJ 361, 300
Foster, R.S. & Cordes, J.M. 1990, ApJ 364, 123
Foster, R.S., Fairhead, L. & Backer, D.C. 1991, ApJ 378, 687
Fruchter, A.S. et al. 1990, ApJ 351, 642
Fukushima, T. 1994, A&A (submittted)
Guinot, B. & Petit, G. 1991, A&A 248, 292
Helfand, D.J., Manchester, R.N. & Taylor, J.H. 1975 ApJ 198, 661
Hellings, R.W. 1990, in *Proceedings Workshop on Impact of Pulsar Timing on Relativity & Cosmology*, D. Backer (Ed.), p. k1
Hu, W., Romani, R.W. & Stinebring, D.R. 1991, ApJ 366, L33
Kaspi, V.M. 1994, in *Millisecond Pulsars – A Decade of Surprise*, A.S. Fruchter, M. Tavani & D.C. Backer (Eds.), ASP Conf. Proc. (in press)
Kaspi, V.M., Taylor, J.H. & Ryba, M.F. 1994, ApJ 428, 713
Lestrade, J.-F. 1994, in *Millisecond Pulsars – A Decade of Surprise*, A.S. Fruchter, M. Tavani & D.C. Backer (Eds.), ASP Conf. Proc., (in press)
Lyne, A.G. et al. 1988, Nat 332, 45
Nice, D.J. & Taylor, J.H. 1994, ApJ (submitted)
Phillips, J.A. & Wolszczan, A. 1992, ApJ 385, 273
Phinney, E.S. 1992, Phil. Trans. Roy. Soc. Lond. A 341, 39
Phinney, E.S. & Kulkarni, S.R. 1994, ARA&A 32, 591
Rasio, F.A. 1994, ApJ 427, L107
Rajagopal, M. & Romani, R.W. 1994, ApJ (submitted)
Romani, R.W. 1988, Phys. Lett. B 215, 477
Ryba, M.F. & Taylor, J.H. 1991, ApJ 371, 739
Seidelmann, P.K. & Fukushima, T. 1992, A&A 265, 833
Shapiro, I.I. 1964, Phys. Rev. Lett. 13, 789
Sigurdsson, S. 1993, ApJ 415, L43
Standish, E.M. Jr. 1990a, A&A 233, 252
Standish, E.M. Jr. 1990b, in *Proceedings Workshop on Impact of Pulsar Timing on Relativity & Cosmology*, D. Backer (Ed.), p. R1
Standish, E.M. Jr. 1990c, A&A 233, 272
Suleymanova, S.A. & Shitov, Y.P. 1994, ApJ 422, L17
Taylor, J.H. 1992, Phil. Trans. Roy. Soc. Lond. A 341, 117
Taylor, J.H. et al. 1992, Nat 355, 132
Thorsett, S.E., Arzoumanian, Z. & Taylor, J.H. 1993a, ApJ 412, L33
Thorsett, S.E. et al. 1993b, ApJ 405, L29
Vilenkin, A. 1981, Phys. Lett. B107, 47
Weisberg, J.M., Romani, R.W. & Taylor, J.H. 1989, ApJ 347, 1030
Wolszczan, A. 1994, Science 264, 538

PULSAR VELOCITIES

MATTHEW BAILES

Australia Telescope National Facility, CSIRO
P.O. Box 76
Epping, N.S.W.
2121 Australia

Abstract. Lyne & Lorimer (1994) recently demonstrated that revisions to the pulsar distance scale, coupled with new interferometric measurements of pulsar proper motions and a better treatment of selection effects, indicate that typical pulsar velocities are of the order $450\,\mathrm{km\,s^{-1}}$. This is between a factor of 2–4 greater than most estimates made over the last decade. This paper looks at the implications of these higher velocities for the various theories about their origin. An extremely simple argument is used to place a fairly rigid upper limit for the rate at which neutron star pairs merge of $10^{-5}\,\mathrm{yr^{-1}}$ in the Galaxy. It appears inevitable that an extremely large fraction of binaries containing neutron stars coalesce during the common-envelope stage of massive binary evolution.

1. Introduction

Early work on the observed pulsar distribution led Gunn & Ostriker (1970) to propose that pulsars are high-velocity objects, with typical velocities of $\sim 100\,\mathrm{km\,s^{-1}}$. In the 1980s many pulsars had their velocities measured using interferometric (Lyne, Anderson & Salter 1982) or scintillation techniques (Cordes 1986). These measurements confirmed that pulsars often possess velocities of several hundred $\mathrm{km\,s^{-1}}$, much greater than that of their progenitors, the OB stars, whose typical velocities are a few tens of $\mathrm{km\,s^{-1}}$. How do pulsars achieve such high velocities?

2. The Observational Data

For many years, the Lyne et al. (1982) interferometric study provided the only reasonable sample of pulsar proper motions and hence velocities. The study measured the proper motions or meaningful upper limits for 26 pulsars. However, as much information was derived from the direction of the pulsar proper motion vectors as from their magnitude, and these data provided the basis for a number of fundamental conclusions concerning radio pulsars, their evolution and kinematics. These are listed below:

- Radio pulsars have large transverse velocities with an rms of $\sim 170\,\mathrm{km\,s^{-1}}$, and a mean of $130\,\mathrm{km\,s^{-1}}$.
- Most pulsars are leaving the Galactic plane.
- The "kinetic ages" of pulsars are usually less than the spin-down age, indicating that pulsar magnetic fields decay.
- A correlation exists between the transverse velocity and magnetic field strength (Anderson & Lyne 1983).
- There is no preferential orientation of the pulsar spin and velocity vectors.

Cordes (1986) used the scintillation properties of pulsars to estimate the velocities of 69 pulsars. The mean velocity he obtained for pulsars was closer to $100\,\mathrm{km\,s^{-1}}$. In Fig. 1, the estimated distribution of pulsar velocities at several epochs are shown. In an important paper, Harrison & Lyne (1993) demonstrated that the difference between the mean velocities of pulsars with velocities measured by interferometric and scintillation techniques could be understood if the scattering screen was not assumed to be midway between the pulsar and the Sun, but rather considerably closer to the Galactic plane, as might be expected. This explains the difference between the velocity distribution obtained from the two techniques.

In the late 1980s there were three independent pulsar proper-motion surveys commenced (Bailes et al. 1990; Fomalont et al. 1992; Harrison, Lyne & Anderson 1993), more than doubling the number of pulsars with known proper motions; this was largely due to the Harrison et al. (1993) survey. Pulsars for such proper-motion surveys were generally chosen on the basis of distance and flux, the former allowing measurement of proper motion on a reasonable time scale, and the latter providing adequate signal-to-noise ratio. These more recent surveys have measured the proper motions of pulsars more distant from the Sun, and consequently further from the Galactic plane than the earlier surveys. Restricting the sample to local objects in effect excludes old, high-velocity pulsars. Compensation for this increases the mean speed by a factor of 1.2. This effect was first recognised by Cordes (1986) and recently quantified by Lyne & Lorimer (1994).

The latest pulsar distance model of Taylor & Cordes (1993) is much more elaborate than that of Lyne, Manchester & Taylor (1985). The need for a new distance model was emphasized by the discovery of the globular cluster pulsars, for which the existing model significantly underestimated distances. The new model also attempts to address the enhanced free electron density associated with spiral arms, and the tendency for the old model to underestimate the distances to nearby pulsars. Lyne & Lorimer have demonstrated that adopting the new pulsar distance model increased the mean velocity of pulsars by a factor of 1.6.

The two effects mentioned above account for the increase in mean transverse velocity shown in Fig. 1. Lyne & Lorimer (1994) argue that the observed pulsar velocity distribution is biased by the inclusion of old, low-velocity pulsars. Old, high-velocity pulsars do not generally appear in pulsar proper motion surveys as they reside far from the Galactic plane. By removing all pulsars with ages >3 Myr we end up with the "unbiased" transverse velocity distribution shown in Fig. 1d. This distribution has a mean of $370 \, \text{km s}^{-1}$, almost a factor of four greater than the mean implied by the Cordes (1986) study.

How believable is this new velocity distribution? It is disturbing to see the assumed velocity distribution change so drastically over just a few years. I believe that there are several factors which might be biasing the velocity distribution toward higher velocities. One is that any simple distance model will inevitably overestimate the distances to some pulsars and underestimate the distance to others. It may be that no simple distance model is appropriate to our Galaxy, and that several pulsars have their distances and hence velocities (derived from proper motions), greatly overestimated. Another factor is that proper motions measured are the vector addition of the true proper motion, the measurement error and any systematic errors. The vector nature of these quantities usually makes the derived proper motion greater than the true proper motion. Nevertheless, unless the new distance model is severely in error, it appears that radio pulsars often have large velocities, and that when we take into account the unknown radial component of velocity, a mean of $400 \, \text{km s}^{-1}$ is not unreasonable. Frail (these proceedings), using the displacement of pulsars associated with supernova remnants, obtains a similar mean velocity to that derived by Lyne & Lorimer.

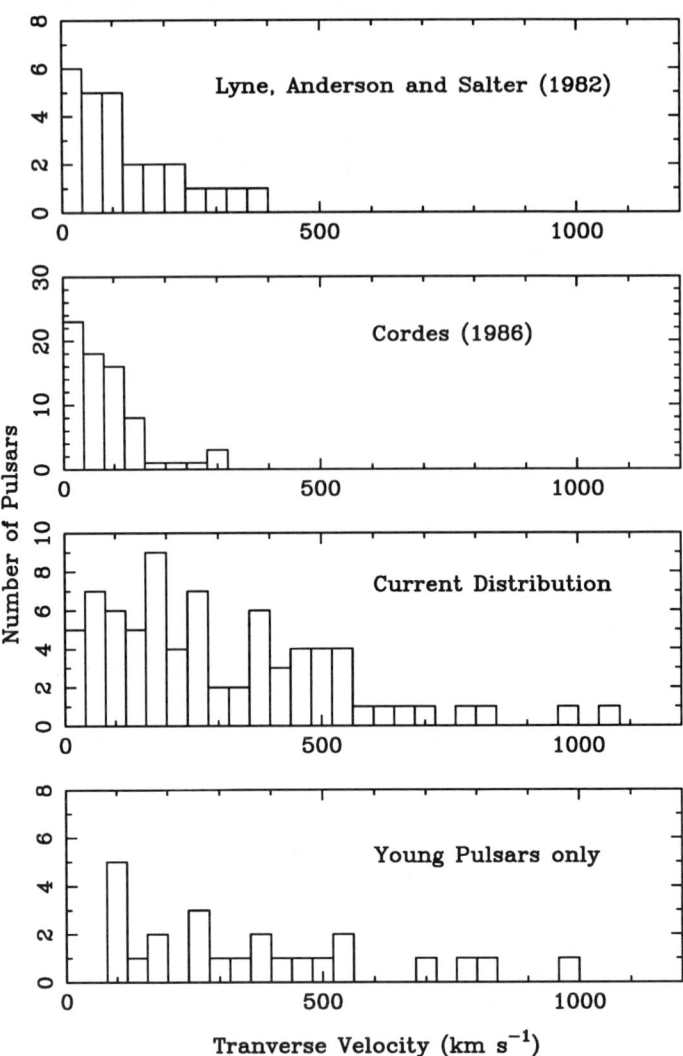

Figure 1. The changing distribution of pulsar transverse velocities. From the top: the distribution derived from the interferometric observations of Lyne, Anderson & Salter (1982), the scintillation velocities of Cordes (1986), the most recent distribution using the latest pulsar distance model, and the distribution if we restrict the sample to pulsars younger than 3 Myr.

3. The Origin of the Velocities

It has been proposed that the large velocities of radio pulsars occur because neutron stars often appear in tight binaries which disrupt during the supernova explosion that produces the pulsar. The velocities are therefore just a product of the pre-supernova conditions. This elegant idea, put forward by Gott, Gunn & Ostriker (1970), was largely championed by V. Radhakrishnan and co-workers in the 1980s. Unfortunately, the vast majority of pulsars are not members of binary systems, and hence what constitutes a typical pre-supernova binary is hard to determine.

In an early paper, before the appearance of much reasonable proper motion data, Radhakrishnan & Srinivasan (1981) recognised the important role that a companion could play in the final appearance of a pulsar. They used evolutionary arguments to identify pulsars which may have been "processed" in a binary system. The binary pulsar PSR 1913+16 is generally accepted to have accreted matter from a companion, and this is thought to explain its weaker than average magnetic field strength and shorter spin period. Radhakrishnan & Srinivasan used some fundamental physical arguments to demonstrate that the first-born pulsars in massive binaries should have space velocities greater than that of the second-born pulsars if there were no additional "kicks" imparted to the pulsars at birth. PSR 1913+16 was used as an example of a first-born pulsar. They identified two other pulsars, PSR 1804−08 and PSR 1541−52, as weak-field, and hence the first-born pulsars in binaries which disrupted at the time of the second supernova explosion. They bravely predicted that these pulsars should have large proper motions.

Ultimately it was shown that weak-field pulsars appeared to have lower velocities than strong-field ones (Anderson & Lyne 1983), and even though neither PSR 1804−08 nor 1541−52 had meaningful upper limits on their velocities, the appearance of the velocity-magnetic moment correlation caused extensive revision to be made to the binary break-up model. Radhakrishnan (1984) abandoned the idea that massive binaries produce a fast, weak-field pulsar and a slow, strong-field pulsar. Instead, he introduced the idea that some binaries containing a neutron star and a main-sequence star become "completely tidally disrupted", leaving behind a solitary neutron star, with a low field strength and velocity. In this revised model, massive binaries produced two high-velocity pulsars, and lower-mass binaries formed slower, weak-field pulsars. This could therefore reproduce the correlation observed between velocity and magnetic moment.

Dewey & Cordes (1987) made very detailed simulations of the pulsar population to examine whether the observed pulsar velocity distribution could be explained without the use of some acceleration mechanism or "asymmetric kick" such as that proposed by Shklovskii (1970). Their conclusion was that it could not, because there were too many low-velocity pulsars in any synthesized population compared with that of the real pulsars. They chose the velocity distribution well fitted by Cordes' scintillation measurements, with a mean velocity of only $100\,\mathrm{km\,s^{-1}}$. Now that we believe that this underestimates the true velocities of pulsars by a factor of ~ 4, the conclusions of Dewey & Cordes are strongly reinforced. The real issue here though is what fraction of pulsars are assumed to come from single stars. Dewey & Cordes assumed that the fraction was close to 50%. It is trivial to show that 50% of pulsars do not have velocities as small as the OB stars. If we hypothesize that a much larger fraction of pulsars comes from binaries than simple measurements of the binary fraction of OB stars suggest, then the argument against the binary break-up model is not so simple. More compelling evidence that pulsars receive some sort of velocity kick comes from recent evidence that the spin axis is misaligned with the orbital angular-momentum axis in PSR 1913+16 (Weisberg, Romani & Taylor 1989), and that some young pulsars have extremely high velocities of the order of $1000\,\mathrm{km\,s^{-1}}$ (Cordes, Romani & Lundgren 1993; Frail & Kulkarni 1991). Radhakrishnan & Shukre (1985) have shown that it is difficult to accelerate a pulsar using binaries to velocities much greater than $500\,\mathrm{km\,s^{-1}}$ without the system becoming too compact before the final supernova explosion.

Since the paper of Dewey & Cordes (1987), most of the effort in the pulsar velocity debate has gone into explaining the correlation between velocity and magnetic moment. Bailes (1989) determined the velocities of the first- and second-born neutron stars in binaries similar to those Dewey & Cordes had assumed, to demonstrate that, even with kicks, tight binaries would produce an anti-correlation between velocity and magnetic moment and far too many binary pulsars. My conclusion at the time was that most binaries disrupt whilst still wide, probably by avoiding the common-envelope phase. Bhattacharya & Van den Heuvel (1992) suggested an elegant solution to the problem in which they postulated that all pulsars received kicks, but many binaries remain bound after the first supernova explosion and coalesce *during* the common-envelope phase. This model has many attractive features. First, it avoids the production of too many binary pulsars. Second, the common-envelope phase does not have to be avoided. Third, it is still possible to get low-velocity pulsars which have accreted matter and may have slightly weaker fields as a result, producing a very weak correlation between velocity and magnetic field strength, as is observed.

Recently, Camilo, Nice & Taylor (1993) reported the discovery of a distant 60 ms pulsar, with remarkably similar spin period and field strength to that of PSR 1913+16. The z-height of this pulsar is large ($\sim 1\,\mathrm{kpc}$) and it seems very likely that it has a large velocity to have attained such a height. It appears therefore, that with the discovery of a weak-field, high-velocity pulsar, the original prediction of Radhakrishnan & Shukre (1985) has been finally fulfilled. It is fascinating to see how the correlation led Radhakrishnan (1984) to propose the coalescence idea to preserve the binary break-up model only to have it incorporated into the rival model.

An offset dipole could cause a radio pulsar to accelerate to speeds of up to several $100\,\mathrm{km\,s^{-1}}$ from asymmetric radiation of the spin-down energy (Harrison & Tademaru 1975). The spin vector and proper motion vector should be aligned if this mechanism is responsible for pulsar acceleration. Anderson & Lyne (1983) dismissed the rocket mechanism as there was no correlation between the spin and velocity vectors. In the light of our new-found doubts about field decay (Bailes 1989; Bhattacharya et al. 1992), the true ages of radio pulsars can easily exceed the gravitational oscillation period of the disk. Thus, when comparing the alignment between the spin axes and proper motions of pulsars, old pulsars in which the Galactic potential has had time to act should be excluded. A revised study, which omits all old pulsars could be the final word on whether the rocket mechanism could be the pulsar acceleration mechanism.

One interesting thing to come out of the recent velocity measurements is that the apparent correlation between velocity and magnetic field strength has weakened. Although it is still statistically significant, the scatter in the data is enormous. Lorimer (1994) has shown that there is about a 7% chance that it is totally spurious. A small class of low-field, low-velocity pulsars could entirely explain the observed "correlation".

4. The Paucity of Neutron Star Binaries

Although there are now a few dozen binary pulsars known, most of these contain millisecond pulsars. There are only three convincing neutron star pairs in the Galactic disk, PSRs 1913+16, 1534+12 and 2303+46. It is possible that PSR 1820–11 contains two neutron stars but this is hard to establish. It seems incredible that such a large fraction of massive stars are members of binaries and yet so few neutron star pairs exist. If half of all massive stars are members of binaries, then over 99% of systems must either disrupt or coalesce to avoid the production of too many binary pulsars. With two supernova explosions, one might imagine that disruption probabilities of 90% per explosion might suffice, but even with large kick velocities such probabilities are hard to attain. A further complication is that it is vital to retain enough massive binaries to explain the incidence of massive X-ray binaries.

The problem with kicks is that the fraction of pulsars we expect to be members of binaries becomes very large when the kick velocity is of similar magnitude to the relative orbital velocity of any pre-supernova binary (Bailes 1989). This is because many pulsars get kicked into bound orbits. The only way to avoid this is to keep binaries very wide, or to have them coalesce during the common-envelope phase.

We only know of two systems which are capable of coalescing in less than a Hubble time, PSRs 1913+16 and 1534+12. In both cases the spin periods and field strengths of the pulsars place them far from the general pulsar population on the magnetic-field, spin-period diagram - at the location we would expect from the effects of mass transfer. It is with some confidence therefore that we identify them as the first-born pulsars in the binary, which was subsequently spun up during the giant phase of the companion. If we believe the unbiased transverse-velocity distribution shown in Fig. 1, and there are no "kicks", almost all radio pulsars must originate from very tight binaries prior to the second explosion. Since the only binary pulsars which would have been tight enough to produce *two* high-velocity neutron stars have, at most, a few single pulsars similar to them in the observed sample, we can produce the following argument against symmetric supernova explosions:

If almost all pulsars emerge from tight binaries similar to the progenitors of PSRs 1913+16 and 1534+12 and over 99% of them disrupt to achieve the correct fraction of binary pulsars, why is the single pulsar population not inundated with high-velocity pulsars with similar spin periods and magnetic fields to the two pulsars mentioned above?

The only satisfactory solution is that pre-supernova binaries that lead to 1913+16-like systems must be extremely rare. I conclude that close binaries at the time of the second supernova must be avoided by the complete spiral-in of neutron stars into their massive companions. R. Taam reaches similar conclusions from entirely different reasoning in these proceedings.

How rare are 1913+16-like systems and their progenitors? The answer is crucial, because these systems are the only guaranteed source of gravitational waves strong enough to be detected by the next generation of detectors. Several authors have tried to answer this question by extrapolating the population of the two known neutron star-neutron star binaries, first to the rest of the Galaxy, and then to the universe (Phinney 1991; Narayan, Piran & Shemi 1991). These studies obtain "conservative" values for the coalescence rate of neutron star pairs of 10^{-6} yr^{-1} for the Galaxy. Both PSR 1913+16 and PSR 1534+12 have weak fields and long radio lifetimes. Recently Van den Heuvel (1992) suggested that these rates are far too conservative and should be multiplied by a factor of 100 to take into account all of the neutron star binaries with much shorter radio lifetimes than the recycled pulsars we see. He claims that the coalescence rate should be nearer 10^{-4} yr^{-1}. The implication is that we can observe only one in every hundred neutron star pairs.

The easiest way to counter such an enormous population of invisible neutron star pairs is to consider, not the recycled pulsar in the binary, but rather the second-born or "normal" pulsar. Fortunately, the second-born pulsar has no peculiar characteristics, and should simply resemble the 650 or so other normal pulsars we know of. The statistics of these are much more certain than those of the two known neutron star binaries. Lorimer et al. (1993) estimate the birth rate of normal pulsars to be 1/125–1/250 yr^{-1}. A salient fact exists. The birth rate of normal pulsars in neutron star binary pairs must equal the birth rate of recycled pulsars in similar systems. We do not know of a single "normal" pulsar in a neutron star binary that will coalesce in a Hubble time. Therefore the birth rate of relativistic binary pulsars is less than one in 650 times the maximum pulsar birth rate, or

$$\frac{1}{650} \times \frac{1}{125} \sim 10^{-5} \, \text{yr}^{-1}. \tag{1}$$

This is a factor of ten less than the birth rate suggested by Van den Heuvel and is based upon 650 objects, not two.

If we take the above conservative upper limit on the birth rate of coalescing neutron star binaries and combine it with the conservative lower limits of previous workers, the coalescence rate R is starting to be reasonably well constrained with 10^{-6} yr^{-1} < R < 10^{-5} yr^{-1}. However, the earlier estimates were based upon the fact that a certain fraction of the Galaxy had been searched for such objects and only two had been found. Since then, the amount of the Galaxy that has been extensively surveyed for short-period pulsars has increased dramatically, with *no* more relativistic binaries discovered, despite a five-fold increase in the number of millisecond pulsars. The first-order correction to the birth rate estimates of neutron star pairs is to decrease the birth rate of relativistic binaries by a factor of five to $\sim 2 \times 10^{-7}$ yr^{-1}.

We conclude that, although the birth rate of relativistic binary pulsars in the Galaxy is uncertain, it is probably between 2×10^{-7} and 10^{-5} yr^{-1}.

5. Conclusions

Radio pulsars have large space velocities which are probably obtained as the result of asymmetry in the explosion which produced the pulsar. It is difficult to disrupt binaries routinely where the kick is of the same order as the relative velocity of the binary constituents, and the paucity of neutron star binaries therefore dictates that very few binaries can resemble the progenitors of PSRs 1913+16 and 1534+12 before the final explosion. If massive binaries with neutron star companions usually coalesce during the common-envelope phase of their evolution the small fraction of binary pulsars can be understood. The end-product of such evolution is open to debate, but may produce solitary millisecond pulsars, slow weak-field pulsars or completely invisible neutron stars. The coalescence rate of neutron star binary pairs can be constrained by the complete lack of any "normal" pulsars known in such systems.

Acknowledgments. I thank V. Radhakrishnan, E.P.J. van den Heuvel, S.R. Kulkarni, A.G. Lyne, R.J. Dewey, J. Cordes and R.N. Manchester for many stimulating discussions over the years on this subject.

References

Anderson, B. & Lyne, A.G. 1983, Nat 303, 597
Bailes, M. 1989, ApJ 342, 917
Bailes, M. et al. 1990, MNRAS 247, 322
Bhattacharya, D. & Van den Heuvel, E.P.J. 1991, Phys. Reports 203, 1
Bhattacharya, D. et al. 1992, A&A 254, 198
Camilo, F., Nice, D.J. & Taylor, J.H. 1993, ApJ 412, L37
Cordes, J.M. 1986, ApJ 311, 183
Cordes, J.M., Romani, R.W. & Lundgren, S.C. 1993, Nat 362, 133
Dewey, R.J. & Cordes, J.M. 1987, ApJ 321, 780
Fomalont, E.B. et al. 1992, MNRAS 258, 497
Frail, D.A. & Kulkarni, S.R. 1991, Nat 352, 785
Gott, J.R., Gunn, J.E. & Ostriker, J.P. 1970, ApJ 160, L91
Gunn, J.E. & Ostriker, J.P. 1970, ApJ 160, 979
Harrison, E.R. & Tademaru, E. 1975, ApJ 201, 447
Harrison, P.A. & Lyne, A.G. 1993 MNRAS, 265, 778
Harrison, P.A., Lyne, A.G. & Anderson, B. 1993, MNRAS 261, 113
Lorimer, D.R. 1994, PhD thesis, The University of Manchester
Lorimer, D.R. et al. 1993, MNRAS 263, 403
Lyne, A.G. & Lorimer, D.R. 1994, Nat 369, 127
Lyne, A.G., Anderson, B. & Salter, M.J. 1982, MNRAS 201, 503
Lyne, A.G., Manchester, R.N. & Taylor, J.H. 1985, MNRAS 213, 613
Narayan, R., Piran, T. & Shemi, A. 1991, ApJ 379, L17
Phinney, E.S. 1991, ApJ 380, L17
Radhakrishnan, V. 1984, in *Millisecond Pulsars*, S.P. Reynolds & D.R. Stinebring (Eds.), NRAO (Green Bank), p. 130
Radhakrishnan, V. & Shukre, C.S. 1985, in *Supernovae, Their Progenitors and Remnants*, G. Srinivasan & V. Radhakrishnan (Eds.), Indian Academy of Sciences (Bangalore), p. 155
Radhakrishnan, V. & Srinivasan, G. 1981, in *Proc. 2nd Asian–Pacific Regional Meeting of the IAU*, B. Hidayat & M.W. Feast (Eds.), Tira Pustaka (Jakarta), p. 423
Shklovskii, I.S. 1970, Astr. Zh. 46, 715
Taylor, J.H. & Cordes, J.M. 1993, ApJ 411, 674
Van den Heuvel, E.P.J. 1992, in *X-ray Binaries and Recycled Pulsars*, E.P.J. van den Heuvel & S.A. Rappaport (Eds.), Kluwer (Dordrecht), p. 233
Weisberg, J.M., Romani, R.W. & Taylor, J.H. 1989, ApJ 347, 1030

Discussion

J. van Paradijs: Nick White and I (1995, ApJ, submitted) have recently calculated the galactic distribution of LMXB for several evolutionary scenarios, both with and without kick velocities (over and above that expected from sudden mass loss). We find that this extra kick is necessary to account for the distance of LMXB from the galactic plane; the Lyne and Lorimer (1994) velocity distribution – taken as kick velocity distribution – given a reasonable agreement with the observed LMXB distribution.

M. Bailes: There are a large number of assumptions one needs to make in order to follow the evolution of the system from post-explosion to LMXB, I suspect that within the parameter space there is room for kicks from Lyne and Lorimer (1994), to say a factor of two lower than this.

V. Lipunov: What is the 3σ accuracy of the mean velocity?

M. Bailes: The Lyne and Lorimer (1994) estimate is $450\pm90\,\mathrm{km\,s^{-1}}$ where $90\,\mathrm{km\,s^{-1}}$ is the statistical uncertainty only (1σ). If you wanted to say that the distance scale is wrong by 30% you could increase this 1σ uncertainty to $\sim 120\,\mathrm{km\,s^{-1}}$. Extending this to 3σ gives an enormous value of $450\pm360\,\mathrm{km\,s^{-1}}$.

A REVIEW OF GALACTIC MILLISECOND PULSAR SEARCHES

A. G. LYNE
The University of Manchester,
Nuffield Radio Astronomy Laboratories,
Jodrell Bank, Macclesfield,
Cheshire, SK11 9DL, UK.

Abstract. Since the discovery of the first millisecond pulsar in 1982, the main discoveries of more of these exotic objects have come in two bursts. The first followed the realization late in the decade that they might be formed readily in globular clusters and resulted in the discovery of about two dozen objects in these clusters between 1987 and 1990. The second burst started shortly after this as improved computer technology permitted widespread searches of the Galaxy with high sensitivity and is continuing at the present time. This paper describes the main successful surveys for these galactic millisecond pulsars of which 29 are now known. 21 of these are in circular orbits with low-mass companion stars. The large number of objects now available allows a number of firm statistical trends to be seen which all point to an evolutionary phase of these systems involving the spin-up of old neutron stars during the accretion of matter from less evolved companions during their giant phase. Surveys of the present generation are still very insensitive to short period pulsars or highly accelerated pulsars in close or massive binary systems, limitations which should be removed in future years.

1. Introduction

In 1982, the first millisecond pulsar (MSP) was discovered by Backer *et al.* In spite of intense efforts during the next few years, only 2 more had been discovered by 1987. PSR B1953+29 was found by Boriakoff *et al.* (1983) and PSR B1855+09 by Stokes *et al.* (1986). At the time, the lack of discoveries suggested that they were rare objects in the sky.

The main reason for the small number in fact was that the searches were severely limited by the data storage and computer technology available at that time. The problem lies in the dispersion of the pulsed signals by the ionized component of the interstellar medium, preventing the search process being simply a periodicity search of a time sequence. Because the pulses at low frequencies are delayed relative to higher frequencies, recording a single, wide receiver band, which is required to provide high sensitivity, results in narrow pulses being broadened, possibly to such an extent that they cover more than the whole pulse period so that no periodicity would be observed at all. The most effective way of combating this is to split the receiver band into many narrow channels and to record the output of all of these in order to permit subsequent searches over many values of dispersion measure. Higher time resolution requires the use of narrower filters, so that, roughly speaking, in order to cope with dispersion, the size of the data sets and the magnitude of the processing tasks both increase as the square of the sampling rate. Thus, moving from the earlier searches for normal pulsars, which sampled at about 100 Hz, to the millisecond pulsar searches sampling at 3 kHz required 3 orders of magnitude increase in computing resources.

Typical sampling rates in modern searches are 3–4 kHz for each of several hundred frequency channels, amounting to approximately 1 million data samples/second. For instance a 2.5 minute observation of the present Parkes survey described below would result in a data set of 128 million data points to be stored and processed. It would have taken in excess of one day to process such a dataset on a Vax 780, the standard computer available to astronomers 10 years ago. The small number of millisecond pulsars that were found in these early years were mostly discovered with the aid of super computers.

Because of these limitations, large scale surveys of the sky were impossible in the 1980s. However, it was realised that the origin of the high rotation rate of MSPs might lie in the spin up of old, ordinary pulsars during mass transfer in the accretion of material from a less evolved companion star in its giant phase (Fabian, Pringle & Rees 1975; Alpar et al. 1982). Such objects are believed to be witnessed in the low-mass X-ray binary systems (LMXBs). The large number of LMXBs seen in globular clusters suggested the possibility of many spun-up neutron stars in these clusters. Subsequently, in 1987, the first MSP was found in the core of a globular cluster, M28 (Lyne et al. 1987), and in the next few years a total of 34 pulsars were found in globular clusters. These searches were successful because only a few dozen telescope pointings were required and the necessary computations were relatively modest. Most of these discoveries were MSPs, the majority of which were in binary systems, supporting the spin-up hypothesis. While these MSPs provided much information about stellar interactions within clusters and provided a number of interesting systems to study, they could not be used for their great intrinsic precision as clocks since they were by no means in inertial frames and they gave no indication of how widespread the MSP phenomenon might be in the Galaxy. See Lyne (1994) for a review of globular-cluster searches which will not be discussed further here.

2. New All Sky Surveys

By 1992.0, only 5 MSPs were known outside the globular-cluster system, but this figure has now (1994.5) risen to 29. The reasons for this significant advance are two-fold. Firstly the discovery of PSR B1257+12 by Wolszczan & Frail (1992) at high galactic latitude suggested that MSPs were more isotropic than normal pulsars (Johnston & Bailes 1991), so encouraging searches away from the galactic plane, where the earlier limited searches had been conducted. Secondly, computing resources available to astronomers had increased by about 2 orders of magnitude over the decade since PSR B1937+21 was discovered. Groups at the three main observatories of Arecibo, Parkes and Jodrell Bank have been active in the recent discoveries. There are also two other surveys of the northern hemisphere being carried out by the Princeton group, using telescopes at Cambridge and Green Bank.

The groups active at Arecibo all used essentially the same observing system, consisting of receivers using an 8 MHz bandwidth centered on 430 MHz. The surveys at Parkes and Jodrell Bank were similar, using cryogenic receivers centered on 436 and 410 MHz respectively. Filterbanks were used in all cases and the systems are summarized in Table 1.

The sensitivities of these surveys are complicated functions of pulsar period and dispersion measure (DM). For periods in excess of 10 ms and for zero DM, the Arecibo and Parkes surveys have sensitivities of 0.9 and 3 mJy on cold sky (i.e., at high galactic latitude). However, even for zero DM, instrumental broadening results in reduction in sensitivity by factors of between about 4 and 10 for objects with period of 1 ms. For higher dispersion measures or shorter periods the sensitivity reduction is even greater. See Camilo (1994) for a fuller technical review of these searches.

Between them, the whole of the sky is being covered by the 3 main surveys. Roughly speaking, the Parkes survey is scheduled to cover the whole sky south of the Equator, the Arecibo experiments to encompass declinations between 0 and +35 degrees while the Jodrell Bank survey covers the region north of about +30 degrees. These surveys are all still in progress but have already proved to be most successful. So far, a total of 29 pulsars have been found and are summarised in Tables 2 and 3. 21 of the 29

TABLE 1. Parameters of 5 Surveys for Millisecond Pulsars.

	Arecibo	Jodrell Bank	Parkes	Cambridge	Green Bank
Telescope	305-m	76-m	64-m	36000-m^2	43-m
Frequency (MHz)	430	408	436	81.5	370
Integration Time (sec)	32.8	314	157	100	134
Bandwidth (MHz)	2x8	2x8	2x32	1	2x40
Channel Width (MHz)	0.25	0.125	0.125	0.004	0.078
Sampling Rate (kHz)	4.0	3.3	3.3	1.3	4
System Temp (K)	62	50	55	–	–
Min flux density (mJy)	0.9	4	3	90	10
Area of Sky (deg^2)	12700	10300	20600	20000	20600
No of beams in area	582000	22000	44000	17000	11300
Area surveyed to 1994.5 (deg^2)	3000	3000	18000	?	?
Fast pulsars found	11	1	16	0	0

objects are in circular orbits with low-mass companions, consistent with the binary accretion process being responsible for the rapid pulsar rotation. Of the 11 pulsars discovered at Arecibo, 5 were discovered in searches prior to 1992, and 5 in the surveys described above. The eleventh, PSR J0751+18, like PSR B1953+29, was found in a survey of the error boxes of a number of gamma-ray sources. It is not clear whether either of these sources is in fact related to a gamma-ray source.

A number of notable pulsars lie within these lists of discoveries. Firstly PSR 1257+12 is an object which appears to have at least 3 bodies of planetary mass in orbit around it (Wolszczan & Frail 1992). Another, PSR J0437-4715 is a 5.75 millisecond pulsar in a circular 5.7-day orbit with a companion of $\sim 0.14\,M_\odot$ (Johnston et al. 1993). It is by far the brightest of all known MSPs and high-quality single pulses have been observed, it is also the closest known millisecond pulsar to the Earth, with a distance of about 150 pc. The proximity of the binary system to Earth offers new opportunity for studies at all wavelengths. Indeed, the companion has been optically identified (Bell, Bailes & Bessell 1993), and X-rays have been detected from the surface of the neutron star (Becker & Trümper 1993). Two of these pulsars show eclipses by the atmosphere of the companion stars, PSRs B1957+20 and J2052-08. The former was the first such system to be found and optically identified (Fruchter, Stinebring & Taylor 1988). Occulted by greater periods at lower frequency, it seems that the occultation is due to absorption in a wind from the companion star which is being ablated by high-energy radiation from the pulsar (Ruderman, Shaham &

TABLE 2. Millisecond Pulsars Discovered at Arecibo.

PSR	P_{rot} ms	P_{orb} days	e	M_{min} $M_\odot$	d kpc	z kpc	$\log(\tau)$ $\log(yr)$	$\log(B)$ $\log(G)$	V_t km s^{-1}	Dis Ref
J0751+18	3.48	0.3	*0.0	.13	2.00	.72				1
J1025+10	16.45	7.8	*0.0	.72						2
B1257+12	6.22				.62	.60	8.95	8.93	276	3
J1713+0747	4.57	67.8	.000075	.28	.89	.38	9.95	8.30	31	4
B1855+09	5.36	12.3	.000022	.25	1.00	.05	9.68	8.49	28	5
B1937+21	1.56				3.58	−.02	8.36	8.62	10	6
B1953+29	6.13	117.3	.000330	.18	5.39	.04	9.52	8.63		7
B1957+20	1.61	0.4	.000040	.02	1.53	−.13	9.17	8.23	206	8
J2019+2425	3.93	76.5	.000111	.32	.91	−.11	9.95	8.26	102	9
J2317+1439	3.44	2.5	.000001	.18	1.89	−1.27	10.3	8.00		10
J2322+2057	4.81				.78	−.47	8.89	8.34	94	9

* - The eccentricity of these systems is small, but not yet measured.

Reference keys:
1: Lundgren, Zepka & Cordes 1995; 2: Camilo 1995; 3: Wolszczan & Frail 1992; 4: Foster, Wolszczan & Camilo 1993; 5: Stokes, Taylor & Dewey 1985; 6: Backer et al. 1982; 7: Boriakoff, Buccheri & Fauci 1983; 8: Fruchter, Stinebring & Taylor 1988; 9: Nice, Taylor & Fruchter 1993; 10: Camilo, Nice & Taylor 1993.

Tavani 1989; Rasio, Shapiro & Teukolsky 1989). This is supported by the observation that the optical brightness of the companion star varies with orbital phase, being brightest as the side illuminated by the pulsar faces the Earth (Kulkarni, Djorgovski & Fruchter 1988). Three other known occulting low-mass systems are all in globular clusters. Both PSR B1957+20 and PSR J0437−4715 show Hα wind nebulae where the pulsar high-energy radiation interacts with the interstellar medium (Kulkarni & Hester 1988; Bell et al. 1995).

3. Statistical Overview

The findings of the two main surveys are roughly consistent with the same space density of MSPs in the local galactic plane. Taking into account the regions of sky so far covered, we find that the Parkes survey is discovering pulsars at a rate of 1 MSP/1000 deg^2 above the limiting flux density of 3 mJy. Because of the greater sensitivity, the Arecibo surveys are finding

TABLE 3. Millisecond Pulsars Discovered at Parkes and Jodrell Bank.

PSR	P_{rot} ms	P_{orb} days	e	M_{min} $M_\odot$	d kpc	z kpc	$\log(\tau)$ $\log(yr)$	$\log(B)$ $\log(G)$	V_t km s^{-1}	Dis Ref
Parkes Millisecond Pulsars										
J0034−0534	1.88	1.6	<.0001	.14	.98	−.91	9.64	8.04		1
J0437−4715	5.76	5.7	.000020	.14	.14	−.09	9.20	8.76	72	2
J0613−0200	3.06	1.2	<.000022	.13	2.19	−.35	9.69	8.26		3
J0712−68	5.49				1.02	−.40				4
J1025−07	5.16				.35	.23				4
J1045−4509	7.47	4.1	.000019	.16	3.24	.69	9.79	8.58		1
J1455−3330	7.99	76.2	.000167	.26	.74	.28	>10.3	<8.3	60	3
J1604−72	14.84	6.3	*0.0	.26	1.65	.43				
J1643−1224	4.62	147.0	.000506	.12	4.86	1.76	9.34	8.60		3
J1730−2304	8.12				.51	.05	>9.6	<8.7	22	3
J1745−11	4.07				.17	.03				
J1804−27	9.34	11	*0.0	.20	1.17	−.06				4
J2052−08	4.51	0.099	*0.0	.03	1.23	−.63				
J2124−3358	4.93				.24	−.17	>9.1	<8.7		4
J2129−57	3.73	6.6	*0.0	.13	2.54	−1.76				
J2145−0750	16.05	6.8	.000021	.43	.50	−.34	>10.1	<8.8	31	1
Jodrell Bank/Caltech Millisecond pulsars										
J0218+4232	2.32	2.0	<0.00002	.17	5.85	−1.76	9.58	8.18		
J1012+5307	5.26	0.6	*0.0	.11	.53	.41				5

Reference keys:
1: Bailes et al. 1994; 2: Johnston et al. 1993; 3: Lorimer et al. 1995; 4 :Lorimer 1994; 5: Nicastro et al. 1995.

pulsars at a rate of about 4 MSP/1000 deg^2 above the limiting flux density of 0.9 mJy. We would expect the observed surface density of pulsars observed by Arecibo to exceed that observed by Parkes by a factor of 3.3 if the pulsars are distributed in an essentially planar population or by a factor of 6 for an isotropic distribution. The observed factor of 4 is within this range and is consistent with what can be seen from the distribution of the pulsars in Tables 3 and 4, that MSPs clearly have a large scale height, albeit with

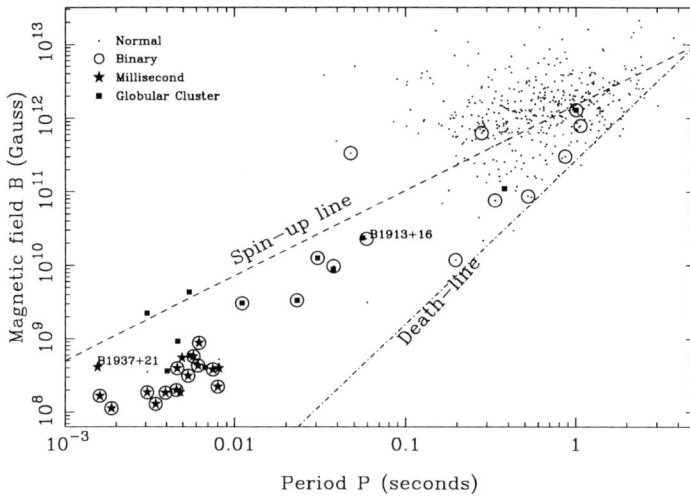

Figure 1. Pulsar magnetic field against rotational period for all known pulsars with period derivative measurements.

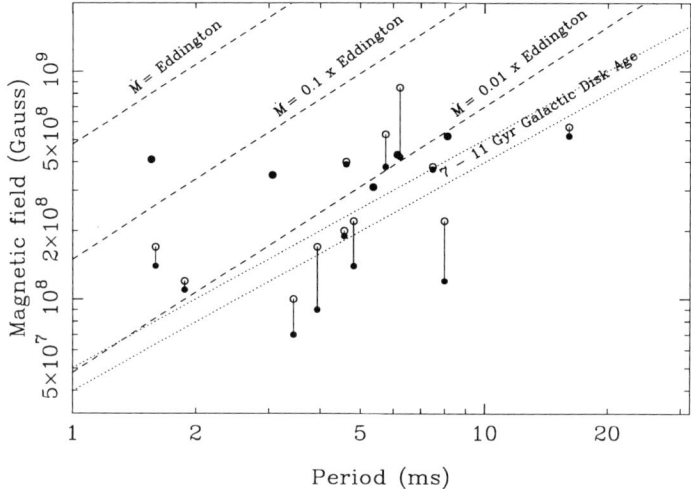

Figure 2. Pulsar magnetic field against rotational period for millisecond pulsars. The open symbols are the fields derived from the measured values of period derivative while the filled ones are derived after correction for the Shklovskii effect (see text). The dotted lines are isochrones delimiting the likely age of the Galaxy.

some concentration towards the Galactic Plane (Lorimer 1995).

All the MSPs have small surface magnetic fields and high ages, as determined from their period derivatives in the normal way (Manchester & Taylor 1977). Fig. 1 shows their position in the $B - P$ diagram relative to the rest of the pulsar population. However, it has been pointed out by

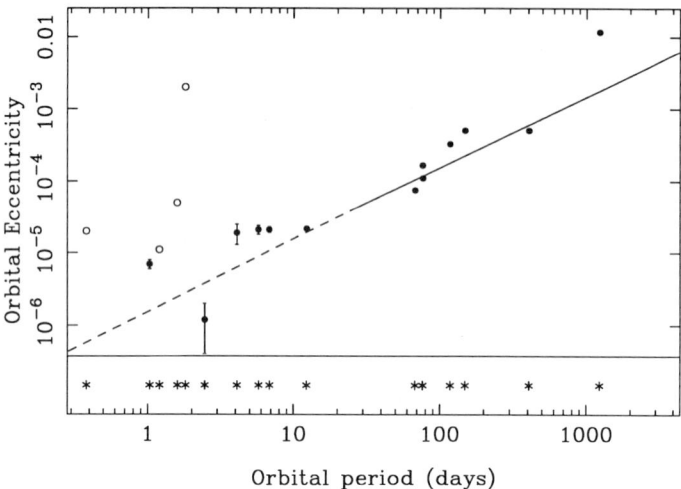

Figure 3. Orbital eccentricity against orbital period for low-mass pulsar binary systems. The open symbols represent orbits with only upper limits to the eccentricity measurement. Along the bottom of the diagram is the distribution in orbital period.

Camilo et al. (1994) that the observed values of derivative, $\dot{P}$, are contaminated by the Shklovskii effect (Shklovskii 1970) which arises from the pulsar transverse speed, V_t, and increases them by an amount $\Delta \dot{P} = P \times V_t^2/cd$, where d is the distance and c is the velocity of light. After correction for this effect, the ages of several pulsars exceed 10^{10} years, in excess of the age of the Galaxy as can be seen in Fig. 2 which displays the lower left-hand corner of Fig. 1 together with corrected values of B for those pulsars for which estimates of their transverse speeds are available (Lorimer 1994). The great ages suggest that either the present periods are similar to those at birth, immediately after spin-up, or else their magnetic fields may have decayed.

There is good evidence that the fast rotation rates are due to processing in binary systems. The fact that the large majority are in circular orbit with low-mass companions supports this hypothesis as does the loose relationship between the spin period and orbital period predicted for binary pulsars (Bailes et al. 1994), similar to the Corbet diagram (Corbet 1984) for high-mass X-ray binaries. Moreover, the evolution of the mass-exchanging systems in which spin up takes place is expected to result in a relation between the final pulsar orbital period and its white-dwarf companion mass (Joss, Rappaport & Lewis 1987). Despite the uncertainty in determining these masses due to the usually unknown orbital inclination, most pulsars in circular low-mass systems do obey this law.

Expected eccentricities are expected to be $\sim 10^{-43}$ or less, due to the

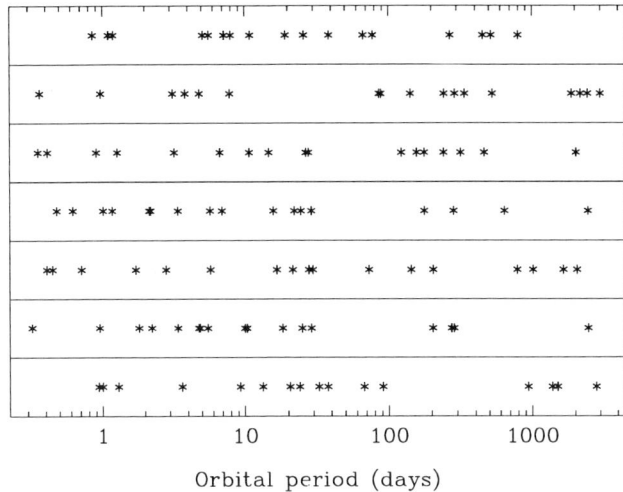

Figure 4. The orbital period distributions resulting from seven consecutive Monte Carlo simulations obtained assuming uniform probability between 0.3 and 3000 days. Most of these show gaps of similar width to that observed in Fig. 3.

circularising effect of the accretion process (Phinney 1993). These values are tiny compared with the observed values which all lie between 10^{-6} and 10^{-2}. Phinney suggests that this might arise from convection in the giant star prior to termination of accretion. He predicts a linear relationship between the eccentricity and orbital period, which is shown in Fig. 3 together with the observed values which follow the prediction extremely well (Lorimer et al. 1995). This provides further good circumstantial evidence for the spin-up scenario for these objects.

The distribution of orbital period is also shown at the bottom of Fig. 3, showing a gap over a factor of about 6, centered on 30 days (Camilo 1994). In the context of the orbital period/companion mass relationship (Joss, Rappaport & Lewis 1987), this would imply a gap in the masses of the companions. A simple Monte Carlo simulation demonstrates that the gap is not significant. Fig. 4 shows 7 successive simulations which assume a uniform probability distribution of 17 periods between 0.3 and 3000 days. In each of these, there is a gap over a similar factor somewhere in the distribution. Clearly, much larger numbers of systems will be required to establish whether the proposed gap in orbital period and hence in companion mass really exists.

4. Conclusion

The surveys discussed above have been remarkably successful at detecting these very weak objects. The MSPs show a large range of extreme astrophysical phenomena and, as clocks, they show incredible rotation stability and promise to act as superb probes of gravitation fields. However, despite these successes, the surveys have very poor sensitivity to any MSPs which have rotational periods of a millisecond or less or any which may be highly accelerated in close or massive binary systems. In the next 10 years or so, computing resources are expected to increase enough to allow us to find out whether such exotic systems exist.

References

Alpar, M.A. et al. 1982, Nat 300, 728
Backer, D. C. et al. 1982, Nat 300, 615
Bailes, M. et al. 1994, ApJ 425, L41
Becker, W. & Trümper, J. 1993, Nat 365, 528
Bell, J.F., Bailes, M. & Bessell, M.S. 1993, Nat 364, 603
Bell, J.F. et al. 1995, ApJ 440, L81
Boriakoff, V., Buccheri, R. & Fauci, F. 1983, Nat 304, 417
Camilo, F. 1994, In: *The Lives of the Neutron Stars* (NATO ASI Series), M.A. Alpar, Ü. Kızıloğlu & J. van Paradijs (Eds.), Kluwer (Dordrecht), p. 243
Camilo, F. 1995, PhD thesis, Princeton University
Camilo, F., Nice, D.J. & Taylor, J.H. 1993, ApJ 412, L37
Camilo, F., Thorsett, S.E. & Kulkarni, S.R. 1994, ApJ 421, L15
Corbet, R.H.D. 1984, A&A 141, 91
Fabian, A.C., Pringle, J.E. & Rees, M.J. 1975, MNRAS 172, 15P
Foster, R.S., Wolszczan, A. & Camilo, F. 1993, ApJ 410, L91
Fruchter, A.S., Stinebring, D.R. & Taylor, J.H. 1988, Nat 333, 237
Johnston, S. & Bailes, M. 1991, MNRAS 252, 277
Johnston, S. et al. 1993, Nat 361, 613
Joss, P.C., Rappaport, S. & Lewis, W. 1987, ApJ 319, 180
Kulkarni, S.R. & Hester, J.J. 1988, Nat 335, 801
Kulkarni, S.R., Djorgovski, S. & Fruchter, A.S. 1988, Nat 334, 504
Lorimer, D.R. 1994, PhD thesis, The University of Manchester
Lorimer, D.R. 1995, these proceedings
Lorimer, D.R. et al. 1995, ApJ 439, 933
Lundgren, S.C., Zepka, A.F. & Cordes, J.M. 1995, ApJ (submitted)
Lyne, A.G. 1994, in *Millisecond Pulsars - A Decade of Surprises*, A. Fruchter, M. Tavani, & D.C. Backer (Eds.), Astronomical Society of the Pacific, (in press)
Lyne, A.G. et al. 1987, Nat 328, 399
Manchester, R.N. & Taylor, J.H. 1977, *Pulsars*, Freeman (San Francisco)
Nicastro, L. et al. 1995, MNRAS (in press)
Nice, D.J., Taylor, J.H. & Fruchter, A.S. 1993, ApJ 402, L49
Phinney, E.S. 1993, Phil. Trans. Roy. Soc. A341, 39
Rasio, F.A., Shapiro, S.L. & Teukolsky, S.A. 1989, ApJ 342, 934
Ruderman, M., Shaham, J. & Tavani, M. 1989, ApJ 336, 507
Shklovskii, I.S. 1970, SvA 13, 562
Stokes, G.H., Taylor, J. & Dewey, R.J. 1985, ApJ 294, L21
Wolszczan, A. & Frail, D.A. 1992, Nat 355, 145

THE LOCAL LOW-MASS BINARY PULSAR POPULATION

D.R. LORIMER
The University of Manchester,
Nuffield Radio Astronomy Laboratories,
Jodrell Bank, Macclesfield,
Cheshire, SK11 9DL, U.K.

Abstract. Using a fully self-consistent approach to account for known survey selection effects, we constrain the number and scale height of low-mass binary pulsars (LMBPs) in the local solar neighbourhood. Our results show that the local surface density of LMBPs with luminosities above $2.5\,\mathrm{mJy\,kpc}^2$ is $\sim 20\,\mathrm{kpc}^{-2}$. Assuming that these are long-lived ($\lesssim 10^{10}\,\mathrm{yr}$) objects, their local birth rate is at least $2\,10^{-9}\,\mathrm{kpc}^{-2}\,\mathrm{yr}^{-1}$. Whilst this is in excellent agreement with the birth rate of their proposed progenitors, the low-mass X-ray binaries, there are several uncertainties involved which could significantly increase our derived birth rate, perhaps by an order of magnitude. Models in which the scale height of LMBPs above the galactic plane exceeds 500 pc are found to be most consistent with the data. The mean space velocity at birth required to produce scale heights of this order in $10^{10}\,\mathrm{yr}$ is found to be $\gtrsim 80\,\mathrm{km\,s}^{-1}$.

1. Introduction

The discovery of binary and millisecond pulsars with distinctly different properties (periods typically $P \lesssim 30$ ms and inferred magnetic field strengths $10^8\,\mathrm{G} \lesssim B \lesssim 10^{10}\,\mathrm{G}$) compared to the normal pulsars poses fundamental questions as to their origin, evolution and galactic population. In the standard binary pulsar formation model, the neutron star is formed during the supernova explosion of the initially more massive star in the binary system and remains bound to its companion. During the first $10^{7-8}\,\mathrm{yr}$ or so, it may be active as a normal radio pulsar spinning down to periods $\gtrsim$ few s. If the companion star is sufficiently massive to evolve into a giant and overflow its Roche lobe, the now old spun down neutron star can gain a new lease of life as a pulsar by spinning up as it accretes mass from the companion (Alpar et al. 1982). This is often referred to as the "recycling" scenario (Radhakrishnan 1982). Such a spin up process is already known to be happening in the X-ray binaries (see for example Rappaport & Joss 1983) making these systems attractive progenitors for the recycled pulsars. Within this framework, the double neutron star binaries like PSR B1913+16 evolve from the massive X-ray binaries (Smarr & Blandford 1976); in contrast, the observationally more numerous millisecond and low-mass binary pulsars (LMBPs) evolve from the low-mass X-ray binaries (LMXBs). In the latter case, the companions are believed to be low-mass ($\lesssim 0.5\,M_\odot$) white-dwarf stars. In this paper, we shall restrict our attention to LMBP statistics; the interested reader should read the review by Bhattacharya (these proceedings) for more details on the formation scenarios.

Despite the general acceptance of the recycling model amongst the pulsar community, one basic link that can (in principle) be tested by pulsar statistics remains poorly established: equality of LMXB-LMBP birth rates. This well known "birth rate problem" was first pointed out by Kulkarni & Narayan (1988), who found that the LMBP birth rate appears to be at least an order of magnitude larger than the LMXBs. The difficulty in accepting this result is small-number statistics (Kulkarni & Narayan's analysis was based on only four LMBPs and was virtually dominated by one millisecond pulsar, B1855+09). The effects of small-number statistics were well demonstrated by Johnston & Bailes (1991), who repeated the analysis with the benefit of a revised distance scale and the completion of two high-frequency searches along the galactic plane. The effect of the new distance scale was to increase the inferred distance and hence our perceived luminosity for PSR B1855+09. Since more luminous pulsars are potentially visible out to larger distances compared with faint ones, the overall effect of B1855+09 on the LMBP birth rate was significantly reduced.

Even in the light of small-number statistics, the birth rate problem has often been advocated as evidence for alternative LMBP formation scenarios such as the accretion induced collapse of a white dwarf (Michel 1987; Grindlay & Bailyn 1988). The recent spate of millisecond pulsar discoveries (see the review by Lyne, these proceedings) means that we now have the prospect of a more reliable evaluation of the underlying population which is the aim of this paper.

2. Model Calculations

There are a large number of selection effects at play in the observed pulsar sample. Most importantly, intrinsically faint pulsars cannot be detected out to very large distances compared to their more luminous counterparts due to the inverse square law and are therefore under-represented in a flux-limited sample. In addition, there are several more subtle deleterious effects on pulsar searches which cause deviations from a simple flux limit: dispersion smearing, sky background noise, interstellar scattering etc. (see Narayan 1987 for further details). In this paper, we are interested in the size and underlying distribution of the local LMBP population and require a correction or "scale factor" to account for those pulsars missed by the surveys as a result of these selection effects.

The method we shall adopt was originally used in an analysis of the normal pulsar population (Phinney & Blandford 1981; Vivekanand & Narayan 1981) and involves computing the volume V in which a pulsar is potentially observable by the present surveys and comparing this with the effective volume V_{max} of the Galaxy. The ratio V_{max}/V serves as a good statistical estimator for the number of "similar" pulsars that exist in the volume V_{max} whose beams intersect our line of sight and we refer to this as the pulsar's scale factor ξ. In practice, we calculate the scale factors numerically, using a detailed computer model of the galactic pulsar population and the major surveys. A full account of the model is given in an earlier paper (Lorimer et al. 1993)

It is interesting to note that the size of the observed sample of LMBPs is approaching that used for the pioneering analysis of the normal pulsars by Gunn & Ostriker (1970) almost 25 years ago. Given the large amount of controversy still inherent in studies of normal-pulsar statistics, we restrict the calculations in this paper to the local solar neighbourhood which is defined by a circle of radius 3 kpc projected on the galactic plane with the Sun at the centre. Since the present all-sky surveys are most sensitive to detecting millisecond pulsars in this region, this has the effect of minimizing any systematic errors present in our calculations. The relevant parameters for the sample of 16 LMBPs that satisfy this criterion are summarised in Table 1.

3. The Local Distribution and Number of LMBPs

The underlying distribution of pulsars that we seed our model galaxy with is initially unknown, forcing us to make an initial guess at the distribution. However, because our scale factor calculation involves recording the parameters of the synthetic pulsars that were theoretically *detectable*, we can make an important consistency check as to the validity of the assumed distributions by comparing the model and true observed pulsars. For simplicity, we have opted for a uniform planar distribution together with an exponential function to model the decrease in pulsar density with increas-

TABLE 1. Assumed parameters for the 16 LMBPs.

PSR	P (ms)	$|z|$ (pc)	L_{436} (mJy kpc^2)	α	W/P (%)	ξ_{500}
J0034−0534	1.88	910	16	−3.0	64	90
J0437−4715	5.76	95	11	−1.5	43	50
J0613−0200	3.06	350	100	−1.6	55	8
J0751+18	3.47	720	40	−2.0	30	12
B0820+02	864.87	520	62	−2.4	4	5
B1257+12	6.22	605	7.8	−2.4	23	40
J1455−3330	7.99	280	6.4	−3.0	32	70
J1713+0747	4.57	380	27	−2.0	9	10
J1730−2304	8.12	55	11	−1.8	25	20
B1831−00	520.95	170	105	−1.5	4	5
B1855+09	5.36	55	15	−1.6	13	15
B1957+20	1.61	125	47	−3.4	4	20
J2019+2425	3.93	105	17	−1.9	20	20
J2145−0750	16.05	335	13	−1.3	30	18
J2317+1439	3.44	1300	80	−2.0	20	7
J2322+2057	4.81	475	2.5	−2.0	12	130

ing height z above the plane. On comparing the model and observed data, we found that the precise form of planar distribution is unimportant to our results (as one would expect for a local population in a steady state); the exponential scale height h_z of pulsars above the galactic plane, however, significantly affects matters. This is well demonstrated in Fig. 1 which shows the variation in Kolmogorov-Smirnov (KS) probability between the model and observed z distributions as a function of scale height h_z.

From Fig. 1, we infer that the most favourable scale height of LMBPs is $\gtrsim 500$ pc. Because of the difficulty of detecting pulsars at large z heights and the uncertainty in the distance scale beyond $z \sim 1$ kpc (Taylor & Cordes 1993), we cannot rule out the possibility of scale heights up to and beyond ~ 1 kpc. Therefore, strictly speaking, this result should be regarded as a lower limit.

Assuming for now that $h_z \sim 500$ pc, in Table 1 we give the scale factor of each LMBP. Summing all these scale factors, leads to an estimate for the local surface density $\Sigma \sim 20$ LMBPs kpc^{-2}. Assuming that LMBPs are long-lived objects with lifetimes $\sim 10^{10}$ yr (Camilo, Thorsett & Kulkarni 1994), we infer a local birth rate of 2 10^{-9} LMBPs kpc^{-2} yr^{-1}, entirely consistent with the local LMXB birth rate of 4 10^{-9} LMXBs kpc^{-2} yr^{-1} (Coté & Pylyser 1989). Taking the uncertainty in h_z into consideration,

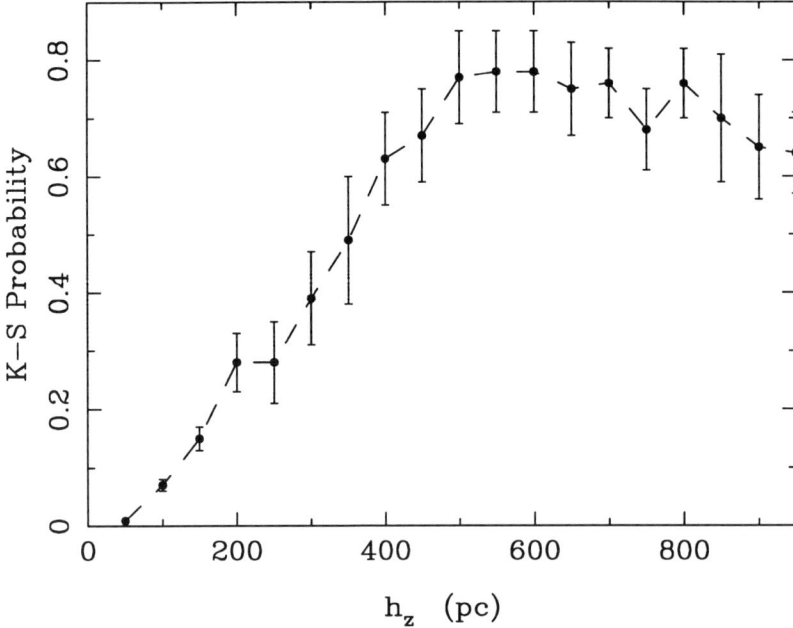

Figure 1. Results of Kolmogorov-Smirnov tests of galactic models of LMBPs with the observed z distribution, as a function of the assumed scale height h_z.

we have found that the inferred local surface density from the scale factor calculations follows a simple linear relationship with h_z: $\Sigma \simeq (h_z/75\,\mathrm{pc})+13$ LMBPs kpc^{-2}.

Whilst our results give strong support for the LMXB-LMBP connection and the standard formation model, there are a number of caveats to be aware of. As discussed previously (Lorimer 1994) the random errors in the method are large (up to a factor of 3), due primarily to distance errors which affect the inferred luminosity and hence scale factor of each pulsar. Systematic errors can occur in several ways. Firstly, the calculation does not take into account the number of pulsars with luminosities *below* the minimum value L_{min} in the observed sample. In the present sample $L_{min} = 2.5\,\mathrm{mJy\,kpc^2}$ set by PSR J2322+2057. The number of pulsars with $L < L_{min}$ is uncertain, but could be comparable to the number with $L > L_{min}$. Secondly, we have made the simplifying assumption that all LMBPs beam to the entire sky. Given the large amount of scatter between different beaming models (see Lorimer et al. 1993) this conclusion should be regarded as provisional. A "worst case" scenario could be a beaming factor of 5. Finally, as mentioned above, there is the possibility that the scale height of LMBPs is $\gtrsim 1$ kpc. From our relationship between the number of LMBPs and h_z, this could increase the scale factors by 30%. All these systematics serve to increase the number and birth rate of LMBPs by quite uncertain factors and it is not impossible that the real LMBP birth rate could be more than an order of magnitude larger than calculated here, making it difficult to see how all LMBPs can form from LMXBs.

4. The Birth Velocity of LMBPs

From our analysis of the present sample, we have demonstrated that the scale height of LMBPs above the galactic plane is $\gtrsim 500$ pc. If we make the assumption that the LMBP were born close to the galactic plane, at a z distance of, say, $\lesssim 100$ pc, they *must* have received a substantial kick velocity at birth in order to attain the present scale height. To estimate the size of the kick velocity, we performed Monte Carlo simulations in which 1000 LMBPs were assigned a one-dimensional (1D) kick velocity away from the galactic plane from a Gaussian distribution and allowed to oscillate in a realistic model of the galactic gravitational potential for up to 10^{10} yr. The z distribution of the steady-state population was then fitted to an exponential whose scale height was compared to $h_z = 500$ pc. Since the true scale height of LMBPs is almost certainly greater than or equal to 500 pc, this allows us to set a lower limit on their birth velocities.

From the results of these simulations we find that, for an initial scale height of 100 pc together with a Gaussian 1D velocity kick with an rms of $50 \,\mathrm{km\, s^{-1}}$, the scale height produced is in excellent agreement with $h_z = 500$ pc. Simulations with lower initial kick velocities produce too many LMBPs at low z heights. Assuming, for simplicity, that the velocity distribution is Maxwellian, the required 3D mean birth velocity is $\gtrsim \sqrt{8/\pi} \times 50 = 80 \,\mathrm{km\, s^{-1}}$. This is relatively low when compared to normal pulsars, for which $V_{\mathrm{birth}} \sim 450 \,\mathrm{km\, s^{-1}}$ (Lyne & Lorimer 1994), but consistent with expectations that LMBPs occupy the low end of the velocity spectrum. Much larger velocities become less likely since they decrease the binary survival probability, reducing the number observed. Our results do not rule out the possibility that LMBPs form via AIC in which kick velocities of $\sim 100 \,\mathrm{km\, s^{-1}}$ are entirely plausible (Woosley, Timmes & Baron 1992). Indeed, given the uncertainties in the LMBP birth rates, we should not discount the idea that a substantial fraction of LMBPs are formed in this way.

Acknowledgements. I thank Matthew Bailes for many useful discussions and Jan van Paradijs for pointing out the possibilities of substantial kick velocities via AIC.

References

Alpar, M.A. *et al.* 1982, Nat 300, 728
Camilo, F., Thorsett, S.E. & Kulkarni, S.R., 1994, ApJ 421, L15
Coté, J. & Pylyser, E.H.P., 1989, A&A 218, 131
Grindlay, J.E. & Bailyn, C.D. 1988, Nat 336, 48
Gunn, J.E. & Ostriker, J.P. 1970, ApJ 160, 979
Johnston, S. & Bailes, M. 1991, MNRAS 252, 277
Kulkarni, S.R. & Narayan, R. 1988, ApJ 335, 755
Lorimer, D.R. 1994, in *Lives of the Neutron Stars*, M.A. Alpar, Ü. Kızıloğlu & J. van Paradijs (Eds.), Kluwer Academic Publishers, (in press)
Lorimer, D.R. *et al.* 1993, MNRAS 263, 403
Lyne, A.G. & Lorimer, D.R. 1994, Nat 369, 127
Michel, F.C., 1987, Nat 329, 310
Narayan, R. 1987, ApJ 319, 162
Phinney, E.S. & Blandford, R.D. 1981, MNRAS 194, 137
Radhakrishnan, V. 1982, Contemp. Phys. 23, 207
Rappaport, S. & Joss, P.C. 1983, in *Accretion-Driven Stellar X-Ray Sources*, W.H.G. Lewin & E.P.J. van den Heuvel (Eds.), Cambridge University Press, p. 1
Smarr, L.L. & Blandford, R. 1976, ApJ 207, 574
Taylor, J.H. & Cordes, J.M. 1993, ApJ 411, 674
Vivekanand, M. & Narayan, R. 1981, J. Astrophys. Astr. 2, 315
Woosley, S.E., Timmes, F.X. & Baron, E. 1992, in *X-ray Binaries and Recycled Pulsars*, E.P.J. van den Heuvel & S.A. Rappaport (Eds.), Kluwer Academic Publishers, p. 167

MODELS FOR THE FORMATION OF BINARY AND MILLISECOND PULSARS

D. BHATTACHARYA
Raman Research Institute
Bangalore 560080. India.

1. Introduction

Under the name "binary and millisecond pulsars" are grouped a number of radio pulsars, forming roughly 10% of the known pulsar population, which distinguish themselves on three counts:

- Short spin period
- Low magnetic-field strength
- Presence of an evolved binary companion

Although in a few cases not all the three characteristics are present, this class brings together those pulsars whose spin history has involved a stage of spin up on accretion of matter from a binary companion, a process popularly known as "recycling". Another name of this group of pulsars is thus "recycled pulsars".

Understanding the formation of these pulsars mainly concerns understanding each of the above characteristics – namely (i) why the spin periods of the majority of them are short, (ii) why do most of them have much lower magnetic fields than isolated pulsars (see Fig. 1), and (iii) what evolutionary path led to the specific orbital characteristics and companion masses that individual pulsars of this class have. In this article we shall mainly concentrate on the third issue, and make only a few remarks about the first two.

2. Spin Periods

It is quite remarkable that most pulsars with binary companions have very short spin periods (see Fig. 1). As mentioned above, this is attributed to the

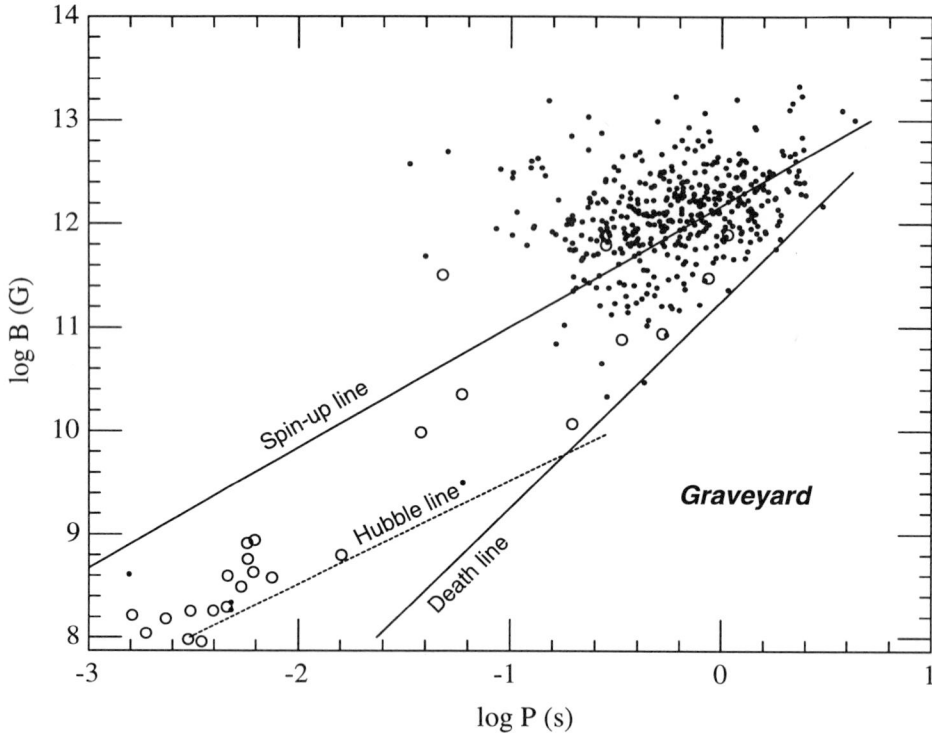

Figure 1. The magnetic fields and periods of known pulsars in the galactic disc. Filled dots indicate isolated pulsars and open circles binaries. The spin up line is the relation defined in eq. (1), with $\dot{M} = \dot{M}_{\rm Edd}$. Pulsar activity ceases to the right of the death line, and the spin down age ($\equiv P/2\dot{P}$) of a pulsar equals 10^{10} yr on the Hubble line.

spin up of the neutron star during accretion, as is seen to be happening in a number of X-ray binaries. A rough description of this process is as follows. During the mass transfer from the companion to the magnetized neutron star, matter approaches the neutron star through an accretion disc, in quasi-Keplerian orbits. As it approaches the "magnetospheric radius" $r_{\rm m}$ of the neutron star, the dynamics of the matter flow comes to be dominated by the magnetic field of the neutron star, and the matter is forced to corotate with the neutron star. If the corotation speed exceeds Keplerian speed at this point, the neutron star loses spin angular momentum to the incoming matter, and slows down. If, on the other hand, the Keplerian speed exceeds the corotation speed, part of the orbital angular momentum of the matter is transferred to the neutron star till corotation is achieved, spinning the neutron star up as a result. Given enough time a spin equilibrium would be achieved, where the Keplerian speed and the corotation speed would match at the magnetospheric radius. This leads to an "equilibrium spin period"

$$P_{eq} = 1.9 \,\text{ms} \left(\frac{B_s}{10^9 \,\text{G}}\right)^{6/7} \left(\frac{\dot{M}}{\dot{M}_{\text{Edd}}}\right)^{-3/7} \tag{1}$$

where B_s is the dipole field strength at the surface of the neutron star, $\dot{M}$ is the rate of mass accretion and $\dot{M}_{\text{Edd}}$ is the Eddington accretion rate. From the rate of accretion of angular momentum, it can also be estimated that spin up to P_{eq}, starting from a much longer spin period, would require the accretion of a minimum mass

$$\Delta M \simeq 0.1 \, M_\odot \, (P_{eq}/1.5 \,\text{ms})^{-4/3} \tag{2}$$

Admittedly, the picture described above is rather simplified (see Ghosh, these proceedings, for a more detailed discussion), but it does produce correct order-of-magnitude estimates.

One of the successes of this spin up (recycling) hypothesis for the origin of short spin periods of binary and millisecond pulsars is that most of these pulsars lie to the right of the critical "spin up line" in the $(B - P)$ diagram (Fig. 1), obtained by setting $\dot{M} = \dot{M}_{\text{Edd}}$ (the maximum accretion rate) in Eq. (1). An important point to note is that if a neutron star has to be spun up to a period as short as a few milliseconds in an Eddington-limited accretion, it would take a persistent accretion phase lasting $\sim 10^7$ yr or more, since the Eddington limit for a neutron star is $\sim 10^{-8} \, M_\odot \, \text{yr}^{-1}$. In addition, it would also need the magnetic field of the neutron star to be low, $\lesssim 10^9$ G.

3. Magnetic-Field Strength

Most pulsars processed in interacting binaries have considerably lower field strengths than garden-variety isolated pulsars (Fig. 1). The reason for this is not very well understood, but several physical models have been proposed. For about two decades following the discovery of pulsars it was believed that the magnetic fields of neutron stars undergo spontaneous Ohmic decay with a time scale of a few million years. In this picture, the recycled pulsars would have lower magnetic-field strengths simply because they are older. For a variety of reasons, however, this is no longer a preferred hypothesis [see Bhattacharya & Srinivasan (1995) for a review]. The idea that processing of a neutron star in a binary system is directly responsible for lowering its field strength (Bisnovatyi-Kogan & Komberg 1974; Taam & Van den Heuvel 1986; Bailes 1989) has of late gained currency. Physical models proposed for this kind of field evolution fall in two major categories, namely (a) Spindown-induced field evolution, and (b) Mass accretion-induced field evolution.

Models of category (a) propose an intimate relation between the spin period of the star and its magnetic field strength. At least two different physical mechanisms have been considered to accomplish this. One suggestion involves the interaction of the Abrikosov fluxoids in the superconducting protons carrying the magnetic flux in the neutron star core with the quantized neutron superfluid vortices which carry the angular momentum. Spin down (prior to spin up due to accretion) of the neutron star causes neutron vortices to move outwards, carrying the fluxoids with them. This "expelled" flux then undergoes Ohmic decay in the crust. Srinivasan et al. (1990) argued that this manner of evolution is in good qualitative agreement with observed distribution of isolated and binary pulsar fields. Quantitative modelling of low-mass binary systems in this scenario has also yielded encouraging results (Jahan Miri & Bhattacharya 1994). The second type of models in this category associate magnetic field evolution with crustal plate tectonics (Ruderman 1991). According to this picture spin down causes plates with frozen magnetic field configurations to move towards the stellar equator, where opposite poles can combine and destroy much of the magnetic field.

Category (b) is an even larger mixed bag of models, but practically all of them assume an initial field configuration confined entirely to the outer crust of the neutron star. The physical effects proposed here have been (i) screening of the stellar magnetic field by incoming diamagnetic plasma (Bisnovatyi-Kogan & Komberg 1974), (ii) an inverse thermoelectric battery that destroys existing field as heat flows up the density gradient (Blondin & Freese 1986), (iii) advection of the field with the accreted matter into the deeper layers of the crust, where the plasma expands sideways and reconnection occurs (Romani 1990), (iv) reduction of crustal conductivity due to heating resulting from accretion and consequent enhanced ohmic decay of the magnetic field (Urpin & Geppert 1994) and (v) compression of the current-carrying layers causing reduction in length scale of the field distribution and hence faster Ohmic decay (Konar et al. 1994). These models require further development before direct comparison with observations can be made.

In short, a number of models have been proposed, and are being currently worked upon, which attempt to explain the low magnetic fields of recycled pulsars as a direct result of the interaction of the neutron star with its binary companions. However at present we have neither sufficient knowledge about these models, nor enough observational constraints to unequivocally choose between them.

4. Orbital Characteristics

We now come to the major part of this paper, namely the discussion of the orbital characteristics of binary and millisecond pulsars, the nature of their companions, and the probable evolutionary history leading up to the present systems.

Binary and millisecond pulsars come with a great variety of companion types, which can be roughly classified into the following categories:

1. None
2. Evaporating secondary with very low mass
3. White dwarf with mass $< 0.45 M_\odot$
4. White dwarf with mass $> 0.45 M_\odot$
5. Neutron star
6. Massive main-sequence star
7. Low-mass main-sequence star (?)
8. Planets

Of these, pulsars with companion types 1–3 above are called low-mass binary pulsars (LMBPs), and those with companion types 4 and 5 high-mass binary pulsars (HMBPs). Companions of type 6 and 7 are unevolved, the pulsars are therefore not "recycled", and will hence be excluded from the present discussions. One millisecond pulsar in the galactic disc (PSR 1257+12, see Wolszczan, these proceedings) and perhaps one in a globular cluster (PSR 1620−26, Thorsett et al. 1993) have been reported to have planetary-sized companions. It is not clear, however, whether they belong to the low-mass or the high-mass class (see Section 7).

What are the progenitors of these pulsars? Since they have undergone mass accretion at some point in their lives, clearly their progenitors are to be looked for among the accreting binaries with neutron star components. Two major classes of such binaries are known – the high-mass X-ray binaries (HMXBs), and the low-mass X-ray binaries (LXMBs). The low-mass binary pulsars are usually thought to have descended from low-mass X-ray binaries, and the high-mass binary pulsars from high-mass X-ray binaries.

According to the conventional classification, a low-mass X-ray binary consists of a neutron star and a donor star with mass less than the mass of the neutron star (although Her X-1 with a donor mass $\sim 2\,M_\odot$ is also usually grouped with this category). The traditional HMXBs, on the other hand, contain donors more massive than $\sim 8\,M_\odot$. There is a clear gap in the donor masses between these two categories – the "intermediate-mass" secondaries do not seem to be represented among the visible X-ray binaries. This does not mean, however, that binaries containing intermediate-mass companions to neutron stars do not exist in nature. Our failure to detect them may be caused by a very short X-ray lifetime of these objects. It is well known that the Roche-lobe overflow mass transfer is unstable if the donor mass exceeds that of the neutron star, and this phase is probably marked by a shrouding of the accreting X-ray source by a large quantity of outflowing matter, keeping X-rays out of view. This is also true of high-mass binaries, of course – but high-mass stars have strong stellar winds, sufficient to generate detectable X-rays in HMXBs *before* the Roche-lobe overflow occurs. In intermediate mass binaries the weak stellar winds are unlikely to give rise to a detectable accreting X-ray source. We must therefore be aware that the intermediate-mass binaries, as a class, may well exist and contribute to recycled pulsar population, although they may not be represented among the accreting X-ray binaries. We shall have more to say about this below.

5. The Evolution of High-Mass X-ray Binaries and the Origin of HMBPs

According to the standard evolutionary scenario, high-mass X-ray binaries produce X-rays in the two early phases – when the neutron star accretes the stellar wind of the companion, and when only the atmosphere of the donor overflows the Roche lobe. Beyond this point, very heavy mass transfer due to Roche-lobe overflow ensues and the neutron star spirals into the massive secondary. If the initial orbital period is less than ~ 1 year, a complete spiral-in is likely, with the neutron star entering the centre of the massive companion, and a Thorne-Żytkow object is produced. The final outcome of this is unclear – if the Thorne-Żytkow object is able to lose most of its envelope either in stellar winds or due to magneto-rotational effects a recycled pulsar may be left, otherwise a black hole appears to be the inevitable result (see Podsiadlowski, these proceedings).

If the initial orbit is wide enough, with orbital period larger than ~ 1 year, then according to the conventional picture the neutron star would escape complete spiral in, and will be left in a tight binary with the helium core of the secondary. This helium star, if it is massive enough, would in time explode in a supernova leaving a new-born neutron star. If the binary survives the explosion, one expects to see a recycled pulsar and a young neutron star in an eccentric orbit, similar to PSR 1913+16 and PSR 1534+12 systems. On the other hand if the binary disrupts the final result would be an isolated young pulsar and an isolated recycled pulsar. If the helium star is not massive enough to produce a neutron star, it would expand and transfer mass to the neutron star companion (lighter helium stars expand more). A second spiral-in may occur and eventually a recycled pulsar and a heavy white dwarf in a compact, circular orbit will be left – similar to the observed system PSR 0655+64.

While the above scenario appears reasonable for the production of dou-

TABLE 1. HMBP orbital periods

Pulsar Name	Binary components	Orbital period
B1913+16	NS + NS	8 h
B1534+12	NS + NS	10 h
B2303+46	NS + NS	12 d
B0655+64	NS + WD	1 d
J1023+10	NS + WD	8 d
J2145−0750	NS + WD	7 d

NS = Neutron Star; WD = White Dwarf

ble neutron star and neutron star + heavy white-dwarf binaries, it does depend heavily on the poorly understood evolutionary phase of spiral-in. Recently, Chevalier (1993) and Brown (1994) have questioned the scenario on this ground. These authors show that in the case of a spherically symmetric mass transfer, a neutron star spiralling into a hydrogen-rich giant would experience a "hypercritical accretion", with transfer rates exceeding $0.1\,M_\odot\,yr^{-1}$, when the Eddington limit is no longer valid since most of the cooling is done by neutrinos. Such a neutron star will end up accreting an enormous quantity of material and would invariably collapse to a black hole. If this assertion proves to be correct, then one can never hope to obtain a double neutron star system or a compact neutron star-white dwarf binary through the conventional spiral-in.

Indications from the observations are, however, that spiral-in products do exist. Systems listed in Table 1 all have short binary periods, and rather heavy compact remnants – the initial orbit of the progenitor system must have been considerably wider, since the present orbits are not wide enough to accommodate even one of the progenitor stars in the main sequence or a slightly evolved phase.

What does the existence of these systems tell us? Either, the manifestly non-spherical geometry of the spiral-in process manages somehow to avoid the predicament of hypercritical accretion [although both Chevalier (1993) and Brown (1994) argue that taking angular momentum into consideration should not change their result], or one must find their progenitors among hitherto unconsidered class of objects. Brown (1994) suggests that double helium star binaries might be the progenitors of some double neutron star binaries. Finding suitable progenitors of neutron star + white-dwarf binaries might be even more difficult. Clearly this question deserves careful attention and further study.

6. Low-Mass X-ray Binaries and the Origin of LMBPs

The standard evolution of low mass X-ray binaries follows one of two classical tracks, depending on the starting orbital period P_0: for $P_0 \lesssim 0.5$ day the evolution is driven by angular momentum loss by gravitational radiation and magnetic braking. These systems come into contact while the donor is on the main sequence and continues to shrink as mass transfer proceeds. The orbital period is expected to pass through a minimum of ~ 80 min, at which point the secondary would become degenerate and the orbit would widen again, albeit very slowly. It seems likely that the full course of this evolution is never followed, and at quite an early stage irradiation of the companion might significantly modify the evolutionary track (Podsiadlowski 1991; Harpaz & Rappaport 1991; Tavani 1991). The nature and extent of this modification is, however, hotly debated (see Ritter, these proceedings), and a clear picture is yet to emerge.

Evolution of LMXBs with wider initial orbits ($P_0 \gtrsim$ a few days) seems to be better understood – nuclear evolution drives the expansion of the secondary and hence the mass transfer, and the orbit widens as a result. Typically an increase of orbital period by about an order of magnitude is expected between the beginning and the end of the mass transfer (Webbink, Rappaport & Savonije 1983). The mass transfer ends when the white-dwarf core of the donor is left. This evolutionary sequence leaves a helium white dwarf, with a mass $\lesssim 0.45 \, M_\odot$, orbiting the recycled pulsar in a circular orbit. The wider the initial orbital period, the more evolved is the donor at the time of contact and hence the heavier is the final white dwarf remnant. This defines a relation between the final secondary mass and the final orbital period (Joss, Rappaport & Lewis 1987), and most low-mass binary pulsars seem to obey this relationship, within observational uncertainties, strengthening the case for their origin from LMXBs [see Phinney & Kulkarni (1994) and Lorimer (1994) for a comparison of the observed data with the predicted relation]. It must, however, be pointed out that it has not yet been possible to fully reconcile the birth statistics of LMXBs and LMBPs of different categories (see Lorimer, these proceedings).

A curious point regarding the orbital-period distribution of LMBPs has been noted by Camilo (1994): there appears to be a "gap" in the period distribution between 12d and 65d where no LMBP is seen (see Fig. 2). No known selection effect would create such a gap – so if the gap is real and not a statistical fluke the reason behind it must be evolutionary. However, there is no known evolutionary effect leading to gaps in this range, either. Enhanced magnetic braking effects for LMXBs with initial orbital periods up to a few days have been considered (Pylyser & Savonije 1988), and might produce a similar gap between ~ 1–10 days in the final orbital period, which

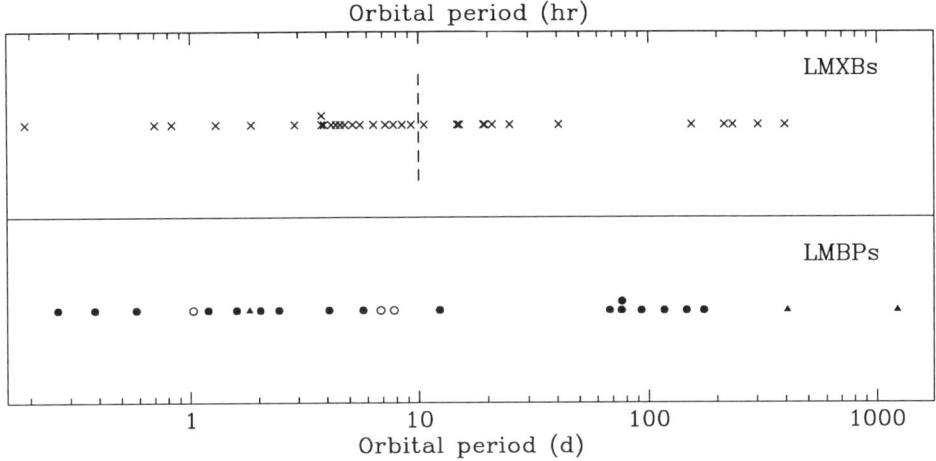

Figure 2. The distribution of the orbital periods of known low-mass X-ray binaries (top) and low-mass binary pulsars (bottom). Note the difference in scale between the two panels. Evolution of the systems to the left of the vertical dashed line in the top panel is strongly affected by angular momentum loss and the secondary comes into contact in the main sequence. Systems with larger orbital periods are likely to have evolved donors. In the bottom panel filled symbols are systems with secondary mass $\lesssim 0.45\,M_\odot$, while the open symbols denote systems with higher secondary masses, which are probable descendants of intermediate-mass binaries. Systems with eccentricity higher than 0.001 are shown as triangles. There appears to be a "gap" in the LMBP period distribution between 12 and 65 days. It may be that this is a reflection of a corresponding gap in the period distribution of LMXBs, in the range between 40 h and 5 d, as seen in the top panel. (Figure courtesy Fernando Camilo.)

is still much smaller than the gap periods referred to above. On the other hand, this observed gap might reflect the original period distribution of LMXBs – indeed among the observed LMXBs there seems to be a dearth of systems with orbital periods between 40 h and 5 days (Fig. 2) – roughly the same range that might evolve to the region of the observed gap in the final evolutionary products.

7. Millisecond Pulsars

Millisecond pulsars, right since their discovery, have been considered to be products of LMXB evolution (Alpar *et al.* 1982). This is because their ultra-rapid spin ($P \lesssim 10$ ms) can be obtained by recycling only if the near-Eddington accretion phase is prolonged – something that obtains only in LMXBs, particularly in wide ones. It is no surprise, therefore, that $\sim 75\%$ of the known disc population of millisecond pulsars *are* LMBPs. What is surprising, however, is that quite a few millisecond pulsars are single, i.e., with no companion. It would be fair to say that we do not really understand the formation of single millisecond pulsars. If, like other millisecond pulsars, these too were formed in LMXBs then their companions had to be got rid of in some way. The foremost suggestion in this regard has been that of Ruderman et al. (1989), who argued that the radiation from a fast-spinning millisecond pulsar can ablate its companion away, thus predicting the existence of systems like PSR 1957+20 with evaporating secondaries. It appears, however, that the vaporization of the companion of 1957+20 may not be vigorous enough to completely destroy it (Fruchter et al. 1990; Ryba & Taylor 1991), and this pulsar being amongst the most powerful millisecond pulsars known casts some doubt on the effectiveness of this mechanism. However, a recently proposed variant of this scenario, involving irradiation and tidal dissipation, has been argued to be capable of destroying the companion of PSR 1957+20 within $\sim 10^8$ yr (Applegate & Shaham 1994).

A second suggestion has been that the degenerate secondary, towards the end of the evolution of "close" LMXBs, might undergo unstable Roche lobe overflow as it expands on loss of mass, and be completely disrupted to form a heavy disc around the neutron star (Ruderman & Shaham 1985). A serious objection to this idea was raised by Jeffrey (1986), who pointed out that even if such a disc were to form, slow accretion from it onto the neutron star would spin the neutron star down to a very long period well before the disc disappears.

Another way of losing the companion of a neutron star is the "ionization" of a binary due to an encounter with a third passing star. This may happen at times in globular clusters (see Rappaport, Putney & Verbunt 1989), but would be an extremely rare occurrence in the galactic disc.

If these models are found to be inadequate to explain the occurrence of single millisecond pulsars, one must look for their origin outside the standard LMXB group. A view taken by some is that single millisecond pulsars are just born with short periods and low magnetic fields, and one need not invoke recycling to explain their origin (Michel 1987). While it can never be conclusively *proven* to be otherwise, it would seem odd that the single millisecond pulsars blend in very well with the rest of the millisecond pulsar population, which are clearly recycled, given their binary nature. Moreover, none of the millisecond pulsars in the galactic disc, including the single ones, lies above the "spin up line" in Fig. 1.

It has been pointed out by Bonsema & Van den Heuvel (1985) that the merger of a white dwarf of mass exceeding $\sim 0.66\,M_\odot$ with a neutron star may well produce a millisecond pulsar and leave no companion for it. This remains a viable route, although the statistics of it are yet to be worked out. The progenitor in this case would be very similar to PSR 0655+64, but with a tighter orbit (orbital period of a few hours). This kind of progenitor system would, in turn, be a descendant of a high-mass X-ray binary, or perhaps an intermediate-mass one.

An even more exotic scenario involving high-mass binaries would envisage the formation of a single millisecond pulsar as the end product of a Thorne-Żytkow object (Wijers et al. 1992). The degree of spin up that can occur during and after the spiral-in of the neutron star into the companion's envelope is not entirely clear. In case of a complete spiral-in, a fraction of the orbital angular momentum resides in the stellar envelope – so one expects the Thorne-Żytkow stars to be rapidly spinning. Whether or not accretion of some of this material onto the central neutron star, the magnetic field of which might have been considerably weakened during the course of the evolution, can spin it up to millisecond periods needs to be quantitatively worked out.

At this point it is appropriate to state that every scenario for the formation of single millisecond pulsars that I described above, with the exception of that involving ionization, has also been used in the context of formation of planets around millisecond pulsars (see Podsiadlowski 1993 for a review). The generic idea in these models is that once the secondary is destroyed, its debris can eventually condense into planets. As a result, the situation is now slightly confused – it is not possible to tell which of these routes might really contribute to planet formation and which might leave genuinely isolated millisecond pulsars.

8. Millisecond Pulsars from Intermediate-Mass Binaries

To conclude this discussion, I would like to draw the attention of the reader to two recently discovered near-millisecond pulsars – PSR 2145–0750 and PSR 1023+10. Both have spin periods of 16 ms. The orbital periods are short, 6.8 days and 7.8 days, respectively. Most importantly, the masses of their white-dwarf companions are moderately high – $0.5\,M_\odot$ and $0.8\,M_\odot$, respectively. Since helium white dwarfs cannot grow to masses beyond $0.45\,M_\odot$, these remnants must be C-O dwarfs. This means that the progenitors of these remnants must have gone through helium burning, and hence mass transfer must have taken place in a rather late (AGB) phase – implying a rather wide initial orbit and a subsequent spiral in (Van den Heuvel 1994). According to the calculations of Iben & Tutukov (1985, 1993) the original secondary masses in these cases should have been in the range $1-3\,M_\odot$ for PSR 2145–0750 and $4-6\,M_\odot$ for PSR 1023+10. For PSR 2145–0750 an original secondary mass towards the heavier end of the above range seems preferable, since PSR 0820+02, starting with a donor of $\sim 1\,M_\odot$ and an orbital period of $\gtrsim 200$ days has *expanded* its orbit to the present period of 1232 days, and has left a $\sim 0.3-0.4\,M_\odot$ helium white-dwarf remnant. It seems likely, therefore, that the spiral-in products like PSR 2145−0750 and PSR 1023+10 descend from systems with secondary masses higher than $\sim 1\,M_\odot$.

The important point that emerges from this is that these pulsars seem to be descendants of intermediate-mass binaries, with secondaries in the range 3–6 $M_\odot$ (see also Kulkarni 1994). Indeed the pulsar PSR 0655+64 should also be classified in this category (Van den Heuvel & Taam 1984). However, these two new pulsars have demonstrated that spin up in this class of binaries can proceed to very nearly the millisecond range. This could be the result of an effect pointed out by Iben & Tutukov (1993) – that the post-spiral-in remnants of stars with original mass in the range 4–6 $M_\odot$ continue to undergo Roche lobe overflow and transfer mass at the rate of 10^{-8} to $10^{-6}\,M_\odot\,\mathrm{yr}^{-1}$ for about 10^6 yr (see also Van den Heuvel 1994).

In sum, these recent discoveries have established intermediate-mass binaries as substantial contributors to recycled pulsar population – and shown that many of these pulsars could be spinning very rapidly, at nearly millisecond periods. Intermediate-mass binaries have remained in relative oblivion so far because they do not show up as X-ray sources. These recycled pulsars have thus opened up a new and exciting avenue for the study of this class of binary systems.

Acknowledgements. I am indebted to Ed van den Heuvel, Philipp Podsiadlowski, Fernando Camilo and Duncan Lorimer for many discussions and access to unpublished information during the preparation of this talk, much of which was done during a very useful month spent at the Nuffield Radio Astronomy Laboratories at Jodrell Bank. A travel grant by the International Astronomical Union and hospitality of the University of Utrecht are gratefully acknowledged.

References

Alpar, M.A., Cheng, A.F., Ruderman, M.A. & Shaham, J. 1982, Nat 300, 728
Applegate, J.H. & Shaham, J. 1994, ApJ 436, 312
Bailes, M. 1989, ApJ 342, 917
Bhattacharya, D. & Srinivasan, G. 1995, in *X-ray Binaries*, W.H.G. Lewin, J. van Paradijs & E.P.J. van den Heuvel (Eds.), Cambridge University Press, (in press)
Bisnovatyi-Kogan, G.S. & Komberg, B.V. 1974, SvA 18, 217
Blondin, J.M. & Freese, K. 1986, Nat 323, 786
Bonsema, P.F.J. & Van den Heuvel, E.P.J. 1985, A&A 146, L3
Brown, G.E. 1994, ApJ (in press)
Camilo, F. 1994, in *Lives of the Neutron Stars*, M.A. Alpar, Ü. Kızıloğlu & J. van Paradijs (Eds.), Kluwer Academic Publishers, (in press)
Chevalier, R. 1993, ApJ 411, L33
Fruchter, A.S., Berman, G., Bower, G. *et al.* 1990, ApJ 351, 642
Harpaz, A. & Rappaport, S.A. 1991, ApJ 383, 739
Iben, I. & Tutukov, A.V. 1985, ApJS 58, 661
Iben, I. & Tutukov, A.V. 1993, ApJ 418, 343
Jahan Miri, M. & Bhattacharya, D. 1994, MNRAS 269, 455
Jeffrey, L.C. 1986, Nat 319, 384
Joss, P.C., Rappaport, S.A. & Lewis, W. 1987, ApJ 319, 180
Konar, S., Bhattacharya, D. & Urpin, V.A. 1994, in *Proceedings of the 6th Asia-Pacific Regional Meeting of the IAU*, V.K. Kapahi & N. Dadhich (Eds.), (in press)
Kulkarni, S.R. 1994, in: *Proceedings of the Aspen Conference on Millisecond Pulsars: A Decade of Surprises*, (in press)
Lorimer, D. 1994, *The Galactic Population of Millisecond and Normal Pulsars*, Ph.D. Thesis, University of Manchester
Michel, F.C. 1987, Nat 329, 310
Phinney, E.S. & Kulkarni, S.R. 1994, ARA&A (in press)
Podsiadlowski, P. 1991, Nat 350, 136.
Podsiadlowski, P. 1993, in: *Planets Around Pulsars*, J.A. Phillips, S.E. Thorsett & S.R. Kulkarni (Eds.), Astron. Soc. of the Pacific (San Fransisco)
Pylyser, E.H.P. & Savonije, G.J. 1988, A&A 191, 57
Rappaport, S.A., Putney, A. & Verbunt, F. 1989, ApJ 345, 210
Romani, R.W. 1990, Nat 347, 741
Ruderman, M. 1991, ApJ 366, 261
Ruderman, M. & Shaham, J. 1985, ApJ 289, 244
Ruderman, M., Shaham, J. & Tavani, M. 1989, ApJ 336, 507
Ryba, M.F. & Taylor, J.H. 1991, ApJ 380, 557
Srinivasan, G., Bhattacharya, D., Muslimov, A.G. & Tsygan, A.I. 1990, Curr. Sci. 59, 31
Taam, R.E. & Van den Heuvel, E.P.J. 1986, ApJ 305, 235
Tavani, M. 1991, ApJ 366, L27
Thorsett, S.E., Arzoumanian, Z. & Taylor, J.H. 1993, ApJ 412, L33
Urpin, V. & Geppert, U. 1994, A&A (submitted)
Van den Heuvel, E.P.J. 1994, A&A (in press)
Van den Heuvel, E.P.J. & Taam, R.E. 1984, Nat 309, 235
Webbink, R.F., Rappaport, S.A. & Savonije, G.J. 1983, ApJ 270, 678
Wijers, R.A.M.J., Van den Heuvel, E.P.J., Van Kerkwijk, M.H. & Bhattacharya, D. 1992, Nat 355, 593

Discussion

S.R. Kulkarni: The period gap may be spurious, but I would like to draw your attention to a curious point. Pylyser & Savonije (1988, A&A 191, 57) considered the evolution of compact LMXBs (accretion driven by angular momentum loss). Several authors (Taam, Webbink et al.) have considered the evolution of extended (Roche-lobe fed) LMXBs. These models disagree in the final orbital period range where they overlap (see Coté & Pylser 1989, A&A 218, 131). This suggests that in the final period range around about 10 days the evolution is extremely sensitive to the input physics. There could well be a discontinuity in this period range.

D. Bhattacharya: I agree that evolution dominated by angular momentum loss for short period binaries could be the cause of the period gap. However, even as strong an angular momentum loss considered by Pylyser and Savonije (1988) would be unable to place the gap where it is. Perhaps an even stronger angular momentum loss can do so.

PULSAR/SUPERNOVA REMNANT ASSOCIATIONS

D. A. FRAIL
National Radio Astronomy Observatory
P.O. Box 0, Socorro, NM, 87801, USA

1. Introduction

If one is interested in issues related to the birth and evolution of neutron stars, a promising avenue of research begins with the study of the byproducts of a Type II supernova event, namely a pulsar (PSR) and its supernova remnant (SNR). However, it has long been a matter of some embarrassment that out of the more than 600 known PSRs and the 175 catalogued SNRs there are few associations. In fact, for nearly 15 years following the discovery of PSRs only the Crab and Vela PSRs had an associated remnant. This failure has spawned something of a cottage industry to "explain away" the low incidence of PSR/SNR associations. Beaming (Frail & Moffett 1993; Narayan & Schaudt 1988), injection (Narayan 1987), field growth (Blandford, Applegate & Hernquist 1983) and black-hole formation (Chevalier 1994) have been invoked in this context at one time or another.

In the last decade there has been tremendous progress in this field. These advances have come from two approaches: (1) High-frequency PSR searches of the Galactic plane (e.g., Johnston *et al.* 1992) which have overcome the high brightness temperatures and interstellar scattering that plagued previous surveys, allowing us to see deeper into the plane of the Galaxy, reaching the young, luminous PSRs, and (2) radio imaging of the fields of these young PSRS using the latest in low-frequency imaging algorithms (Cornwell 1993) has resulted in the discovery of extended, low surface brightness SNRs (e.g., Kassim & Weiler 1990).

The last comprehensive review on PSR/SNR associations was by Helfand & Becker (1984). Since then the incremental progress in this area has been reviewed by Kulkarni (1990) and Bailes & Johnston (1993). In the spirit of these later mini-reviews we will describe a number of recent PSR/SNR associations and summarize some preliminary conclusions reached from a study of all the PSR/SNR associations by Frail, Goss & Whiteoak (1994).

2. Pulsar-Supernova Remnant Associations

2.1. PSR 1338–62 AND G 308.8–0.1

This PSR/SNR pair is a good illustration of how improvements in both image fidelity and PSR parameters have led to new associations. Both the PSR (Manchester, D'Amico & Tuohy 1985) and the SNR (Caswell, Milne & Wellington 1981) had been previously known but the association was not proposed until Caswell et al. (1992) made a sensitive radio image of G 308.8–0.1 and Kaspi et al. (1992) made timing observations of the PSR to refine the position of the PSR and establish its young age (12 kyrs). The near central location of the PSR within G 308.8–0.1, and the close agreement in the ages argue persuasively for a real association.

Other recent examples where improved imaging has led to newly proposed associations of known PSRs include PSR 1757–24 and G 5.4–1.2 (Frail & Kulkarni 1991), PSR 1800–21 and W 30 (Kassim & Weiler 1990), PSR 1706–44 and G 343.1–2.3 (McAdam, Osborne, & Parkinson 1993).

2.2. PSR 1643–43 AND G 341.2+0.9

PSR 1643–43 has a period P of 232 msec, a characteristic age τ_c of 32.6 kyrs and a dispersion measure based distance of 6.9 kpc (Johnston et al. 1994). Recent work by Frail et al. (1994) shows that PSR 1643–43 lies in an extended ($22' \times 16'$), non-thermal radio source which is likely a supernova remnant G 341.2+0.9. The Σ-D distance of G 341.2+0.9 is between 8.3–9.7 kpc (Milne 1979; Clark & Caswell 1976; Allakhverdiyev et al. 1983). In addition to the shell-type emission which brightens on the side closest to PSR 1643–43, there is diffuse emission in the immediate vicinity of the PSR. This includes a $4'$ nebulosity just east of the PSR which is joined to the PSR by a "bridge" of emission. The appearance of this structure is consistent with an implied westward motion of the PSR.

Frail et al. (1994) argue that the similar distances of the PSR and the SNR in combination with the morphological evidence, are strong indicators that a real physical association exists between the two. Other PSR/SNR associations studied by Frail et al. (1994) include PSR 1706–44/G 343.1–2.3 and PSR 1727–33/G 354.1+0.1.

3. The Distribution of Pulsar Velocities at Birth

All known PSRs with characteristic ages less than 60,000 years are included in the table below. Of the 18 PSRs in this table there are 15 potential PSR/SNR associations. One of the most striking characteristics of these associations is that in the majority of cases the PSR is offset from the center of its SNR. Following Shull et al. (1989) we quantify this by the parameter $\beta = \theta_p/\theta_s$ where θ_p is the angular displacement of the PSR from the geometric center of the remnant, θ_s is the angular radius of the supernova remnant. Note the values of β in the table are rarely close to zero, and in fact in a few cases $\beta > 1.0$ (i.e., the PSR is outside the remnant). If the PSR were born at the center of the remnant then we can straightforwardly calculate an *implied* PSR velocity V_{PSR} by dividing its transverse displacement from the center of the remnant by its age τ_c.

The median of these transverse velocities is 480 km s^{-1}. Twelve of the 15 young PSRs have velocities above the mean of 217 km s^{-1} for the Harrison et al. (1993) proper-motion sample. Such large V_{PSR} values have been noted before in discussions of individual associations (e.g., Kassim & Weiler 1990; Manchester et al. 1991), but it has only recently been recognized (Frail et al. 1994; Caraveo 1993) that this may be a general property of young PSRs as a whole. In an independent study, Lyne & Lorimer (1994) corrected for well-known selection effects in proper-motion samples (Cordes 1986; Helfand & Tademaru 1977) and used the newer Taylor & Cordes (1993) distance model to derive a mean PSR birth velocity of 450±90 km s^{-1}.

TABLE 1. Young Pulsars and Proposed Associated Supernova Remnants

PSR Name	P (s)	D (kpc)	Age (10^3 yrs)	SNR Name	β	V_{PSR} (km s^{-1})
0531+21	0.033	2.0	1.3	Crab Nebula		150
1509−58	0.150	4.4	1.5	MSH 15−52	0.25	3100
0540−69	0.050	49.4	1.7	SNR 0540−693	0.06	260
1610−50	0.232	7.3	7.4	Kes 32	1.7	3600
0833−45	0.089	0.5	11	Vela XYZ	0.02	120
1338−62	0.193	8.7	12	G 308.8−0.1	0.30	750
1757−24	0.125	4.6	15	G 5.4−1.2	1.3	1800
1800−21	0.134	3.9	16	W 30	0.73	1100
1706−44	0.102	1.8	17	G 343.1−2.3	1.4	670
1853+01	0.267	3.3	20	W 44	0.58	420
1046−58	0.124	3.0	20			
1737−30	0.607	3.3	21			
1823−13	0.101	4.3	21			
1727−33	0.139	4.2	26	G 354.1+0.1		460
1643−43	0.232	6.9	32	G 341.2+0.9	0.73	480
1930+22	0.144	9.8	40	G 57.1+1.7	0.45	1600
2334+61	0.495	2.5	41	G 114.3+0.3	0.0	<50
1758−23	0.416	3.0	59	W 28	1.3	300

3.1. A CAVEAT OR TWO

There are some potential difficulties with this method of determining PSR velocities from their offsets from the SNR center.

(1) Not all the associations in the table are secure and much effort needs to be made to test whether these associations are real or not. However, eliminating questionable associations does not significantly change the result (Frail et al. 1994). We also know that only a few percent of the disk is covered by SNRs and thus chance superpositions are rare. 85% of the PSRs in the table are found in or next to an SNR, compared with only 3% for PSRs with $\tau_c > 10^5$ yrs. Furthermore, there is no indication that we are searching any deeper around young PSRs than in general purpose low frequency interferometric surveys of the Galactic plane (e.g., Taylor, Wallace & Goss 1992).

To properly demonstrate that a PSR/SNR association is a *true* physical association and not just a *chance* superposition of objects on the sky, agreement is required between the ages and distances of both objects. For the most part the distances and ages for the PSRs and SNRs in the table agree but they are poorly known, reflecting the uncertainty in the methods that are being used (H I absorption, DM distance, Σ-D, Sedov ages, etc.). Morphological evidence, such as a flat spectrum plerionic nebula surrounding the PSR or signs of increased brightness in the radio SNR shell closest to the PSR, suggests an interaction and hence is useful. Finally, a measured PSR proper motion is a useful and final discriminant for distinguishing real and chance superpositions. Unfortunately, Crab and Vela are the only PSRs in the table to have had such measurements made and as such they represent the most firm associations.

(2) This method rests on the assumption that we can identify the birthplace of the PSR as the center of the SNR. This issue is discussed in more depth by Frail et al. (1994). In short, it is necessary to have a high quality image of the SNR before determining PSR velocities by this method.

The Vela association is a good illustration of the dangers of using this method when the shape of the SNR is poorly known. Based on an offset of the PSR from the center of the Vela SNR the inferred proper motion of the PSR was thought to be $800\,\mathrm{km\,s^{-1}}$ and yet proper-motion measurements yielded a considerably lower value of $120\,\mathrm{km\,s^{-1}}$ (Bailes et al. 1989). However, the true shape of the remnant has only recently been defined by low-frequency radio observations (Dwarakanath 1991) and X-ray observations (Aschenbach 1992) revealing a larger and more circular remnant than had been previously suspected. Now if we use the new center of the remnant the inferred velocity of the PSR is the same as its proper motion value.

4. Implications

The implications of the existence of large numbers of high-velocity PSRs is far-reaching; it has an impact on the mechanisms that give rise to PSR velocities at birth, and it affects how they interact with their surroundings and what the final distribution of PSRs in the Galaxy will be. If this result holds up we will need to re-examine the birth rate of PSRs, their survival in binary systems and their escape from globular clusters and the Galaxy (Lyne & Lorimer 1994). Future efforts should concentrate on measuring proper motions to test the veracity of the associations. Both the magnitude and the direction of the velocity vector are useful in this regard. The former will test whether the high velocities are real and the latter will test the association (i.e., the PSR must originate from the remnant).

If a significant population of high-velocity PSRs does exist then they could escape the disk, forming a halo population of old neutron stars. This extended halo population has been postulated to exist for many years in order to explain γ-ray bursts as a galactic phenomenon (e.g., Li & Dermer 1992) but critics have argued that an "ad hoc" population of Galactic objects is less preferable to a cosmological distribution of sources (Paczyński 1993). Now that evidence is accumulating in favor of a real high-velocity neutron star population and it has been shown that they are able to reproduce the high degree of angular isotropy seen by current γ-ray instruments (Podsiadlowski, Rees & Ruderman 1994) there is cause to revisit the extended halo models. To produce the observed distribution of γ-ray bursts the high-velocity PSRs must be endowed with a special property that distinguish them from the low-velocity PSRs, and this may be related to the origin of the bursts. High magnetic fields (10^{14}–10^{15} G) have been evoked in some models of bursters (e.g., Duncan & Thompson 1992) but the dipole fields of the PSRs in the table are only slightly above the mean $\log B$ of 12.51 (Stollman 1987).

With $V_{\rm PSR} \simeq 500\,{\rm km\,s^{-1}}$ a PSR catches up to its SNR in only 40 000–70 000 years (Shull et al. 1989). Such PSRs will act as a "fountain of youth" injecting fresh relativistic particles and field into the compressed shell of the aging remnant. Thus ambient density may not be the dominant factor influencing the radio lifetimes of supernova remnants. The distinction between shell-type remnants and pulsar-powered nebulae has been blurred, creating a new class of supernova remnants called "interacting composites". Examples may include a number of objects mentioned by Shull et al. (1989) like G 5.4−1.2, W 28 and G 57.1+1.7 as well as G 114.3+0.3 (Kulkarni et al. 1993), G308.8−0.1 (Kaspi et al. 1992), and MSH 15−52 (Caswell, Milne & Wellington 1981).

References

Allakhverdiyev, A.O., Amnuel, P.R., Guseinov, O.H. & Kasumov, F.K. 1983, ApSS 97, 261
Aschenbach, B. 1992, in *Highlights of Astronomy*, Vol. 9, J. Bergeron (Ed.), Kluwer (Dordrecht), p. 223
Bailes, M. & Johnston, S. 1993 in *Review of Radio Science (URSI)*, W. Ross Stone (Ed.), Oxford University Press, London, p. 677
Bailes, M. et al. 1989, ApJ 343, L53
Blandford, R., Applegate, J.H. & Hernquist, L. 1983, MNRAS 204, 1025
Caraveo, P.A. 1993, ApJ 415, L111
Caswell, J.L., Milne, D.K. & Wellington, K.J. 1981, MNRAS 195, 89
Caswell, J.L. et al. 1992, ApJ 399, L151
Chevalier, R.A. 1994, in IAU Coll. 145, R. McCray & Z. Wang (Eds.), Cambridge U. Press (London), (in press)
Clark, D.H. & Caswell, J.L. 1976, MNRAS 174, 267
Cordes, J.M. 1986, ApJ 311, 183
Cornwell, T.J. 1993, VLA Scientific Memorandum 164, National Radio Astronomy Observatory
Duncan, R.C. & Thompson, C. 1992, ApJ 392, L9
Dwarakanath, K.S. 1991, J. Astrophys. Astr. 12, 199
Frail, D.A. & Kulkarni, S.R. 1991, Nat 352, 785
Frail, D.A. & Moffett, D.A. 1993, ApJ 408, 637
Frail, D.A., Goss, W.M. & Whiteoak, J.B.Z. 1994, ApJ (in press)
Harrison, P.A., Lyne, A.G. & Anderson B. 1993, MNRAS 261, 113
Helfand, D.J. & Becker, R.H. 1984, Nat 307, 215
Helfand, D.J. & Tademaru, E. 1977, ApJ 216, 842
Johnston, S. et al. 1992, MNRAS 255, 401
Johnston, S. et al. 1994, A&A (in press)
Kaspi, V. M. et al. 1992, ApJ 399, L155
Kassim, N.E. & Weiler, K.W. 1990, Nat 343, 146
Kulkarni, S.R. 1990, in *Neutron Stars and Their Birth Events*, W. Kundt (Ed.), Kluwer Academic Publishers, p. 59
Kulkarni, S.R., Predehl, P., Hasinger, G. & Aschenbach, B. 1993, Nat 362, 135
Li, H. & Dermer, C.D. 1992, Nat 359, 514
Lyne, A.G. & Lorimer D.R. 1994, Nat 369, 127
Manchester, R.N., D'Amico, N. & Tuohy, I.R. 1985, MNRAS 212, 975
Manchester, R.N. et al. 1991, MNRAS 253, P7
McAdam, W.B., Osborne, J.L. & Parkinson, M.L. 1993, Nat 361, 516
Milne, D.K. 1979, Aus. J. Phys. 32, 83
Narayan, R. 1987, ApJ 319, 162
Narayan, R. & Schaudt, K.J. 1988, ApJ 325, L43
Paczyński, B. 1993, in *Compton Gamma Ray Observatory*, M. Friedlander, N. Gehrels & D.J. Macomb (Eds.), AIP (New York), p. 981
Podsiadlowski, P., Rees, M.J. & Ruderman, M. 1994, MNRAS (preprint)
Shull, J.M., Fesen, R.A. & Saken, J.M. 1989, ApJ 346, 860
Stollman, G.M. 1987, A&A 178, 143
Taylor, J.H. & Cordes, J.M. 1993 ApJ, 411, 674
Taylor, A.R., Wallace, B.J. & Goss, W.M. 1992, AJ 103, 931

PERIASTRON OBSERVATIONS OF THE PSR B1259–63/SS 2883 BINARY SYSTEM

SIMON JOHNSTON
Research Centre for Theoretical Astrophysics
University of Sydney, NSW 2006, Australia.

1. Introduction

PSR B1259–63 is a 47-millisecond pulsar which was discovered in a high frequency survey of the galactic plane (Johnston et al. 1992a) and was subsequently found to be in a highly eccentric orbit with a main-sequence Be star known as SS 2883 (Johnston et al. 1992b). Radio observations of the pulsar led to a phase connected timing solution which predicted the epoch of periastron to be 1994 January 9 (MJD 49361.2); optical observations of the Be star led to a determination of its mass and of the size of its circumstellar disk (Johnston et al. 1994a): the star is of approximate spectral type B1e, with mass $10\,M_\odot$ and radius $6\,R_\odot$. If this mass is correct and the pulsar has a mass of $1.4\,M_\odot$, then the inclination angle of the plane of the orbit with respect to the sky is 35°. This pulsar has an unusually flat radio spectrum compared to most pulsars, which makes it easily detectable up to 8.4 GHz. The narrow pulse permits dispersion and scattering measurements for studying the ionized plasma in the system. Moreover, the pulses are highly linearly polarized and permit determination of the rotation measure (RM), allowing measurements of the magnetic field along the line of sight. The 3.5-yr orbit of the pulsar around its companion thus provides us with an excellent probe of the stellar wind of the Be star over a wide frequency range.

Be stars are generally defined as non-supergiant B-type stars which show one or more of the Balmer lines in emission. These emission lines generally show a double-peaked structure, and it was realized (Struve 1931) that this could be interpreted as originating from a rotating disk of material in the equatorial region of the star. These disks have been observed at optical, infrared, millimeter and radio wavelengths and are generally assumed to consist of high-density, slowly moving material which is more or less confined to the equatorial plane of the star. Be stars are also bright UV sources, and observations of UV resonance lines seem to indicate the presence of a more global wind, with low density and high velocity.

2. Observations and Data Analysis

All observations of the pulsar were made using the Parkes 64-m radio telescope of the Australia Telescope National Facility in New South Wales, Australia. Four separate dual-channel cryogenic receiver systems were used at frequencies of 1.5, 2.3, 4.8 and 8.4 GHz. Each receiver was sensitive to two orthogonal planes of linear polarization and had a probe injecting a linearly polarized calibration signal at 45 degrees to the signal probes. Flux density calibration was carried out on Hydra A (PKS 0915–118) for which we assumed flux densities of 42, 24, 14 and 8 Jy at each of the four observing frequencies, giving system equivalent flux densities of 40, 80, 80 and 130 Jy on cold sky, respectively. The back end consisted of a filter bank for each polarization of 64 individual contiguous sub-bands each 5 MHz wide. The outputs of each sub-band were separately detected to produce the intensities of the two orthogonal polarizations and also multiplied with phase shifts of $\pm 45°$ to provide all four Stokes' parameters. The outputs from each frequency channel were high- and low-pass filtered, sampled every 0.6 ms, one-bit digitised and recorded on magnetic tape for off-line analysis. Typical integration times were 30 minutes, after which the entire feed and receiver package were rotated through 90 degrees; this permits the removal of certain instrumental effects from the polarization data. Calibration of the polarization was carried out with a pulsed calibration noise source and, as a further check, by observations of pulsars with known polarization characteristics.

3. Results

In broad terms, our results showed that the pulses became depolarized, at least at low frequencies, as early as 1993 October. The pulsar was not detectable in timing data taken on 1993 November 27 at 1.5 GHz but had re-appeared, although scattered, at this frequency on 1993 December 11. The pulsar then became increasingly scatter-broadened at even the highest frequency, and the last detection of the pulses prior to periastron was made (at 4.8 GHz) on 1993 December 20. During this period there was evidence for an increase in the dispersion measure (DM) of the pulsar. In spite of extensive observations throughout 1994 January, the pulses remained undetectable. The first detection of the pulses following periastron occurred on 1994 February 2, when the pulses were unscattered even at 1.5 GHz.

The pulse period was close to that predicted from the binary parameters, although phase could not be unambiguously connected across periastron passage. Around this time, the pulses were depolarized at 1.5 GHz and partially polarized at the higher frequencies. By mid April 1994, the DM, the polarization and the RM had returned to the values observed in 1993 August.

The pulsar has a DM of 146.7 cm^{-3} pc for most of its orbital period. The first evidence of a change in DM is from data taken at 1.5 GHz on 1993 November 18. These data show a larger than expected residual from the phase connected solution implying an increase in the value of DM of about 0.4 ± 0.2 cm^{-3} pc. On 1993 December 9 from observations at 2.3 GHz and 5 GHz, the DM was 150.4 ± 0.2 cm^{-3} pc, an increase of 3.7 cm^{-3} pc over the nominal DM. The last observations of the pulsed emission from PSR B1259–63 before periastron, taken at 5 GHz on 1993 December 20 showed the DM to have increased to 158.0 ± 0.2 cm^{-3} pc. Following periastron, multi-frequency observations on 1994 February 4, showed the DM to be 146.9 ± 0.2 cm^{-3} pc, an enhancement of at most 0.4 cm^{-3} pc. By 1994 March 4, the DM had returned to its nominal value of 146.7 cm^{-3} pc.

The pulsar was observed in 1993 August as part of a study of the polarization properties of pulsars (Manchester & Johnston 1994). The pulsar shows about 80 per cent linear polarization at all frequencies from 1.5 to 8.4 GHz. The RM at that epoch was $+19$ rad m^{-2}. The pulsar was next observed in polarization mode on 1993 October 8 at 1.5 GHz, when the linear polarization amounted to less than 2 per cent of the total flux. At this epoch no observations were made at higher frequencies. It was further observed at 2.3, 4.8 and 8.4 GHz from 1993 December 9–11 and no linear polarization was evident at these frequencies. This continued throughout December in spite of an extensive search of the RM parameter space.

Following periastron, and the re-emergence of the pulses, depolarization was still in evidence at 1.5 GHz on 1994 February 2. Four sets of observations were made at 8 GHz on 1994 February 3. The first two have low signal-to-noise ratios and no polarization can be measured. The final two observations revealed a rather stronger signal, and there is evidence for a very large and negative RM. The two observations show similar polarization position angles of 70° and 15° for the main pulse and the interpulse. This can be compared with the position angles for the data taken on 1994 March 4 (35° and $-20°$). In spite of the 180° position angle ambiguities, by looking at the percentage of linear polarization at different RMs we conclude that the best value from the position angle changes and the linear polarization is -6780 rad m^{-2}. What makes these observations more remarkable is that there is apparently an increase of only 0.2 cm^{-3} pc in the DM of the 1.5 GHz data taken at the same epoch. From the ratio of the RM to the DM we can get an estimate of the magnetic field parallel to the line of sight. Using an RM of $-6800\,\mathrm{rad\,m^{-2}}$ and a DM of $0.2\,\mathrm{cm^{-3}\,pc}$, the magnetic-field strength is $0.2\,\mathrm{G}$ at a distance of $\sim 45\,R_*$.

The 5 GHz data taken only 5 days (February 8) later show remarkable changes in both the the RM and the percentage of linear polarization on short time scales. Twelve 30 minute integrations were made in total. The first half of the data shows no polarization down to a limit of 5 per cent of the total intensity. During the second half of the data, the fractional linear polarization rose steadily and in the final integration was as high as 60 per cent. In the first of the integrations that shows some linear polarization, the RM was $-1630\,\mathrm{rad\,m^{-2}}$. This then increased by about $200\,\mathrm{rad\,m^{-2}}$ before settling at a value around $-1400\,\mathrm{rad\,m^{-2}}$ for the rest of the integration. Observations with such a large change in RM in such a short time scale are unprecedented. There were no DM changes measurable in these observations.

On 1994 March 4 polarization observations were again made at three frequencies. At 5 and 8 GHz, the pulsar appeared to have the same fractional linear polarization as in 1993 August. In the 1.5 GHz data, however, the linear component was only about 50 per cent. The RM had changed to $-69\,\mathrm{rad\,m^{-2}}$ as measured across the bandwidth at 1.5 GHz, and this change was confirmed by the change in the absolute position angle at 1.5, 5 and 8.4 GHz when compared with the 1993 August data. During these observations there were no changes in the RM over short time scales. By 1994 April 12, linear polarization had been fully restored at 1.5 GHz (measurements were made at this frequency only). The RM was $25\,\mathrm{rad\,m^{-2}}$. Further observations on 1994 May 10 gave an RM of $22\,\mathrm{rad\,m^{-2}}$.

4. Discussion

We will attempt to model the equatorial wind or disk of the Be star using a simple model. Waters (1986) derived a numerical model based on IRAS observations of Be stars. He assumed that the electron number density, n_e, as a function of radius varied as

$$n_e(r) = n_0(r/R_*)^{-\beta} \qquad (1)$$

where r the distance from the star and n_0 is the density at the radius, R_*, of the stellar surface. Waters found that β typically had values between 2 and 4. Optical observations of the Balmer emission lines from a sample of Be stars (Slettebak et al. 1992) showed that these disks extended to at least a few tens of stellar radii and that n_0 had values between 10^{11} and $10^{13}\,\mathrm{cm^{-3}}$. He further assumed that the disk had an opening angle of about $15°$ and that the disk was truncated at some radius. Recent radio observations of Be stars (Dougherty et al. 1991; Dougherty & Taylor 1992) showed that the emission can extend as far as a few hundreds of stellar radii and in the case of ψ Persei extends to thousands of stellar radii. While it is thus clearly unphysical to assume an abrupt cut-off to the disk, we assume here that outside the disk the gaseous material contributes only a insignificant amount to the integrated electron density along the line of sight. Throughout the modelling process we assume an orbital inclination of 35° and that the equatorial plane of the Be star lies in the orbital plane.

We will assume that the magnetic field, B, in the wind has the form

$$B(r) = B_0(r/r_*)^{-1} \qquad (2)$$

where B_0 is the magnetic field strength at the stellar surface. The field thus has a $1/r$ dependence; this is probably true beyond 5–10 R_* although inside this radius the field may be dipolar. However, at all orbital phases at which there are meaningful data the line of sight does not come within 10 R_* so the exact nature of the magnetic field structure within this radius is not important. We will further assume that the galactic contribution to the RM is $\sim$25 rad m^{-2} which is consistent with the values measured in 1993 August and after 1994 April.

We can use the polarization data to place limits on the extent of the circumstellar disk. Depolarization occurred in 1993 October and the RM did not stabilize until 1994 April. At these dates, the separation of the pulsar from its companion was about 150$\dot{R}_*$. We thus take this to be the extent of the stellar disk. This is much larger than the extent derived from the optical emission lines (Johnston et al. 1994a), however the cooler, less dense outer portions of the disk are better probed at longer wavelengths such as infra-red. The DM data only show measurable changes much closer to the star, at 100 R_* before periastron and 60 R_* after periastron. Such asymmetry is expected because the line of sight traverses a greater region of the disk prior to periastron. Due to the low inclination of the binary orbit, however, this only applies if the opening angle of the disk is large. Even so, the rather small observed DM changes imply a large value of β in Eq. (2). The best fit to the DM changes has $n_0 = 4.5\ 10^{12}$ cm^{-3} and $\beta = 4.2$. This electron density compares well to the numbers derived on a large sample of Be stars by Waters et al. (1987) but the value of β is higher than most of the stars in their sample. If we assume a slow radial outflow to the disk of $\sim$5 km s^{-1} then this leads to a derived mass loss rate in the disk of 5 10^{-8} M$_\odot$ yr^{-1}. This is a typical value for early-type B stars. However even with an r^{-4} dependence on the electron density, it is hard to produce little or no DM change on 1994 February 2. One possible explanation for this apparent strong asymmetry in the DM changes is that the equatorial plane of the star is *not* in the same plane as the pulsar orbit.

Although it is hard to constrain the magnetic field given the very few values of RMs that we obtained it is possible to reproduce the general evolution of the RM with time. For the simple field structure described above we might expect the RM contribution from the system to change sign when the pulsar crosses the line of sight to its companion and indeed the data support this. Also, the very steep increase in the RM in early February is satisfactorily modelled despite the small DM changes at this time. Using the values of n_0 and β obtained above, we derive a magnetic field strength at the stellar surface of 0.8 G. If this model is correct, then it implies a very large value of RM, in excess of 25000 rad m^{-2}, in 1993 December. As we have seen, strong depolarization occurs at RMs of 1500 and 7000 rad m^{-2} at 5 and 8 GHz respectively, so the lack of polarization in the pulses in the 1993 December data is not greatly surprising.

Given this model, and assuming a smooth wind (i.e., no clumping) we are in a position to calculate the optical depth as a function of orbital phase. The optical depth, τ, for free-free radiation is give approximately by

$$\tau = 8.2 \ 10^{-2} T^{-1.35} \nu^{-2.1} \int_0^L n_e^2 dl \qquad (3)$$

where T is the temperature in Kelvin, ν the observing frequency in GHz. If we assume that the temperature in the wind has a constant value of $\sim 10^4$ K (Waters 1986) then we find that the optical depth is about unity at 1.5 GHz on 1993 December 17 and unity at 5 GHz on 1993 December 23, close to the dates of the last detections at these frequencies. Given our simple model, these values agree well with the observations. On 1993 December 10, the flux density at 1.5 GHz was lower than average and following this date the pulsar was undetectable at 1.5 GHz. The flux density at 5 GHz on 1993 December 21 was very low and this was the last detection of the pulsar prior to periastron. Thus it would appear as if both free-free absorption and pulse scatter broadening were responsible for the lack of detection of the pulses through periastron. The model also fits the post-periastron data though not as well as the pre-periastron data. The optical depth decreases below unity on 1994 January 16 at 5 GHz and on 1994 January 22 at 1.5 GHz. The observations show that the pulsar was undetectable, as predicted, at 1.5 GHz on January 21 but was not detected at 5 GHz the following day in spite of the fact that the predicted optical depth is only ~ 0.1. The pulsar was then detectable at both frequencies on 1994 February 3. The discrepancy between the model and the observations can be explained by a moderate amount of clumping in the wind. This may also explain the non-detection of pulsed emission on 1993 November 18 at which time the optical depth from the model was low.

5. Conclusions

Between 1993 August and 1994 May, a time span which encompasses the periastron passage on 1994 January 9, we have carried out extensive multi-frequency radio observations of PSR B1259-63. Using a basic model for the structure and form of a Be star's disk (Waters 1986), which assumes a finite disk with an opening angle θ and a power law radial dependence on the electron density and magnetic field, we have been able to model the large-scale features of the observations. A full report on these observations and their implications will be given in the papers by Johnston *et al.* (1994b), Manchester *et al.* (1994) and Melatos *et al.* (1994).

References

Dougherty, S.M. & Taylor, A.R. 1992, Nat 359, 808
Dougherty, S.M., Taylor, A.R. & Waters, L.B.F.M. 1991, A&A 248, 175
Johnston, S. *et al.* 1992a, MNRAS 255, 401
Johnston, S. *et al.* 1992b, ApJ 387, L37
Johnston, S. *et al.* 1994a, MNRAS 268, 430
Johnston, S. *et al.* 1994b, MNRAS (submitted)
Manchester, R.N. & Johnston, S. 1994, ApJ (submitted)
Manchester, R.N. *et al.* 1994, ApJ (submitted)
Melatos, A., Johnston, S. & Melrose, D.B. 1994, MNRAS (submitted)
Slettebak, A., Collins, G.W. & Truax, R. 1992, ApJS 81, 335
Struve, O. 1931, ApJ 73, 94
Waters, L.B.F.M. 1986, A&A 162, 121
Waters, L.B.F.M., Coté, J. & Lamers, H.J.G.L.M. 1987, A&A 185, 206

TIMING OBSERVATIONS OF THE SMC BINARY PSR J0045−7319

V. M. KASPI
IPAC/Caltech/Jet Propulsion Laboratory
Pasadena, CA 91125, U.S.A.

R. N. MANCHESTER AND M. BAILES
ATNF/CSIRO
Epping, Australia

AND

J. F. BELL
ANU/Mount Stromlo and Siding Spring Observatories
Canberra, Australia

Abstract. We describe radio timing observations of the binary pulsar PSR J0045−7319 made over the past 3.3 yr. We show that a simple timing model involving a standard Keplerian orbit describes the data well; however, significant low-level systematic residuals never before seen in other binary pulsars remain. We consider various possible origins of the residuals.

1. Introduction

PSR J0045−7319 (PSR B0042−73) was discovered in a systematic search of the Magellanic Clouds for radio pulsars (McConnell et al. 1991). The pulsar's association with the SMC is evidenced by its high dispersion measure of 105 pc cm^{-3}, since models of the galactic electron distribution account for no more than $\sim$25 pc cm^{-3} along that line of sight (Taylor & Cordes 1993). PSR J0045−7319 is the only known radio pulsar in the SMC, and is the most distant pulsar known.

Radio timing observations of the pulsar by Kaspi et al. (1994) have shown that it is in a 51 day binary orbit with a companion having a minimum mass of 4 M$_\odot$. Although the pulsar is a faint source (its flux density at 430 MHz is only $\sim$1 mJy), its large distance makes it easily the most

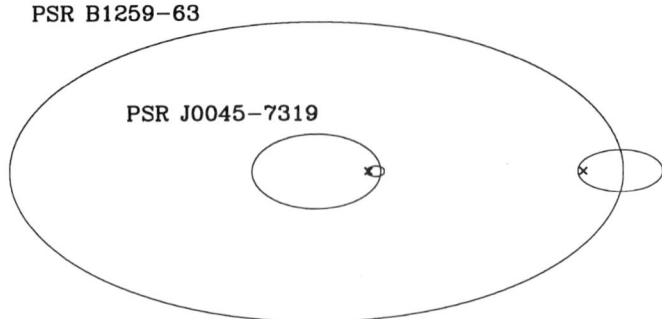

Figure 1. Relative sizes of the PSRs B1259–63 and J0045–7319 orbits, assuming 10 M$_\odot$ companions in both cases. The orbits of both binary components are shown. Crosses mark locations of the binary centers of mass.

luminous binary radio pulsar known. Optical observations in the direction of the pulsar have revealed a V = 16 mag B1 star at the timing position. The probability of chance alignment being small, we conclude that the PSR J0045–7319 binary system is the second in a new class of binary pulsars having massive, non-degenerate companions. The first example of such a system was PSR B1259–63, reported by Johnston *et al.* (1992) and also discussed in this volume by Johnston. A schematic drawing of the relative sizes of the two orbits is given in Fig. 1, where we have assumed 10 M$_\odot$ companions in both systems, consistent with estimates from optical photometry and spectroscopy.

2. Timing Properties

We have made timing observations of PSR J0045–7319 for 3.3 yr at the 64-m Parkes radio telescope. We obtained a total of 155 pulse arrival times from February 1991 through June 1994. Most observations were made at 430 MHz; however, a few were made at 660 and 1520 MHz. The standard pulse timing analysis was done using the TEMPO software package (Taylor & Weisberg 1989). The best-fit timing model including astrometric, spin, and Keplerian orbital parameters is given in Table 1; the post-fit residuals after subtraction of this model are shown in Fig. 2 versus date and in Fig. 3 versus orbital phase.

3. Discussion

The residuals in Figs. 2 and 3 clearly show significant deviations from the straightforward spin-down and Keplerian orbit model. We consider here possible origins of the deviations. We note, however, that the deviations,

TABLE 1. Astrometric, Spin, and Orbital Parameters for PSR J0045−7319 assuming a Keplerian orbit.

Right ascension, α (J2000)	$00^h\ 45^m\ 35^s.09 \pm 0^s.08$
Declination, δ (J2000)	$-73°\ 19'\ 03''.1 \pm 0''.3$
Period P	$0.926275835356(15)$ s
Period Derivative, $\dot{P}$	$4.486(1)\ 10^{-15}$
Epoch of period	MJD 48964.2000
Orbital Period, P_b	$51.169226(3)$ days
Projected semi-major axis, $a_p \sin i$	$174.2540(8)$ lts
Longitude of periastron, ω	$115°.2537(7)$
Eccentricity, e	$0.807995(5)$
Epoch of periastron	MJD 49578.56742(3)
r.m.s. timing residual	15 ms

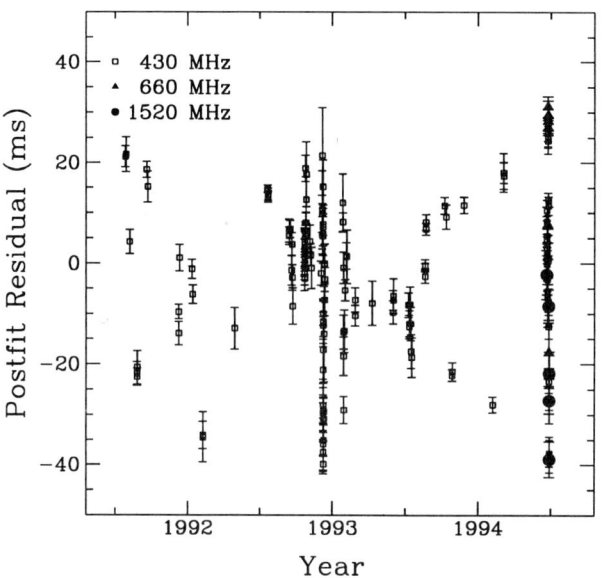

Figure 2. Residuals versus time using the model given in Table 1.

though significant, are low-level, with maximum excursions under 5% of the pulse period. Thus, we are effectively probing the details of the dynamics of this main-sequence star binary system with unprecedented precision.

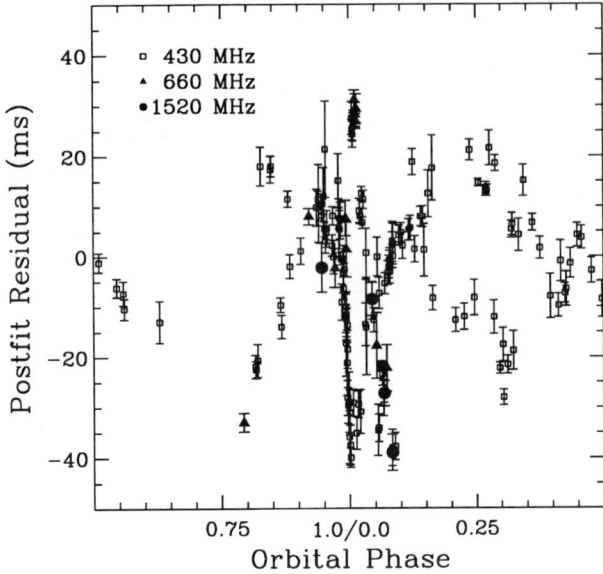

Figure 3. Residuals versus orbital phase using the model given in Table 1.

3.1. TIMING NOISE?

Deviations from simple spin down models amounting to significant fractions of the pulsar period are frequently seen in timing data from isolated pulsars (e.g., Arzoumanian *et al.* 1994, Kaspi 1994), and are usually ascribed to rotational instabilities inherent to the pulsar itself. The timing residuals for PSR J0045−7319, however, do not share the salient characteristics of standard "timing noise". Timing noise typically has a "red" spectrum, that is, has more power at long time scales, with spectral indexes ranging from −2 to −6 (Groth 1975). By contrast, the residuals for PSR J0045−7319 as shown in Fig. 2 show no sign of long-term modulation. Further, timing noise is a phenomenon most common to young pulsars having characteristic ages $\simeq 10^5$ yr, and has been shown to be correlated with $\dot{P}$ (Cordes & Downs 1985). Here, however, the pulsar's characteristic age is $3 \; 10^6$ yr; its $\dot{P}$ is correspondingly small. Thus, standard pulsar timing noise seems an unlikely source of the residuals.

3.2. ADDITIONAL ORBITAL COMPANION?

Unexplained deviations of pulse arrival times from simple spin-down models have also been used as evidence for an unseen orbital companion (Backer, Foster & Sallmen 1993; Thorsett, Arzoumanian & Taylor 1993). From Figs. 2 and 3, it is clear that any additional orbital companion must have an orbital period much shorter than the length of the data span. However, a Fourier transform of the residuals reveals no obvious periodicities.

3.3. APSIDAL ADVANCE?

Inspection of the residuals as a function of orbital phase (Fig. 3) reveals no obvious trends such as might be expected for a secular variation of a single Keplerian parameter, such as the longitude of periastron. Apsidal advance at the level of a few hundredths of a degree per year is expected in this system because of the tidal deformation of the companion near periastron, as well as from rotational deformation of the companion and general relativistic effects (Kaspi *et al.* 1994). Such an effect would be at a similar level to the observed deviations from the Keplerian orbit, but would have an obvious signature.

3.4. DISPERSION MEASURE VARIATIONS?

In Fig. 4 we show two well-sampled residual time series, both taken around periastron. The upper panel shows the residuals as a function of orbital phase (which is also time since we are looking at a single orbit) at periastron on MJD 48964. All data at this epoch were taken at 430 MHz. The obvious systematic trend could be due to an orbital-phase-dependent variation in the line-of-sight integrated electron density, the dispersion measure, due to an ionized wind from the B star. A variation in the integrated electron density will result in time delays or advances of the pulsar signal because radio waves are dispersed by free electrons according to the standard $1/f^2$ plasma law, where f is the observing frequency.

To test this hypothesis, we repeated the detailed periastron observation 11 orbits later, this time, observing at three different radio frequencies, 430, 660, and 1520 MHz. The results of these June 1994 observations are shown in the lower panel of Fig. 4. It is clear that arrival times at all frequencies are consistent within the measurement uncertainties, which argues strongly against the stellar wind as an origin of the timing residuals. These results can be used to set an interesting and uniquely direct limit on the B star mass loss; the implied upper limit on $\dot{M}$ is $\sim 10^{-11}$ $M_\odot$ yr^{-1} for a wind velocity of 1000 km s^{-1}.

3.5. TIDAL INTERACTION?

In the absence of alternatives, a remaining possibility for the origin of the PSR J0045−7319 residuals lies in a gravitational interaction between the components that results in a deformation of the companion's field. However such possibilities have been considered in detail in recent work by Kumar,

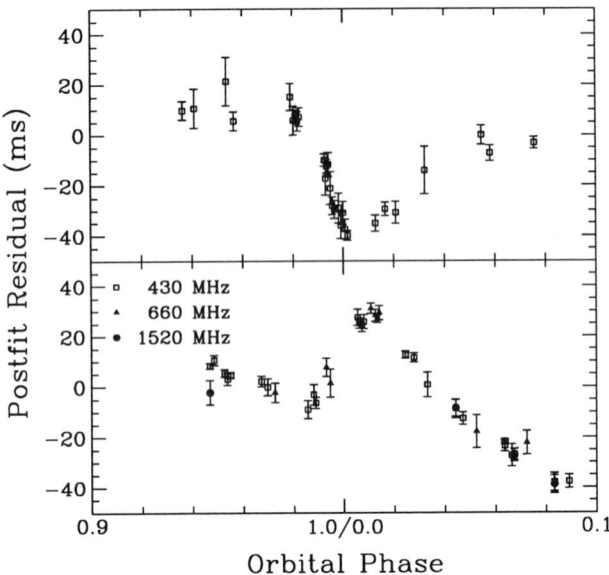

Figure 4. Residuals versus time at two periastron epochs. The upper panel is around periastron on MJD 48964, while the lower panel is around periastron of MJD 49527, 11 orbits later. Periastron is at orbital phase 0.0/1.0, and the timing model used is given in Table 1. The duration of the orbital phase interval shown in these figures is 10 days.

Ao & Quataert (1994), who conclude that apart from apsidal advance, no deviations from a Keplerian orbit should be detectable.

4. Conclusions

Timing observations of the PSR J0045–7319 binary system over a 3.3 yr period have revealed systematic deviations of pulse arrival times from a simple Keplerian orbit. We have considered several possible origins of these deviations, however their origin remains unclear. In the process, however, we have set a limit on mass loss from the B star that is low, but not incompatible with other highly model-dependent determinations of B star mass loss rates (Snow 1982). The systematic variation in the residuals near periastron suggests they are due to a dynamical interaction between the neutron star and B star, rather than to the neutron star itself. Thus, the above timing results provide strong evidence against a proposed black hole/pulsar model of the PSR J0045–7319 system (Lipunov, Postnov & Prokhorov 1994) (independent of the persuasive optical observations) since a neutron star/black hole binary should be an extremely clean system, similar to the well-studied neutron star/neutron star binaries (e.g., Taylor & Weisberg 1989).

References

Arzoumanian, Z. et al. 1994, ApJ 422, 671
Backer, D.C., Foster, R.S. & Sallmen, S. 1993, Nat 365, 817
Cordes, J.M. & Downs, G.S. 1985, ApJS 59, 343
Groth, E.J. 1975, ApJS 29, 443
Johnston, S. et al. 1992, ApJ 387, L37
Kaspi, V.M. 1994, PhD thesis, Princeton University
Kaspi, V.M. et al. 1994, ApJ 423, L43
Kumar, P., Ao, C.O. & Quataert, E.J. 1994, (preprint)
Lipunov, V.M., Postnov, K.A. & Prokhorov, M.E. 1994, ApJ (in press)
McConnell, D. et al. 1991, MNRAS 249, 654
Snow, T.P. 1982, ApJ 253, L39
Taylor, J.H. & Cordes, J.M. 1993, ApJ 411, 674
Taylor, J.H. & Weisberg, J.M. 1989, ApJ 345, 434
Thorsett, S.E., Arzoumanian, Z. & Taylor, J.H. 1993, ApJ 412, 33

Discussion

V.M. Lipunov: I think that the B star is not connected to a binary pulsar, but a black hole (see Lipunov, Postnov & Prokhorov, 1994, ApJ, in press). The main arguments are the absence of a radio eclipse and the absence of a high space (radial) velocity of the B star.

THE MASSES OF THE NEUTRON STARS IN M15C

W.T.S. DEICH
Netherlands Foundation for Research in Astronomy
Postbus 2, 7990 AA Dwingeloo, The Netherlands

AND

S.R. KULKARNI
California Institute of Technology

Abstract. Several years of timing the pulsar in the binary neutron star system M15C have yielded the masses of both stars: the total mass is $M_T = 2.7121(6)\,M_\odot$; the companion mass is $m_c = 1.36(4)\,M_\odot$; and the pulsar mass is $m_p = 1.35(4)\,M_\odot$. We argue that this system is not likely to have formed through accretion-induced collapse (AIC), and that the standard model also has problems in explaining the formation.

1. Introduction

Pulsars were discovered in M15 in 1988, as part of an extensive program for searching for millisecond pulsars in globular clusters (Wolszczan et al. 1989; Anderson et al. 1990). There are now eight known pulsars in M15 (Wolszczan et al. 1989; Anderson et al. 1990; Anderson 1993; Middleditch, private communication). Fig. 1 shows the locations of the pulsars in the cluster. The binary pulsar M15C is well outside the core of the cluster, from which it was probably ejected as a result of the collision that formed the system (Phinney & Sigurdsson 1991).

We report here the results of timing the binary pulsar PSR 2127+11C (M15C). We have measured the masses of the pulsar and its neutron star companion: both are close to $1.35\,M_\odot$, which is inconsistent with the hypothesis that most or even all of the pulsars in globular clusters were formed through accretion-induced collapse (AIC) of massive white dwarfs (Grindlay & Bailyn 1988; Bailyn & Grindlay 1990).

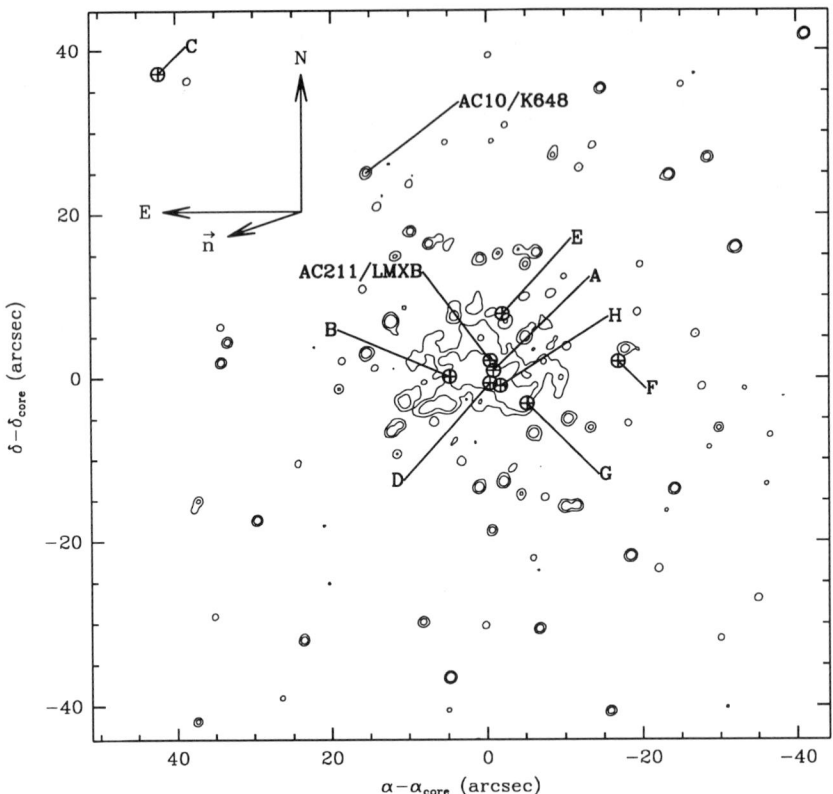

Figure 1. The pulsars superimposed on a VLA image of M15. From Prince et al. (1991).

2. Observations

All of the M15 observations were carried out with the Arecibo Observatory's 305-m radio telescope, using dual circularly-polarized line feeds. The center frequencies were either $\nu_0 \sim 430\,\text{MHz}$ or $\nu_0 \sim 1400\,\text{MHz}$ (only PSR 2127+11A is bright enough to obtain useful timing data at 1400 MHz), using bandwidths $B_{430} = 10\,\text{MHz}$ and $B_{1400} = 40\,\text{MHz}$.

The Observatory's three-level, 40 MHz autocorrelation spectrometer was used in two modes. In the first mode ("XCORCHN"), it was divided into 2 polarizations × four subcorrelators of 32 lags each, for an effective spectral resolution of 2 polarizations × 128 channels across the bandpass. In each subcorrelator, a 32-lag autocorrelation function (ACF's) was formed every 506.625 μs.[1] The correlator data were dumped to tape for off-line analysis,

---

[1] In practice, the autocorrelator samples all lags in 500 μs, but suffers dead time while it is occupied dumping a block of ACF's to the computer that writes the data to tape. The mean time between lags is precisely 506.625 μs.

TABLE 1. PSR 2127+11C Timing Parameters.

Pulse Period	P	30.5292951283 (3)	(ms)
Period Derivative	$\dot{P}$	4.9818 (80)	10^{-18} s s^{-1}
Dispersion Measure	DM	67.132(1)	cm^{-3} pc
Right Ascension	α	$21^h 30^m 1^s.20548$ (36)	J2000 coordinates
Declination	δ	$12°10'38''.1926$ (83)	J2000 coordinates
Orbital Period	P_b	28968.36974 (2)	s
Projected Semi-Major Axis	$a_1 \sin i$	2.51857 (21)	ls
Eccentricity	e	0.681388 (5)	
Epoch of Periastron	T_0	2447633.02	JD
Longitude of Periastron	ω_0	$316°.3629$	
Apsidal Motion	$\dot{\omega}$	4.4636	deg y^{-1}
Einstein Delay	γ	0.00467	ms
Orbital Period Decay	$\dot{P}_b$	-3.937	10^{-12} s s^{-1}
Total Mass	M	2.7121 (6)	M$_\odot$
Companion Mass	M_C	1.363 (40)	M$_\odot$

allowing the same data set to be used for simultaneous observations of all the pulsars in the beam.

In the second mode ("XCORHI"), 128-lag ACF's of the input voltage were folded synchronously with a Doppler-corrected pulsar period, using 256 pulse phase bins. This was only done for observations of PSR 2127+11A, yielding an effective time resolution of ∼0.43 ms. The integration time was 5 minutes for all of the XCORHI observations.

The observations reported here span the interval 21 February 1988 to 7 November 1993.

3. The masses of PSR 2127+11C and its companion

PSR 2127+11C, which is much more distant from the core than any of the other seven known pulsars (Fig. 1), is a binary pulsar with a neutron star companion (Anderson 1992). Long-term timing has yielded a measurement of the orbital period decay and the masses of the pulsar and its companion. This is the first accurate mass measurement of a pulsar in a globular cluster, and the fourth accurate binary neutron star mass measurement.

All of the PSR 2127+11C pulse time of arrival (TOA) measurements are derived from observations at 430 MHz. The modest amounts of data available at 1400 MHz did not yield satisfactory TOA's, due to the steep spectral index of this pulsar. Timing from Arecibo continues with monthly

TABLE 2. PSR 2127+11C Timing Solution Comparison. A comparison of the timing solution to the 1.5 y of data used in Anderson et al. (1992) with the present timing solution.

Parameter, X	$\Delta(X)/\sigma(X)$
Pulse Period, P	0.0
Period Derivative, $\dot{P}$	0.4
Dispersion Measure, DM	0.04
Right Ascension, α	0.2
Declination, δ	0.7
Orbital Period, P_b	0.1
Projected Semi-Major Axis, $a_1 \sin i$	0.2
Eccentricity, e	0.1
Longitude of Periastron, ω_0	0.1
Total Mass, M	0.02
Companion Mass, M_C	0.2

observations of M15 at 430 MHz. The dispersion was estimated by Anderson (1992) from pulse arrival times across the 10-MHz receiver bandwidth. The timing parameters are presented in Table 1.

We have compared our results with those from the last published timing solution (Anderson 1992), and find good agreement in all parameters. Table 2 compares the fitted parameters from Anderson with the same parameters from our data. The first column gives the parameter, and the second column gives the difference between the Anderson results and ours in units of the uncertainty reported by Anderson. Note that the timing solution reported here includes most of the same data used by Anderson, and that Anderson's uncertainties are conservative, being the formal 3σ errors. Thus one should not wonder that for all parameters X, $\Delta(X)/\sigma(X) < 1$.

On secular time scales, the apparent period of the pulsar varies due to acceleration of the pulsar and centrifugal contributions. Particularly for pulsars in clusters, the observed parameters can be dominated by external accelerations; for example, PSR 2127+11A has a negative period derivative, which is due to acceleration in the cluster potential. We therefore consider these influences on PSR 2127+11C:

$$\frac{\dot{P}}{P} = \frac{\dot{P}_0}{P_0} + \frac{a\dot{n}}{c} + \frac{V_\perp^2}{cD}$$

where P_0 is the intrinsic period, **a** is the pulsar acceleration due to sources outside the binary system, **n** is a unit vector from the Earth to the pulsar, c is the speed of light, and D is the pulsar's distance. The observed value for PSR 2127+11C is $\dot{P}/P = 1.6 \, 10^{-16} \, \text{s}^{-1}$. The second term includes accelerations due to passing stars, the cluster potential, and the Galaxy. The mean cluster potential dominates the term, which Phinney (1992) has estimated to be $0.6 \, 10^{-17} \text{s}^{-1}$. The third term is an order of magnitude smaller: $V_\perp = 474 \frac{\mu_{\text{tot}}}{10 \, \text{mas y}^{-1}} \frac{D}{10 \, \text{kpc}} \, \text{km s}^{-1}$, where μ_{tot} is the proper motion of the pulsar, and $V_\perp^2/cD = 2.5 \, 10^{-18} \, \text{s}^{-1}$.

Therefore the observed period and period derivatives are a good measure of the intrinsic values, and we conclude that the masses of PSR 2127+11C and its companion are $1.350(40) \, \text{M}_\odot$ and $1.363(40) \, \text{M}_\odot$, respectively.

4. Discussion

Although the binary systems M15C and PSR B1913+16 are superficially similar, with similar orbital period, spin period, and eccentricity, they have very different origins. PSR B1913+16 is a classic product of a high-mass X-ray binary, in which the progenitors were two massive stars in a binary system (Flannery & Van den Heuvel 1975; Burrows & Woosley 1986).

In contrast, the M15C progenitor stars came from different systems and formed a binary *after* the initial supernova. This can be seen from two separate lines of reasoning: first, the characteristic age of PSR 2127+11C is $\tau_c = P/2\dot{P} = 208 \, \text{My}$, yet there could not have been any massive stars remaining in the cluster as recently as 200 My ago. Thus PSR 2127+11C was not spun up by accretion from its present companion.

Second, had the progenitors of the pulsars been in a single binary system, the second explosion would have given the system a large enough kick to escape the cluster. The orbit would have been circular just before the second explosion, and the mass of the helium core that exploded must have been in the range 4–6 $\text{M}_\odot$ (to produce a companion mass $M_C = 1.33 \, \text{M}_\odot$; see Burrows & Woosley 1986). To produce a bound final system with $M_T = 2.71 \, \text{M}_\odot$ and eccentricity $e = 0.68$ would require an asymmetric explosion whose minimum recoil speed is (Burrows & Woosley 1986; Flannery & Van den Heuvel 1975)

$$V_{\min} = 710 \, (r^{0.5} - (1+e)^{0.5}) \, \text{km s}^{-1} = 80 - 250 \, \text{km s}^{-1}$$

where r is the ratio of the post-explosion to pre-explosion total system mass and e the post-explosion eccentricity.

In the "standard model" for millisecond pulsar production (see for example Bhattacharya & Van den Heuvel 1991) neutron stars in globular clusters were born from massive stars with $M \geq 6-8 \, \text{M}_\odot$. Most were ejected in the explosion, but some fraction of the retained neutron stars underwent tidal capture and produced low-mass X-ray binaries (LMXBs), which evolved to produce the "recycled" spun-up pulsars. The most serious difficulty with the standard model is the so-called birth rate problem, namely that the LMXB lifetime would have to be very short, $\sim 10^7$ yr, in order to produce the inferred number of millisecond pulsars.

The competing AIC scenario starts with an accreting binary white-dwarf system. When the accreting white dwarf exceeds the Chandrasekhar limit it collapses directly to a millisecond-period neutron star. A series of papers (e.g., Nomoto 1987; Nomoto & Kondo 1991) has shown that carbon-oxygen white dwarfs can readily produce neutron stars via AIC. The AIC model has the virtue of avoiding the birth rate problem entirely.

Nonetheless, the masses of PSR 2127+11C and its companion do *not* fit easily into the AIC scenario. When a white dwarf collapses to a neutron star, it loses 10–15% of its gravitational binding energy in neutrinos, and $\Delta M \leq 0.02\,M_\odot$ in the explosion. The final mass will be in the range 1.23–1.3 $M_\odot$. The new neutron star must then accrete about 0.1 $M_\odot$ to reach the observed present mass of PSR 2127+11C $= 1.35\,M_\odot$. (The companion is not lost, but the orbit grows and becomes eccentric with $e \sim 0.1$.) Yet accreting 0.1 $M_\odot$ should lead to a spin period an order of magnitude smaller than observed, and (by comparison with other spun-up pulsars) its magnetic field should also be an order of magnitude less than the inferred 10^{10} Gauss.

If the pulsar's mass is taken to be 1.30 $M_\odot$ (1.5 σ below the best-fit value), then it could have been produced by AIC without requiring any accretion after forming the neutron star. But in that case, the pulsar's companion would have to be 1.41 $M_\odot$, and *it* must have undergone significant accretion after becoming a neutron star.

In summary, AIC cannot account for both neutron stars in M15C without at least one of the members accreting $\sim 0.1\,M_\odot$ after collapse to a neutron star.

The standard model also experiences discomfort explaining M15C. Here the prime difficulty is that B is large. If spin-up occurred via a low-mass companion, one would expect B to be as small as other pulsars that have been spun up this way. On the other hand, if PSR 2127+11C was simply born spinning slow, then the characteristic age of just 200 My implies a very high birth rate of these pulsars.

Acknowledgements. We are grateful to S.B. Anderson, A. Wolszczan, and T.A. Prince for allowing us to use their first 1.5 yrs of M15 time series data in our analysis.

References

Anderson, S.B. et al. 1990, Nat 346, 42
Anderson, S.B., 1993, Ph.D. Thesis, California Institute of Technology
Bailyn, C.D. & Grindlay, J.E. 1990, ApJ 353, 159
Bhattacharya, D. & Van den Heuvel, E.P.J. 1991, Phys. Rept. 203, 1
Burrows, A. & Woosley, S.E. 1986, ApJ 308, 680
Flannery, B.P. & Van den Heuvel, E.P.J. 1975, A&A 39, 61
Grindlay, J.E. & Bailyn, C.D. 1988, Nat 336, 48
Nomoto, K. 1987 in *The Origin and Evolution of Neutron Stars*, IAU Symp. 125, D.J. Helfand & J.H. Huang (Eds.), Reidel, p. 281
Nomoto, K. & Kondo, Y. 1991, ApJ 367, L19
Phinney, E.S. 1992, Phil. Trans. Roy. Soc. Lond. A341, 39
Phinney, E.S. & Sigurdsson, S. 1991, Nat 349, 220
Prince, T.A. et al. 1991, ApJ 374, L41
Wolszczan, A. et al. 1989, Nat 337, 531

Discussion

J. van Paradijs: Your rejection of the standard evolutionary scenario (which succesfully can explain PSR 1913+16) in the case of M15C depends on the assumption, that the helium star that produced the $1.35\,M_\odot$ neutron star, had a mass in the range 4–$6\,M_\odot$. However, the work of Habets (1986, A&A 167, 61) has shown that helium stars with masses as low as $2.2\,M_\odot$ may form a neutron star; therefore the system may remain bound as the neutron star is formed.

5

X-ray Binaries

HIGH-MASS X-RAY BINARIES: RECENT DEVELOPMENTS

F. NAGASE

The Institute of Space and Astronautical Science
3-1-1 Yoshinodai, Sagamihara, Kanagawa 229, Japan

1. Introduction

There are about a dozen extensively investigated high-mass X-ray binaries (HMXBs), including LMC X-4, Cen X-3, 4U 1700–37, SMC X-1, Cyg X-1, and Vela X-1. Bhattacharya & Van den Heuvel (1991) compiled a list of "standard" HMXBs (see table 8 of their review article) and most of them, except for Cyg X-1 and 4U 1700–37, are accreting X-ray pulsars with an early-type or a Be star companion. Cyg X-3 was long considered to be a low-mass X-ray binary (LMXB). It was, however, recently revealed from infrared observations that the companion star has characteristics of a Wolf-Rayet star and it may be a fairly massive helium star (Van Kerkwijk et al. 1992; Van Kerkwijk 1993). I shall review here some recent progress in observational studies of the "standard" HMXBs and Cyg X-3.

I will concentrate on two topics and review mainly observational results from GINGA and ASCA. The first topic is a brief summary of measurements of the orbital period changes in X-ray binary pulsars. A few results have been added recently from GINGA observations, in addition to the well established results on the orbital period decay in the Cen X-3 binary system (Murakami et al. 1983; Kelley et al. 1983). The orbital period changes of LMXBs will be briefly compared with those of HMXBs. The second topic deals with results obtained from recent high-resolution spectroscopic observations with the ASCA SIS detectors (CCD cameras). Since the spectra obtained with ASCA from black-hole candidates, LMXBs and the jet-like object SS 433 are reviewed in these proceedings by Inoue, I here present results obtained for X-ray pulsars and Cyg X-3, and discuss line emission from plasmas in photo-ionization equilibrium.

2. Measurements of Orbital Period Change

Studies of orbital period changes in HMXBs provide insight into their present state, e.g. the presence of tidal torques and the rotation rate of the optical counterparts, and their evolution, including their lifetime as bright X-ray sources as the Roche lobe moves through the stellar atmosphere. Before GINGA observations, Cen X-3 was the only X-ray binary pulsar for which a finite rate of orbital period change had been measured (Murakami et al. 1983; Kelley et al. 1983).

GINGA observations between 1987 and 1992 extended the observational baseline for several X-ray pulsars, with maximum intervals of nearly 20 years between UHURU and GINGA. This extension of the baseline enables us to examine in detail the rate of change of the orbital periods which in the case of X-ray pulsars can be supplemented by accurate determinations of the orbital parameters through fitting of the Doppler shifted pulse arrival times. In the case of Cyg X-3 the orbital period (P_{orb}) and its change rate ($\dot{P}_{orb}$) are derived from the modulation of the X-ray light curve. However, accurate measurements of the P_{orb} and $\dot{P}_{orb}$ are possible in Cyg X-3 because of the long baseline of observations, stable modulation of the X-ray light curve and the extremely short (0.2 d) binary period.

A negative rate of period change $\dot{P}_{orb}/P_{orb} = (-1.78 \pm 0.08)\ 10^{-6}\ \mathrm{yr}^{-1}$ was first obtained for Cen X-3 using the eclipse data obtained with UHURU, SAS-3 and HAKUCHO (Kelley et al. 1983). The orbital decay was confirmed by Nagase et al. (1992) by adding the TENMA and GINGA determinations of mid-eclipse time using the orbital Doppler analysis of pulse arrival times; the rate of change was improved to $\dot{P}_{orb}/P_{orb} = (-1.738 \pm 0.004)\ 10^{-6}\ \mathrm{yr}^{-1}$ (see Fig. 1). It is worth pointing out that there are significant deviations of the residuals from the best-fit ephemeris in the early UHURU data. This is unlikely to be intrinsic to the nature of the binary evolution, and Deeter (private communication) suggests from his reexamination of the UHURU data that they can possibly be fit consistently to the currently improved orbital parameters.

Although Her X-1 is a LMXB, precise determination of the orbital parameters is possible for this X-ray pulsar, utilizing pulse arrival time analysis to the orbital Doppler effect. It was found by Deeter et al. (1991) that the mid-eclipse epoch of Her X-1 determined from the GINGA data occurs about 40 s earlier than expected from the extrapolation of the previously determined ephemeris. By combining the GINGA ephemeris of Her X-1 with that of previous measurements, they obtained a finite rate of orbital period decrease of $\dot{P}_{orb}/P_{orb} = (-1.32 \pm 0.16)\ 10^{-8}\ \mathrm{yr}^{-1}$. This value is about two order of magnitude smaller than that for Cen X-3 and is the smallest finite value so far measured for either HMXBs or LMXBs.

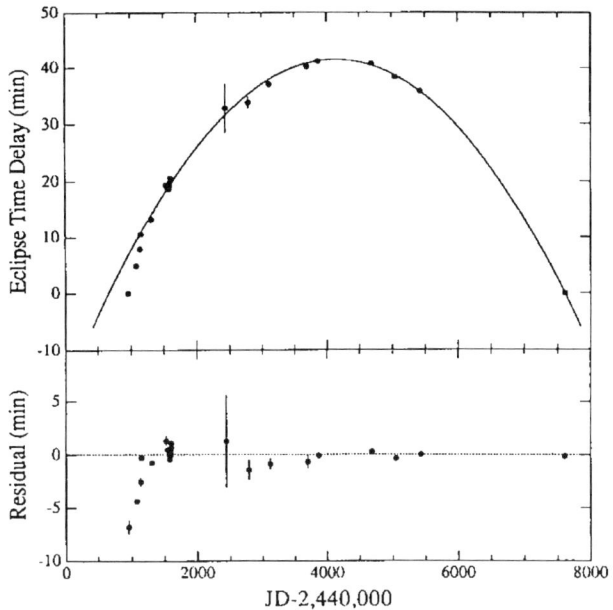

Figure 1. Delays of mid-eclipse time of Cen X-3. *Upper panel*: the observed-minus-calculated eclipse times are plotted with respect to a linear ephemeris. *Lower panel*: residuals of the observed eclipse times from the best-fitting quadratic ephemeris. (Adopted from Nagase et al. 1992).

Levine et al. (1993) analyzed the SMC X-1 data obtained with GINGA in 1987 and 1988, and found a decay of the orbit with $\dot{P}_{orb}/P_{orb} = (-3.36 \pm 0.02) \, 10^{-6} \, \text{yr}^{-1}$, by combining the GINGA results with previously measured ephemerides. This rate of orbital period change is a factor of two larger than that of Cen X-3. These two binaries posses quite similar features in many respects. Based on the result, Levine et al. have discussed the relations between the time scales for stellar evolution, orbital decay and neutron star spin up in the SMC X-1 system.

Levine et al. (1991) also analyzed the GINGA data of LMC X-4, a system similar to Cen X-3. They obtained $\dot{P}_{orb}/P_{orb} = (1.1 \pm 0.8) \, 10^{-6} \, \text{yr}^{-1}$ by combining their results with previously measured ephemerides, which gives a 2σ confidence range for $\dot{P}_{orb}/P_{orb}$ of $-0.5 \, 10^{-6} \, \text{yr}^{-1}$ to $+2.7 \, 10^{-6} \, \text{yr}^{-1}$. Combining the GINGA results with previous ephemeris measurements, Corbet et al. (1993) derived a rate $\dot{P}_{orb}/P_{orb} = (3.3 \pm 4.0) \, 10^{-6} \, \text{yr}^{-1}$ for 4U 1538−52, suggesting a 2σ upper limit of $8 \, 10^{-6} \, \text{yr}^{-1}$. Similarly, Nagase (1992) derived a rate $\dot{P}_{orb}/P_{orb} = (-0.1 \pm 2.1) \, 10^{-6} \, \text{yr}^{-1}$ for Vela X-1, yielding a 2σ upper limit of $4 \, 10^{-6} \, \text{yr}^{-1}$.

Cyg X-3 shows a clear and stable intensity modulation with a 4.8 hr period, which is generally believed to be the orbital period of the binary sys-

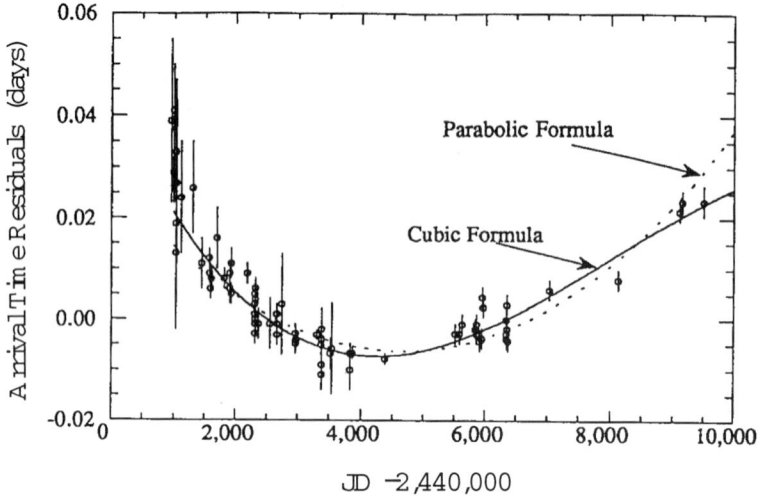

Figure 2. Arrival time residuals of the orbital modulation of Cyg X-3 with respect to a best-fit linear ephemeris. (Adopted from Kitamoto et al. 1995).

tem. This orbital period is typical for LMXBs. However, if the companion star is a massive Wolf-Rayet star (Van Kerkwijk et al. 1992; Van Kerkwijk 1993), this would be an extraordinary close system among HMXBs. An increase of the 4.8 hr period (i.e., orbital growth) was suggested by Van der Klis & Bonnet-Bidaud (1981) and this was further established by follow-up work (e.g., Kitamoto et al. 1987, 1992; Van der Klis & Bonnet-Bidaud 1989). Van der Klis & Bonnet-Bidaud (1989) evaluated the second derivative term $\ddot{P}_{orb} = (-1.6 \pm 0.4)\, 10^{-10}\, \text{yr}^{-1}$ from a cubic fit, implying a large decrease in $\dot{P}_{orb}$, and hence $\dot{P}_{orb}$ is expected to turn negative at $\sim$ 1987. New ephemerides were obtained recently from ASCA observations, and it was found that the fit to the arrival time data including these ASCA points yields significantly smaller values of $\ddot{P}_{orb}$, and even a parabolic ephemeris cannot be rejected (see Fig. 2 for a comparison of the parabolic fit and the cubic fit). The global rate of period change fit to all available data for Cyg X-3 is $\dot{P}_{orb}/P_{orb} = (1.17 \pm 0.44)\, 10^{-6}\, \text{yr}^{-1}$ (Kitamoto et al. 1995).

In addition to the above, changes of the orbital period have been reported for three LMXBs, 4U 1820–30, X 1822–371 and EXO 0748–676. Combining the latest ROSAT observations of 4U 1820–30 in 1993, Van der Klis et al. (1993) derived an orbital period decrease of $\dot{P}_{orb}/P_{orb} = (-5.3 \pm 1.1)\, 10^{-8}\, \text{yr}^{-1}$. In contrast, an orbital period increase in X 1822–371 of $\dot{P}_{orb}/P_{orb} = (3.7 \pm 1.6)\, 10^{-8}\, \text{yr}^{-1}$ was reported by Hellier et al. (1990). EXO 0748–676 is a unique eclipsing LMXB which was discovered with EXOSAT and exhibits frequent bursts and dips. The history of the study of orbital period change in this LMXB is also interesting. Parmar et al. (1991) first reported an orbital decay at a rate $\dot{P}_{orb}/P_{orb} \sim -2\,10^{-7}\,\mathrm{yr}^{-1}$ by using the EXOSAT and the first GINGA observation. Thereafter, adopting the second and third GINGA observations, Asai et al. (1993) suggested that the secular tendency is rather an orbital growth at a rate $\dot{P}_{orb}/P_{orb} \sim 0.9\,10^{-7}\,\mathrm{yr}^{-1}$. They also suggested that an ephemeris varying with a sinusoidal function provide a better fit to all the EXOSAT and GINGA data. Corbet et al. (1994) further confirmed this sinusoidal ephemeris by adopting recent ASCA observations, and suggested the possibility that this system is a hierarchical triplet.

TABLE 1. Measurements of Orbital Period Changes

Source Name	P_{orb} (days)	$\dot{P}_{orb}/P_{orb}$ (yr^{-1})	M_{opt} ($M_\odot$)	References
Cen X-3	2.09	$(-1.738 \pm 0.004)\,10^{-6}$	20 ± 4	1, 2, 3
SMC X-1	3.89	$(-3.36 \pm 0.02)\,10^{-6}$	17 ± 4	4
LMC X-4	1.41	$(1.1 \pm 0.8)\,10^{-6}$ $(-0.5\,10^{-6} \le \dot{P}_{orb}/P_{orb} \le +2.7\,10^{-6})$	15 ± 4	5
Vela X-1	8.96	$(-0.1 \pm 2.1)\,10^{-6}$ $(|\dot{P}_{orb}/P_{orb}| \le 4\,10^{-6})$	23 ± 2	6
4U 1538−52	3.73	$(3.3 \pm 4.0)\,10^{-6}$ $(|\dot{P}_{orb}/P_{orb}| \le 8\,10^{-6})$	20 ± 4	7
Cyg X-3	0.02	$(1.17 \pm 0.44)\,10^{-6}$	(~ 10)	8, 9, 10
Her X-1	1.70	$(-1.32 \pm 0.16)\,10^{-8}$	2.0 ± 0.2	11
4U 1820−30	0.008	$(-5.3 \pm 1.1)\,10^{-8}$	(~ 0.06)	12
4U 1822−371	0.23	$(3.7 \pm 1.6)\,10^{-7}$	(~ 0.3)	13
EXO 0748−67	0.16	(sinusoidal variation?)	(~ 0.2)	14, 15, 16

References: 1; Murakami et al. 1983, 2; Kelley et al. 1983, 3; Nagase et al. 1992, 4; Levine et al. 1993, 5; Levine et al. 1991, 6; Nagase 1992, 7; Corbet et al. 1993, 8; Kitamoto et al. 1987, 9; van der Klis and Bonnet-Bidaud 1989, 10; Kitamoto et al. 1995, 11; Deeter et al. 1991, 12; van der Klis et al. 1993, 13; Hellier et al. 1990, 14; Parmar et al. 1991, 15; Asai et al. 1992, 16; Corbet et al. 1994.

The derived rates of orbital period change are summarized in Table 1, together with the orbital periods and estimates of the companion masses. The rates are also plotted in Fig. 3 against the estimated companion mass. For X-ray pulsar systems, the masses of the companion star are calculated from measurements of the orbital Doppler delay curve of the X-ray sources (neutron stars) and of the optical radial-velocity curve of the companion stars. Values are adopted from Nagase (1989), except for 4U 1538−52 for which an improved value is adopted from recent measurements of the

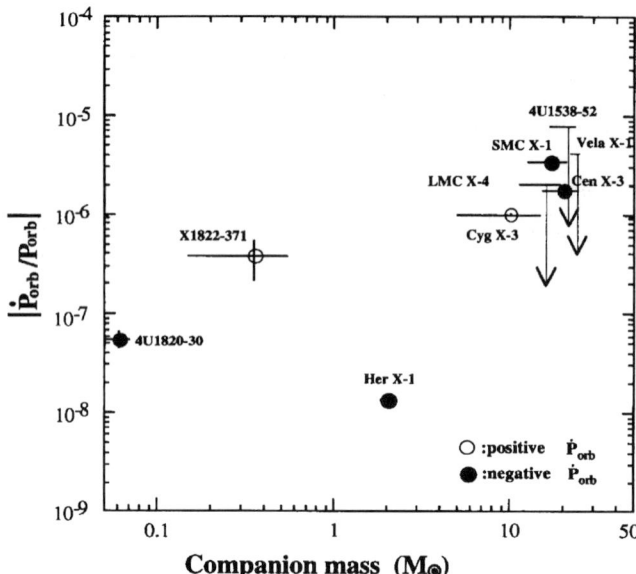

Figure 3. The magnitudes of the rate of change of the orbital period in X-ray binaries plotted against the mass of companion stars.

companion radial-velocity curve (Reynolds et al. 1992). For others sources, nominal masses estimated for the particular classes of companion stars are cited.

Although there are several mechanisms that cause the orbital period to change (see e.g., Bhattacharya & Van den Heuvel 1991), gravitational radiation loss from the binary system and mass transfer from a massive donor star to the neutron star cannot extract enough angular momentum from the system to explain the orbital decays measured in these massive binary pulsars. If the massive companion is not co-rotating with the orbit, tidal friction due to asynchronism becomes an effective mechanism of orbital period change (Kelley et al. 1983). Both orbital decay and growth are possible by this model depending on the ratio of the rotation frequency of the companion star and the orbital frequency. Levine et al. (1991, 1993) have interpreted the observed rate of change in Cen X-3, SMC X-1 and LMC X-4 all consistently with this scheme. They further suggest that Her X-1 is located in the regime of stable tidal equilibrium (i.e., synchronous rotation).

Since a naive mass transfer model does not explain the observed rate of orbital period change in Her X-1, Deeter et al. (1991) adopted a model of magnetically channeled flow. If Her X-1 possesses a wind with mass flow rate comparable to the mass accretion rate, and the companion star has a magnetic field of several tens of Gauss at the stellar surface, then the material is channeled to a large distance of the Alfvén radius and the torque exerted by the material explains the observed rate of change in Her X-1.

Mass loss via a stellar wind could be a plausible mechanism for causing an observable rate of change in the orbital period, if the mass loss rate of the companion star is large. This mechanism may explain the observed orbital growth in Cyg X-3, since the mass loss rates of Wolf-Rayet stars are generally large (Van Kerkwijk et al. 1992). From the observed rate of change in orbital period, a mass loss rate $\dot{M}_w = \dot{P}_{orb}M/2P_{orb} \sim 6\ 10^{-6}\ M_\odot\ yr^{-1}$ is derived, assuming the companion star mass to be $10\ M_\odot$ (Kitamoto et al. 1995). Thus, the interpretation of angular momentum loss via a stellar wind is plausible, since the estimated mass loss rate is in the typical range for Wolf-Rayet stars.

3. X-Ray Spectroscopy of HMXBs with ASCA

About a dozen typical HMXBs were observed during the initial 6 month performance verification phase of ASCA. High resolution X-ray spectra were obtained from these observations using the SIS detectors (CCD cameras). Spectroscopic results obtained from Vela X-1, Cen X-3 and Cyg X-3 are presented and discussed in this section.

Typical HMXBs have an evolved early-type companion star with a strong stellar wind. A neutron star orbiting the massive primary star captures the wind and X-ray emission is produced in the vicinity of the neutron star. The X-ray photons irradiate and ionize the surrounding wind, thus forming a photo-ionized sphere surrounding the neutron star. Since the direct beam from the neutron star is blocked by the companion star during eclipse phases, reprocessed X-ray emission from the photo-ionized stellar wind can be observed without contamination by the stronger direct beam. At the phase after ingress of the neutron star we can probe the trailing side of the photo-ionized sphere, then the outer region at the mid-eclipse phase, and finally the leading side of the sphere before egress from eclipse. Thus, the summed eclipse spectrum yields X-ray emission from the whole photo-ionized region.

3.1. VELA X-1

Vela X-1 is an eclipsing wind-fed X-ray binary pulsar. It is a relatively wide system with an 8.96 d orbital period, an eclipse duration of 1.7 d, and a long pulse period of 283 s. This pulsar was observed with ASCA in June and July 1993, covering an entire eclipse transition by the two observations. The spectrum obtained during an eclipse phase, when the direct beam from the vicinity of the neutron star disappears, is shown in Fig. 4.

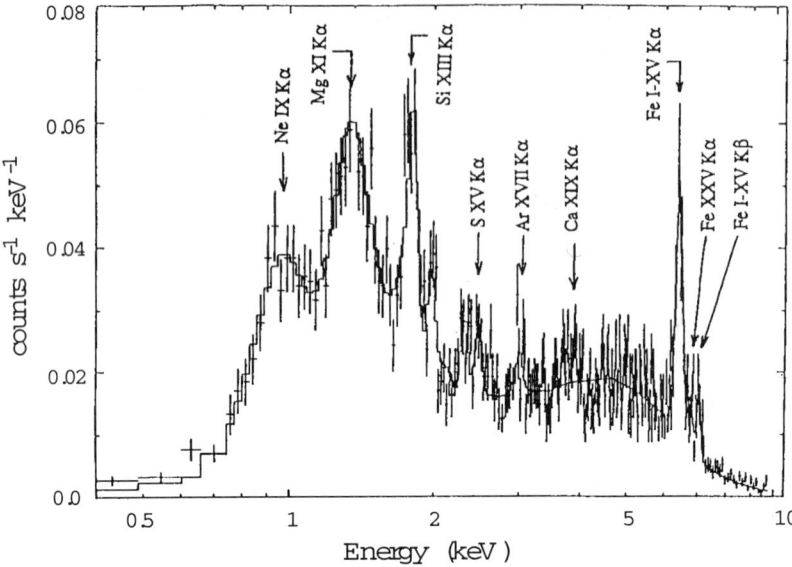

Figure 4. Energy spectrum of Vela X-1 obtained with the SIS0 detector (CCD camera) on board ASCA during the eclipse phase. A model spectrum convolved from the best-fit parameters is compared to the data by a histogram.

Remarkably, the spectrum consists of many intense lines from 0.9 keV to 7 keV that are superposed on the flat continuum. The energies of these lines and their tentative identifications are indicated in the figure. In addition to the iron $K\alpha$ line at 6.4 keV and the $K\beta$ line at 7.05 keV, which are reprocessed by iron in low-ionization states, one can see $K\alpha$ lines from highly ionized He-like ions of Ne, Mg, Si, S, Ar, Ca and Fe. Although the lines at 0.9 keV and 1.3 keV are apparently wide, these widths may be not intrinsic, but due to blending of H-like $K\alpha$ lines of Mg and Si, and of L-shell transition lines of Fe with the He-like $K\alpha$ lines. This spectral feature suggests that the stellar wind surrounding the neutron star is a multi-zone structure with different degrees of ionization.

The features dominated by the He-like lines are interpreted as the radiative recombination $K\alpha$ lines followed by cascades (Hatchett et al. 1976). As calculated by McCray et al. (1984), the stellar wind near the neutron star can be easily ionized to high ionization stages by irradiation of hard X-rays from the neutron star. Thus, zones of different ionization are formed surrounding the neutron star. Contrary to the thermal-equilibrium plasma, the electron temperature of this photo-ionization plasma is relatively low, typically about 100 eV (Kallman & McCray 1982). Thus radiative recombination followed by cascades becomes a dominant process. If this is the case, the line-like emission above the recombination edge due to the free-bound transitions should also be considerable. By coincidence, the edge energies of He-like and H-like oxygen lie near the Kα line of He-like Ne, and the edges of Ne lie around the Mg line, etc. It is difficult, unfortunately, to resolve these edge emission structures and the transition lines in the present spectrum because of the limited resolution and statistics.

3.2. CENTAURUS X-3

Cen X-3 is another typical eclipsing X-ray binary pulsar with an orbital period of 2.09 d, an eclipse duration of 0.4 d and a pulse period of 4.84 s^{-1}. This pulsar is very luminous and the companion is an O-type star, hence it is believed that an accretion disk and stellar wind co-exist in this system. This X-ray pulsar Cen X-3 was observed with ASCA in June 1993, throughout an entire eclipse transition from a pre-eclipse dip phase to eclipse egress.

Kα emission lines of the H-like ions of Mg and Si are visible in the spectra obtained both during the pre-eclipse and mid-eclipse phases of Cen X-3, in addition to the iron emission lines. This is in contrast with the Vela X-1 eclipse spectrum, in which Kα lines from He-like ions dominate. This may be caused by the fact that the luminosity of Cen X-3 is more than an order of magnitude larger than that of Vela X-1; therefore, the ionization of the surrounding stellar wind in this system will be more complete as calculated by Hatchett & McCray (1977).

Three emission lines of Fe can be resolved at 6.4, 6.7 and 6.9 keV, in addition to an absorption edge at 7.1 keV in both the pre-eclipse and mid-eclipse spectra (Ebisawa et al. 1995). As seen in Fig. 5, the intensity of the 6.4 keV fluorescent line is prominent at the pre-eclipse phase but is extremely reduced during the eclipse phase. However, the intensities of the 6.7 keV line due to He-like ions of Fe and the 6.9 keV line due to H-like ions of Fe remain about the same in the two phases. These results suggest that the fluorescent 6.4 keV line is produced by cold matter located relatively close to the neutron star, whereas the 6.7 keV and 6.9 keV lines are attributed to the hot highly ionized plasma spread out over the size of companion star. Similar results were suggested previously by Nagase et al. (1992) by analyzing the GINGA data. However, in their analysis they assumed *a priori* two narrow lines at 6.4 keV and 6.7 keV and derived the intensities of the two lines from the shift of the center energy in the observed line, which is apparently broad due to the poor resolution of the GINGA proportional counters. In contrast the ASCA observations resolve all three lines.

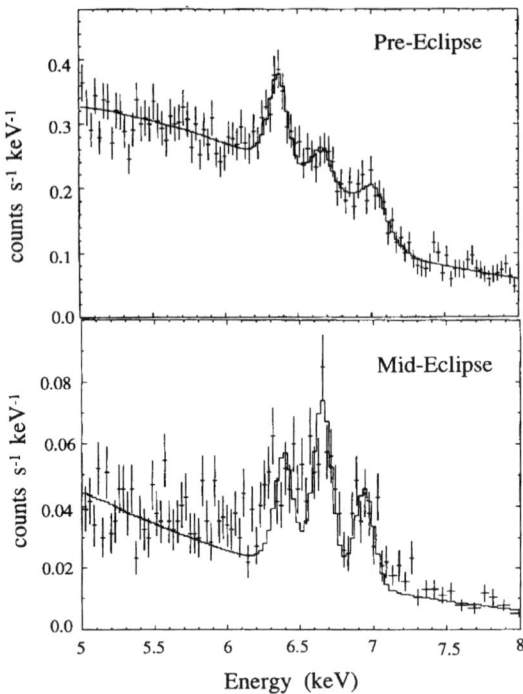

Figure 5. Energy spectra of Cen X-3 in the range of 5-8 keV obtained with the ASCA SIS0 detector. The three iron lines resolved at 6.4 keV, 6.7 keV and 6.9 keV are compared to the spectrum obtained during pre-eclipse phase and that obtained during mid-eclipse phase. Model spectra convolved from the best-fit parameters are compared to the data by a histogram.

3.3. CYGNUS X-3

Cyg X-3 was observed with ASCA during the PV phase on June 11, 1993 in its high-intensity state. The spectrum obtained by the CCD camera is shown in Fig. 6, averaged over the 4.8 hr orbital period. Using this ASCA spectrum Kitamoto *et al.* (1994) resolved for the first time the broad Fe emission line into three components at 6.4 keV, 6.7 keV and 6.9 keV They also found that the 6.7 keV and 6.9 keV lines are modulated with the 4.8 hr orbital period with a phase lag relative to the continuum; the line intensities are at maximum when the continuum intensity is at minimum.

In addition, Kα lines of both the He-like and H-like ions of Si, S, Ar, and Ca are clearly visible in the spectrum (see Fig. 6). The prominent H-like Kα lines of Si and S do not show the 4.8 hr orbital modulation unlike the

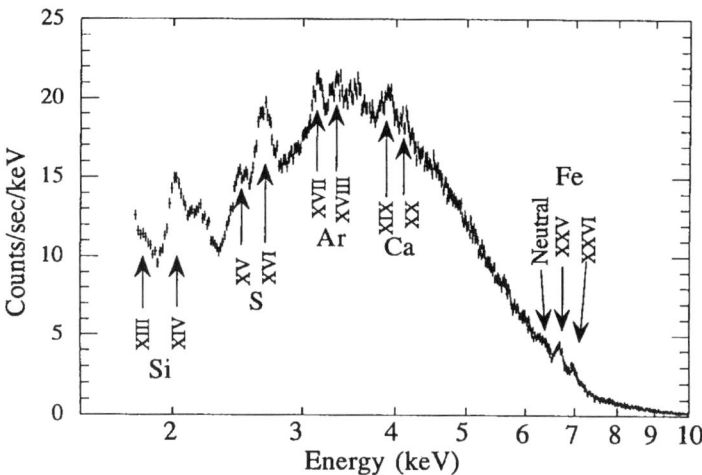

Figure 6. Energy spectrum of Cyg X-3 obtained with the ASCA SIS0 detector. Identifications of prominent lines are indicated by arrows. (Adopted from Kitamoto et al. 1994).

iron line. Those lines from lighter elements cannot originate from the same region that emits the He-like and H-like Kα lines of iron. Nevertheless, the plasma is in thermal equilibrium or in photo-ionization equilibrium, the spectrum can hardly be fitted by a simple plasma model. These emission lines and their orbital phase dependences may provide crucial clues to the understanding of the X-ray emission from the enigmatic X-ray binary Cyg X-3.

Acknowledgements. The author would like to express his thanks to all the ASCA team members. Particular thanks are due to Drs. S. Kitamoto and K. Ebisawa for providing data prior to publication. He is also grateful to Dr. S. Skinner for his careful review of the manuscript.

References

Asai, K. et al. 1992, PASJ 44, 633
Bhattacharya, D. & Van den Heuvel, E.P.J. 1991, Phys. Rep. 203, 1
Corbet, R.H.D., Woo, J.W. & Nagase, F. 1993, A&A 276, 52
Corbet, R.H.D. et al. 1994, ApJ 436, L15
Deeter, J.E. et al. 1991, ApJ 383, 324
Ebisawa, K. et al. 1995, (in preparation)
Hatchett, S. & McCray, R. 1977, ApJ 211, 552
Hatchett, S., Buff, J. & McCray, R. 1976, ApJ 206, 847
Hellier, C. et al. 1990, MNRAS 244, 39P
Kallman, T.R. & McCray, R. 1982, ApJS 50, 263

Kelley, R.L. et al. 1983, ApJ 268, 790
Kitamoto, S. et al. 1987, PASJ 39, 259
Kitamoto, S. et al. 1992, ApJ 384, 263
Kitamoto, S. et al. 1994, PASJ 46, L105
Kitamoto, S. et al. 1995, (in preparation)
Levine, A. et al. 1991, ApJ 381, 101
Levine, A. et al. 1993, ApJ 410, 328
McCray, R. et al. 1984, ApJ 282, 245
Murakami, T. et al. 1983, ApJ 264, 563
Nagase, F. 1989, PASJ 41, 1
Nagase, F. 1992, in *Frontiers of X-Ray Astronomy*, Y. Tanaka & K. Koyama (Eds.), Uni. Acad. Press, Inc., Japan, p. 79
Nagase, F. et al. 1992, ApJ 396, 147
Nagase, F. et al. 1994, ApJ 436, L1
Parmar, A.N. et al. 1991, ApJ 366, 253
Reynolds, A.P., Bell, S.A. & Hilditch, R.W. 1992, MNRAS 256, 631
Van der Klis, M. & Bonnet-Bidaud, J.M. 1981, A&A 95, L5
Van der Klis, M. & Bonnet-Bidaud, J.M. 1989, A&A 214, 203
Van der Klis, M. et al. 1993, A&A 279, L21
Van Kerkwijk, M.H. 1993, A&A 276, L9
Van Kerkwijk, M.H. et al. 1992, Nat 355, 703

Discussion

R.A.M.J. Wijers: What is the accuracy with which the positions of emission line centers can be determined using ASCA? Is there any hope of measuring velocities of the line forming region?

F. Nagase: The nominal accuracy is 20 eV for the position of a line center. The instrumental energy resolution is 120 eV at 6 keV.

Note to answer by R. Wijers: This means one can measure a velocity of $\frac{v}{c} \simeq \frac{20\,\mathrm{eV}}{5\,\mathrm{keV}} = 4\,10^{-3}$, i.e. $v \simeq 1200\,\mathrm{km\,s^{-1}}$. This is probably not small enough to measure orbital velocities in realistic cases, except perhaps if one uses 10 lines in one source and the line forming process is well-understood and simple (then $\Delta E \sim 20\,\mathrm{eV}/\sqrt{10} \sim 6\,\mathrm{eV}$, and $v \sim 350\,\mathrm{km\,s^{-1}}$. An example of this procedure is the work done with ASCA on SS433.

LOW-MASS X-RAY BINARIES—RECENT DEVELOPMENTS

M. VAN DER KLIS
*Astronomical Institute "Anton Pannekoek"
and Center for High-Energy Astrophysics
Kruislaan 403, 1098 SJ Amsterdam, The Netherlands*

Abstract. Recent developments in the field of low-mass X-ray binaries are briefly reviewed, with particular emphasis on a comparison between the systems that contain accreting low magnetic-field neutron stars and those that contain black-hole candidates. The possibility that inclination effects play a role in black-hole candidate phenomenology is explored.

1. Introduction

Low-mass X-ray binaries (LMXB) are defined as X-ray binary systems in which the mass donor stars have a mass $M < 1\,M_\odot$. The donor star mass is very significant from the point of view of binary evolution, but from the point of view of the compact objects the LMXB are a mixed bag. They include persistent and transient black-hole candidates, low magnetic-field neutron stars such as bright bulge sources, persistent as well as transient bursters and dippers, and even a few accretion powered pulsars. For this paper, in the spirit of my assignment to review recent developments in the field of LMXB, I shall mostly ignore the pulsars and focus on a comparison between systems containing low magnetic-field accreting compact objects, i.e., black-hole candidates and low magnetic-field neutron stars, as some of the more exciting recent developments have to do with the comparison between neutron stars and black-hole candidates. A very recent development is that maybe we are beginning to understand some of the effects of binary inclination in the non-dipping LMXB; I will briefly summarize the status of this in Section 5. For more extensive reviews I refer to Van der Klis (1994b,d).

In the process of accretion onto compact objects, the X-ray spectrum and the rapid X-ray variability originate in the same physical region (near the compact object), so that these X-ray properties are expected to be coupled. The hypothesis that the mass flux $\dot{M}$ towards the compact object governs both the X-ray spectrum and the power spectrum, so that when $\dot{M}$ varies these observables will show correlated variations, works well in explaining the data. Stellar-mass black holes and neutron stars have similar masses and dimensions, and therefore the phenomena accompanying accretion onto them may be expected to show similarities. Indeed, in practice, similarities have emerged that indicate that a unified description of these accretion phenomena may be possible. The fact that a phenomenon is seen in both neutron star and black-hole candidate systems is in itself very revealing, as it shows that the phenomenon can not be due to any property that is unique to either neutron stars or black holes, such as the presence or absence of a surface, or the presence or absence of a strong non-aligned magnetic field. In studying the similarities of neutron stars and black holes, furthermore, some characteristics have emerged that may indeed be unique to black holes.

The power spectra of accreting compact objects can be described in terms of a small number of simple shapes (see Van der Klis 1994b,d). *Power law noise* has a power distribution $\propto \nu^{-\alpha}$, *band limited noise* one that steepens towards high ν and flattens towards low ν. Band limited noise that has a maximum at $\nu > 0$ is called *peaked*; if the maximum is at $\nu = 0$ the component is called *flat-topped*. The same power spectral component can be at one time flat-topped and at another time peaked. *Quasi-periodic oscillations* (QPO) are a type of peaked noise. Usually, the term QPO is reserved for relatively narrow peaks.

2. Z and Atoll Sources

Z and atoll sources (Hasinger & Van der Klis 1989, hereafter HK89) are low magnetic-field neutron stars. They have been extensively described previously (Van der Klis 1989, 1994b and references therein), and only a summary of their properties is presented here. The X-ray spectral changes are usually subtle, and *colour-colour diagrams* (CDs) and *hardness-intensity diagrams* (HIDs), plots of X-ray hardness ratios *vs.* each other or *vs.* count rate are used to describe the X-ray spectral variations. The sources produce a characteristic track in the CDs and HIDs, and source position in the track is used as an indication for $\dot{M}$.

Six *Z sources* are known. They produce Z-shaped tracks in X-ray CDs and HIDs. $\dot{M}$ is inferred to increase following the Z track from upper left to lower right. Z source power spectra show three broad noise components, *very-low-frequency noise* (VLFN), *low-frequency noise* (LFN) and *high-frequency noise* (HFN), and two QPO components, *horizontal-branch oscillations* (HBO) and *normal- and flaring-branch oscillations* (N/FBO).

VLFN is 1–6% amplitude power law noise that gets stronger with $\dot{M}$. HBO and LFN are a QPO and a band limited noise component that appear and disappear together, and are likely physically related. They are strongest at low $\dot{M}$ and disappear at high $\dot{M}$. HBO frequency (13–55 Hz) and LFN cut-off frequency (2–20 Hz) increase with $\dot{M}$. LFN can be flat topped or peaked, depending on the source. N/FBO have a preferred frequency near 6 Hz. In Sco X-1 and GX 17+2, their frequency has been observed to increase from $\sim$6 to $\sim$20 Hz when $\dot{M}$ increases.

The most successful HBO model is the *magnetospheric beat frequency model* (Alpar & Shaham 1985; Lamb et al. 1985), which requires Z sources to have a magnetosphere. Some pulsars show QPO that may be caused by a similar mechanism (Angelini et al. 1989; Finger et al. in these proceedings). In most models for the N/FBO, *radiation pressure* plays the key role (Van der Klis et al. 1987; Hasinger 1987; Lamb 1989; Fortner et al. 1989; Miller & Lamb 1992; Alpar et al. 1992). Z sources have near-Eddington luminosities, and the NBO frequency is roughly similar in each Z source, in accordance with the idea that the frequency is determined by the Eddington critical luminosity L_{Edd}. Lamb (1991) proposed a comprehensive model for the QPO and X-ray spectral properties of Z sources that uses the above ingredients.

A dozen *atoll sources* are known (HK89; Van der Klis 1994b). They show one curved branch in the CD, often fragmented due to observational effects. $\dot{M}$ increases from left to right along the branch. Their power spectra show two broad noise components called *very-low-frequency noise* (VLFN) and *high-frequency noise* (HFN). Atoll source VLFN is power law noise similar to that in Z sources. Atoll source HFN has a cut-off frequency of 0.3–20 Hz and depends strongly on $\dot{M}$. At low $\dot{M}$ it is strong (up to 22%); when $\dot{M}$ increases this decreases to <2% while the cut-off frequency increases (Yoshida et al. 1993; Prins et al. 1994). Atoll HFN is sometimes flat-topped and sometimes peaked.

HK89 proposed that the neutron stars in atoll sources have lower magnetic field strengths than Z sources, and are constrained to lower mass fluxes $\dot{M}$. The lower field explains why the (magnetospheric) HBO are not seen in atoll sources, and the lower $\dot{M}$ why the same is true for the (near-Eddington) N/FBO. The implied relation between $\dot{M}$ and magnetic field strength may have an evolutionary origin (Van der Klis 1991). Predictions are that an atoll source that becomes bright will show Z source high-$\dot{M}$ properties (N/FBO and appropriate spectral branches), but never HBO, and that a Z source that becomes faint will show millisecond pulsations.

The properties of Cir X-1 fit the first prediction. This source is a low magnetic-field neutron star (it shows type 1 X-ray bursts; Tennant et al. 1986a, b) with a complex phenomenology that most likely originates in the large variations in mass transfer that the system undergoes as a function of its 17-d period. Sometimes (at intermediate brightness levels and away from periastron) its power-spectrum and CD behaviour are very similar to those of an atoll source on the banana branch (Oosterbroek et al. 1994). When the source becomes very bright, at periastron, it sometimes shows 6–20 Hz QPO and spectral branches that are reminiscent of Z source N/FBO behaviour (Tennant 1987; Makino et al. 1992; Oosterbroek et al. 1994). The source is apparently an example of an atoll source that can reach $\dot{M}_{\rm Edd}$ (van der Klis 1991; Oosterbroek et al. 1994). As will be discussed in Section 4, Cir X-1 also shares some characteristics with black-hole candidates.

3. Black-Hole Candidates

Three source states are distinguished in black-hole candidates (Tananbaum et al. 1972; Oda et al. 1976; Miyamoto et al. 1991). In the *low state* (LS) the X-ray spectrum is a flat power law with photon spectral index 1.5–2. In the *high state* (HS) the 1–10 keV flux is much higher due to a soft component; the power law is sometimes "sticking out" from under the soft component at higher energies. In the *very high state* (VHS) the X-ray spectrum is similar to that in the high state (at higher 1–10 keV flux), with perhaps an additional hard power law component. The VHS is mainly distinguished from the HS by the properties of its rapid X-ray variability.

Fig. 1 summarizes the power spectra in the three states. The LS power spectrum shows strong (30–50% amplitude) band-limited noise with $\nu_{\rm cut}$ between 0.03 and 0.3 Hz. This LS noise is usually flat-topped, but sometimes peaked (Vikhlinin et al. 1994). The level of the flat top and, in anti-correlation with this, the cut-off frequency $\nu_{\rm cut}$ sometimes vary, whereas the power spectrum above $\nu_{\rm cut}$ remains approximately unchanged (Belloni & Hasinger 1990; Miyamoto et al. 1992a). In the HS power law noise with $\alpha \sim 1$ and an amplitude of a few % is present. Sometimes LS noise is present in the hard X-ray spectral component seen in the HS. Slow QPO with frequencies similar to the LS noise cut-off frequencies (~ 0.08–0.8 Hz; Motch et al. 1983; Ebisawa et al. 1989; Grebenev et al. 1991) and possibly related to peaked LS noise sometimes occur in LS and HS. The rare VHS shows 3–10 Hz QPO and rapidly variable broad-band noise. The QPO show second harmonics and possible subharmonics. The noise in the VHS alternates in shape, sometimes within 1 s, between band-limited ($\nu_{\rm cut} \sim 1$–10 Hz), and power law shaped ($\alpha \sim 1$). CD/HID branches occur in the VHS, and the power spectral parameters seem to depend on position in the branches, but

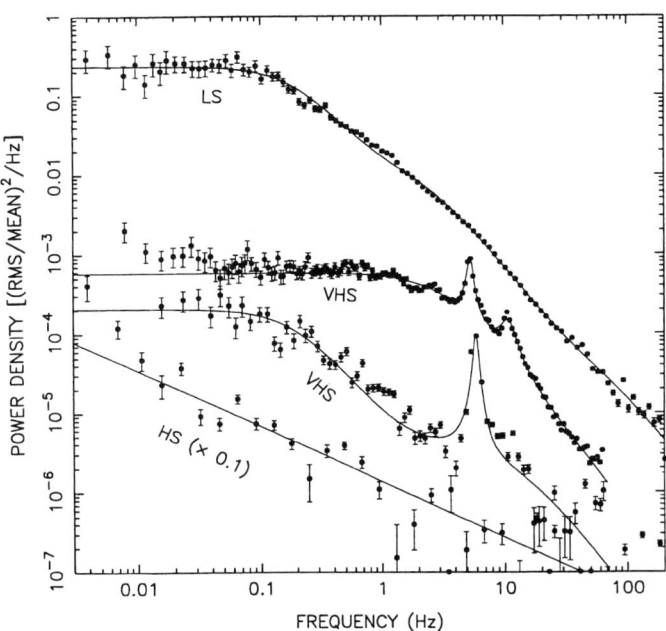

Figure 1. Power spectra from GINGA data of black-hole candidates in the low (LS; Cyg X-1), high and very high (HS and VHS; GS 1124−68) states.

these branch structures are not very similar from one epoch to the next ("messy" branches). The LS and VHS band limited noise cut-off frequency and amplitude fit one relation (Van der Klis 1994c), suggesting that they form one phenomenon. The VHS power law noise is similar to that in the HS.

The transient black-hole candidate GS 1124−68 (Nova Mus '91) in its decay went through all three states (Miyamoto *et al.* 1992b; Kitamoto *et al.* 1994), strongly suggesting that the states directly follow $\dot{M}$. In the bright low magnetic-field neutron-star systems the assumption of strict dependence on $\dot{M}$ worked very well. Recent evidence indicates that some black-hole transients, even when they are very luminous, remain in the "low" state; see Section 5.

4. Similarities between Black-Hole Candidates and Low Magnetic-Field Neutron Stars

There is a number of striking similarities between black-hole candidate and neutron star phenomenology (see Van der Klis 1994a,d).

The black-hole candidate LS is very similar to the atoll source low $\dot{M}$ ("island") state. Both states occur at the lowest 1–10 keV count rates and inferred $\dot{M}$ levels. Both are dominated by strong (several 10%) band limited noise (LS noise and atoll HFN) which is sometimes flat-topped and sometimes slightly peaked. When an atoll source becomes really faint, the power spectra are nearly indistinguishable from those of black-hole candidates in the low state, and the 1–20 keV (Langmeier et al. 1987; Yoshida et al. 1993) and 13–80 keV (Van Paradijs & Van der Klis 1994) X-ray spectra become hard, just as in black-hole candidates in the LS. Even the inverse correlation between cut-off frequency and flat-top level, characteristic for black-hole candidates in the LS, was seen in an atoll source, 4U 1608−52, at low $\dot{M}$. Z source LFN fits in with black-hole candidate LS noise and atoll HFN: it is also stronger at lower $\dot{M}$, disappears at higher $\dot{M}$, can be peaked and flat-topped, and has a higher ν_{cut} at higher $\dot{M}$. The absence of a similar band-limited noise component in pulsars, and also the beat-frequency model as applied to Z sources, suggest that such noise arises through inhomogeneities in the inner, radiation pressure dominated part of the disk, which in pulsars is disrupted by magnetic stresses.

The black-hole candidate VHS has strong similarities to the Z source high $\dot{M}$ ("normal/flaring branch") state. Both occur at the highest inferred $\dot{M}$ levels, and both show QPO, with similar frequencies (6–20 Hz in the neutron star systems, 3–10 Hz in the black-hole candidates), that depend on the position of the source in branched tracks in the HID/CDs. Clearly different is the harmonic content of the QPO (black-hole candidate VHS QPO show strong harmonics, Z source N/FBO do not) and the character of the HID/CD branches (much "messier" in BHCs). Another difference is that Z sources do not show the fast changes in broad-band noise shape seen in black-hole candidates.

The properties of Cir X-1 provide a further link between neutron stars and black holes. In some of its high states (Tennant 1987; Makino et al. 1992; Oosterbroek et al. 1994), this source shows a mix of characteristics of Z sources and black-hole candidates in high $\dot{M}$ states (see Fig. 2). It shows QPO with frequencies between 6 and 20 Hz and no second harmonics (both Z source characteristics) in combination with messy branches in the CD/HID and fast changes in the shape of the broad-band noise (BHC characteristics). The reason, then, that Cir X-1 sometimes resembles a black hole in its rapid variability characteristics, as was noted by Toor (1977) and Samimi et al. (1979), while its X-ray bursts show it to be a

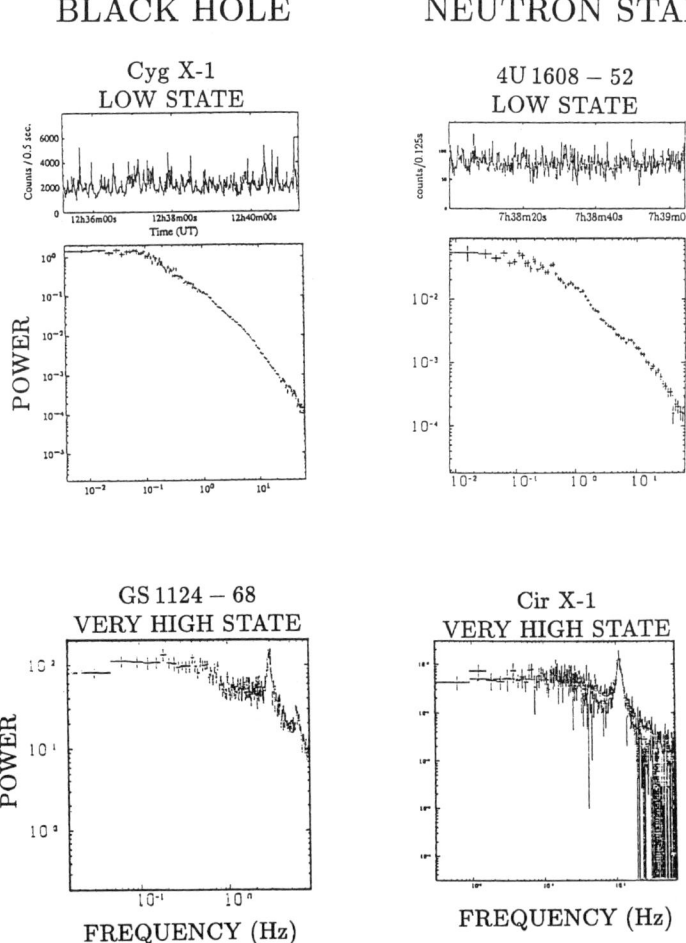

Figure 2. Power spectra from the black-hole candidates Cyg X-1 (*top left*) and GS 1124−68 (*bottom left*) in the low state and the very high state, respectively, and from the low magnetic-field neutron stars 4U 1608−52 (*top right*) and Cir X-1 (*bottom right*) in the atoll island state and a very high X-ray brightness state, respectively, illustrating the similarity between neutron star and black-hole candidate low and very high states. Compiled from Inoue (1992), Takizawa et al. (1994) and Makino et al. (1991).

neutron star, is that it is the only neutron star that we know that has a magnetic field as low as in atoll sources that sometimes accretes at near- or super-Eddington rates. Cir X-1 is therefore a key object as it can help to distinguish between phenomena that are characteristic for accretion onto any compact object that has no appreciable magnetic field, and phenomena that are truly characteristic for accretion onto a black hole. Following this line of reasoning, one concludes that a high harmonic content of the high

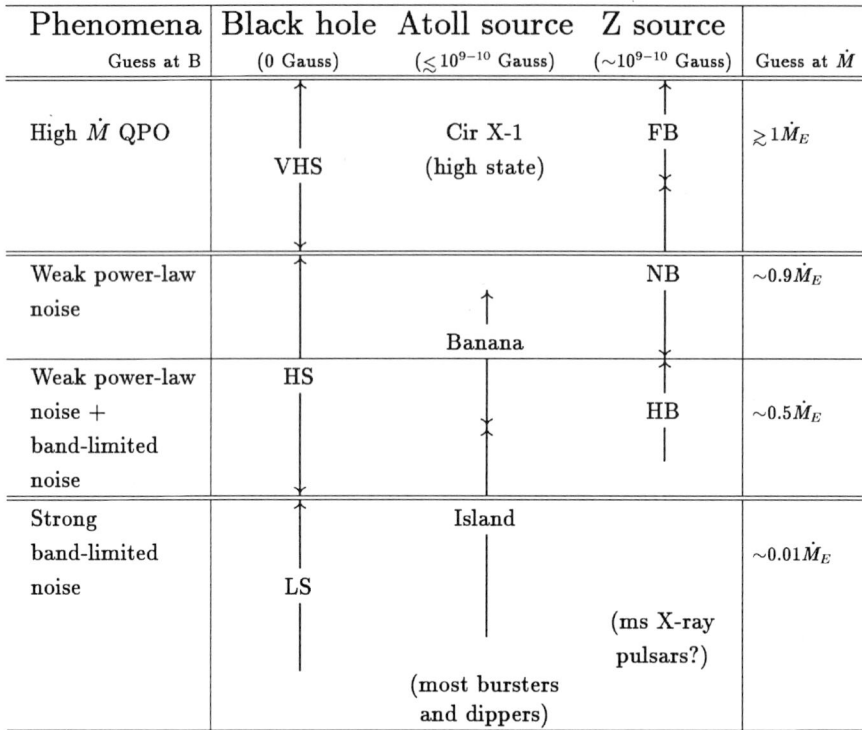

Figure 3. Proposed classification scheme for X-ray binary source states. There are three states that are common to neutron stars and black holes; in a given source the mass transfer rate $\dot{M}$ towards the compact object determines the state. The power spectral shapes that are characteristic of each state are indicated in the leftmost column. The correspondence between the source states of each source type is indicated. Magnetic field strengths and mass fluxes are rough indications only. In particular, other source parameters might affect the $\dot{M}/\dot{M}_{\rm Edd}$ levels at which state transitions occur.

$\dot{M}$ QPO may be a black-hole signature, whereas variable broad-band noise and messy branches are not.

On the basis of this array of similarities, it can be concluded that the phenomenology of the black-hole candidates and low magnetic-field neutron stars may be described in terms of three $\dot{M}$-driven states that are common to accreting low magnetic-field neutron stars and accreting black holes (Van der Klis 1994a). Fig. 3 presents a line-up of the three common states of black-hole candidates and low magnetic-field neutron stars.

It was proposed recently that the power law noise components (VLFN) seen in accreting neutron stars might be due to unsteady nuclear burning on the neutron star surface (Bildsten 1993). If correct, then the amplitude of this noise is constrained by the ratio of nuclear burning to accretion energy (times a correction factor dependent on the wave form of the noise). For hydrogen, this is ~0.04; for helium only ~0.01. Note that 4U 1820−30, which is believed to accrete hydrogen-poor matter, sometimes shows VLFN with a strength of 4.5% (HK89), in apparent violation of this. The power law noise of black-hole candidates in the HS could not be caused by the same mechanism.

5. Inclination Effects

Detailed examination of the properties of Z sources, in particular in their flaring branches, has led Kuulkers & Van der Klis (1994) to propose that obscuration by a geometrically thick inner accretion disk plays a role in Z source phenomenology. The disk swells when $\dot{M}$ increases, and for higher inclination i obscuration effects set in at lower $\dot{M}$. A similar model could explain some of the differences seen between black-hole candidates (Van der Klis 1994a). Some black-hole transients, such as GS 2023+338, show only a hard power-law X-ray spectral component, even when they are very bright, and in for example GX 339−4 the observable energy flux in the 1–200 keV band is higher in the low state than in the high state (see Fig. 3 in Grebenev et al. 1993).

The reason for the disappearance of the hard LS X-ray spectral component in the HS may be obscuration of a central, hot and rapidly variable region by matter in, e.g., a puffed-up accretion disk. For a pole-on viewing geometry no obscuration would occur and the system would show only the hard, rapidly variable X-ray spectral component at all $\dot{M}$ levels; this might explain the behaviour of GS 2023+338. The increasing concentration of the hard X-rays towards the (rotation) polar axes with increasing $\dot{M}$ would in this scenario explain why the apparent 1–200 keV luminosity of GX 339−4 in its LS seems to be (at least sometimes) higher than in its HS and VHS: most of the energy would be leaving the system in the HS and VHS along the polar axis and not be seen by us (see Fig. 4).

In the low magnetic-field neutron stars the X-ray flux is an unreliable indicator of $\dot{M}$; the same might turn out to be the case in the black-hole candidates. Note, that the mass flux $\dot{M}$ that by hypothesis determines the state is that *towards* the compact object, just as is the case in the Z sources; at near- and super-Eddington rates, not all of this matter may actually be accreted; jets might for example be formed when $\dot{M}$ becomes high enough.

HARD

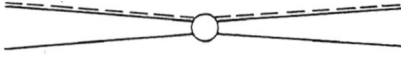

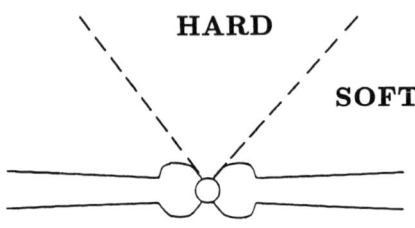

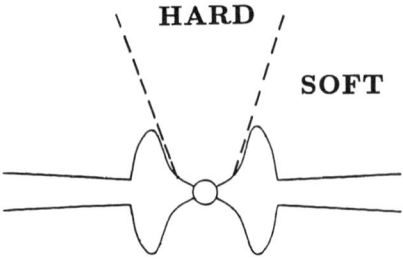

Figure 4. A geometrically thick inner disk could obscure a central, hot and rapidly variable emitting region for observers at high inclination.

6. Conclusion

A picture may be emerging in which the millisecond fluctuations in black holes, and in neutron stars with high, low and very low magnetic fields can be understood in common terms. Just two structures determine the basic physics of the accretion process, namely the magnetosphere and the inner (radiation pressure dominated) disk. Z sources have the most complex phenomenology, showing HBO and N/FBO, as well as LFN and HFN, because their magnetic field is weak enough to allow the presence of a mostly undisturbed inner accretion disk (like in black holes and atoll sources) and strong enough to allow the presence of a small magnetosphere (like in pulsars). X-ray pulsars (not discussed here) have a magnetosphere and no inner disk and therefore show only an HBO-like and an HFN-like component, black holes and atoll sources have an inner disk and no appreciable magnetosphere and therefore only show an N/FBO-like and an LFN-like component. Inner disk structure causes anisotropic emission and thereby inclination effects are introduced in the phenomenology.

Acknowledgements. This work was supported in part by the Netherlands Organization for Scientific Research (NWO) under grant PGS 78-277.

References

Alpar, M.A. & Shaham, J. 1985, Nat 316, 239
Alpar, M.A. et al. 1992, A&A 257, 627
Angelini, L., Stella, L. & Parmar, A.N. 1989, ApJ 346, 906
Bildsten, L. 1993, ApJ 418, L21
Belloni, T. & Hasinger, G. 1990, A&A 227, L33
Ebisawa, K., Mitsuda, K. & Inoue, H. 1989, PASJ 41, 519
Fortner, B., Lamb, F.K. & Miller, G.S. 1989, Nat 342, 775
Grebenev, S.A. et al. 1991, SvAL 17(6), 413
Grebenev, S. et al. 1993, A&AS 97, 281
Hasinger, G. 1987, A&A 186, 153
Hasinger, G. & Van der Klis, M. 1989, A&A 225, 79 [HK89]
Inoue, H. 1992, in *Accretion disks in compact Stellar Systems*, J.C. Wheeler (Ed.), (ISAS RN 518)
Kitamoto, S. et al. 1994, in preparation
Kuulkers, E. & Van der Klis, M. 1994, A&A (submitted)
Lamb, F.K. 1989, *Proceedings 23rd ESLAB Symposium*, ESA SP-296, 215
Lamb, F.K. 1991, in *Neutron Stars, Theory and Observation*, J. Ventura & D. Pines (Eds.), NATO ASI Vol. 344, p. 445
Lamb, F.K. et al. 1985, Nat 317, 681
Langmeier, A. et al. 1987, ApJ 323, 288
Makino, Y., Kitamoto, S. & Miyamoto, S. 1991, poster presented at the 28th Yamada Conference, Nagoya, Japan, April 8–12 1991
Makino, Y., Kitamoto, S. & Miyamoto, S. 1992, in *Frontiers of X-ray Astronomy*, Y. Tanaka & K. Koyama (Eds.), Universal Academy Press, Tokyo, p. 167
Miller, G.S. & Lamb, F.K. 1992, ApJ 388, 541
Miyamoto, S. et al. 1991, ApJ 383, 784
Miyamoto, S. et al. 1992a, ApJ 391, L21
Miyamoto, S. et al. 1992b, *GINGA Memorial Symposium* (ISAS, Tokyo, 1992), F. Makino & F. Nagase (Eds.), p. 37
Motch, C. et al. 1983, A&A 119, 171
Oda, M. et al. 1976, Ap&SS 42, 223.
Oosterbroek, T. et al. 1994, A&A (in press)
Prins, S. et al. 1994, A&A (in preparation)
Samimi, J. et al. 1979, Nat 278, 434
Takizawa, M. et al. 1994, (in preparation)
Tananbaum, H. et al. 1972, ApJ 177, L5
Tennant, A.F. 1987, MNRAS 226, 971
Tennant, A.F., Fabian, A.C. & Shafer, R.A. 1986a, MNRAS 219, 871
Tennant, A.F., Fabian, A.C. & Shafer, R.A. 1986b, MNRAS 221, 27P
Toor, A. 1977, ApJ 215, L57
Van der Klis, M. 1989, ARA&A 27, 517

Van der Klis, M. 1991, in *Neutron Stars, Theory and Observation*, J. Ventura & D. Pines (Eds.), NATO ASI Vol. C344, p. 319
Van der Klis, M. 1994a, ApJS 92, 511
Van der Klis, M. 1994b, in *X-Ray Binaries*, W.H.G. Lewin, J. van Paradijs & E.P.J. van den Heuvel (Eds.), Cambridge University Press, (in press)
Van der Klis, M. 1994c, A&A 281, L17
Van der Klis, M. 1994d, in: *The Lives of the Neutron Stars*, M.A. Alpar, Ü. Kızıloğlu & J. van Paradijs (Eds.), NATO ASI Vol. C450, p. 301
Van der Klis, M. et al. 1987, ApJ 316, 411
Van Paradijs, J. & Van der Klis, M. 1994, A&A 281, L17
Vikhlinin, A. et al. 1994, (preprint)
Yoshida, K. et al. 1993, PASJ 45, 605

Discussion

S.R. Kulkarni: 1.) The previous speaker (dr Nagase) mentioned that the orbital period change in the LMXB 4U 1820–30 to be 5×10^{-8} yr^{-1} and said you would be talking about this source. The recent determination of the center of NGC 664 and the position of the UV counterpart of 4U 1820–30 by HST (I. King et al. 1993, ApJ 413, L117) places the 4U-source within 0.6 arcsec of the cluster center. Thus all of the observed orbital periodicity can be attributed to the cluster potential.

2.) Recently, Bildsten wrote a paper (1993, ApJ 418, L21) attributing much of the LFN to incomplete burning on neutron star surface. If so, there should be considerable difference in the LFN features between LMXBs and black hole systems.

M. van der Klis: 1.) In our A&A Letter (1993, A&A 279, L21) on this source, we conclude that the variations in the shape of the light curve that we observe with ROSAT are just by themselves sufficient to explain most of the observed orbital phase changes. So, it seems we now have *two* independent ways to explain the period changes without requiring binary evolutionary efforts.

2.) Bildstens idea is that the VLFN (the power law that eliminates the power spectra at the lowest frequencies) is caused by nuclear burning, not the LFN. This is a very interesting thought and we are following up on this. The measurement of the VLFN is difficult, as it is sometimes too steep to measure with standard Fourier techniques, and also because on the relevant time scales ($>10^2$–10^3 s) the data often show gaps. However, at this stage it is already clear that black hole candidates *do* sometimes show VLFN-like power laws, which in the low state might be mashed by the presence of the strong shot noise component down to relatively low frequencies. A problem for the nuclear burning model might be the core of the atoll source 4U 1820–30. If this source is accreting H-depleted material, the energy available from nuclear burning may be too little to explain the observed VLFN amplitudes in that source.

APERIODIC FLUX VARIABILITY IN A 0535+262

MARK H. FINGER
Compton Observatory Science Support Center/USRA
Goddard Space Flight Center

ROBERT B. WILSON AND B. ALAN HARMON
NASA/Marshall Space Flight Center

AND

WILLIAM S. PACIESAS
University of Alabama Huntsville

Abstract. A "giant" outburst of A 0535+262, a transient X-ray binary pulsar, was observed in 1994 February and March with the Burst and Transient Source Experiment (BATSE) onboard the Compton Gamma-Ray Observatory. During the outburst power spectra of the hard X-ray flux contained a QPO-like component with a FWHM of approximately 50% of its center frequency. Over the course of the outburst the center frequency rose smoothly from 35 mHz to 70 mHz and then fell to below 40 mHz. We compare this QPO frequency with the neutron star spin-up rate, and discuss the observed correlation in terms of the beat frequency and Keplerian frequency QPO models in conjunction with the Ghosh-Lamb accretion torque model.

1. Introduction

A 0535+262 is a 103 s X-ray pulsar in a binary system with the Be star HDE 245770. Since its initial discovery in 1975 (Rosenberg et al. 1975), the source has been frequently observed to undergo transient outbursts. The outbursts show a range of peak intensities, with the largest reaching 3 Crab in the 2–10 keV band. For a review of previous observations see Giovannelli & Graziati (1992).

A major outburst of A 0535+262 occurred in 1994 February and March (Wilson et al. 1994a). Hard X-ray observations with BATSE were made continuously during the 50 day duration of the outburst. At the peak of the outburst the intrinsic spin up rate determined from pulse timing was approximately 1.2×10^{-11} Hz s^{-1} (Wilson et al. 1994b), clearly indicating the presence of an accretion disk. The formation of a transient accretion disk during "giant" outbursts has previously been inferred from optical and UV observations (Motch et al. 1991).

During 27 days of the outburst a broad quasi-periodic oscillation (QPO) like feature appeared in Fourier power spectra of the flux (Finger et al. 1994b). One possible explanation for this feature is the beat frequency model (Alpar & Shaham 1985; Lamb et al. 1985). In this model blobs of matter, in the process of being entrained in the neutron star's magnetic field, orbit the neutron star at approximately the Keplerian frequency of the inner edge of the accretion disk, accreting at a rate that is modulated by the magnetic field. This produces a peak in the power spectra at the beat frequency between the Keplerian and the pulsar spin frequencies. Another possible model is that the inner edge of the accretion disk contains structures that persist for a few cycles around the neutron star, and modulate the observed flux by obscuration (Van der Klis et al. 1987). In this case, the power spectral feature should be located near the Keplerian frequency of the disk inner edge.

Both of these models predict a simple relationship between the QPO frequency and the rate of mass accretion through the disk. The accretion rate also determines the torque on the neutron star. We compare the observed relationship between the neutron star spin up rate and the QPO frequency with the predictions of the beat frequency or Keplerian frequency models combined with the Ghosh & Lamb (1979) accretion torque model. The beat frequency model does best at predicting the trend of the data, but the Keplerian frequency model is not ruled out by the observations.

2. Observations and Analysis

BATSE (Fishman et al. 1989) has eight unshielded planar scintillation detectors oriented in the corner directions of a cube. Its full sky field of view allows the continuous monitoring of transient outbursts.

Pulsed hard X-ray emission from A 0535+262 was first detected with BATSE on January 28th (TJD 9380). The flux initially remained at a low level, but then on February 3 the flux began to rise quickly (Finger et al. 1994a), peaking on February 18 (Wilson et al. 1994a,b). The last BATSE pulsed flux detection was on March 20, fifty days after the first detection. The light curve of the February-March outburst as determined by the Earth occultation method is shown in panel A of Fig. 1.

A pulse timing analysis was performed to determine the intrinsic spin-up rate of A 0535+262. This analysis used the binary orbital parameters

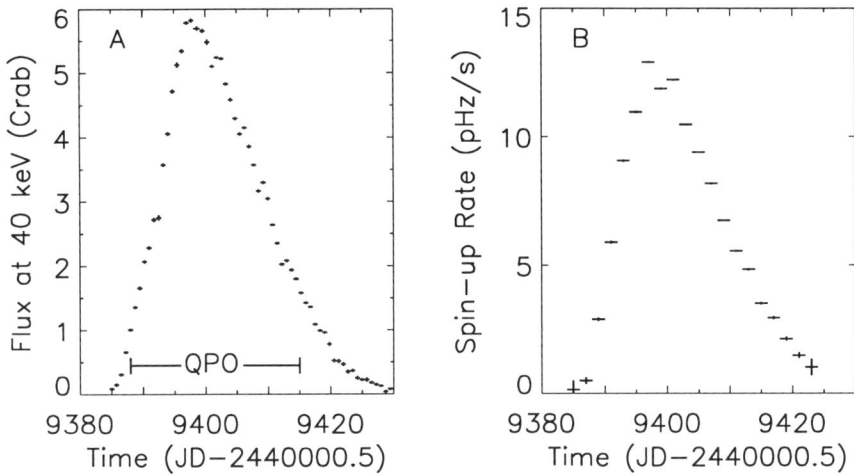

Figure 1. A) shows the A 0535+262 flux history determined from Earth occultations. Also shown is the interval of QPO detections. B) shows the intrinsic spin-up of A 0535+262 during the outburst.

recently determined by Finger et al. (1993), based on a series of weak outbursts of A 0535+262 that occurred near periastron passage in the three orbits previous to the February-March outburst. The spin-up rate during the outburst is shown in panel B of Fig. 1.

Aperiodic variability in the source flux was noticed early in the outburst (Finger et al. 1994b). Daily average power spectra were made from the DISCLA channel 1 rates (20–50 keV, 1.024 s resolution) after subtraction of the daily mean pulse profile. The power spectrum for February 19 is shown in panel A of Fig. 2. These power spectra consist of an approximately $1/f$ power-law component extending from at least 5 mHz to 0.5 Hz, and a significant concentration of noise power in a bump centered in the 35–70 mHz range. For convenience we will call this bump in the power spectra a QPO, although the bump typically has a FWHM/center frequency of 50% or slightly greater, and therefore does not strictly meet the QPO definition used in connection with LMXBs. Typical fractional r.m.s. amplitudes of the coherent, power-law (10–500 mHz), and QPO components were 20%, 15%, and 9%, respectively. The QPO was detectable for an interval of 27 days indicated in panel A of Fig. 1. Panel B of Fig. 2 shows the history of the QPO center frequency, obtained by fitting the daily power spectra with the sum of Gaussian and power-law models. As the outburst progressed the center frequency rose smoothly from 35 mHz to 70 mHz near the peak of the outburst, and returned slowly to 35 mHz.

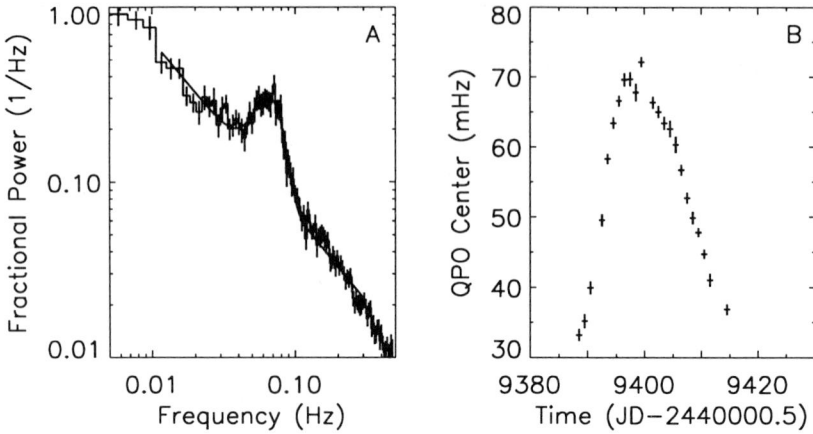

Figure 2. A) shows the mean power spectrum for February 19. The best fit model is superposed within the fit interval. B) shows the center frequency of the Gaussian QPO component of the power spectral model.

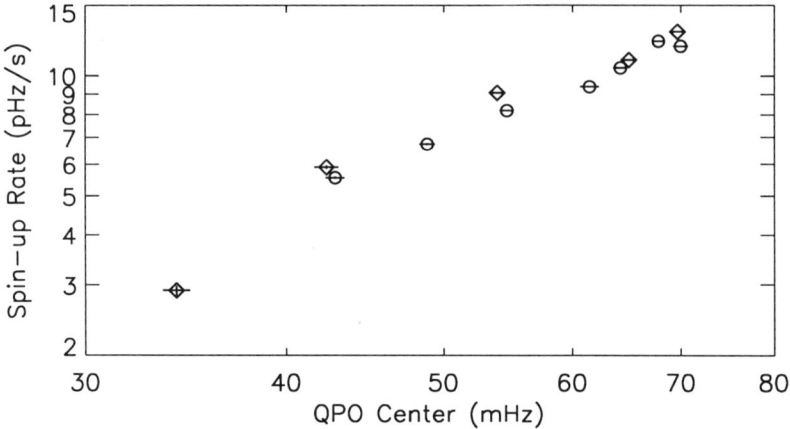

Figure 3. Comparison of intrinsic spin-up rate of A 0535+262 during the outburst and the QPO center frequency. Diamond symbols are used during the rise of the outburst and circles during the decline.

Fig. 3 compares the QPO center frequency and the neutron star spin up rate. For the plot, pairs of center frequencies were interpolated to the midpoint of the spin up rate measurements. The QPO frequency and the spin up rate are seen to be highly correlated. The tracks from the rise and the fall of the outburst are amazingly close, with spin up on the rise being only slightly higher than spin up on the fall at the same QPO frequency.

3. Discussion

Simple theoretical predictions of the relationship between the QPO center frequency and the spin up rate can be obtained by combining either the beat frequency model, or the Keplerian frequency model with the Ghosh & Lamb (1979) accretion torque model. The Ghosh & Lamb model gives the accretion torque as

$$N = \dot{M}\sqrt{GMr_0}\, n(\omega_s) \tag{1}$$

where $\dot{M}$ is the accretion rate, M is the neutron star mass, and r_0 is the radius of the inner edge of the accretion disk. The dimensionless torque function $n(\omega_s)$ depends only on the fastness parameter ω_s which is the ratio of the neutron star spin frequency ν_{ns} to the Keplerian frequency at the inner edge of the accretion disc ν_K. The radius of the inner edge of the accretion disk is given by

$$r_0 = (GM)^{1/3}(2\pi\nu_K)^{-2/3} = \eta\mu^{4/7}(2GM)^{-1/7}\dot{M}^{-2/7} \tag{2}$$

where η is a geometry-dependent constant that Ghosh & Lamb computed to be 0.52, and μ is the neutron star magnetic moment. If we assume that the torque N acts on the solid body moment of inertia of $2/5MR^2$, the spin-up rate may be written as

$$\dot{\nu}_{ns} = \aleph n(\omega_s)\nu_K^2 \text{ where } \aleph = \frac{5\pi\eta^{3.5}\mu^2}{\sqrt{2GM^2R^2}}. \tag{3}$$

The dimensionless constant $\aleph$ has a value of $2.1\,10^{-9}$ for the representative values of $\eta = 1$, $\mu = 10^{31}\,\mathrm{G\,cm^3}$, $M = 1.4\,M_\odot$, and $R = 10^6$ cm.

A plot of $\dot{\nu}_{ns}/\nu_K^2$ versus the fastness parameter ω_s will therefore yield a measurement of $\aleph n(\omega_s)$. This is shown in Fig. 4 for both the beat frequency and Keplerian frequency models. Also shown on the plots is the theoretical relationship for several values of $\aleph$ using the approximation for $n(\omega_s)$ given by Ghosh & Lamb (1979). For both the beat frequency model and the Keplerian frequency model the approximate scaling of $\dot{\nu}_{ns} \propto \nu_K^2$ is born out by the data. For the beat frequency model, the observations are consistent with $\aleph n(\omega_s)$ being a slowly varying function that decreases with increasing ω_s (decreasing $\dot{M}$), although detailed agreement between observation and model prediction is not achieved. For the Keplerian frequency model, the observed form of $\aleph n(\omega_s)$ rises at low ω_s and then falls at higher ω_s. The agreement between observations and model prediction is worse than in the beat frequency model case. The observations therefore favour the beat frequency model, for which we estimate a value of $\aleph = (2.5 \pm 0.5)\,10^{-9}$.

For assumed values of $\eta = 1$, $R = 10^6$ cm, and $M = 1.4\,M_\odot$, we compute a polar magnetic field of $B = 2\mu/R^3 = 2\,10^{13}$ G. Observations with

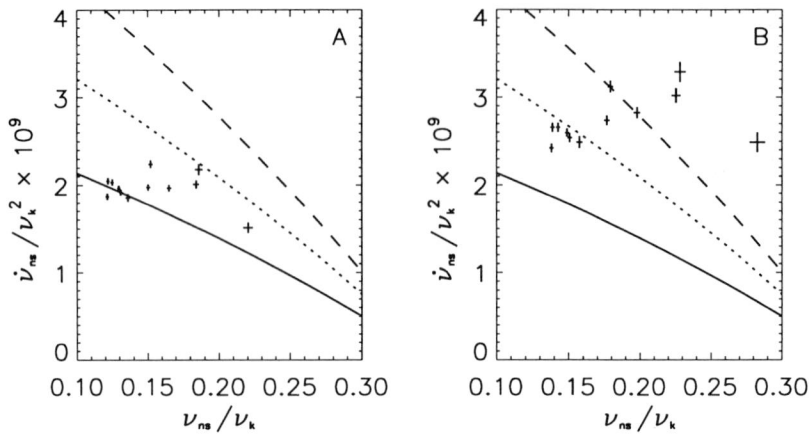

Figure 4. The ratio of the neutron star spin-up rate to the Keplerian frequency squared for the beat frequency (panel A) and Keplerian frequency (panel B) models. The curves give the Ghosh-Lamb prediction, for values of the dimensionless constant ℵ of $2.0\ 10^{-9}$ (solid), $3.0\ 10^{-9}$ (dotted), and $4.0\ 10^{-9}$ (dashed).

OSSE during this same outburst of A 0535+262 revealed a cyclotron line at an energy of 110 keV (Grove *et al.* 1994), or a polar magnetic field of $B = 9.5\ 10^{12}$ G. Given the number of poorly known parameters, this rough agreement is encouraging.

Acknowledgements. M.H.F. would like to thank J. van Paradijs, M. van der Klis, and L. Bildsten for useful discussions.

References

Alpar, M.A. & Shaham, J. 1985, Nat 316, 239
Finger, M.H. *et al.* 1993, in *Evolution of X-ray Binaries*, S.S. Holt & C.S. Day (Eds.), AIP Conf. Proc. Vol. 408, p. 459
Finger, M.H., Wilson, R.B. & Hagedon, K.S. 1994a, IAU Circ. 5931
Finger, M.H., Wilson, R.B. & Harmon, B.A. 1994b, IAU Circ. 5934
Fishman, G.J. *et al.* 1989, *Proceedings GRO Science Workshop*, GSFC, p. 2-39
Ghosh, P. & Lamb, F.K. 1979, ApJ 234, 296
Giovannelli, F. & Graziati, L.S. 1992, Space Sci. Rev. 59, 1
Grove, E. *et al.* 1994, ApJ (in press)
Lamb, F.K. *et al.* 1985, Nat 317, 681
Motch, C. *et al.* 1991, ApJ 369, 490
Rosenberg, F.D. *et al.* 1975, Nat 256, 628
Van der Klis, M. *et al.* 1987, ApJ 313, L19
Wilson, R.B., Harmon, B.A. & Finger, M.H. 1994a, IAU Circ. 5933
Wilson, R.B. *et al.* 1994b, IAU Circ. 5945

Discussion

P. Ghosh: Your determination of $\frac{d \ln \nu_K}{d \ln M}$ (which is nominally 2/7, as verified by early QPO work on LMXB) is very likely going to be the best measurement so far. This has very important implications for the state of the inner accretion disk in X-ray pulsars. The value is in agreement with the usual one-temperature and gas-pressure dominated disk, and not with some other dark models.

A. Alpar: The observation of the beat frequency does not depend on the existence of beaming or polar caps if the field is weak enough, as in LMXBs.

M. Finger: This is a HMXB with beaming and the observation of ν_{rotation} is expected.

A. Alpar: To see the inclination dependence comparatively and to discuss ν_{rotation}, we hope for QPO observations of this quality from other HMXBs.

J. van Paradijs: This is a response to the remark by P. Ghosh. One can derive relations between $\dot{\nu}$ and ν_K for different disk models. It turns out that the only disk model (from a sample of models published by P. Ghosh) that fits the approximately quadratic dependence of $\dot{\nu}$ on ν_K is that of a classical gas-pressure dominated disk; a radiation-pressure dominated disk, or a two-temperature disk, lead to very different $\dot{\nu}(\nu_K)$ relations, inconsistent with the one observed for A 0535+262.

ASCA OBSERVATIONS OF WHITE DWARFS, NEUTRON STARS AND BLACK HOLES

H. INOUE

Institute of Space and Astronautical Science
3-1-1, Yoshinodai, Sagamihara, Kanagawa 229, Japan

1. Introduction

ASCA, the fourth Japanese X-ray astronomy satellite, was launched by the Institute of Space and Astronautical Science (ISAS) on 1993 February 20. ASCA is designed to be a high-capability X-ray observatory (Tanaka et al. 1994). It is equipped with nested thin-foil mirrors which provide a large effective area over a wide energy range from 0.5 to 10 keV. Two different types of detectors, CCD cameras (SIS) and imaging gas scintillation proportional counters (GIS) are employed as the focal plane instruments.

The ASCA instruments cover the most important energy band for plasma diagnostics, because the K lines and the K absorption edges from oxygen through iron (and also the L lines of iron) at various ionization stages all lie within this band. On the other hand, the previous high-sensitivity imaging missions, the EINSTEIN Observatory and ROSAT, are limited to narrower energy bands than ASCA: the EINSTEIN Observatory is limited to <4 keV, and ROSAT to <2 keV. The ASCA SIS can individually resolve all major lines (except the L line complex around 1 keV). Motion of plasma of the order or greater than $1000 \mathrm{~km~s^{-1}}$ can be measured significantly from Doppler shift of the line energies. Also, ASCA has a much larger effective area, hence a much larger photon collection power, than the EINSTEIN Observatory and ROSAT, which is an advantage for detailed line spectroscopy requiring large enough numbers of photons for meaningful statistics. These capabilities of ASCA allow diagnostics of accreting matter around compact objects through studies of emission and absorption features. This review shows some recent ASCA results on studies of accreting matter around white dwarfs, neutron stars and black holes.

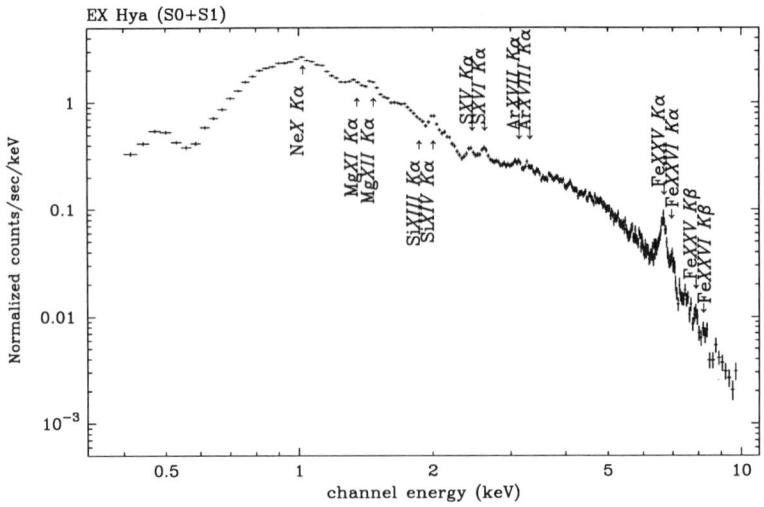

Figure 1. The spectrum of EX Hya obtained with the ASCA SIS. Helium-like and hydrogen-like Kα lines of Mg, Si, S, Ar, and Fe are clearly seen (Ishida et al. 1994).

2. Emission Lines from EX Hydra

EX Hydra is one of the brightest intermediate polars and shows two periodicities in its light curve: one is the orbital period of 98.3 min and the other is the rotational period of the white dwarf of 67.0 min.

ASCA observed this source on July 16, 1993 (Ishida et al. 1994). Fig. 1 shows the ASCA SIS spectrum averaged over the observational period. In this figure, helium-like and hydrogen-like Kα lines of various elements are clearly seen. If the plasma emitting these lines is in ionization equilibrium, the ratio of the intensity of the H-like line to that of the He-like line of a particular element indicates the temperature of the plasma responsible for the line emission from the element. Fig. 2 shows the theoretical ratio of the line intensities for Mg, Si, S, Ar and Fe as a function of the temperature and the allowable range of the temperature from the observation is shown for each of the elements. Clearly, if ionization equilibrium holds, then a multi-temperature plasma is needed.

A cooling flow along an accretion column onto a magnetized white dwarf could be the origin of the multi-temperature plasma. The temperature and the density structure of the accretion column behind the shock have been analytically solved by Aizu (1973). A rough estimate assures that ionization equilibrium holds in the accretion column of EX Hydra. Then, the line ratio can be calculated for each of the elements as a function of the shock temperature and we can obtain the shock temperature by comparing the

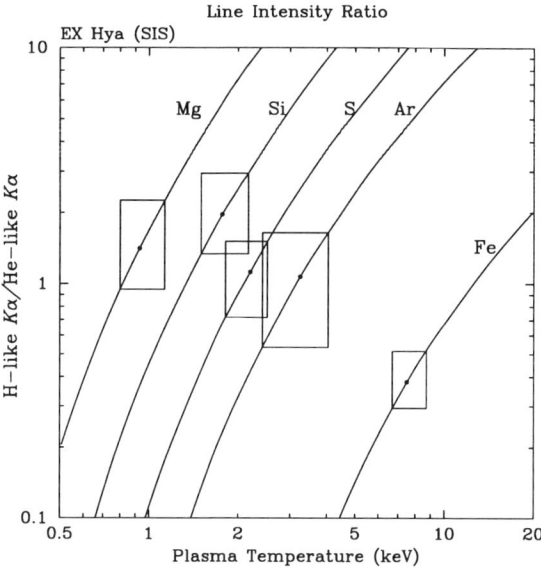

Figure 2. The allowable temperature ranges obtained from the intensity ratios of H-like to He-like Kα lines for Mg, Si, S, Ar and Fe, observed in the EX Hya spectrum. Solid curves are the theoretical intensity ratios for these elements as a function of the plasma temperature.

observed line ratios with the theoretical curves as shown in Fig. 3 (Fujimoto *et al.* 1994).

Since the height of the shock above the white-dwarf surface is considered to be small compared to the radius of the white dwarf, the shock temperature is proportional to the ratio of the mass to the radius of the white dwarf. As a result, the mass of the white dwarf is estimated to be $0.5\pm0.2\,M_\odot$ if we assume the theoretical mass-radius relation of white dwarfs.

The above discussion suggests that matter in the accretion column cools from above 10 keV just behind the shock to below 1 keV along the flow, since emission lines from such light elements as Mg or Si should come from a plasma with a temperature of about 1 keV. Most of the gravitational energy seems to be radiated away from the optically thin accretion column in this case, which is currently considered to be carried into the optically thick atmosphere of the white dwarf in AM Her stars (see, e.g., Frank *et al.* 1985). We need further X-ray spectroscopic studies of the polar-type sources.

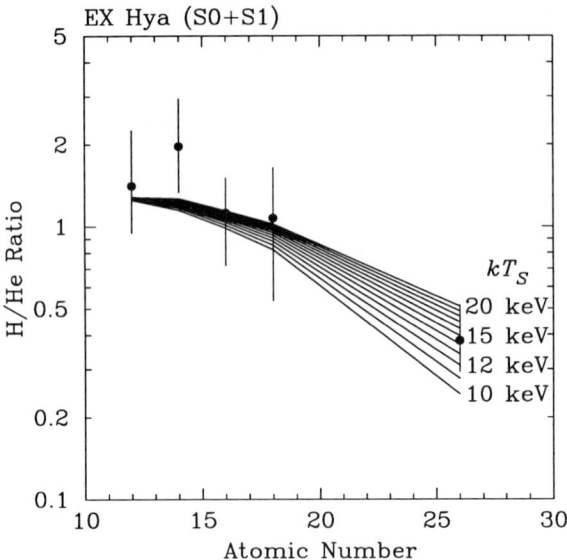

Figure 3. The line intensity ratios expected from the post-shock accretion flow together with the observed ratios. The temperature and the density profiles have been taken from Aizu (1973).

3. Very Dim Phase of 1608–52 and Cen X-4

The low-mass X-ray binaries 1608–52 and Cen X-4 are known to be recurrent transients [see, e.g., Tsunemi (1991) for 1608–52; Matsuoka et al. (1980) for Cen X-4). Both sources are also X-ray bursters (e.g., Nakamura et al. 1989; Matsuoka et al. 1980), which strongly suggests that these sources harbour weakly magnetized neutron stars in their centers of activity.

ASCA observed 1608–52 and Cen X-4 on August 12, 1993, and Feb. 27, 1994, respectively, and both sources were very dim.

The flux from 1608–52 was $7\ 10^{-13}\,\mathrm{erg\,cm^{-2}\,s^{-1}}$ (0.5–10 keV) and its luminosity was estimated to be $10^{33}\,\mathrm{erg\,s^{-1}}$, assuming a distance of 3.6 kpc (Nakamura et al. 1989). The X-ray spectrum was very soft and can be reproduced with either a single power law model, an optically thin thermal emission model, a thermal bremsstrahlung or a blackbody with a photoelectric absorption. The best-fit temperature is ~ 0.5 keV for a thermal bremsstrahlung model, and ~ 0.2 keV for a blackbody.

The flux from Cen X-4 was $1.4\ 10^{-12}\,\mathrm{erg\,cm^{-2}\,s^{-1}}$ (0.5–10 keV) and the luminosity was estimated to be $2\ 10^{32}\,\mathrm{erg\,s^{-1}}$ on the assumption that the distance is 1.2 kpc (McClintock & Remillard 1990). The X-ray spectrum can again be reproduced with either a single power law, an optically thin thermal emission model, thermal bremsstrahlung or a blackbody. The best-fit temperature is $\sim 0.7\,\mathrm{keV}$ for thermal bremsstrahlung and $\sim 0.3\,\mathrm{keV}$ for a blackbody.

ROSAT observations of the transient low-mass X-ray binary Aql X-1 revealed that its quiescent luminosity is also as low as several times $10^{32}\,\mathrm{erg\,s^{-1}}$, with a spectrum that could be described by a blackbody of temperature $\sim 0.3\,\mathrm{keV}$ (Verbunt et al. 1994).

An important question is where the X-rays come from in the very dim phase of weakly magnetized neutron star sources. These sources commonly exhibit luminosities as low as $10^{32-33}\,\mathrm{erg\,s^{-1}}$ and very soft spectra which can be reproduced typically with a blackbody of temperature 0.2–0.3 keV.

One possibility is that mass overflow from the companion star completely ceases in this dim phase and the X-rays come from the coronal activity of the companion star. However, the mass of the companion star is thought to be less than a solar mass and hence the stellar luminosity from the nuclear burning at its center should be less than $10^{33}\,\mathrm{erg\,s^{-1}}$. Since the energy release rate in coronal activity should be a small fraction of the nuclear energy generation rate, coronal activity with a luminosity of $10^{32-33}\,\mathrm{erg\,s^{-1}}$ seems to be too large.

A second possibility is that mass overflow from the companion star takes place but the accretion disk does not extend to the neutron star surface due to the very low efficiency of angular momentum transfer. In this case, gravitational energy release at the outer side of the disk is expected to produce the X-rays. The luminosity can be explained by the gravitational energy release of matter with an accretion rate of $10^{16-17}\,\mathrm{g\,s^{-1}}$ at a distance of $10^{10}\,\mathrm{cm}$ from the neutron star. However, if the emission is optically thick, the temperature should be of the order of $10^4\,\mathrm{K}$; if the emission is optically thin, a temperature of the order of $10^6\,\mathrm{K}$ seems to be too low compared to the temperature corresponding to the gravitational potential at that distance from the neutron star.

It should also be noted that radio pulsar activity is expected in the above two cases but no radio emission has been detected (Kulkarni et al. 1992; see also the poster contribution to this symposium by Kulkarni et al.).

A third possibility is that matter is accreted by the neutron star at a rate of $10^{12-13}\,\mathrm{g\,s^{-1}}$. In this case, a significant fraction of the radiation will heat the surface of the neutron star, irrespectively of how the gravitational energy is converted to radiation, and will be re-radiated as blackbody emission from the neutron star surface. If we estimate the surface area responsible for the blackbody emission from the observed luminosity and the temperature with the help of the Stefan-Boltzmann constant, the result is roughly consistent with emission from a fraction of the neutron star surface.

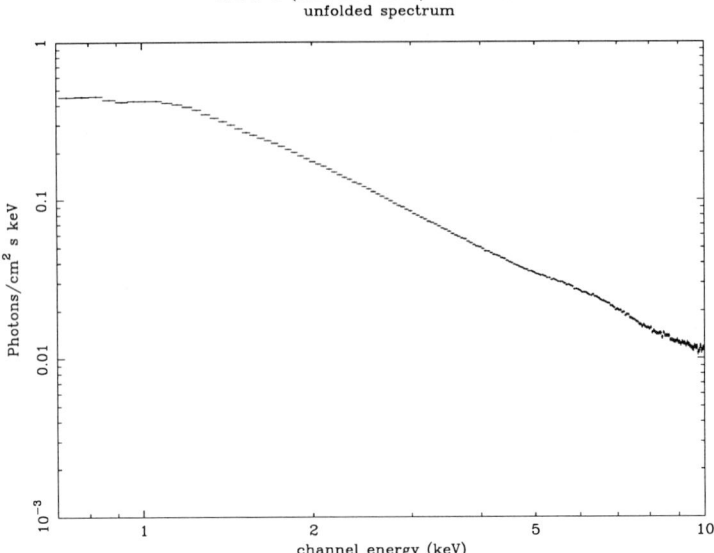

Figure 4. The unfolded photon spectrum of Cyg X-1 obtained with the ASCA GIS.

However, the allowable range in the relation between the strength of the magnetic field at the surface of the neutron star and its rotational period is very limited in this case. In order for mass accretion to take place at the Alfvén radius, the gravitational force should be stronger than the centrifugal force at the Alfvén radius. This condition can be written as

$$P > 63 \ (B/10^8 \text{ G})^{6/7} (\dot{M}/10^{13} \text{ g s}^{-1})^{-3/7} (M/M_\odot)^{-5/7} (R/10^6 \text{ cm})^{18/7} \text{ ms}, \tag{1}$$

where P, B, M, R and $\dot{M}$ are the rotational period, the surface magnetic field, the mass and the radius of the neutron star, and the accretion rate onto the neutron star, respectively. As seen from this equation, very weakly magnetized or very slowly rotating neutron stars are necessary in this case. If this possibility holds true, these sources cannot be progenitors of millisecond pulsars.

4. Iron Lines from Black-Hole Candidates

It is widely known that black-hole candidates generally have two states, the soft (high) state and hard (low) state (see, e.g., Inoue 1993).

The spectrum of Cyg X-1 in the hard state obtained with ASCA GIS is shown in Fig. 4. No prominent emission lines characteristic of thin thermal

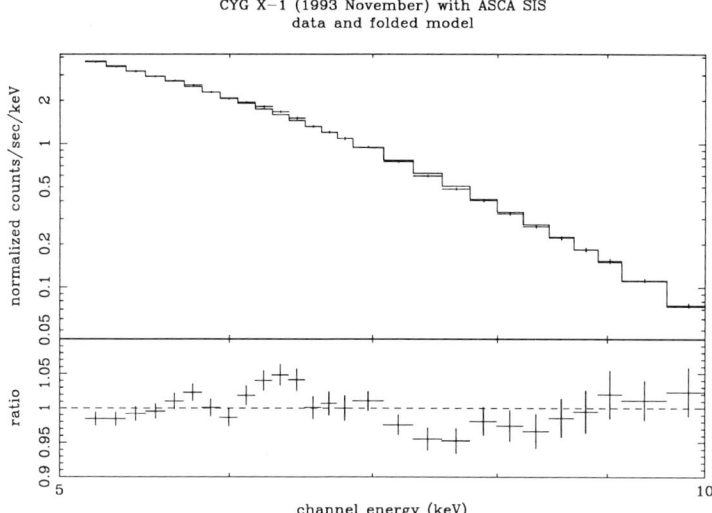

Figure 5. The Cyg X-1 spectrum obtained with the ASCA SIS (crosses) and the best-fit power law spectrum (histograms) (upper panel); the ratio of the observed to the model spectrum (lower panel).

emission are seen. However, we find a slight but significant excess over a single power law around 5–7 keV as seen in Fig. 4. This broad excess can be interpreted in terms of a reflection component superposed on the power-law component (Tanaka 1989). If the reflection component really exists in the spectrum, it must be accompanied by a fluorescent iron line. In fact, if we fit a single power law to the SIS spectrum in the 5–10 keV range, the ratio of the observed spectrum to the power law model shows a clear line feature around 6.3–6.4 keV as seen in Fig. 5. By fitting a Gaussian profile to this emission line feature, it is found that the line is consistent with a fluorescent iron line at 6.4 keV and the line width is less than 200 eV. Hence, this line does not seem to come from a relativistic region near the central object. However, the equivalent width of this narrow line is about 10 eV and this value is much weaker than expected from the intensity of the reflection component. A significant fraction of the broad excess around 5–7 keV might be due to a broad emission line from the relativistic region (Fabian et al. 1989).

ASCA obtained a typical soft-state spectrum from GRS 1009–45. This source was first detected with the GRANAT/WATCH experiment (Lapshov et al. 1993; Harmon et al. 1993). Following the first detection, ASCA observed this source on Nov. 10, 1993 (Tanaka et al. 1993). Fig. 6 shows

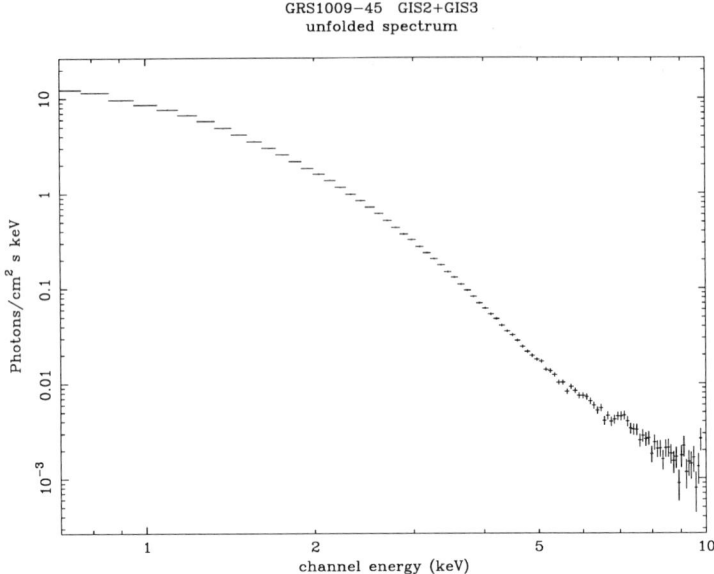

Figure 6. The unfolded photon spectrum of GRS 1009–45 obtained with the ASCA GIS.

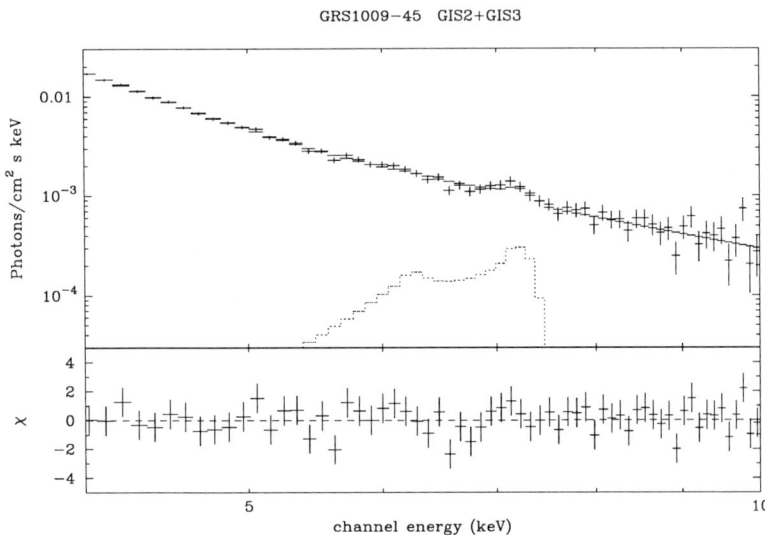

Figure 7. The GIS spectrum of GRS 1009–45 fitted with a line profile from a relativistic accretion disk (dashed line). The detector response has been unfolded. The best-fit line energy is determined with other parameters fixed: $r_{\rm in} = 10\,r_{\rm s}$, $r_{\rm o} = 100\,r_{\rm s}$, $q = -2$, $i = 40°$ (for definitions, see Fabian *et al.* 1989). (Tanaka 1994).

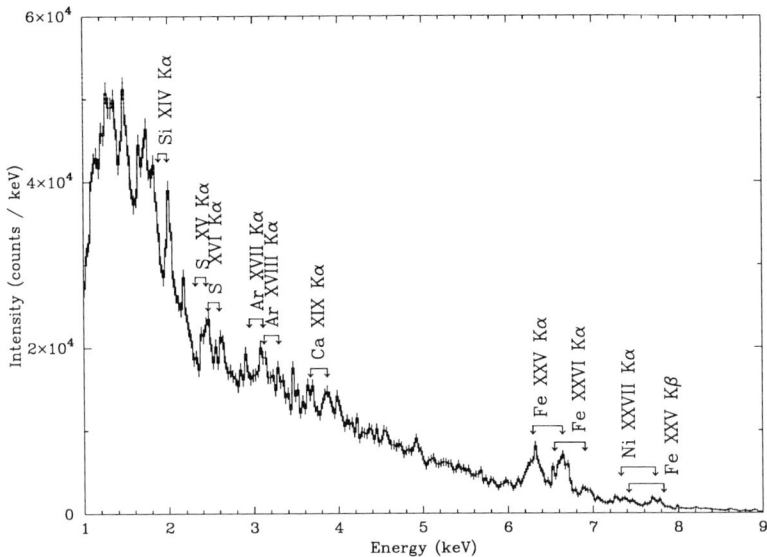

Figure 8. The X-ray count spectrum of SS 433 obtained with the ASCA SIS on 1993 April 23. The candidates for emission lines and possible identifications are shown by arrows. (Kotani et al. 1994).

the ASCA GIS spectrum of GRS 1009-45. It is clearly ultrasoft, and accompanied by a hard tail. A remarkable finding is a line feature near 7 keV. If we fit a Gaussian profile to the feature, the best fit line center energy is 7.1 keV, which does not correspond to any atomic lines without a Doppler effect. The rotational motion in the accretion disk may be the origin of the Doppler effect. In fact, a line profile from a relativistic accretion disk can reproduce the line feature as shown in Fig. 7 (Tanaka 1994).

5. Various Emission Lines from SS 433

SS 433 is a well-known X-ray binary system ejecting bipolar jets. The jets have an ejection velocity of about a quarter of the light velocity and precess with a period of 163 days.

Fig. 8 shows the ASCA SIS spectrum of SS 433 on April 23, 1993, and the presence of various emission lines is obvious (Kotani et al. 1994). These lines can be identified with pairs of emission lines from silicon to iron and all the lines are consistent with coming from either of the two jets. If we compare the degrees of the Doppler shift of the two jets best reproducing the various line energies with the simultaneously obtained Doppler shift at the optical band, we will be able to see whether or not the velocity of the X-ray emission region is the same as the optical emission region. This analysis is in progress.

Coexistence of He-like and H-like Kα lines from silicon to iron suggests the presence of temperature structure in the X-ray emission region as discussed earlier for EX Hydra. The ratios of the intensity of He-like to H-like lines of various elements are again the best indicator of the temperature structure. If we assume that all the physical parameters can be expressed as a power of the radial distance, the line intensity can be obtained by integrating $T^\alpha \Lambda(T)$ over T from 0 to $T_{\max}$, where $\Lambda(T)$ is emissivity at temperature T. Then, the line intensity ratio can be calculated for various elements in terms of the power-index α and compared with the observation. If we assume that $v = $ const., $S \propto r^2$ and $P \propto \rho^{5/3}$, we find $\alpha = -0.25$. (Here, v, S, P and ρ are the velocity, cross-section, pressure and density of matter in a jet, respectively.) This preliminary result shows that the observed line intensity ratios of silicon, sulphur and iron are significantly smaller than those for the adiabatic flow. This may suggest the importance of radiative cooling in the X-ray emitting region of jets.

Acknowledgements. The author is indebted to Dr. M. Ishida and Mr. R. Fujimoto, Dr. K. Asai and Dr. T. Dotani, Prof. Tanaka, Mr. Y. Ueda and Mr. T. Sonobe, and Mr. T. Kotani for the results presented in Sections 2, 3, 4 and 5 of this review paper, respectively. I would like to thank all of them. Thanks are also due to Dr. S. Skinner for his careful reading of the manuscript.

References

Aizu, K. 1973, Prog. Theo. Phys. 49, 1184
Fabian, A.C. et al. 1989, MNRAS 238, 729
Frank, J. et al. 1985, *Accretion Power in Astrophysics*, Cambridge Univ. Press
Fujimoto, R. et al. 1994, (in preparation)
Harmon, B.A. et al. 1993, IAU Circ. 5864
Inoue, H. 1993, in *Accretion Disks in Compact Stellar Systems*, J.C. Wheeler (Ed.), World Scientific (Singapore), p. 303
Ishida, M. et al. 1994, PASJ 46, L81
Kotani, K. et al. 1994 PASJ 46, L141
Kulkarni, S.R. et al. 1992, in *X-Ray Binaries and Recycled Pulsars*, E.P.J. van den Heuvel & S.A. Rappaport (Eds.), Kluwer (Dordrecht), p. 99
Lapshov, I. et al. 1993, IAU Circ. 5864
Matsuoka, M. et al. 1980, ApJ 240, L137
McClintock, J.E. & Remillard, R.A. 1990, ApJ 350, 386
Nakamura, N. et al. 1989, PASJ 41, 617
Tanaka, Y. 1989, in *Two Topics in X-Ray Astronomy*, Proc. 23rd ESLAB Symp., p. 3
Tanaka, Y. 1994, in *New Horizon of X-Ray Astronomy*, F. Makino & T. Ohashi (Eds.), Univ. Acad. Press (Tokyo), p. 37

Tanaka, Y. & ASCA Team 1993, IAU Circ. 5888
Tanaka, Y. et al. 1994, PASJ 46, L37
Tsunemi, H. 1991, in *Frontiers of X-Ray Astronomy*, Y. Tanaka & K. Koyama (Eds.), Univ. Acad. Press (Tokyo), p. 677
Verbunt, F. et al. 1994, A&A 285, 903

Discussion

P. Charles: In your explanation for the quiescent X-ray emission from the neutron star transients 4U 1608–52 and Cen X-4 you dismissed the possibilities of coronal emission from the companion star, because it is too weak. I would like to point out that, since almost all these transients appear to have approximately K0 secondaries, they are analogous with the X-ray active RS CVn systems. Admittedly, typical RS CVn luminosities are still below those observed in the transients, but their binary periods are shorter and so the coronal activities could be higher, making it a significant contribution. This could be tested by future higher sensitivities observations by searching for coronal X-ray line emission.

ROSAT OBSERVATIONS
OF SOFT X-RAY TRANSIENTS IN QUIESCENCE

FRANK VERBUNT
Astronomical Institute, Utrecht University,
Postbox 80.000, 3508 TA Utrecht, The Netherlands

Abstract. Four soft X-ray transients, two with a neutron star and two with a black hole, have been detected at quiescence with ROSAT. Blackbody fits to their spectra give temperatures of 160–300 eV, and surface areas of $<1\,\mathrm{km}^2$. The small surface area suggests that the actual spectrum may be optically thin. The companion star does not contribute significantly to the X-ray luminosity, except perhaps in the case of A 0620−00. From the observation that accretion continues at luminosity levels of $\sim 10^{33}\,\mathrm{erg\,s^{-1}}$ it is concluded that the neutron stars in Aql X-1 and Cen X-4 have a weak magnetic field and rotate rather slowly.

1. Introduction

Occasionally, on average perhaps once a year, the X-ray sky changes due to the sudden appearance of a very bright new source, which after reaching maximum within a week, declines again to its pre-outburst level on a time scale of months. During outburst maximum, these sources are very similar to the permanently bright low-mass X-ray binaries, and they are thought similarly to be neutron stars or black holes that accrete mass from a low-mass companion. Upon cessation of mass transfer, the neutron star in a low-mass X-ray binary may turn into a recycled pulsar. (For reviews see White et al. 1984; Van Paradijs & Verbunt 1984; Tanaka & Lewin 1995; for evolution see, e.g., Verbunt 1993.)

A study of the X-ray emission of soft X-ray transients in quiescence is interesting because it may shed light on the mechanism which causes the accretion onto the compact object in these systems to be intermittent, and because the magnetic field and rotation period of the neutron star may

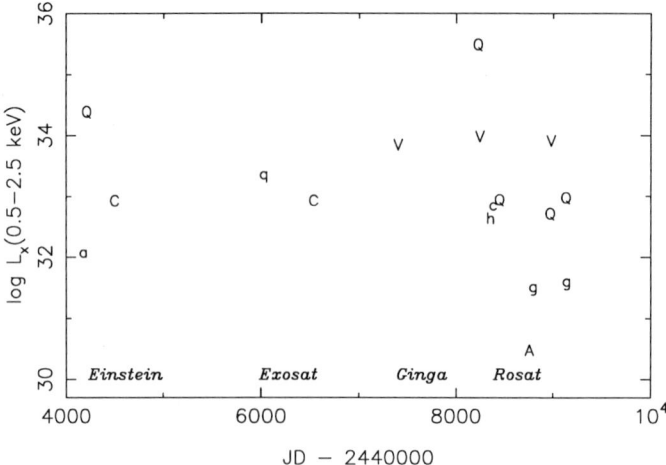

Figure 1. A history of observations of soft X-ray transients in quiescence. X-ray luminosities are for the range 0.5–2.5 keV. Each letter indicates an observation of Q = Aql X-1, A = A 0620−00, C = Cen X-4, V = V404 Cyg (=GS 2023+338), G = GS 2000+25, and H = H 1705−25. Upper case is a detection, lower case an upper limit. The satellites used are indicated near the bottom. (For details, see text and Verbunt et al. 1994.)

be revealed at low accretion rates. Field strength and period may then be compared with those of the recycled radio pulsars with low-mass white-dwarf companions. In Section 2 of this paper I discuss a history of the X-ray observations of soft X-ray transients in quiescence, and in Section 3 the results obtained on the X-ray spectrum at low accretion rates. A possible contribution of the donor star to this X-ray flux is discussed in Section 4, and the magnetic field and rotation period of the neutron star in Section 5.

2. X-ray Observations in Quiescence

Before the launch of ROSAT, there were only two secure detections of a soft X-ray transient in quiescence, both of Cen X-4, at a level of 10^{32-33} erg s^{-1}, with EINSTEIN and with EXOSAT (Van Paradijs et al. 1987). An upper limit at a comparable level was obtained with GINGA (Kulkarni et al. 1992). (Here and in the following the X-ray luminosities are quoted for the range of 0.5–2.5 keV.) Detections at $\sim 10^{34}$ erg s^{-1} with uncertain identification due to inaccurate position were reported for Aql X-1 with EINSTEIN (Czerny et al. 1987) at a level three times above an upper limit obtained with EXOSAT (Van Paradijs et al. 1987), and for V404 Cyg with GINGA (Mineshige et al. 1992). A very low upper limit, at $\sim 10^{32}$ erg s^{-1}, was obtained with EINSTEIN for A 0620−00 (Long et al. 1981).

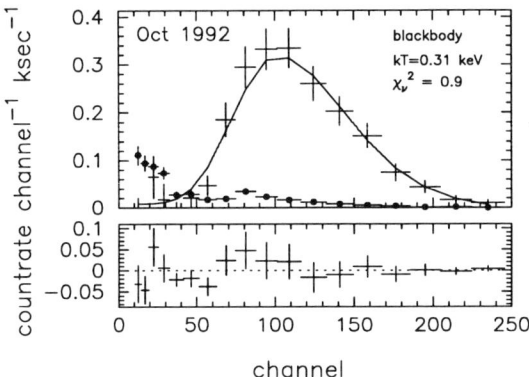

Figure 2. The ROSAT PSPC countrates of Aql X-1 obtained in October 1992, together with a blackbody fit (upper panel) and the difference between fit and observation (lower panel). Channel n corresponds roughly to $n \times 0.01$ keV. The background level is indicated with •. (From Verbunt et al. 1994.)

The dramatic improvement in sensitivity with ROSAT has transformed the field (see Fig. 1). Aql X-1 was detected at various flux levels, down to $4~10^{32}$ erg s^{-1} (Verbunt et al. 1994). V404 Cyg was detected twice at similar count rates; the derived luminosity depends critically on the assumed absorption column: for $N_{\rm H} = (0.3 - 1.5)~10^{22}$ cm^{-2}, the X-ray luminosity is $(1\text{-}8)~10^{33}$ erg s^{-1} (Verbunt et al. 1994; Wagner et al. 1994). A 0620−00 was detected at $2.5~10^{30}$ erg s^{-1} (McClintock et al. 1995). Upper limits were obtained for H 1705−25 at $4~10^{32}$ erg s^{-1} for $N_{\rm H} = 3~10^{21}$ cm^{-2}, and for GS 2000+25 at $3~10^{31}$ erg s^{-1} for $N_{\rm H} = 10^{22}$ cm^{-2} (Verbunt et al. 1994). Further results are now coming in from ASCA (see Inoue, these proceedings).

Thus, we now have quiescent detections of two transients with a neutron star, Cen X-4 and Aql X-1, and of two with a black hole, A 0620−00 and V404 Cyg. The flux level of all of these is too low to be compatible with models that explain the transient mass transfer as the consequence of irradiation of the donor star (see also Lasota, these proceedings).

3. Spectra

The ROSAT XRT-PSPC combination has a limited spectral resolution, providing about six independent flux points in the 0.1–2.4 keV range. We show the spectrum of Aql X-1 at its lowest observed flux level in Fig. 2 together with the best fitting blackbody spectrum, which has a temperature of 0.31 keV. This temperature, although somewhat lower than the characteristic temperature at higher flux levels, is such that the spectrum peaks in the ROSAT PSPC sensitivity range: the low observed flux therefore corresponds to a low bolometric flux.

Low temperatures have also been found for the detections with the

ROSAT PSPC of V404 Cyg (210 eV, Wagner et al. 1994), of A 0620−00 (160 eV, McClintock et al. 1995), and with ASCA of Cen X-4 and H 1608−522 (Inoue, these proceedings).

The interpretation of the spectra as blackbody spectra leads to rather small radii of the emitter. The bolometric luminosities at minimum are about $4.25 \ 10^{30}$ erg s^{-1} for A 0620−00 and $5 \ 10^{32}$ erg s^{-1} for Aql X-1, corresponding to emitting areas of < 1 km^2 in both cases, rather too small to be from the accretion disk. In the case of Aql X-1 we may be seeing a ring on the neutron star, the boundary layer, which is heated by continued accretion. Against this interpretation is the observation of a similar emitting area in the black hole system. Perhaps it is more likely then that we do see the spectrum of the accretion disk, which is optically thin. If so, the conversion of observed to bolometric flux is uncertain.

4. Contribution of the Companion

At the very low X-ray luminosities now found of soft X-ray transients in quiescence, one has to investigate the possibility that the donor star contributes to the observed flux. To do this, I use the survey of RS CVn type systems by Dempsey et al. (1993), which has the advantage of being made with the same instrument, the ROSAT PSPC, used to detect the quiescent transients. Here the definition of RS CVn systems is taken widely, and includes binaries consisting of main-sequence stars, but with increased magnetic activity due to rapid rotation. In Fig. 3 I show the X-ray luminosity of RS CVn's as a function of the radius of the active star. The luminosity in the 0.5–2.5 keV range is assumed to be half of the L_X tabulated by Dempsey et al., and the stellar radii have been derived from V, $B - V$ and the distance as tabulated by Dempsey et al. using the relations of $B - V$, T_{eff} and bolometric correction in Tables 3.3 and 3.5 in Mihalas & Binney (1981). No reddening corrections have been applied.

In the same figure, I plot the detected luminosities of the soft X-ray transients, using radii of $0.6 \text{R}_\odot$ for Cen X-4 (McClintock & Remillard 1990), $0.8 \text{R}_\odot$ for A 0620−00 (Shahbaz et al. 1994), and $9 \text{R}_\odot$ for V404 Cyg (Wagner et al. 1992). I further assumed a radius for Aql X-1 slightly larger than that for Cen X-4. It should be noted that all these radii are very uncertain. It is seen that the companion may contribute significantly only in A 0620−00. The companion is not expected to contribute significantly to Aql X-1, Cen X-4, or V404 Cyg. As an aside, note that Fig. 3 indicates that the X-ray flux of cataclysmic variables with $L_X < 3 \ 10^{30}$ erg s^{-1} may be affected by flux from the donor star.

The comparison of the soft X-ray transients is made with RS CVn binaries rather than with rapidly rotating single stars, because the binaries are

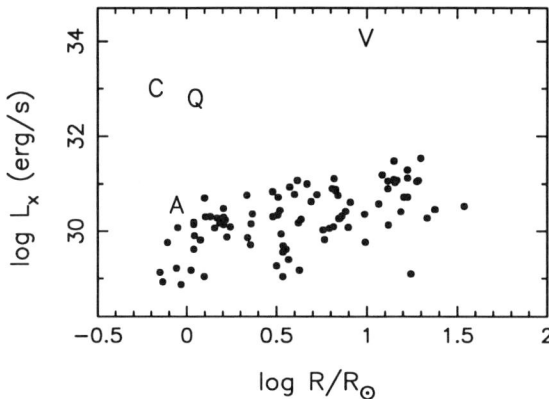

Figure 3. 0.5–2.5 keV X-ray luminosities of RS CVn binaries, adapted from Dempsey et al. (1993), plotted as a function of the radius of the active star (•). The position of some X-ray transients in this diagram is also given, with the same symbols as in Fig. 1.

known to be more active than single stars rotating equally rapidly (Schrijver & Zwaan 1991). HD 22403 is an example of a binary consisting of two low-mass main-sequence stars, a G2 V and a K V star, which is a remarkably bright X-ray source, at $L_X \simeq 0.5\ 10^{31}\ \mathrm{erg\,s^{-1}}$ (Dempsey et al. 1993).

5. Magnetic Field and Pulse Period

The observation that soft X-ray transients with a neutron star are still accreting at very low rates provides a strong limit on the magnetic field and rotation period of the neutron star (Stella et al. 1994; Verbunt et al. 1994). At the surface of the magnetosphere the disk matter is forced into corotation with the neutron star. When the magnetic field B is too strong (i.e. the magnetospheric surface too far from the neutron star), or the rotation period P too short, the centrifugal force caused by the corotation exceeds the gravitational attraction by the neutron star, and the matter is flung out, rather than accreted. Thus, accretion requires

$$P \gtrsim (0.4\ \mathrm{s})\ B_9^{6/7} \dot{M}_{13}^{-3/7} \qquad (1)$$

where B_9 is the surface magnetic field of the neutron star in units of 10^9 G, and $\dot{M}_{13}$ the accretion rate in units of $10^{13}\,\mathrm{g\,s^{-1}}$.

The ROSAT observation of Aql X-1 at its lowest quiescent level implies $\dot{M} \sim 3\ 10^{12}\,\mathrm{g\,s^{-1}}$, so that $P \gtrsim (0.6\,\mathrm{s})\ B_9^{6/7}$. A somewhat less stringent limit may be derived for Cen X-4. It is clear from these conditions that the neutron stars in Cen X-4 and in Aql X-1 will require a lot of additional spin-up if they ever are to become millisecond pulsars upon cessation of mass transfer.

6. Conclusions

The ROSAT PSPC observations have provided a wealth of unprecedented observations of low-mass X-ray binaries in quiescence.

From these it is clear that accretion continues, albeit at a very low rate, during the quiescent intervals between outbursts. Blackbody spectra fitted to the observed data give surface areas of the order of $1\,\mathrm{km}^2$. This suggests that the spectra are optically thin, and may affect the conversion of observed flux into bolometric flux, and thus luminosity.

The continued accretion onto the neutron stars in Cen X-4 and Aql X-1 at rates $\dot M \sim 10^{13}\,\mathrm{g\,s}^{-1}$ indicates that these neutron stars rotate rather slower than expected for the progenitors of millisecond recycled pulsars.

Acknowledgements. I thank Luigi Stella and Shri Kulkarni for discussions about the quiescent soft X-ray transients, and Jeffrey McClintock et al. for allowing me to quote their results on A 0620−00 before publication. This research is supported by the Netherlands Organization for Scientific Research under grant PGS 78-277.

References

Czerny, M., Czerny, B. & Grindlay, J.E. 1987, ApJ 312, 122
Dempsey, R. et al. 1993, ApJS 86, 599
Kulkarni, S.R. et al. 1992, in *X-ray Binaries and Recycled Pulsars*, E.P.J. van den Heuvel & S. Rappaport (Eds.), Kluwer Academic Publishers, p. 99
Long, K., Helfand, D. & Grabelsky, D. 1981, ApJ 248, 925
McClintock, J.E. & Remillard, R.E. 1990, ApJ 350, 386
McClintock, J.E., Horne, K. & Remillard, R.E. 1995, ApJ (in press)
Mihalas, D. & Binney, J. 1981, *Galactic Astronomy, Structure and Kinematics*, Freeman, 2nd Edition
Mineshige, S. et al. 1992, PASJ 44, 117
Schrijver, C. & Zwaan, C. 1991, A&A 151, 183
Shahbaz, T., Naylor, T. & Charles, P. 1994, MNRAS 268, 756
Stella, L. et al. 1994, ApJ 423, L47
Tanaka, Y. & Lewin, W.H.G. 1995, in *X-ray Binaries*, W.H.G. Lewin, J. van Paradijs & E.P.J. van den Heuvel (Eds.), Cambridge University Press (in press)
Van Paradijs, J. & Verbunt, F. 1984, in *High Energy Transients in Astrophysics*, S.E. Woosley (Ed.), AIP Conf. Proc. Vol. 115, p. 49
Van Paradijs, J. et al. 1987, A&A 182, 47
Verbunt, F. 1993, ARA&A 31, 93
Verbunt, F. et al. 1994, A&A 285, 903
Wagner, R. et al. 1992, ApJ 401, L97
Wagner, R.M. et al. 1994, ApJ 429, L25
White, N.E., Kaluzienski, L.J. & Swank, J.H. 1984, in *High Energy Transients in Astrophysics*, S.E. Woosley (Ed.), AIP Conf. Proc. Vol. 115, p. 31

Discussion

R.W. Romani: Given that the flux in quiescence of these transients is small and that your spectral results suggest that it may be non-thermal, is it possible that this flux is generated via shock acceleration at the hot spot where the accretion stream impacts, at least for the soft X-ray transients?

F. Verbunt: I think that this is unlikely. The gradual change in temperature as the flux decreases suggests that the same process is responsible for the X-ray emission throughout.

R. Sunyaev: In the case of Aql X-1 and Cen X-4 it is possible that the disk exists around the neutron star with a magnetic field but that accretion does not take place (see R.A. Sunyaev & N.I. Shakura 1986, Pis'ma Astron. Zh. 12, 286, translated SvA Lett. 12(2), 117). In that case the disk radiates the energy due to angular momentum transport from the neutron star.

F. Verbunt: Yes, that is a possibility. I remember that W. Priedhorsky put forward a similar mechanism for Sco X-1 (ApJ 306, L97). In my estimates I have used the simplest description.

M. van Kerkwijk: Is it possible that the low X-ray flux that you detected arises at the magnetosphere of the neutron star (i.e. in the inner disk), so that the neutron star could actually spin at a much shorter period than you derive?

F. Verbunt: See my answer to the previous speaker. Perhaps I should add that these mechanisms are not possible in a disk around a black hole, as no magnetosphere is present in that case, whereas the luminosities we derive for the black-hole V404 Cyg are actually rather similar to those for Cen X-4 and Aql X-1, and the spectrum derived is also rather similar. I therefore doubt that a magnetosphere plays an important role.

BLACK-HOLE SYSTEMS:
OPTICAL SPECTROSCOPY AND IR PHOTOMETRY

P.A. CHARLES
Oxford University
Nuclear & Astrophysics Laboratory,
Keble Road, Oxford OX1 3RH, United Kingdom

Abstract. The X-ray transient systems have provided the first opportunities for detailed studies of the mass losing star in low-mass X-ray binaries. During X-ray quiescence the cool star is the dominant light source in the red and near-IR. Optical spectroscopy yields the mass function (itself a lower limit to the compact-object mass), the rotational broadening leads to the mass ratio, q (assuming only that the star fills its Roche lobe), and the IR ellipsoidal light curve gives the system inclination (for high q). In such cases, a complete solution to the system parameters is possible, and this has been performed for A 0620−00 (V616 Mon) and GS 2023+338 (V404 Cyg), leading to the first accurate black-hole masses (which are in the range 10–12 $M_\odot$).

1. Introduction

Barely 10 years ago the only galactic black-hole candidates (BHCs) under consideration were Cyg X-1, LMC X-3 and A 0620−00 (McClintock 1986). Stemming initially from optical observations of the supergiant primary in Cyg X-1, the X-ray signature (spectrum and variability; see Tanaka & Lewin 1995) was then used as a guide to finding further candidates. But massive X-ray binaries such as Cyg X-1 and LMC X-3 are difficult objects for determining accurate masses since there is no velocity information associated with the compact object itself. The primaries are highly evolved and of uncertain mass, and hence only lower limits can then be derived to that of the compact object. (Nevertheless, this limit for Cyg X-1 exceeds 3 $M_\odot$, and it very likely is a black hole.) The low-mass X-ray binaries (LMXBs) should be more straightforward for such measurements (as the companion star is much lighter), but they suffer from the major difficulty that the light is dominated by the much brighter X-ray irradiated accretion disc.

X-ray signatures alone are certainly not an unambiguous indicator of the presence of a black hole, as neutron stars can mimic such behaviour (see McClintock 1991). The signatures are merely indicators that the source is a BHC and that dynamical evidence should be obtained. Observationally, there is *no* evidence for any object with neutron star characteristics to have a mass above $2\,M_\odot$ (Thorsett et al. 1993), and $3\,M_\odot$ has been the dividing line, above which an object is declared to be a BHC. McClintock (1986) declared the *holy grail* in this field to be a source with a mass function $f(M) > 5\,M_\odot$ for which it could be unequivocally stated that the compact object was not a neutron star.

2. Soft X-ray Transients

In the last 10 years the field has been transformed by our ability to obtain optical spectra of the companions to the soft X-ray transients (SXTs).[1] They are LMXBs, but are X-ray luminous for only ~ 6 months in every 10–50 years. During the long X-ray quiescence the accretion disc becomes extremely faint, rendering the companion star visible. But even the brightest SXT only has $V \sim 18$, most are around 19–21, and so it is only recently that the equipment has existed on large telescopes to enable such studies to be made. SXTs now dominate the list of BHCs (see table 2 of White 1994).

With their low-mass secondaries (so far, all are K0–K7) this allows firm lower limits to be set to the compact-object mass on the basis of $f(M)$ alone, as the secondaries can be studied in detail during quiescence. In this class, GS 2023+338 (=V404 Cyg), has the highest $f(M)$ known at $6.1\,M_\odot$ (Casares et al. 1992; Casares & Charles 1994).

In this review I shall concentrate on the spectroscopic work on quiescent SXTs, but with some recent IR photometry in order to give a complete picture of the two most intensively studied objects. For a detailed discussion of the optical and IR photometry and the complexities of their behaviour see the accompanying review by Haswell.

---

[1] Note that even though several of the objects in this class do not exhibit the soft X-ray excess that gave SXTs their name, I shall still use the general term SXT to refer to them. This is so as to distinguish them from the "hard transients" that are associated with accreting neutron stars in Be systems, even though a hallmark of SXTs is a hard power law tail extending to very high energies. SXTs are also sometimes referred to as "X-ray novae", but they bear no relation to classical optical novae which are thermonuclear events on the surfaces of accreting white dwarfs.

3. Dynamical Mass Measurements

High spectral resolution observations were first performed on the SXT prototype A 0620–00 (=V616 Mon), which has a K5V secondary and 7.8 hr period (McClintock & Remillard 1986). With $f(M) = 2.9\,M_\odot$ it was immediately recognised as a strong BHC since, for any reasonable value of M_2, the secondary mass, the compact object significantly exceeds the maximum neutron star mass. In SXTs there is no emission from the black hole itself, although there is strong Hα emission from its surrounding disc. Attempts have been made to detect the orbital motion of the compact object from radial-velocity shifts in the wings of the line (which emanate from the inner disc). Whilst detected (although small, as expected), the disc behaviour is obviously complex as the phasing of this motion does not align with that of the secondary (Orosz et al. 1994).

To determine the masses in SXTs we therefore make use of: (i) the rotational broadening of the secondary star's spectral features; (ii) the ellipsoidal modulation of the secondary. These can be combined with $f(M)$ to yield a complete solution of the binary parameters with only minimal assumptions. With their late-type secondaries, the SXTs therefore provide the only method for obtaining accurate masses in BHCs. Indeed, only one SXT (Cen X-4) displays the properties of a neutron star (it bursts). Determining these masses provides a crucial challenge for understanding the late evolution of massive stars and the formation of supernova remnants (see, e.g., Verbunt & Van den Heuvel, 1995). Here I will summarise the results on the prime targets in this class. For more details on these objects see Van Paradijs & McClintock (1995).

3.1. THE NATURE OF THE COMPANION STAR

Since these stars must be (at least very close to) filling their Roche lobes (given their X-ray activity), their size is given by $R_2/a = 0.46(1+q)^{-1/3}$ where $q = M_X/M_2$. Combining this with Kepler's 3rd Law leads to the well-known result that the secondary's mean density $\rho = 110/P_{hr}^2\,\mathrm{g\,cm^{-3}}$ (see Table 1). As ρ for a MS star of this spectral type is $\sim 5\,\mathrm{g\,cm^{-3}}$ then only A 0620–00 and Nova Mus can contain (relatively) unevolved secondaries. It is on this basis that the luminosity classes are given above, since there are no suitable luminosity discriminants within existing spectra. Computations of the evolution of such a star lead to the concept of a "stripped-giant" and give $0.2\,M_\odot < M_2 < 1.3\,M_\odot$ (King 1993).

TABLE 1. Optical and IR properties of soft X-ray transients

Source	P (hrs)	Sp. Type	f(M) (M$_\odot$)	ρ (g cm^{-3})	E_{B-V}	V	K	$v_{rot}\sin i$ (km s^{-1})	K_2 (km s^{-1})
A 0620−00	7.8	K5V	2.91	1.81	0.35	18.3	6	83	433
Nova Mus	10.4	K0-4V	2.86	1.02	0.29	20.5			399
Cen X-4	15.1	K7IV	0.20	0.48	0.1	18.4	15.0		146
V404 Cyg	155.3	K0IV	6.08	0.0046	1	18.4	12.5	39	208.5

from McClintock & Remillard 1986; Remillard et al. 1992; Orosz et al. 1994; McClintock & Remillard 1990; Casares & Charles 1994

3.2. ROTATIONAL BROADENING

With the size of the secondary restricted, then assuming co-rotation leads to a rotational velocity of $v_{rot}\sin i = K_2 \times 0.46\,(1+q)^{2/3}/q$ (Wade & Horne 1988), and hence yields q directly from the radial-velocity curve (K_2) and $v_{rot}\sin i$. This is technically challenging as typical values of v_{rot} are 30–80 km s^{-1} and so high spectral resolution (≤ 1 Å) is needed. Being faint ($V \geq 18$), then large telescopes are required even for the brightest SXTs. Casares & Charles (1994) used the 4.2 m WHT to determine $v_{rot}\sin i$ for V404 Cyg (Fig. 1) by subtracting different broadened versions (including the effects of rotation and limb darkening) of a K0IV template and performing a χ^2 test on the residuals. This gave $v_{rot}\sin i = 39\pm 1$ km s^{-1} and hence $q = 16.7\pm 1.4$. The absence of eclipses implies $i < 80°$ and hence $7\,M_\odot < M_X < 24\,M_\odot$ (Fig. 2). Clearly, further constraints on M_X, M_2 require knowledge of i. This is possible by exploiting the ellipsoidal modulation of the secondary. A similar study of A 0620−00 gave $v_{rot}\sin i = 83\pm 5$ km s^{-1} and hence $q = 15\pm 2$ (Marsh et al. 1994).

3.3. ELLIPSOIDAL MODULATION

The hallmark double-humped variation (resulting from the tidal distortion of the secondary) has been seen optically in all four SXTs of Table 1. However, there is clear contamination of these light curves, probably due to some combination of the accretion disc (e.g., the stream impact region), X-ray heating (although this should be small) and possible starspots on the secondary (cf. RS CVn systems). The principal contributor is probably the disc since veiling has been measured in all three and can be ∼20–30%, and substantial (∼20%) flickering on timescales $< P_{orb}$ is present in V404 Cyg.

Hence Shahbaz et al. (1993, 1994a,b) undertook a campaign of IR pho-

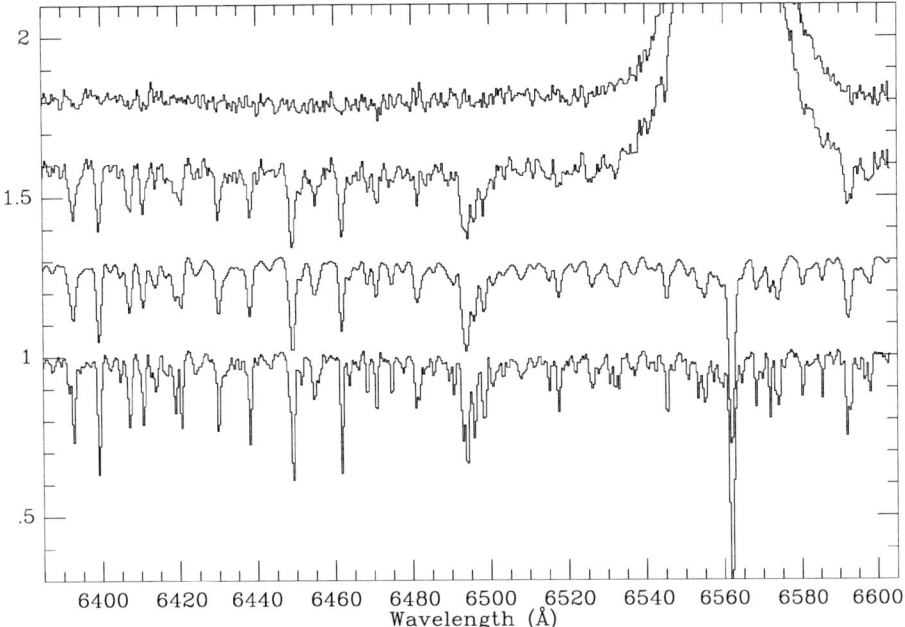

Figure 1. Determining the rotational broadening in V404 Cyg. From bottom to top: the K0IV template star (HR 8857); the same spectrum broadened by 39 km s^{-1}; Doppler corrected sum of V404 Cyg (dominated by intense Hα emission from the disc); residual spectrum after subtraction of the broadened template (from Casares & Charles 1994).

tometry, where the disc contamination would be less, and the limb- and gravity-darkening less affected by uncertainties in T_{eff}. JHK photometry from UKIRT and the AAT yielded the first light curves for these objects (Fig. 3 shows A 0620–00).

Model calculations show that the shape and amplitude are a function of q and i, but they are insensitive to q if $q > 5$. The fits are summarised in Table 2.

TABLE 2. System parameters for three soft X-ray transients

Source	q	i	$R_2(\mathrm{R}_\odot)$	$a(\mathrm{R}_\odot)$	$M_X(\mathrm{M}_\odot)$	$M_2(\mathrm{M}_\odot)$
A 0620–00	15	37	0.8	4.3	10.0	0.6
Cen X-4	(1–30)	31–54	-	-	0.5–2	0.2–0.7
V404 Cyg	16.7	55	5.6	34	12	0.6

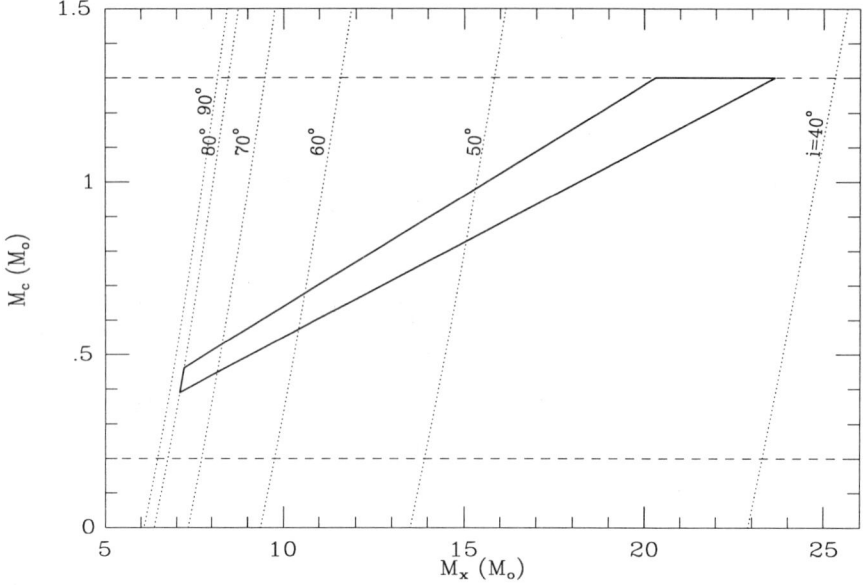

Figure 2. Constraints on M_X and M_2 for a range of values of i in V404 Cyg based on the radial-velocity curve ($f(M)$) and determination of q (from the rotational broadening). It is the limited constraint on i (absence of eclipses) that leads to a wide range of M_X (from Casares & Charles 1994).

The power of this technique is most clearly demonstrated for those systems (V404 Cyg and A 0620–00) in which the rotational broadening has been detected. This measures q directly, and then the ellipsoidal modulation tightly constrains i, giving the first direct mass measurements for black holes in our galaxy. Note that, even though we do not yet have an accurate q for Cen X-4, M_2 must be less than that of a main-sequence star of the same spectral type ($<0.7\,M_\odot$) and hence M_X must be in the range 0.5–2 $M_\odot$. This is in excellent accord with that expected for a neutron star, and provides a useful confirmation of this basic approach.

4. Lithium in the Secondary Star

Perhaps the most surprising discovery in this field is that all three of the SXTs have high lithium abundances. Strong Li I $\lambda 6708$ was discovered by Martín et al. (1992) in V404 Cyg when fitting with its K0IV template. This was a great surprise as Li is found with this abundance only in young stars, where the convective mixing (that leads to Li destruction) has not yet had time to substantially reduce the initial Li content. Li is an important element in galactic chemical abundances because galactic gas is enriched in

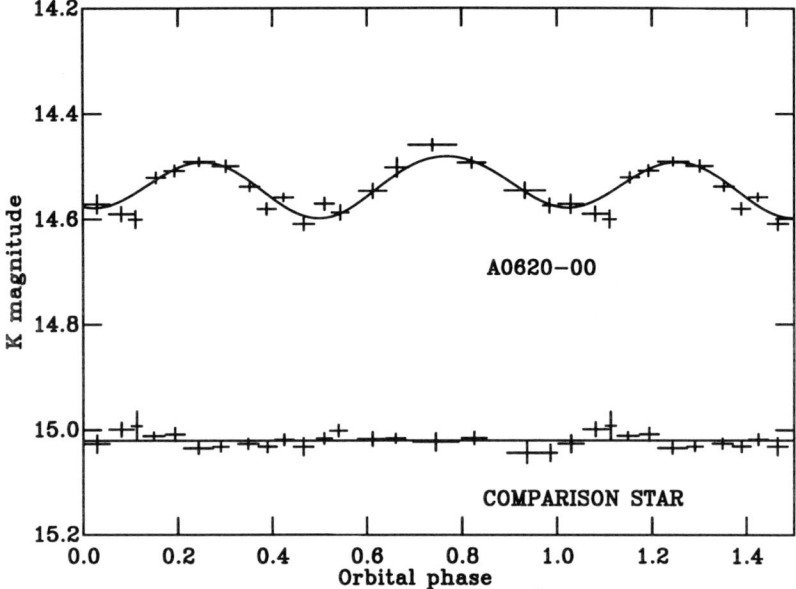

Figure 3. IR light curves of A 0620–00 (from Shahbaz et al. 1994a).

Li relative to the halo, and much effort has been put into locating the source of enrichment. Subsequently, Li was discovered in comparable spectra of A 0620–00 (Marsh et al. 1994) and Cen X-4 (Martín et al. 1994), with comparable abundances of log $N_{Li} \simeq 2.0$–3.3 (see Fig. 4).

It is *very* unlikely that all three of these SXTs are young ($\leq 10^7$ yrs) and so the conclusion must be that Li is being created at these sites. They have a wide range of orbital periods (0.32–6.5 days), two are black holes while one is a neutron star, and the secondaries are of significantly different sizes. The *only* property they share is the huge X-ray outburst that recurs every few decades, and it has been proposed that spallation processes during these outbursts result in the production of Li (Martín et al. 1994). Large mass outflows seen during the outbursts transfer the Li to the secondary.

A possible test for this mechanism has been suggested that is related to the observed γ-ray line at 476 keV in Nova Mus 1991. This was originally interpreted as a gravitationally-redshifted $e^- - e^+$ annihilation line, double-peaked due to Keplerian rotation of the disc. Instead, it could be associated with the 478 keV line of ^{7}Li, which also provides a more natural explanation for the line width. If so, the temporal history of the feature

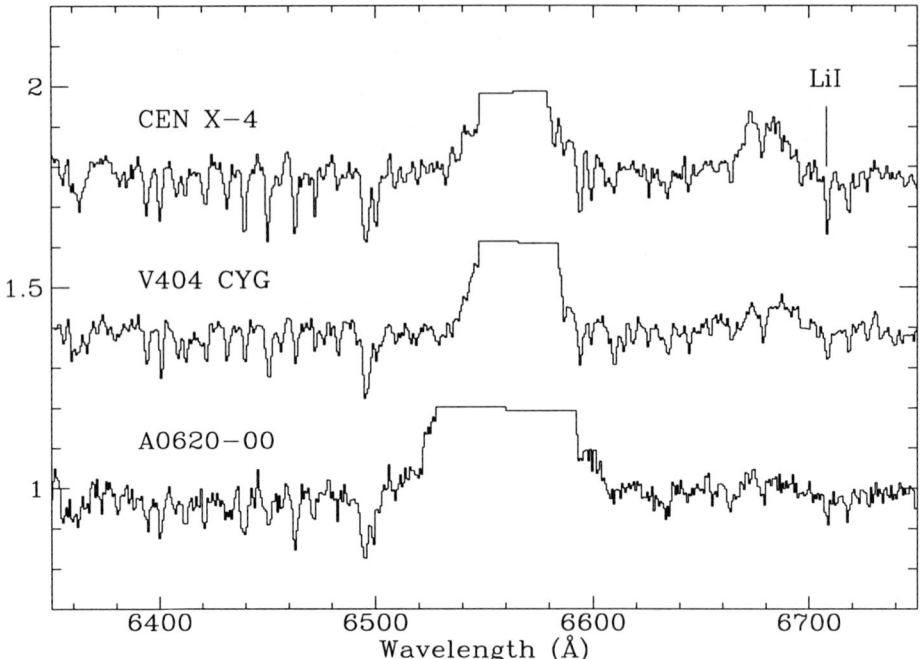

Figure 4. AAT and WHT spectra of Cen X-4, V404 Cyg and A 0620–00 showing the presence of Li I λ6708 (from Charles et al. 1994).

(it only lasted for ~1/2 day) will give information on the acceleration and spallation mechanisms that are taking place.

5. GRO J0422+32: an SXT exhibiting Mini-Outbursts

Nova Per 1992 (GRO J0422+32) was suggested as a BHC on the basis of its hard spectrum (similar to V404 Cyg) and 400–600 keV excess. It has a 5.1 hr optical period (Kato et al. 1992; Chevalier & Ilovaisky, 1993) and a slightly longer "superhump" period (see Haswell's review in these Proceedings). This has also been seen in GS 2000+25 (Charles et al. 1991) and Nova Mus 1991 (Bailyn 1992). GRO J0422+32 was thus considered as potentially an important system because it has low reddening ($E_{B-V} \sim 0.2$), the shortest period in the class, an unusually slow decline ($\sim 0^m.01\,d^{-1}$) and could also be the first high i SXT (dips in the light curve were reported by Kato et al. (1993).

It started declining rapidly 240 d after the peak but in August and December 1993 it exhibited two "mini-outbursts", behaviour that is unprecedented in this class. Spectroscopy with the WHT showed very broad

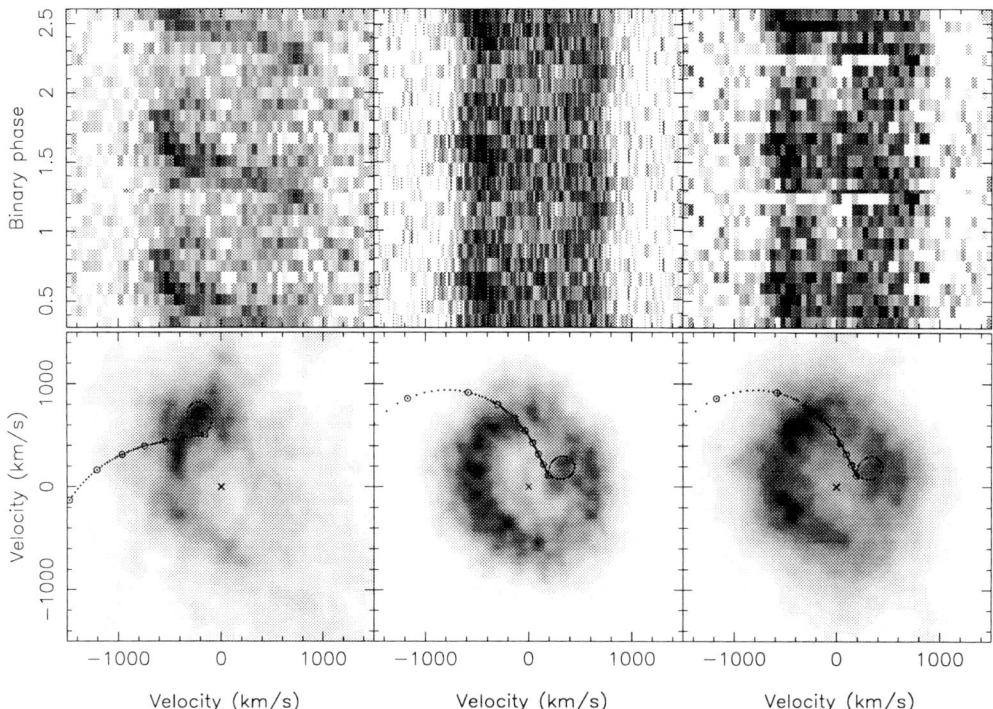

Figure 5. Grey-scale representation of time-resolved spectra of GRO J0422+32 obtained of the He II λ4686, Hα and Hβ profiles. The bottom panels show the resulting Doppler tomograms for each profile in velocity space. The secondary star interpretation is shown for He II, while the disc location is indicated for the others (from Casares et al. 1995).

($\pm \sim 3000\,\mathrm{km\,s^{-1}}$) Balmer absorption, and complex He II and Balmer emission components (Harlaftis et al. 1994). A grey-scale representation of the time-resolved spectra (Fig. 5) shows an S-wave component within He II that has semi-amplitude $\sim 750\,\mathrm{km\,s^{-1}}$ and a period of 5.1 hrs. This is the *first clear velocity modulation* observed in GRO J0422+32, and can be tentatively associated with the orbital period.

The time-resolved spectra can be used as input into a Doppler tomography analysis (Fig. 5). Without a detection of the secondary star, or an ellipsoidal modulation, the precise location of the sharp emission component is impossible to determine. It could be the X-ray heated face of the secondary star, or the hot spot where the stream impacts the disc. The former is favoured due to the correlation between He II λ4686 and the X-rays, but X-ray heating must be low and there is a complete absence of Bowen emission (which was strong during the main outburst). The two possible locations are shown in Fig. 5. If on the secondary star, then the velocities observed imply $f(M) \sim 9\,M_\odot$, whereas the theoretical gas stream path to the hot spot gives $q \sim 5$. Conclusive evidence must await the secondary's radial-velocity curve now that the object seems to have reached true quiescence (Zhao et al. 1994). It is crucial to confirm the 5.1 hr period because, if real, then it means that in future SXT outbursts it will be possible to glean useful dynamical information from high resolution observations of the He II $\lambda 4686$ line profiles.

Acknowledgments. I particularly thank Jorge Casares, Tim Naylor and Tariq Shahbaz for their help in preparing this manuscript.

References

Bailyn, C.D. 1992, ApJ 391, 298
Casares, J. & Charles, P.A. 1994, MNRAS (in press)
Casares, J., Charles, P.A. & Naylor, T. 1992, Nat 355, 614
Casares, J. et al. 1995, MNRAS (submitted)
Charles, P.A. et al. 1991, MNRAS 249, 567
Charles, P.A. et al. 1994, in *The Evolution of X-ray Binaries*, S.S. Holt & C.S. Day (Eds.), AIP Conference Proc. Vol. 308, p. 371
Chevalier, C. & Ilovaisky, S.A. 1993, IAU Circular 5692
Harlaftis, E.T. et al. 1994, in *The Evolution of X-ray Binaries*, S.S. Holt & C.S. Day (Eds.), AIP Conference Proc. Vol. 308, p. 91.
Kato, T., Mineshige, S. & Hirata, R. 1992, IAU Circ. 5676
Kato, T., Mineshige, S. & Hirata, R. 1993, IAU Circ. 5704
King, A.R. 1993 MNRAS 260, L5
Marsh, T.R., Robinson, E.L. & Wood, J.H. 1994, MNRAS 266, 137
Martín, E.L. et al. 1992, Nat 358, 129
Martín, E.L. et al. 1994, ApJ (in press)
McClintock, J.E. 1986, in *Physics of Accretion onto Compact Objects*, K.O. Mason, M.G. Watson & N.E. White (Eds.), Lecture Notes in Physics Vol. 266 (Springer), p. 211
McClintock, J.E. 1991, Ann. N.Y. Acad. Sci. 647, 495
McClintock, J.E. & Remillard, R.A. 1986, ApJ 308, 110
McClintock, J.E. & Remillard, R.A. 1990, ApJ 350, 386
Orosz, J.A. et al. 1994, ApJ (in press)
Remillard, R.A., McClintock, J.E. & Bailyn, C.D. 1992, ApJ 399, L145
Shahbaz, T., Naylor, T. & Charles, P.A. 1993, MNRAS 265, 655
Shahbaz, T., Naylor, T. & Charles, P.A. 1994a, MNRAS 268, 756
Shahbaz, T. et al. 1994b, MNRAS (in press)
Tanaka, Y. & Lewin, W.H.G. 1995, in *X-ray Binaries*, W.H.G. Lewin, J. van Paradijs & E.P.J. van den Heuvel (Eds.), Cambridge Univ. Press, (in press)
Thorsett, S.E. et al. 1993, ApJ 405, L29
van Paradijs, J. & McClintock, J.E. 1995, in *X-ray Binaries*, W.H.G. Lewin, J. van Paradijs & E.P.J. van den Heuvel (Eds.), Cambridge Univ. Press, (in press)
Verbunt, F., van den Heuvel, E.P.J. 1995, in *X-ray Binaries*, W.H.G. Lewin, J. van Paradijs & E.P.J. van den Heuvel (Eds.), Cambridge Univ. Press, (in press)
Wade, R.A. & Horne, K. 1988, ApJ 324, 411
White, N.E. 1994, in *The Evolution of X-ray Binaries*, AIP Conference Proc. 308, 371
Zhao, P. et al. 1994, IAU Circ. 6072

OPTICAL PHOTOMETRY OF BLACK-HOLE CANDIDATES

C. A. HASWELL
Columbia University
Dept. of Astronomy, 538 W. 120th St, New York, NY 10027

Abstract.
The orbital light curves of SXTs are discussed. In principle, the orbital inclination can be determined from the ellipsoidal variations, allowing a complete solution for the component masses. Complications in this analysis are described, and the evidence for superhumps in the light curves is reviewed. Constraints on the compact-object masses derived from superhumps are described.

1. Introduction

This paper will discuss the soft X-ray transient (SXT) class of black-hole candidates which are introduced in the accompanying review by Charles. The long-term optical light curves of these objects form a fascinating topic (e.g., Callanan et al. 1995) which will not be discussed here; instead we will describe orbital and associated periodic photometric modulations, and the consequent system parameter determinations.

Quiescent light curves of SXTs show a double-humped 'ellipsoidal' orbital modulation due to the changing projected area of the Roche lobe filling mass donor. Maxima occur at quadratures, corresponding to the maximum projected area, and are hence expected to be symmetric. Minima occur at the orbital conjunctions; in general they have differing depths due to gravity darkening and heating by the disk. The amplitude of the ellipsoidal modulation depends on the orbital inclination i of the system, so modelling of the observed ellipsoidal light curve can, in principle, constrain i, and hence contribute to an improved solution for the component masses.

There are several complications in the determination of i from the ellipsoidal variations. Inevitably the light from the mass donor is diluted by the flux from the accretion disk; the extent of this diluting component can be assessed using high quality spectroscopic data (e.g., Marsh et al. 1994) but usually some uncertainties remain. Underestimating the diluting flux leads to an underestimate in the orbital inclination. The ellipsoidal light curve may be distorted by other variable contributions to the total light: the bright spot associated with the impact of the mass transfer stream on the edge of the disk (McClintock & Remillard 1990); star spots on the secondary; and variable disk emission, including the "superhump" phenomenon.

Superhumps were first observed in superoutbursts of SU UMa stars, an extreme mass ratio subclass of CVs (Warner 1985). The superhump period, $P_{\rm sh}$, can be expressed as

$$P_{\rm sh} = (1 + \epsilon) P_{\rm orb} \qquad (1)$$

where $P_{\rm orb}$ is the orbital period, and ϵ typically has a value in the range 0.01–0.08 (Skillman & Patterson 1993). There is an empirical relationship between the period excess, ϵ, and the orbital period (Stoltz & Schoembs 1984; Skillman & Patterson 1993)

Theoretical work suggests that superhumps are due to tidal interactions in an extensive, non-circular, precessing disk (Whitehurst 1988; Hirose & Osaki 1990). There are two prerequisites for superhumps. The first is a disk radius which exceeds the radius of the largest tidally stable particle orbit, the 'stability radius'. The disk can extend to the stability radius only for extreme mass ratios, hence the second prerequisite: an extreme mass ratio, $q \equiv M_1/M_2 \gtrsim 4$, where M_1 is the compact object mass, and M_2 is the mass-donor star mass.

Theory is able to reproduce the empirical relationship between ϵ and $P_{\rm orb}$: Osaki (1985) derives

$$\frac{P_{\rm orb}}{P_{\rm prec}} = \frac{3}{4} \frac{1}{\sqrt{q(1+q)}} \left(\frac{R_{\rm d}}{a}\right)^{3/2} \qquad (2)$$

where $P_{\rm prec}$ is the disk precession period, $P_{\rm prec} = \epsilon^{-1} P_{\rm sh}$, $R_{\rm d}$ is the radius of the disk, and a is the binary separation. If either the disk radius or the mass ratio is measured (or assumed), the other quantity can thus be deduced from the period measurements; Mineshige et al. (1992) extend and elaborate on this and consequent constraints on the compact object mass.

Spectroscopic studies imply that SXTs have extreme mass ratios (see the contribution to these proceedings by Charles), so one requirement for a precessing non-circular disk is met in SXTs. Superhumps were detected in the outburst light curves of Nova Muscae 1991 and GS 2000+25 (Charles et al. 1991; Remillard et al. 1992; Bailyn 1992; see Sects. 3 and 4 for further discussion). Recently the detection of persistent superhumps in several extreme mass ratio CVs (Skillman & Patterson 1993; Patterson *et al.* 1993; Patterson & Skillman 1994) shows that an outburst state is not necessary for superhumps to occur. This implies that superhumps may need to be considered in the interpretation of the quiescent light curves of SXTs.

2. A 0620-00: the Prototype and the Problems

McClintock & Remillard's (1986) seminal paper began the study of the ellipsoidal variations in quiescent SXT light curves, and foreshadowed some of the problems. Fig. 1a shows their BV band light curve, it agrees roughly with theoretical expectations, but the asymmetry between the two maxima requires an additional phase-variable contribution to the light curve. Further complications are apparent in subsequent observations: see Figs. 1b and 1c, Bartolini *et al.* (1990), and Haswell *et al.* (1993). Not only are the maxima asymmetric (Haswell 1992; Callanan 1993), but the sense of the asymmetry reverses; worse still, the same statement is true of the minima! In addition, the lower minima in Figs. 1b and 1c appear deeper and sharper than a purely ellipsoidal curve would allow.

The anomalously deep minima in the quiescent light curves have been interpreted as grazing eclipses occurring when the disk radius is large (Karitskaya *et al.* 1988; Haswell *et al.* 1993). Hence both theoretical requirements for superhumps may be satisfied, and persistent superhumps might be expected. In CVs, the persistent superhumps typically have modest amplitudes, so by analogy persistent superhumps in SXTs might be expected to distort, rather than overwhelm, the ellipsoidal modulation in the orbital light curves. The orbital phase at which excess light due to the superhump is observed will depend upon the disk precession phase at which observations are made. Since the light curves shown in Fig. 1 result from averaging together data taken during many orbits, the superhump light would be effectively randomly distributed, and could account for the apparent secular variability in the shape of the orbital light curve (Haswell *et al.* 1993).

Haswell *et al.* (1993) attempted to deduce the orbital inclination from modelling of multi-colour light curves with ellipsoidal variations and a grazing eclipse, assuming a large circular disk; the lack of a deep minima at the opposite conjunction was interpreted as caused by low disk surface brightness. The data shown in Fig. 1c, however, makes this explanation unlikely: a drastic change in disk surface brightness would be required. As the cartoon in Fig. 2 shows, a non-circular disk leads to the possibility that grazing eclipses occur at one conjunction only. As the disk precesses, the grazing eclipse will occur first at one conjunction, then at the other. Hence the mo-

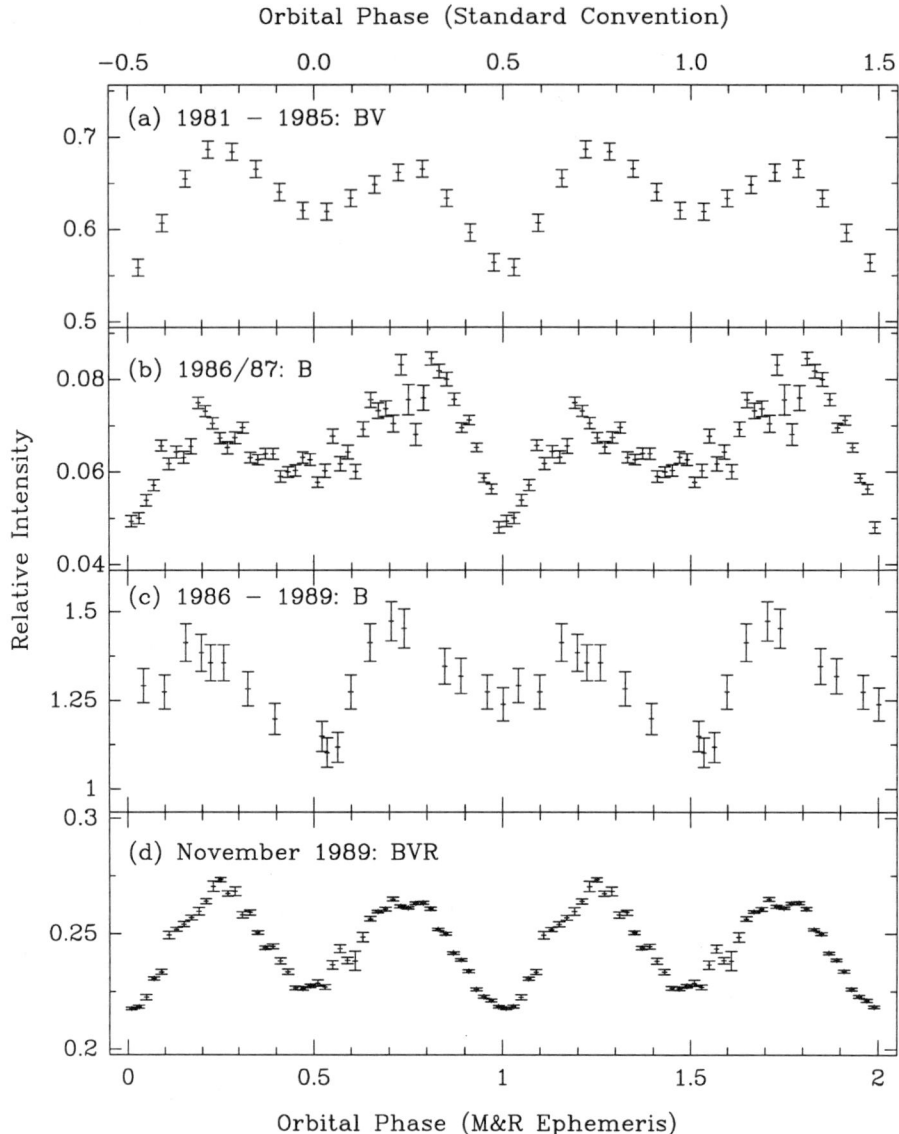

Figure 1. Orbital light curves of A0620–00: (a) from McClintock & Remillard (1986); (b) from Haswell et al. 1993); (c) from Bartolini et al. 1990); (d) from Haswell (1992). Though the double-humped ellipsoidal variation is apparent, it is clearly contaminated. In the individual curves the maxima are distinctly asymmetric; the upper three curves show very prominent differences in the depths of the minima. There is, furthermore, definite secular variability in the shapes of the light curves: the asymmetry between the two maxima reverses between (a) and (b); the asymmetry between the minima reverses between (b) and (c); in (d) neither the maxima nor the minima appear so asymmetric.

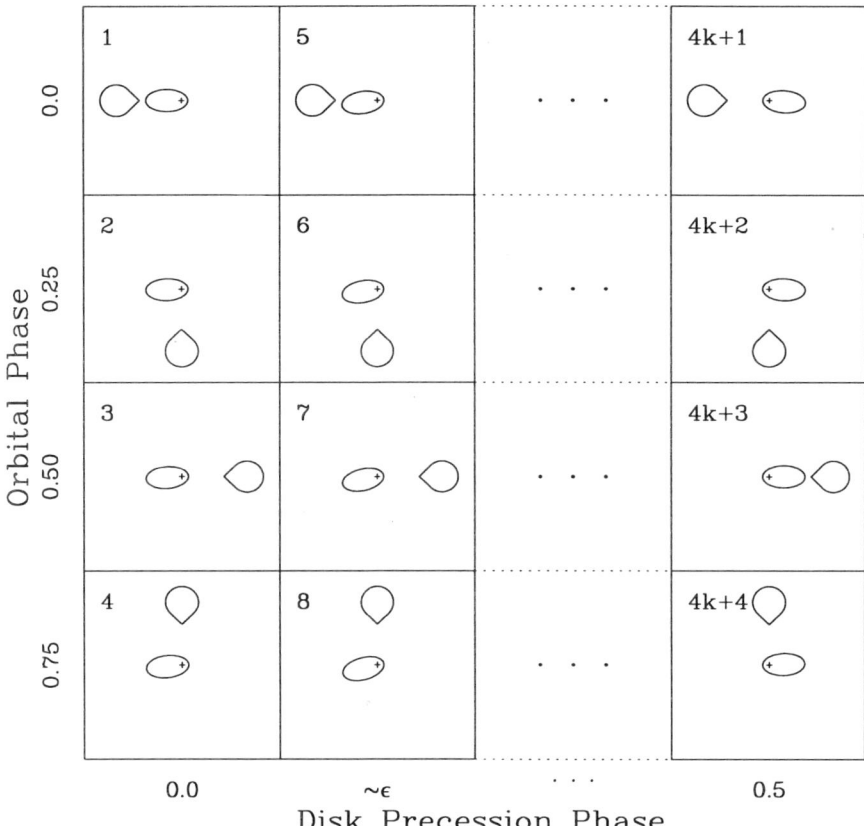

Figure 2. Cartoon depiction of the motion of the mass-donor and non-circular disk in a superhumping binary. Time increases down columns and from left to right, the sequence is indicated by the numbers in the upper left of each panel. Each column represents one binary orbit, and the total time elapsed is one half of the disk precession period. In the first panel the line of apsides of the disk is aligned with the mass-donor at binary phase 0.0; since the disk precession period is long compared to the orbital period, the line of apsides of the disk has moved only slightly by panel 5. After many binary orbits have passed the disk's line of apsides has moved by 180 degrees. If i is about 65 degrees then a grazing eclipse would be observed for the configuration shown in panel 1 (orbital phase 0.0) but not panel 3 (orbital phase 0.5). Conversely in the last column, a grazing eclipse might occur at orbital phase 0.5 but not orbital phase 0.0.

del might simultaneously and self-consistently explain all the anomalies and secular changes shown in Fig. 1.

By November 1989 the photometric modulations were much closer to the expected ellipsoidal variation (see Fig. 1d). The maxima are no longer terribly asymmetric, and the difference between the depths of the minima is small enough to be accounted for by gravity darkening. This suggests that the disk shrank, so that neither grazing eclipses nor superhumps were distorting the light curve at this epoch. Spectroscopic work deduced a disk radius comparable to the size of the primary Roche lobe in 1986 (Johnston et al. 1989), by 1991 the disk radius was approximately half this size (Marsh et al. 1994; see also Orosz et al. 1994).

If the grazing-eclipse hypothesis is accepted, then it allows a simple geometrical constraint to be placed on the orbital inclination. It is reasonable to assume that the disk radius along the line of apsides is no larger than that of the largest coplanar circular disk which can be accommodated in the primary's Roche lobe. Adopting this prescription for the disk radius and demanding grazing eclipses be possible leads to a lower limit on the orbital inclination for a given mass ratio. Taking the mass ratio of $q = 14.94$ from Marsh et al. (1994), we find a maximum disk radius $R_d(\max) = 0.63a$. A system with this configuration will only exhibit grazing eclipses for $i > 62°$; this result is not strongly dependent on the exact value of q until q becomes much smaller than the preferred value. This limiting inclination changes by only $2°$ for the range $10 < q < 20$.

The inclination derived from the grazing-eclipse constraint is in serious disagreement with that determined from the ellipsoidal variations in the K band light curve, $31° < i < 54°$ (Shahbaz et al. 1994a). The weakest part of their analysis is the assumption that the disk contributes negligible flux in the K band, which was justified by extrapolating the accretion disk flux in the optical assuming a power law spectrum. This extrapolation is uncertain, and Haswell (1992), Haswell et al. (1993) and Chevalier & Ilovaisky (1993) show evidence for ellipsoidal amplitudes *decreasing* with increasing wavelength, contrary to the assertion that the flux becomes increasingly dominated by the mass-donor in the infrared. A direct measurement of the accretion disk flux from IR spectroscopy is desirable to resolve this issue.

3. Nova Muscae 1991

The quiescent orbital light curve of Nova Muscae 1991 shows a double-humped ellipsoidal modulation with $P_{\rm orb} = 10.398(14)$ hours (Remillard et al. 1992). Again the ellipsoidal maxima are asymmetric, complicating the interpretation and making a reliable determination of i from the ellipsoidal modulation difficult.

Nova Muscae 1991 showed a 10.5 hour modulation in its optical light curve during the decline from outburst (Bailyn 1992). This modulation is a likely superhump with $\epsilon \approx 0.014$ (Remillard et al. 1992). The error in the determination of ϵ is dominated by the uncertainty in $P_{\rm sh}$: the data showed evidence for a change in $P_{\rm sh}$, but were insufficient to measure the period unequivocally at multiple epochs. This behaviour is consistent with a superhump interpretation: if R_d decreases then $P_{\rm prec}$ will increase (Eq. 2), and P_{sh} decreases. A lower limit for the compact object mass, M_1, can, however, be obtained using a formula derived from Eq. 2 above (Mineshige et al. 1992): $M_1 \gtrsim 0.01\ \epsilon^{-1}\ P_{orb}(\text{hr})\ M_\odot$. Adopting the longest candidate superhump period measured, (10.56(4) hours, see Bailyn 1992), leading to the lowest limit on M_1, we obtain

$$M_1 \gtrsim 6.66^{+2.23}_{-1.30} M_\odot \qquad (3)$$

4. GS 2000+25

In outburst, GS 2000+25 exhibited an 8.33 hour photometric modulation (Charles et al. 1991); in quiescence, a shorter period was observed, indicative of ellipsoidal variations and an orbital period of ~ 8.2 hours (Callanan & Charles 1991). These papers interpreted the outburst modulation as a superhump based on both the shape of the modulation and the period difference, though the latter was only a 3σ result.

Subsequent observations of GS 2000+25 reveal more complicated behaviour: Chevalier & Ilovaisky (1993) determined a period of 8.25836 hours consistent with both their outburst and quiescent data. Evidence is also presented for a 10.015(2) hour period in quiescence, which is attributed to a 'distortion wave'. The beat period of the orbital and distortion wave modulations is 1.96 days; the orbital light curves show one shape for odd JD and another for even JD.

Reanalysing the same data, Shahbaz et al. (1994b) advocate an orbital period of 7.04 hours, dispensing with the distortion wave. However, Fig. 3 shows a high quality I band light curve of GS 2000+25 (Haswell 1995) which strongly supports the 8.26 hour period; the 7.04 hour period is clearly inconsistent with these data.

The superhump and orbital periods can constrain the component masses in this system: $M_1 \gtrsim 6 M_\odot$ was obtained by Mineshige et al. (1992) using the periods reported by Charles et al. (1991) and Callanan & Charles (1991).

5. V404 Cyg

V404 Cyg presents the best case for a stellar black hole, with a compact object mass exceeding $6.08 \pm 0.06 M_\odot$ (Charles et al. 1994). Several groups have measured the ellipsoidal variations in the optical and IR light curves of this system. A best-fit inclination of $i = 56°$ and corresponding compact object mass of 12 $M_\odot$ is derived from the K band light curve (Shahbaz et al. 1994c), assuming a negligible contribution from the disk in that passband. This assumption is motivated by the measured contribution of the disk in

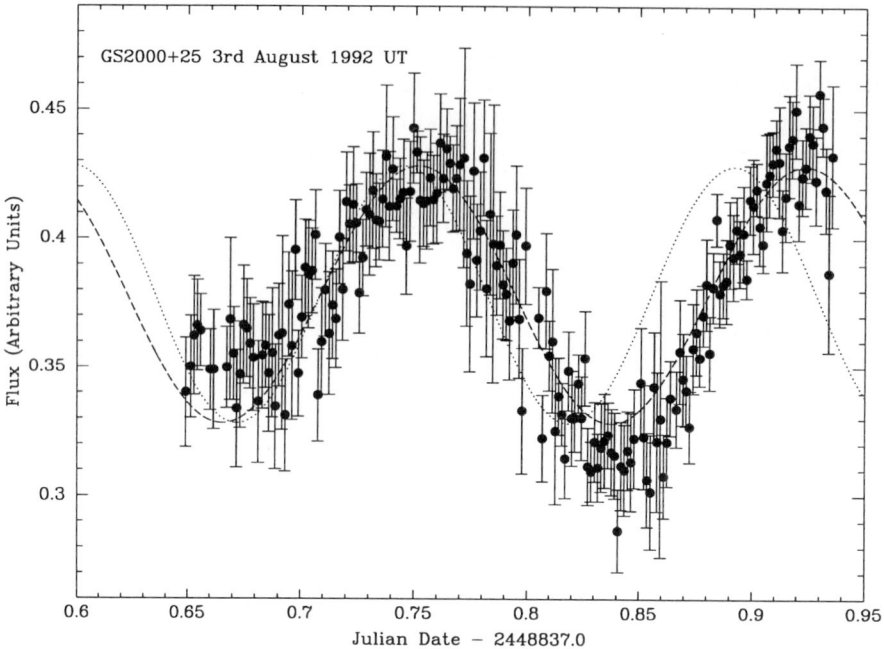

Figure 3. I band light curve of GS 2000+25 taken at the McDonald Observatory 2.7 meter in excellent conditions. This single night of data clearly rules out the proposed 7.04 hour orbital period: a 3.52 hour sine wave is shown by the dotted line. The 8.26 hour period is strongly supported: the dashed line shows a 4.13 hour sine wave.

the optical, which is extrapolated into the IR. The extrapolation may be invalid as it predicts a disk contribution of ∼10% in the R band, while variability of amplitude 20% is observed superimposed upon the R band ellipsoidal modulation (Pavlenko et al. 1994). This extra R band variability has a period of ∼6.2 hours, and likely arises in the accretion flow. Variability superimposed on the ellipsoidal modulation is also observed in the I and H bands (Wagner et al. 1992; Robinson et al. 1994).

A further complication is apparent in the H band data: modelling the ellipsoidal light curve requires a source of X-ray heating of luminosity $\sim 3 \; 10^{36} \, \mathrm{erg \, s^{-1}}$, almost 400 times greater than that observed (Wagner et al. 1994). Independent of the complications, by adopting the most conservative set of assumptions to obtain an extreme lower limit on the orbital inclination, Robinson et al. (1994) deduce an *upper* limit of $M_1 < 12.0 \, M_\odot$ for the mass of the compact object.

6. GRO J0422+32

This object provides the latest opportunity to study the quiescent light curve of an SXT. A 5.1 hour period has been detected in the low-state photometry (Chevalier & Ilovaisky 1994; Garcia et al. 1994; Callanan et al. 1995). It is not yet clear whether this modulation is orbital in origin, though there is some spectroscopic support for this idea (Harlaftis et al. 1994; Charles, these proceedings). The light curves are complex and changing, so it is highly unlikely that the variability is purely ellipsoidal.

If the orbital period were 5.1 hours, then a likely superhump was detected in November 1992 (Kato et al. 1992b), leading to a tentative mass estimate for the compact object (Kato et al. 1992b): $2.9 \lesssim M_1/M_\odot \lesssim 6.2$.

7. Conclusions

The determination of orbital inclinations from modelling of ellipsoidal variations is complicated by other effects which distort the orbital light curves. While observations in the IR may better isolate the ellipsoidal component in the light curve, it is not clear that SXT disks produce negligible flux in the IR. The extensive data on A 0620–00 seem to suggest that the quiescent light curve becomes more purely ellipsoidal as time elapses since the outburst. Observations in late quiescence, with simultaneous spectroscopic determinations of the disk flux at the appropriate wavelength, hold the best promise for a robust determination of i.

Superhumps, while likely responsible for the contamination of the ellipsoidal modulation, offer an alternative approach to the determination of the system parameters in SXTs.

New SXTs are continually being discovered, providing an increasing number of systems to apply these techniques to. Eventually an SXT exhibiting X-ray eclipses will be found, yielding a precise determination of i, and hence providing the opportunity to assess the merits of the techniques discussed in this review.

Acknowledgements. Thanks to Paul Callanan, Phil Charles, Joe Patterson, and Rob Robinson for helpful discussions. I am also grateful to Paul Callanan for comments on this paper and providing data used in Fig. 1.

References

Bailyn, C.D. 1992, ApJ 391, 298
Bartolini, C. et al. 1990, in *Structure and Emission Properties of Accretion Disks*, IAU Colloq. No. 129, C. Bertout, S. Collin, J.-P. Lasota & J. Tran Thanh Van (Eds.), Editions Frontières (Gif-sur-Yvette), p. 371

Callanan, P.J. 1993, PASP 105, 961
Callanan P.J. & Charles P.A. 1991, MNRAS 249, 573
Callanan, P.J. et al. 1995, ApJ (March 1 issue)
Casares, J. & Charles, P.A. 1994, *XXIInd IAU General Assembly Posters*, H. van Woerden (Ed.), p. 83
Charles, P.A. 1995, these Proceedings
Charles, P.A. et al. 1991, MNRAS 249, 567
Chevalier, C. & Ilovaisky, S.A. 1993, A&A 269, 301
Chevalier, C. & Ilovaisky, S.A. 1994, Haute-Provence Preprint No. 83
Garcia, M.R. et al. 1994, BAAS 25, 1381
Harlaftis, E. et al. 1994, in *The Evolution of X-Ray Binaries*, S.S. Holt & C.S. Day (Eds.), AIP Conference Proc. Vol. 308, p. 91
Haswell, C.A. 1992, Ph.D. Thesis, University of Texas at Austin
Haswell, C.A. 1995, (in preparation)
Haswell, C.A. et al. 1993, ApJ 411, 802
Hirose, M. & Osaki, Y. 1990, PASJ 42, 135
Johnston, H.M., Kulkarni, S.R. & Oke, J.B. 1989, ApJ 345, 492
Karitskaya, E.A. & Bochkarev, N.G. 1988, Perem. Zvezdy 22 (6), 943
Kato, T., Mineshige, S. & Hirata, R. 1992a, IAU Circ. 5676
Kato, T., Mineshige, S. & Hirata, R. 1992b, IAU Circ. 5704
Marsh, T.R., Robinson, E.L. & Wood, J.H. 1994, MNRAS 266, 137
McClintock, J.E. & Remillard, R.E. 1986, ApJ 308, 110
McClintock, J.E. & Remillard, R.A. 1990, ApJ 350, 386
Mineshige, S., Hirose, M. & Osaki, Y. 1992, PASJ 44, L15
Orosz, J.A. et al. 1994, ApJ (in press)
Osaki, Y. 1985, A&A 144, 369
Patterson, J. & Skillman, D.R. 1994, C.A.L. preprint number 551
Patterson, J. et al. 1993, ApJS 86, 235
Pavlenko, E.P. et al. 1994, *XXIInd IAU General Assembly Posters*, H. van Woerden (Ed.), p. 86
Remillard, R.A., McClintock, J.E. & Bailyn, C.D. 1992, ApJ 399, L145
Robinson, E.L., Sanwal, D. & Zhang, E. 1994, *XXIInd IAU General Assembly Posters*, H. van Woerden (Ed.), p. 87
Shahbaz T., Naylor T. & Charles P.A. 1994a, MNRAS 268, 756
Shahbaz T. et al. 1994b, MNRAS 268, 763
Shahbaz, T. et al. 1994c, MNRAS (in press)
Skillman, D.R. & Patterson, J. 1993, ApJ 417, 298
Stoltz, B. & Schoembs, R. 1984, A&A 132, 187
Wagner, R.M. et al. 1992, ApJ 401, L97
Wagner, R.M. et al. 1994, (preprint)
Warner, B. 1985, in *Interacting Binaries*, P.P. Eggleton & J.E. Pringle (Eds.), Reidel Publishing Company, p. 367.
Whitehurst, R. 1988, MNRAS 232, 35

Discussion

D. O'Donoghue: MEM eclipse mapping of Z Cha (see O'Donoghue, 1990, MNRAS 246, 29) shows that the superhump light source has a bright feature along the line of centers, precisely where C. Haswell needs a light source to model her infrared light curve variations.

W. Kundt: According to the little I know about superhumps in cataclysmic variables, an eccentric disk is not without problems. For instance, a 180° phaseshift has been seen during transition from outburst to quiescence.

C. Haswell: I'm not familiar with the particular observations you are referring to[1]. As far as I know the precessing non-circular disk model has been spectacularly successful in explaining the observed properties of extreme mass ratio CVs. For example Patterson, Halpern & Shambrook (1993, ApJ 419, 803) recently reported on spectroscopy of AM CVn in which the absorption lines from the disk show pronounced periodic variations in skewness. This feature is very successfully explained by a simple precessing non-circular disk model.

J. van Paradijs: Do you know of attempts to look for the beat between the orbital and super hump periods from the variation in the orbital phase at which the maximum in the perturbation occurs in the light curve, e.g., in the 15 year of data collected on A 0620–00.

C. Haswell: As far as I know this has not been done yet. Obviously it should be done using all the data collected by the many different observers who have studied A 0620–00. In addition to the data I showed, I learned at this conference from Paul Callanan that Jeff McCLintock has some unpublished data which shows a grazing eclipse feature. I suspect features have posed a challenge to interpretation. If the data are extensive enough, we may be able to follow the evolution of the precession period as the disk shrinks.

S.R. Kulkarni: If the secondary is indeed tidally locked, there should be tremendous tidal dissipation which should lead to great chromospheric activity. Is there any observational evidence for this (e.g., the Wilson-Bappu effect, spots, etc.)?

C. Haswell: One of the first explanations we considered for the photometric irregularities in A 0620–00 was migrating star spots on the mass donor (Haswell, Robinson & Horne, 1990, in *Accretion Powered Compact Binaries*, C.W. Muche (Ed.), Cambridge University Press). We now have a body of evidence from A 0620–00 and other systems that a precessing non-circular disk contributes to the photometric irregularities. It is possible, however, that both these phenomena contaminate the light curves. If a precessing non-circular disk is included in the model it is difficult to identify unequivocal evidence for star spots.

---

[1] Note added: In his equation 62, Osaki (1985, A&A 144, 369) discusses this observation in VW Hyi, and gives a possible explanation within the context of the precessing non-circular disk model.

DISCOVERY OF A NEW X-RAY TRANSIENT IN SCORPIUS

GRO J1655-40 ≡ X-ray Nova Scorpii 1994

W.S. PACIESAS
University of Alabama in Huntsville
Huntsville, AL 35899 USA

S.N. ZHANG AND B.C. RUBIN
Universities Space Research Association
Huntsville, AL 35806 USA

AND

B.A. HARMON, C.A. WILSON AND G.J. FISHMAN
NASA/Marshall Space Flight Center
Huntsville, AL 35812 USA

Abstract. A bright transient X-ray source, GRO J1655-40 (X-ray Nova Scorpii 1994) was discovered with BATSE (the Burst and Transient Source Experiment) in late July 1994. More recently, the source also became a strong radio emitter, its rise in the radio being approximately anti-correlated with a decline in the hard X-ray intensity. High-resolution radio observations subsequent to this symposium showed evidence for superluminally expanding jets. Since the hard X-ray emission extends to at least 200 keV and we find no evidence of pulsations, we tentatively classify the source as a black-hole candidate. However, its hard X-ray spectrum is unusually steep (power-law photon index $\alpha \simeq -3$) relative to most other black-hole candidates. In this regard, it resembles GRS 1915+105, the first galactic source to show superluminal radio jets.

1. Introduction

GRO J1655-40 (also designated X-ray Nova Sco 1994) was first detected with BATSE (Zhang et al. 1994a) on 27 July 1994 and located initially to within about 0.3° using Earth occultation imaging (Zhang et al. 1993). A candidate optical counterpart was found by Bailyn et al. (1994) and confirmed spectroscopically by Della Valle (1994). Subsequently, we used additional BATSE data to produce an improved location (Wilson et al.

1994), with a precision of about 0.1°, which agreed well with the optical counterpart.

The radio counterpart to GRO J1655–40 was first detected on 6 August by Campbell-Wilson & Hunstead (1994a) who used the Molonglo Observatory Synthesis Telescope (MOST). From its initial intensity of 370 mJy at 843 MHz, the source continued to flare, reaching more than 5 Jy on 15 August (Campbell-Wilson & Hunstead 1994b). It quickly became apparent that this radio source was unusually bright in comparison with other X-ray transients. The radio spectrum during most of the rise was consistent with optically-thick synchrotron emission (Hjellming 1994a). The early radio observations are described in more detail in an accompanying paper (Hunstead et al. 1995).

Subsequent to this symposium, the radio intensity peaked at ~ 7 Jy around 17–18 August and the spectrum became optically thin (Hunstead et al. 1994). The source was the subject of an intensive campaign of interferometric radio observations. The initial observations showed a double source with the components moving apart (Hjellming 1994b). At an estimated distance of ~ 3.5 kpc (McKay & Kesteven 1994), the apparent separation velocity is superluminal (Reynolds & Jauncey 1994; Hjellming & Rupen 1994a). BATSE observed further X-ray outbursts in September (Paciesas et al. 1994) and November (Zhang et al. 1994b,c), with associated radio activity (Hjellming & Rupen 1994b,c,d). The hard X-ray, radio and optical behavior of X-ray Nova Sco 1994 is summarized elsewhere (Harmon et al. 1995; Hjellming & Rupen 1995; Tingay et al. 1995; Bailyn et al. 1995).

2. Observations

The early history of GRO J1655–40 was remarkable for its unusually fast rise. Fig. 1 shows the hard X-ray intensity derived from Earth occultation measurements using BATSE. We show data from individual source rises and sets in order to provide the best possible time resolution using this technique. The first clear detection of the source occurs around the end of 27 July and the turn-on occurs in less than 0.5 days, and may be as short as 0.25 days. There is a weak indication of emission for several days prior to the turn-on, but this may be a systematic effect, possibly due to interference from nearby sources such as the variable X-ray pulsar OAO 1657–415.

The occultation geometry for several days around the turn-on of GRO J1655–40 was particularly unfavorable for separation of this source from the OAO pulsar. The latter had been detected in a bright state since mid-June by the BATSE daily epoch-folding search. However, the midpoint of its binary eclipse occurred on 27 July and pulsations were not detected by BATSE on either that day or the following day. Furthermore,

DISCOVERY OF A NEW X-RAY TRANSIENT IN SCORPIUS

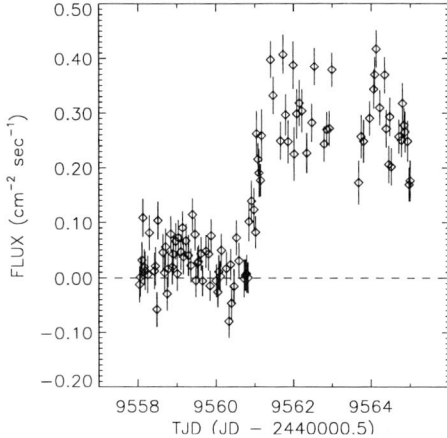 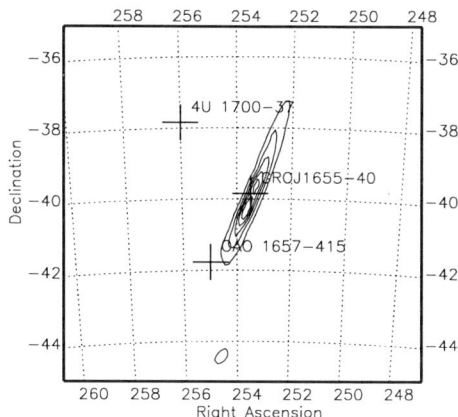

Figure 1. The early intensity history of GRO J1655-40 derived from individual occultation steps. The energy range is 20-430 keV. The rise begins around the end of Truncated Julian Day (TJD) 9560 (27 July) and the turn-on to essentially full intensity occurs in less than 0.5 day.

Figure 2. Occultation transform image for 30 July showing the region surrounding GRO J1655-40. The dominant emission is clearly distinct from the nearby X-ray pulsar OAO 1657-415. Due to the unfavorable occultation geometry the contours are severely elongated along the direction of the Earth's limb.

occultation images of the region showed that the new source was clearly distinct from the OAO source. Fig. 2 shows one such image for 30 July. The effect of the unfavorable occultation geometry is evident in the elongated contours. The OAO source is not visible in the image even though it was detected on this day by the BATSE epoch-folding and Fast Fourier Transform (FFT) searches which are more sensitive than the occultation method for detecting pulsed sources.

The lack of any significant unidentified peaks in the FFT search allows us to place an upper limit of $\sim 10\%$ on the pulsed fraction in the 20-100 keV band at maximum intensity, indicating that the new source is likely to be a black-hole candidate. We also generated power spectra of the data during the times when the source was visible to look for excess red noise or flickering. No obvious evidence of flickering was detected; however, a more detailed investigation is in progress (Crary et al. 1995).

The intensity history of GRO J1655-40 as of the time of this symposium is shown in Fig. 3. Here we have averaged the data in intervals of up to one day and we have omitted data from days when the interference from OAO 1657-415 was most severe. In contrast to most other such transients, the light curve of this source is erratic, showing no clear trend until the rather fast intensity drop around 13 August. Superimposed on Fig. 3 we also show the radio intensity, which increased spectacularly around the time

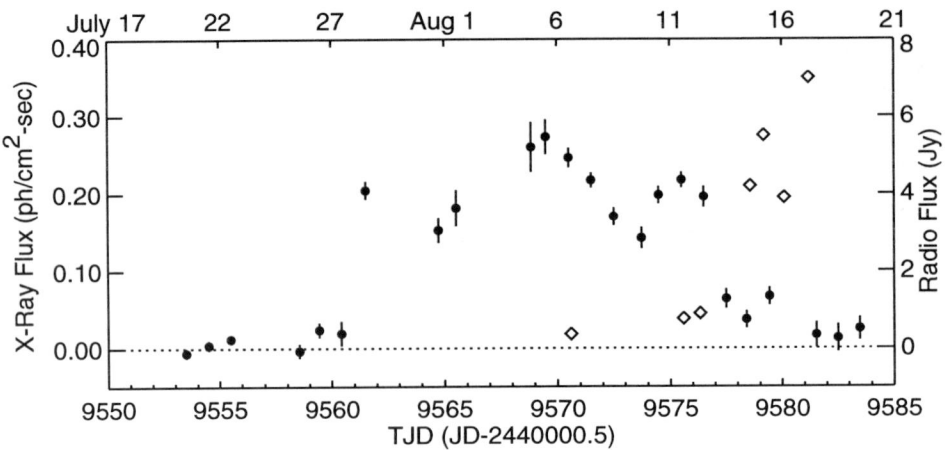

Figure 3. The intensity history of the entire initial outburst of GRO J1655–40. The energy range is 20–100 keV and the data points (solid circles) are one-day averages. The indicated errors are statistical only. Diamonds show the radio intensity at 843 MHz (Campbell-Wilson & Hunstead 1994a,b; Hunstead et al. 1994).

of the hard X-ray drop.

Preliminary analysis showed that the GRO J1655–40 hard X-ray spectrum was fit adequately with a single power law, with some evidence for hardening with time. We obtained values of the photon index $\alpha = -3.38 \pm 0.07$ ($\chi^2 = 18.1$ for 18 d.o.f.) during 1–4 August and $\alpha = -2.95 \pm 0.06$ ($\chi^2 = 66.8$ for 38 d.o.f.) during 4–8 August. We find no evidence for the high-energy cut-off typically seen in the spectra of black-hole candidates (Sunyaev et al. 1991). In fact, nearly contemporaneous observations using the OSSE instrument on CGRO detect emission to at least 600 keV (Kroeger et al. 1994) and in comparison with the BATSE results suggest a spectral flattening at higher energies. In the OSSE energy range ($E > 50$ keV), Kroeger et al. (1994) measured $\alpha = -2.7$ on 4 August and $\alpha = -2.4$ on 9 August.

3. Discussion

Although it is tempting to classify GRO J1655–40 as a black-hole candidate, it is somewhat unusual among unpulsed hard X-ray transients and classification may be premature. The combination of fast rise time, soft power-law spectrum with no high-energy cut off, lack of flickering, and highly variable light curve is not typical of the black hole candidates observed with BATSE (Paciesas et al. 1995). Superluminal radio emission has been observed in only one other galactic source, GRS 1915+105 (Mirabel & Rodriguez 1994) which previously had been recognized as unusual among X-ray transients

(Harmon et al. 1994). Except for its slow rise, the latter source strongly resembles GRO J1655-40 in hard X rays. More careful inter-comparison of these two sources may be useful in understanding their nature.

References

Bailyn, C., Jogee, S. & Orosz, J. 1994, IAU Circ. 6050
Bailyn, C. et al. 1995, Nat (in press)
Campbell-Wilson, D. & Hunstead, R. 1994a, IAU Circ. 6052
Campbell-Wilson, D. & Hunstead, R. 1994b, IAU Circ. 6055
Crary, D. et al. 1995, (in preparation)
Della Valle, M. 1994, IAU Circ. 6052
Harmon, B.A. et al. 1994, in *The Second Compton Symposium*, C.E. Fichtel, N. Gehrels & J.P. Norris (Eds.), AIP Conf. Proc. 304, p. 210
Harmon, B.A. et al. 1995, Nat (in press)
Hjellming, R.M. 1994a, IAU Circ. 6055
Hjellming, R.M. 1994b, IAU Circ. 6060
Hjellming, R.M. & Rupen, M. 1994a, IAU Circ. 6073
Hjellming, R.M. & Rupen, M. 1994b, IAU Circ. 6077
Hjellming, R.M. & Rupen, M. 1994c, IAU Circ. 6086
Hjellming, R.M. & Rupen, M. 1994d, IAU Circ. 6107
Hjellming, R.M. & Rupen, M. 1995, Nat (in press)
Hunstead, R. et al. 1994, IAU Circ. 6062
Hunstead, R. et al. 1995, these Proceedings
Kroeger, R.A. et al. 1994, IAU Circ. 6051
McKay, D. & Kesteven, M. 1994, IAU Circ. 6062
Mirabel, I.F. & Rodriguez, L.F. 1994, Nat 371, 46
Paciesas, W.S. et al. 1994, IAU Circ. 6075
Paciesas, W.S. et al. 1995, in *The Gamma-Ray Sky with COMPTON/GRO and SIGMA*, M. Signore, P. Salati & G. Vedrenne (Eds.), Kluwer Academic Publishers (Dordrecht), (in press)
Reynolds, J. & Jauncey, D. 1994, IAU Circ. 6063
Sunyaev, R.A. et al. 1991, SvAL 17, 409
Tingay, S.J. et al. 1995, Nat 374, 141
Wilson, C.A. et al. 1994, IAU Circ. 6056
Zhang, S.N. et al. 1993, Nat 366, 245
Zhang, S.N. et al. 1994a, IAU Circ. 6046
Zhang, S.N. et al. 1994b, IAU Circ. 6101
Zhang, S.N. et al. 1994c, IAU Circ. 6106

THE RADIO OUTBURST FROM GRO J1655-40

R.W. HUNSTEAD, D. CAMPBELL-WILSON AND T. YE
School of Physics, University of Sydney
NSW 2006, AUSTRALIA

Abstract. Strong variable radio emission from the bright transient X-ray source GRO J1655-40 (X-ray Nova Scorpii 1994) has been detected at 843 MHz with the Molonglo Observatory Synthesis Telescope (MOST). As the hard X-ray intensity from the 1994 August outburst declined the radio output increased rapidly, reaching a peak of nearly 8 Jy some 12 days after the first X-ray peak. VLBI images obtained at this time showed two main components separating with an apparent transverse velocity $>c$. The evolution of the radio spectrum suggests that the time delay between X-ray and radio emissions is due, at least in part, to opacity effects associated with this expansion.

1. Introduction

The discovery of GRO J1655-40 ($\equiv$ X-ray nova Sco 1994) by BATSE and the subsequent optical, radio and X-ray observations are summarised in an accompanying paper (Paciesas et al. 1995). We concentrate here on the discovery and monitoring of the radio emission from this interesting object at 843 MHz, using the Molonglo Observatory Synthesis Telescope (MOST; Robertson, 1991) operated by the University of Sydney.

Following the initial report from BATSE on 1994 August 4 (Zhang et al. 1994) we were able to schedule short ($\sim$3 hour) observations with MOST on August 6.60 and 11.59 UT (mid-exposure). Our field of view, 70' (RA) $\times$ 70'cosec(δ) (Dec), covered the original BATSE error circle of radius 0.3°. The raw images showed a strong point source, $\sim$40' North and 10' West of the BATSE position, which had increased by more than a factor of two between the two epochs. The source position agreed precisely with the position of the proposed optical counterpart reported by Bailyn *et al.* (1994) and the result was reported immediately (Campbell-Wilson & Hunstead, 1994a). It is interesting to note that the radio counterpart of GRO J1655−40 was close to the edge of the MOST synthesised field in these first images, and was discovered only because the BATSE position error was predominantly in declination (more than twice the quoted error), matching the elongation of the MOST field of view.

In this paper we present the MOST light curve through to the end of August. Our more extensive database, including time-resolved measurements when the source was strong, will be published elsewhere (Hunstead *et al.* 1995).

2. Observations

The early observations of GRO J1655−40 with MOST consisted of partial syntheses, and the reported flux densities (Campbell-Wilson & Hunstead, 1994a,b) were measured from raw images obtained at the telescope. We have since carried out a more thorough calibration of all observations, resulting in small adjustments to the reported values. The unexpectedly large increase in flux density seen between August 12.33 (950 mJy) and 14.58 UT (5100 mJy) triggered an intensive monitoring campaign with MOST and provided the impetus for similar campaigns with the Australia Telescope Compact Array (Hunstead *et al.* 1994; McKay & Kesteven 1994), the SHEVE VLBI array (Reynolds & Jauncey 1994; Tingay *et al.* 1995), the VLA (Hjellming 1994a,b; Hjellming & Rupen 1994a,b,c,d) and VLBA (Hjellming & Rupen 1994d).

The MOST 'light' curve for GRO J1655−40, shown in Fig. 1, includes flux density measurements from two separate modes of observing. The data points were obtained either (a) by fitting a theoretical transit beam in real time to short observations (typically 4 minutes), bracketed by 3–5 MOST calibrators (SCAN observing mode: Campbell-Wilson & Hunstead 1994c), or (b) by carrying out a similar fit to individual 24 s data samples from a standard synthesis observation (3–12 hours), with calibration sources observed at the beginning and end. In the latter case the fits were carried out with more sophisticated software which took account of confusing sources in the MOST fan beams; because of source variability we did not attempt to make measurements directly from images. In Fig. 1 we show separately the *mean* flux densities obtained by each method, which emphasises their internal consistency. Points are plotted at the mean epoch of observation and the errors, usually ±3%, come mainly from the uncertainty in calibration. Also shown on Fig. 1 is the BATSE light curve (Paciesas *et al.* 1995) for comparison.

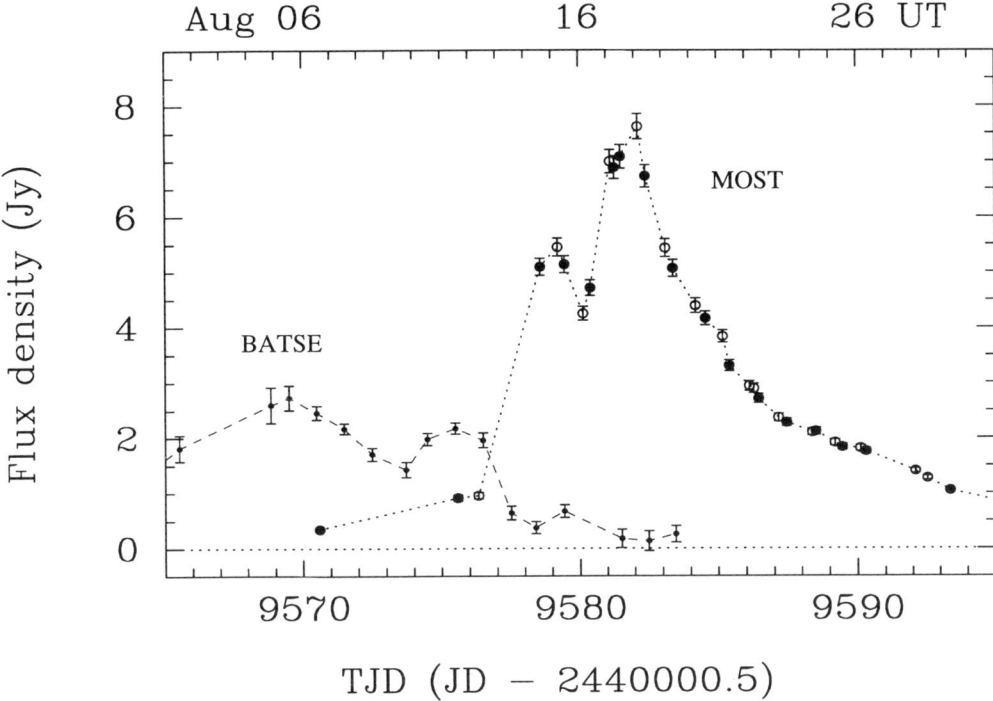

Figure 1. MOST 843 MHz flux density of GRO J1655−40 over the period 1–31 August 1994; open circles refer to SCAN measurements while filled circles are averages spanning each synthesis observation (see text). Points have been joined with a dotted line simply to show the overall pattern of variability. The BATSE one-day averages and statistical errors are plotted as small filled circles joined with a dashed line. The numerical ordinate values correspond to the 20–100 keV flux in photons cm^{-2} s^{-1} multiplied by 10.

3. Discussion

The 2.29 GHz VLBI observations reported by Tingay *et al.* (1995) show a basic double structure with the components moving apart at a rate of 65±5 mas d^{-1}, corresponding to an apparent transverse velocity of 1.5±0.4 c at their estimated distance of 3.0–5.0 kpc. Extrapolation of the outermost radio components back to zero separation, assuming a constant expansion rate, gives an outburst origin time of 1994 August $13.5^{+0.5}_{-0.8}$ UT. It can be seen from Fig. 1 that this epoch corresponds to both a rapid fall in X-ray intensity and a rapid rise in 843 MHz flux density. Unfortunately the sampling of the radio light curve is too coarse to give the true rate of rise; for an exponential increase, the limit on the time constant is $\tau \leq 1.3$ d.

While it is tempting to associate the origin of the VLBI components with the rapid increase in flux density at 843 MHz, it seems likely that the rise may be associated more with a sudden fall in source opacity than with a

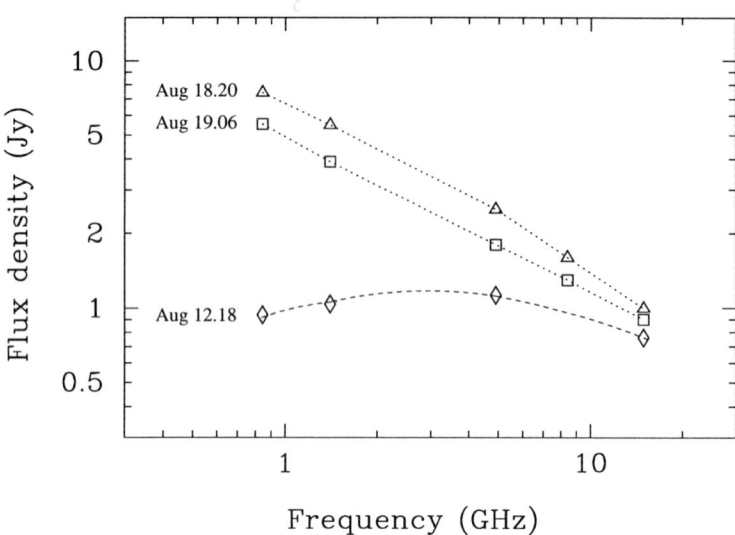

Figure 2. Radio spectra of GRO J1655−40 at three epochs: August 12.18, 18.20 and 19.06. The 843 MHz points are interpolated from Fig. 1 and the higher frequency points are from Hjellming (1994a,b). The dashed line joining the August 12.18 points is an approximate fit by eye to the data.

specific event at August 13.5 in the radio. This point is best demonstrated by Fig. 2 which shows the radio spectrum on August 12.18 (Hjellming 1994a) peaking at ∼3 GHz, with a shape characteristic of synchrotron self absorption. Even at this epoch the radio flux density was substantial at all frequencies, with little increase thereafter at the highest frequencies. The spectra for August 18.20 and 19.06 (Hjellming 1994b), obtained at and just beyond the main 843 MHz peak, show that the source had then become transparent with a spectral index $\alpha \simeq 0.6$ ($S_\nu \propto \nu^{-\alpha}$). These observations are consistent with the simple expanding synchrotron bubble models of Van der Laan (1966) and Hjellming & Johnston (1988). Measurements made at later epochs with the Australia Telescope Compact Array (Hunstead *et al.* 1994) confirm that the radio spectrum remained transparent ($\alpha \sim 0.4$–0.6) during the decline in flux density.

While there is a clear delay between the outbursts at X-ray and radio wavelengths, it is not obvious, because of the opacity effects mentioned above, how the delay should be measured. It has also been suggested that the radio emission may initially have been inhibited by the very mechanism responsible for the X-ray outburst (Tingay *et al.* 1995).

The two peaks in the radio light curve (Fig. 1) cannot readily be explained in terms of Doppler enhancements and time delays associated with approaching and receding jets. However, it is possible that they may be related instead to the overall modulation of the X-ray intensity. The time-averaged BATSE data plotted in Fig. 1 show two clear peaks located at TJD = 9569.5 ± 1 and 9575.5 ± 1. If we relate these to the radio peaks at TJD = 9579.0 ± 0.5 and 9581.8 ± 0.5, the corresponding time delays are 9.5 ± 1 and 6.3 ± 1 d respectively. There is even marginal evidence for a third X-ray peak near TJD = 9580 which may correlate with a weak radio feature at TJD = 9585, a delay of 5 d. While such associations are clearly speculative, they define a trend towards shorter time delays which appears to continue with the later X-ray outburst in 1994 September 6–16 (Paciesas et al. 1994; Harmon et al. 1994; Campbell-Wilson et al. 1994).

GRO J1655–40 has several X-ray and radio properties which single it out from other X-ray transients, and it is likely that a better understanding of its nature and evolution will only be possible when a more complete observational database becomes available.

Acknowledgements. We thank our colleagues for allowing their scheduled MOST observations to be overridden, and Bill Paciesas for communicating the BATSE light curve in advance of publication. The MOST is supported by grants from the Australian Research Council, the University of Sydney Research Grants Committee, and the Science Foundation for Physics within the University of Sydney.

References

Bailyn, C., Jogee, S. & Orosz, J. 1994, IAU Circ. 6050
Campbell-Wilson, D. & Hunstead, R.W. 1994a, IAU Circ. 6052
Campbell-Wilson, D. & Hunstead, R.W. 1994b, IAU Circ. 6055
Campbell-Wilson, D. & Hunstead, R.W. 1994c, Proc. ASA 11, 33
Campbell-Wilson, D., McKay, D.J. & Lovell, J.E. 1994, IAU Circ. 6078
Harmon, B.A. et al. 1994, IAU Circ. 6084
Hjellming, R.M. 1994a, IAU Circ. 6055
Hjellming, R.M. 1994b, IAU Circ. 6060
Hjellming, R.M. & Johnston, K.J. 1988, ApJ 328, 600
Hjellming, R.M. & Rupen, M. 1994a, IAU Circ. 6073
Hjellming, R.M. & Rupen, M. 1994b, IAU Circ. 6077
Hjellming, R.M. & Rupen, M. 1994c, IAU Circ. 6086
Hjellming, R.M. & Rupen, M. 1994d, IAU Circ. 6107
Hunstead, R.W. et al. 1994, IAU Circ. 6062
Hunstead, R.W. et al. 1995, (in preparation)
McKay, D. & Kesteven, M. 1994, IAU Circ. 6062
Paciesas, W.S. et al. 1994, IAU Circ. 6075
Paciesas, W.S. et al. 1995, these Proceedings
Reynolds, J. & Jauncey, D. 1994, IAU Circ. 6063
Robertson, J.G. 1991, Aust. J. Phys. 44, 729
Tingay, S.J. et al. 1995, Nat 374, 141
Van der Laan, H. 1966, Nat 211, 1131
Zhang, S.N. et al. 1994, IAU Circ. 6046

6

Binaries in Globular Clusters

DYNAMICS AND BINARY (TRANS)FORMATION IN GLOBULAR CLUSTERS

PIET HUT
Institute for Advanced Study
Princeton, NJ 08540, U.S.A.

Abstract. Globular clusters form ideal laboratories for studying the interactions between stellar evolution and stellar dynamics. In the past, highly exceptional systems such as X-ray binaries and later millisecond pulsars have provided us with useful diagnostic tools. However, the fate of the bulk of the more normal stars has remained less clear. At present, rapid progress is being made in our understanding of the distributions of normal stars and primordial binaries, as well as their most abundant reaction products: blue stragglers and binaries that are produced through exchange encounters with other single stars or binaries. The complexity of the network of exchange reactions is illustrated through some specific examples, such as a formation scenario for the hierarchical triple system containing the millisecond pulsar PSR B1620−26 in M4, the first triple star system ever detected in a globular cluster.

1. Introduction

During the last decade, a wealth of observational data has been obtained that can help us understand the interactions between stars in dense stellar systems. For example, in globular clusters a large number of millisecond pulsars have been detected, new low-luminosity X-ray sources have been found, and many blue stragglers have been discovered in the inner regions of some of the densest clusters. All these objects suggest a complex history of interactions between single stars and/or binary stars, due to close encounters or even collisions.

Fortunately, during the last decade similar progress has been made in the modeling of dense star clusters. The mechanism of core collapse has become better understood, and the evolution of a star cluster past core collapse has been modeled in considerable detail, using a variety of modeling approaches. The recent discovery that globular clusters contain a significant binary population has further complicated these attempts of cluster modeling. Since the binding energy of all these binaries form a very important factor in the overall energy budget of a cluster, care has to be taken in following the changes in their internal orbital parameters. In turn, to determine these changes, some of the dominant effects of stellar evolution have to be taken into account as well.

Many references to previous work in this general area can be found in a recent review of observational as well as theoretical progress in our understanding of binaries in globular clusters (Hut et al. 1992).

2. Dynamical Transformations of Primordial Binaries

Binaries play a central role in cluster dynamics. If their orbital speed exceeds that of the velocity dispersion of the single stars, the tendency toward energy equipartition during encounters will transfer some of the internal kinetic energy to passing stars. Doing so, energy conservation causes them to shrink, while the negative heat capacity of self-gravitating systems causes them to heat up further, to higher orbital speeds (Lynden-Bell's 'donkey effect': trying to slow down particles in a Kepler orbit speeds them up, and *vice versa*).

The 'gravitational fusion' of single stars into double stars is thus one mechanism that can heat a cluster, in order to balance the energy losses due to the evaporation of stars and the heat flow through the cluster toward the colder halo. Other mechanisms can play a role as well in fueling the central heat engine needed to balance the heat flow from the core to the cluster halo. Mass loss through stellar evolution (especially the much more rapid stellar evolution of the relatively heavier merger remnants) can indirectly heat a cluster through the paradoxical effect of carrying off *kinetic* energy – simply because the *potential* energy carried off per unit mass is much larger, and tilts the balance towards an effective heating. Similarly, the formation of a modest black hole can also cause a heating of the cluster, through the selective disruption of stars on low-energy orbits near the hole (for both mechanisms, see the review by Goodman 1992).

It is far from clear to what extent these various heating mechanisms compete with each other in actual globular cluster cores. Order-of-magnitude estimates indicate that they all can be significant, depending on the precise conditions in the cores, as well as on the nature of the stars. For example, white dwarfs, neutron stars and stellar-mass black holes are likely to produce energy by dynamical binary formation and hardening, while main-sequence stars and giants are likely to suffer physical collisions while attempting to do so.

Whatever the detailed mix of energy sources in individual clusters may turn out to be, it is becoming clear that binaries are crucially important to the energetics of globular clusters. For example, observations of primordial binaries in globular clusters indicate that the binary abundance in globular clusters is not much smaller than that in the Galactic disk and halo (cf., Hut et al. 1992). This suggests that $\gtrsim 10\%$ of the stellar objects in a cluster may be binaries with an orbit of $\lesssim 1$ AU, which implies an average binding energy per binary of $\gtrsim 10$ times that of the average kinetic energy of single cluster stars. This simple reasoning leads to the remarkable conclusion that the internal energy reservoir in binary binding energies may well exceed the total amount of kinetic energy in the cluster as a whole (in the form of center-of-mass motion of single stars and binaries).

With binary stars having locked up the bulk of the energy content of a typical globular cluster, we cannot afford to neglect the transformations in binary properties that take place during the course of normal stellar evolution. The reason is that stellar encounters do not have a monopoly on changing the energy and angular momentum of binaries; isolated binaries, too, have plenty of ways of changing their appearance in complicated ways (cf., Pols & Marinus 1993).

Even a partial list of some of the processes involved in isolated binary evolution gives an idea of the complexity of the physics: tidal capture, magnetic braking, gravitational radiation, runaway mass transfer, and common envelope evolution. Take into account the manifold perturbations and disruptions that can occur during interactions with passing stars or binaries, and the full complexity of the problem becomes clear. The feedback mechanisms between stellar dynamics and stellar evolution in globular clusters play a vital role in the evolution of the cluster as a whole. The term 'ecology', used by Douglas Heggie in his recent 'News and Views' article in Nature (Heggie 1992), indeed captures the essence of this interplay.

3. Fully Automatized Three-Body Scattering Experiments

The first set of binary—single-star scattering experiments were reported by Hills (1975). In these experiments, most encounters took place at zero impact parameter. The first direct determination of accurate cross sections and reaction rates for binary—single-star scattering was made by Hut & Bahcall (1983). For each type of total or differential cross section, a detailed search of impact parameter space was performed as a pilot study, before production runs were started. The problem with the choice of impact parameter (lateral offset from a head-on collision, as measured at infinity) is this: allowing too large an impact parameter can imply a large waste of computer time on uninteresting orbits; while choosing too small an impact parameter will yield a systematic underestimate of some cross sections, since some encounters of interest will be missed.

The first automatic determination of cross sections and reaction rates for binary—single-star scattering has been performed by Hut & McMillan (1994). Rather than relying on human inspection of pilot calculations, their software package includes an automatic feedback system that ensures near-optimal coverage of parameter space while guaranteeing completeness. The basic idea is to maintain a safety zone outside the 'bull's eye' region, defined as the area where 'interesting' reactions take place (where 'interesting' is defined according to the particular interest underlying a particular set of experiments). Allowing for the possibility of dynamically enlarging this safety zone guarantees rapid convergence towards accurate cross sections. One of our first applications has been to determine exchange cross sections for the general problem of binary—single-star scattering with unequal masses (Heggie, Hut & McMillan 1994). Another application is concerned with the formation of triple systems in binary-binary encounters, in the limit where the tightest of the two binaries can be approximated as a single point mass (Rasio, McMillan & Hut 1995), as will be discussed in the next section. References to earlier papers on 3-body scattering can be found in the recent papers by Hills (1992), Heggie & Hut (1993), Hut (1993a), Sigurdsson & Phinney (1993).

The software, developed by Hut & McMillan (1994) to perform the automated scattering calculations, was designed in a multi-layer object-oriented approach and is implemented in C++. At the lowest layer, the orbit integration engine consists of a fourth-order variable–time-step Hermite integrator. What is novel here is the time step criterion used: through one-step iteration, time symmetry is guaranteed to very high accuracy (Hut, Makino & McMillan 1994), which leads to spectacular improvements in long-term stability of the integration scheme, something that is essential for the treatment of long-lasting resonance scattering events.

On top of this lowest layer, there are several layers that contain: 1) checks to determine whether a given scattering experiment has reached its final outgoing state; 2) checks to allow optimization features to be activated, such as analytical integration of inner and outer orbits of hierarchical triple systems in which the outer orbital period vastly exceeds the inner orbital period; 3) diagnostic functions to store information describing the build-up of energy errors; 4) various bookkeeping functions that chart the overall character of the orbits (i.e., democratic vs. hierarchical resonance states); 5) checks for overlap of stellar radii, in which case merging routines are invoked that can replace two colliding stars with a single merger product.

On top of these layers, the first user-accessible layer contains a single-scattering command, with a large number of options. The masses and radii of the stars can be specified, as well as the orbital parameters of the binary, the impact parameter of the encounter, and the relative velocity, asymptotically far before the encounter. The initial distance from which the integration starts is determined automatically (and will be much larger for, say, a $10\,M_\odot$ black hole approaching a given binary compared to a $0.5\,M_\odot$ dwarf). In addition, an overall accuracy parameter gives a handle on the cost/performance tradeoff. In practice, typical relative energy errors can be easily kept as small as 10^{-10}; our production runs usually aim at errors of order 10^{-6}, allowing a speed-up of a factor ten in computer time with respect to the most accurate integrations attainable.

The next layer contains all the management software to conduct a series of scattering experiments. Depending on the type of total or differential cross section requested, the user can choose an appropriate command to activate a 'beam' of single stars aimed at the 'target plate' of binary stars. After a few minutes, a preliminary report appears on the screen, with estimates of all relevant cross sections plus their corresponding error bars. Thereafter, subsequent reports appear, each one after a four times longer interval. Because of the Monte Carlo nature of the orbit parameter sampling, each following report carries error bars that are half the size of the corresponding ones in the previous report. Thus reasonable estimates can often be obtained in ten or fifteen minutes, with more accurate results following in an hour (on a fast modern work station).

Another higher-level layer, on top of the cross section manager, is the Maxwellian rate estimator. After specification of the stellar characteristics, the binary orbit, and the velocity dispersion of the single stars, this estimator computes cross sections at different points under the Maxwellian velocity distribution curve, multiplying the results by the Maxwellian weight factor, and adding those to obtain reaction rates. As before, the longer one is willing to wait, the more accurate the rates become (through an automatic increase in the number of velocity points as well as an continued increase in accuracy of each of cross sections determined at each velocity points).

Further still-higher-level layers can be added with very little additional time investment. With the complexity of orbit integration and scattering management hidden in the various modules, it is relatively straightforward to implement additional levels. For example, one could easily address the inverse scattering problem: given a final system, what is the relative probability that such a system originated from different initials conditions? This could be solved as follows: after specifying the velocity dispersion and other stellar parameters, an automatic tabulation of Maxwellian reaction rates could be performed, while filtering the results to allow only those scattering experiments to be counted that lead to the desired range of final parameters.

Currently, several projects are underway in which we are applying our scattering package to various problems. The initial results are very encouraging, in two respects. First, it has become far easier to carry out many long production runs in so-far unexplored parameter ranges. Second, our package invites real-time experimentation in these new parameter ranges. Not only are new cross sections available in minutes, there are also hooks at all levels to allow graphic output. Actually watching individual resonance scattering orbits in real time has already showed us, for example, that there is a need for an extension of the present classification: apart from the traditional division in democratic and hierarchical resonances, sufficient to cover equal-mass scattering, there are other types that do not fit in these two categories. An example is the class of 'flip-flop' resonances, in which one lighter star is being exchanged several times between repeated encounters of two heavier stars (Heggie, Hut & McMillan 1994).

There are two ingredients that have made these new developments possible. One is the increase in available computer speed of a factor of a thousand, from the earliest scattering experiments performed by Hills (1975) to the ones reported by Hut & Bahcall (1983), with a similar increase of a factor of a thousand between then and now. The second ingredient is the wider availability at present of higher-level computer languages that enable and invite a more flexible and modular approach to programming, making automation far easier.

4. Formation of the PSR B1620–26 Triple System in M4

The hierarchical triple system containing the millisecond pulsar PSR B1620–26 in M4 is the first triple star system ever detected in a globular cluster (Backer, Foster & Sallmen 1993; Thorsett, Arzoumanian & Taylor 1993). Such systems should form in globular clusters as a result of dynamical interactions between binaries (Mikkola 1984; Hut 1992). Specifically, Rasio, McMillan & Hut (1995) have proposed that the triple system containing PSR B1620–26 formed through an exchange interaction between a wide primordial binary and a *pre-existing* binary millisecond pulsar (BMP).

This scenario has the advantage of reconciling the $\sim 10^9$ yr timing age of the pulsar with the much shorter lifetime of the triple system in the core of M4. Using the automated three-body scattering package, we calculated a large number of interactions between the BMP and a primordial binary to determine the triple-formation cross section as well as the predicted characteristics of the triple systems. Some results are shown in Fig. 1.

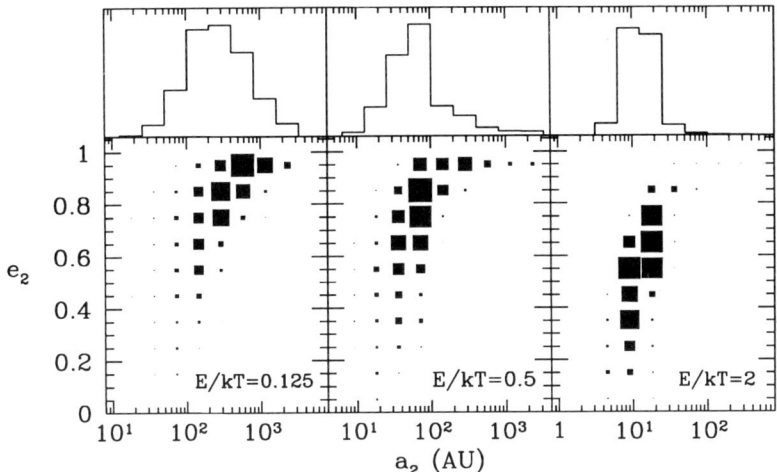

Figure 1. Distribution of the semi-major axis a_2 of the triple's outer orbit (above) and distribution in the (a_2, e_2) plane (below). The size of each square dot is proportional to the differential cross-section for producing a triple system with specific orbital parameters. The incoming primordial binary contained two 0.5 M$_\odot$ stars with semi-major axis $a_{\rm MS} \simeq 50$ AU (left), 13 AU (middle), and 3 AU (right).

Since the eccentricity of the inner BMP in the presently observed system is very small ($e_1 \approx 0.025$), the BMP cannot have been perturbed very much during the exchange interaction. This places a limit on how close either of the two MS stars in the primordial binary can have approached the BMP. In the calculations of exchange interactions, the BMP was treated as a *single mass* and given an effective radius R_e such that close passages with periastron separation $r_p > R_e$ cannot produce an eccentricity larger than observed today.

Calculations of the induced eccentricity in binary-single interactions, also performed in the Starlab environment (Hut 1993b) in which the scattering package was written, indicate that $R_e \approx 3a_1 \simeq 2.4$ AU for this system, where a_1 is the semi-major axis of the BMP. Fig. 1 shows the probability distributions of a_2 and e_2, the semi-major axis and eccentricity of the outer orbit in the triple system for several typical cases. For very soft incoming binaries (much wider than the BMP), the outer orbit tends to be very eccentric (marginally bound), with a wide distribution of separations extending over more than a factor 10 (cf., $E = 0.125$; here E is the binding energy of the incoming primordial binary in units of "kT" in the cluster).

For harder incoming binaries, the average eccentricity gets smaller and the distribution of separations narrower. When the semi-major axis of the incoming binary becomes comparable to that of the BMP, the condition that the BMP not be perturbed forces a more nearly circular outer orbit for the triple (cf., $E = 2.0$). The branching ratio for triple formation peaks at a value of about 50% corresponding to the intermediate case shown in Fig. 1b ($E = 0.5$).

When the numerical results for the triple formation cross section are combined with simple estimates for the rate at which primordial binaries drift into the cluster core by dynamical friction, one obtains the following expression for the predicted number of triple systems containing a detectable pulsar in a cluster like M4,

$$N_{\rm T} \sim 0.1\, N_{\rm BMP} \frac{f_{\rm b}}{0.1} \left(\frac{N_{\rm c}}{10^3}\right)^{-1} \left(\frac{N_{\rm o}}{10^5}\right) \left(\frac{t_{\rm d}}{10^{10}\,{\rm yr}}\right)^{-1} \left(\frac{t_{\rm T}}{10^8\,{\rm yr}}\right). \quad (1)$$

Here $N_{\rm BMP}$ is the number of detectable pulsars in the core, $N_{\rm c}$ is the total number of objects in the core, $N_{\rm o}$ is the number of objects in the outer region of the cluster, $f_{\rm b}$ is the binary fraction there, $t_{\rm d}$ is the dynamical friction timescale, and $t_{\rm T}$ is the lifetime of the triple in the core.

5. Recycling Blue Stragglers to make yet more Blue Stragglers

In the dense cores of globular clusters, a typical star has a significant chance to undergo a collision during a Hubble time. Since the velocity dispersion in globular clusters is one or two orders of magnitude less than the escape velocity at the surface of a typical main sequence star, almost all of the mass of two colliding stars is retained to form a merger product (for a detailed discussion, cf., Lombardi et al. 1995).

In a collision between relatively light stars, the merger remnant may have a mass below that of the main sequence turn-off, $0.8\,{\rm M}_\odot$. In that case, it will be rather difficult to distinguish the merger product from an ordinary star of that mass. For a given candidate, detailed investigation might show unusual abundances, but spotting such stars against the background of normal stars would be akin to seeking for a needle in a haystack.

In a collision where the sum of the masses of the original stars significantly exceeds that of the turn-off, by being larger than $1\,{\rm M}_\odot$ say, the merger remnant will stand out in a H-R diagram, since it will be positioned on or near the main sequence, but bluewards of the turn-off. Such a star, if born at the same time as the rest of the stars of the cluster, should have evolved away from the main sequence. Its presence thus gives the impression of having straggled, and with its relatively bluer color such a star is called a blue straggler. For a recent collection of papers on blue stragglers, see Saffer (1993).

Many, if not most, of the collisions between single stars in star clusters are likely to produce blue stragglers. One reason is that mass segregation will produce an excess of heavier stars in the central regions of a cluster. Other reasons are the larger radii of heavier stars, together with their enhanced gravitational focusing, both of which favor collisions for heavier, rather than lighter stars.

An even more efficient mechanism for inducing collisions is provided by binary–single-star and binary–binary encounters, specially those involving hard binaries, which have a significant chance to lead to resonant encounters where three or four stars are temporarily captured in a small area of space. Similar arguments again predict a larger collision probability for heavier stars. A special feature of 3-body or 4-body channels for blue straggler formation is that the mass of the final blue straggler may well be significantly larger than twice the turn-off mass, if more than two stars are involved in producing a single merger (cf., Davies et al. 1994).

A third way of producing collisions between two stars is through mass overflow from one component of a binary onto the other member, followed by a spiral-in, a process that does not require a high density of neighboring stars.

As a simultaneous illustration of the last two types of blue straggler formation events, a calculation by Simon Portegies Zwart is reproduced in Fig. 2 (originally exhibited as a poster during the Symposium). The calculations were performed within the Starlab environment, in which his simplified binary-evolution models were connected with the automated three-body scattering package described above.

There is nothing special about this particular set of initial conditions. Both the single star and the binary were drawn at random from a prescribed cluster core population of single stars and binaries. At each time step of a thousand years, the probabilities for close encounters between different objects were calculated, and actual collisions were carried out in Monte Carlo fashion by spinning a random number generator to see which stars would actually undergo an encounter. At each time step, those stars and binaries that did not undergo an interaction (the vast majority) were simply evolved with stellar evolution prescriptions: fitting formulas for stellar evolution tracks in the case of single stars, and simplified recipes in the case of binaries (for more details, see Portegies Zwart 1995).

In Fig. 2, an example is given of an interplay between three stars, leading to the formation of two blue stragglers. The first blue straggler is formed in isolation, as the result of binary star evolution, nearly 10 Gyr after its formation. At this time, the $1\,M_\odot$ star attempts to climb the giant track. The first result is that its increased size leads to rapid tidal circularization, erasing the initial eccentricity of 0.3 around $T = 9.97\,\text{Gyr}$. Soon thereafter, the primary fills its Roche lobe, and around $T = 9.98\,\text{Gyr}$ dumps most of its mass on the secondary.

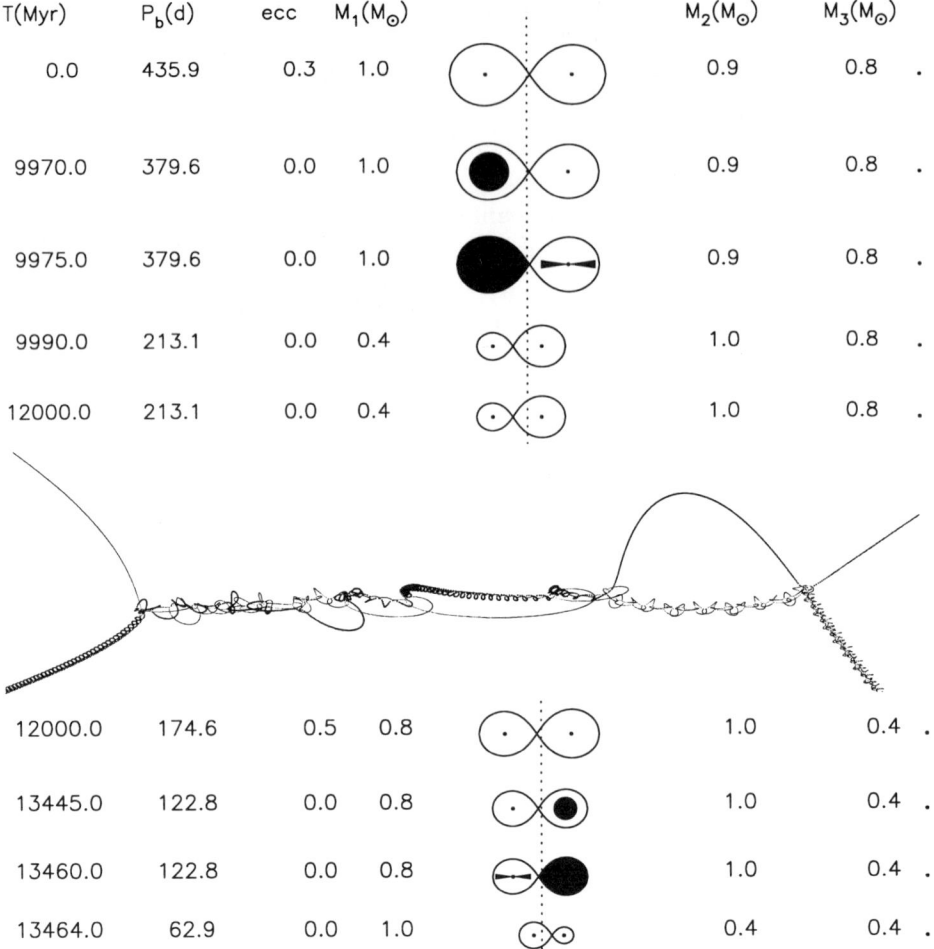

T(Myr)	P_b(d)	ecc	$M_1(M_\odot)$	$M_2(M_\odot)$	$M_3(M_\odot)$
0.0	435.9	0.3	1.0	0.9	0.8
9970.0	379.6	0.0	1.0	0.9	0.8
9975.0	379.6	0.0	1.0	0.9	0.8
9990.0	213.1	0.0	0.4	1.0	0.8
12000.0	213.1	0.0	0.4	1.0	0.8
12000.0	174.6	0.5	0.8	1.0	0.4
13445.0	122.8	0.0	0.8	1.0	0.4
13460.0	122.8	0.0	0.8	1.0	0.4
13464.0	62.9	0.0	1.0	0.4	0.4
17838.0	62.7	0.0	1.0	0.4	0.4

Figure 2. Effect of an exchange reaction on the stellar evolution of a primordial binary, with masses M_1 and M_2, encountering a single star with mass M_3, after having evolved in isolation for 12 Gyr. The orbits of the star during the reaction are plotted in the middle of the figure, with time going from left to right. The total duration of the encounter is about a hundred years. The horizontal time axis for the stellar dynamics event is therefore five orders of magnitude enlarged compared to the vertical time axis displaying the various stellar evolution events.

Each horizontal line applies to a certain age T of the star cluster from which these stars are drawn. This age is given in the first column, in units of 10^6 yr. The next three columns give the orbital period P_b of the binary, in days; the eccentricity ecc of the orbit; and the mass M_1 of the primary. The following column shows a picture of the stars within their Roche lobes, all drawn using the same scale. The next two columns give the mass M_2 of the secondary; and the mass M_3 of the single star. The remaining dot just to the right of the last column indicates the isolated nature of the third star.

However, only $0.1\,M_\odot$ of this supply of matter can be accreted onto the companion, and $0.5\,M_\odot$ leaves the system. The reason for the inefficiency of mass transfer is that the donor star provides mass on its own thermal time scale, while the other star in turn can only accept mass on its thermal time scale, which is longer. By $T = 9.99\,\text{Gyr}$, the primary turns into a low-mass helium-star which then cools to become a white dwarf. From this time on, the secondary shows up as a blue straggler. But before the secondary can reach the turn-off, a third star happens to pass through the system, 2 Gyr later, and is captured into the three-body resonance scattering event, displayed with time moving horizontally from left to right, for a duration of about a hundred years.

The outcome of the event, not surprisingly, leads to the lightest star being ejected, in this case the white dwarf, the end product of the original primary. The incoming $0.8\,M_\odot$ star now takes the place of the primary. The orbit has tightened significantly, as could also be expected (hard binaries tend to get harder; Heggie 1975, Hut 1983). After another 2.4 Gyr, the original secondary begins to climb the giant track, leading to a second phase of tidal circularization, in which the eccentricity induced in the three-body encounter is erased. After another 15 Myr, mass overflow takes place, again $0.6\,M_\odot$ is lost, of which this time $0.2\,M_\odot$ is accepted by exchanged star. This increases the mass of the latter from $0.8\,M_\odot$ to $1.0\,M_\odot$, thereby turning this star into a second blue straggler. The original secondary soon turns into a dwarf, orbiting the $1\,M_\odot$ star. This second straggler phase finally ends around $T = 17.8\,\text{Gyr}$, during the last phase of giant evolution.

The final state is not displayed in this figure. After the second blue straggler will fill its Roche lobe in turn, the binary will shrink significantly, and most likely a common envelope system will be formed. This is likely to lead to the merger of the two cores, resulting in a late type giant of $1.3\,M_\odot$. A bit later, a white dwarf will then be left behind, with a mass of around $0.7\,M_\odot$.

Observationally, the first blue straggler formed by the transfer of mass in a close binary will not be distinguishable from the classical ones. Only after the dynamical exchange interaction, the system becomes clearly different from ordinary blue stragglers: the observation of a blue straggler in an eccentric orbit or in a detached binary with a main-sequence companion is a direct indication for formation by an exchange reaction.

Acknowledgements. I thank Fred Rasio and Simon Portegies Zwart for producing Figures 1 and 2, respectively, and for comments on the manuscript.

References

Backer, D.C., Foster, R.S. & Sallmen, S. 1993, Nat 365, 817
Davies, M.B., Benz, W. & Hills, J.G. 1994, ApJ 424, 870
Goodman, J. 1992, in *Structure and Dynamics of Globular Clusters*, G. Meylan & S. Djorgovski (Eds.), ASP Conf. Proc. Vol. 50, p. 87
Heggie, D.C. 1975, MNRAS 173, 729
Heggie, D.C. 1992, Nat 359, 772
Heggie, D.C. & Hut, P. 1993, ApJS 85, 347
Heggie, D.C., Hut, P. & McMillan, S. 1994, (in preparation)
Hills, J.G. 1975, AJ 80, 809
Hills, J.G. 1992, AJ 103, 1955
Hut, P. 1983, ApJ 272, L29
Hut, P. 1992, in *X-ray Binaries and Recycled Pulsars*, E.P.J. van den Heuvel & S.A. Rappaport (Eds.), Kluwer Academic Publishers, p. 317
Hut, P. 1993a, ApJ 403, 256
Hut, P. 1993b, in *Blue Stragglers*, R.A. Saffer (Ed.), ASP Conf. Proc. Vol. 53, p. 44
Hut, P. & Bahcall, J.N. 1983, ApJ 268, 319
Hut, P. & McMillan, S. 1994, (in preparation)
Hut, P. et al. 1992, PASP 104, 981
Hut, P., Makino, J. & McMillan, S. 1994, (preprint)
Lombardi, J., Rasio, F.A., & Shapiro, S.L. 1995, ApJ (submitted)
Mikkola, S. 1984, MNRAS 208, 75
Pols, O. & Marinus, M. 1993, in *Blue Stragglers*, R.A. Saffer (Ed.), ASP Conf. Proc. Vol. 53, p. 126
Portegies Zwart, S.F. 1995, (in preparation)
Rasio, F.A., McMillan, S. & Hut, P. 1995, ApJ (in press)
Saffer, R.A. (Ed.) 1993, *Blue Stragglers*, ASP Conf. Proc. Vol. 53
Sigurdsson, S. & Phinney, E.S. 1993, ApJ 415, 631
Thorsett, S.E., Arzoumanian, Z. & Taylor, J.H. 1993, ApJ 412, L33

ROSAT OBSERVATIONS OF GLOBULAR CLUSTERS IN THE GALAXY AND IN M31

HELEN M. JOHNSTON AND FRANK VERBUNT
Astronomical Institute, Utrecht University,
Postbus 80.000, NL-3508 TA Utrecht, The Netherlands

GÜNTHER HASINGER
Astrophysikalisches Institut, An der Sternwarte 16,
D-14482 Potsdam, Germany

AND

WOLFRAM BUNK
MPI für extraterrestrische Physik,
D-85748 Garching bei München, Germany

Abstract. X-ray sources in globular clusters fall into two categories: the "bright" sources, with $L_X \sim 10^{36}$–10^{38} erg s^{-1}, and the "dim" sources, with $L_X \lesssim 10^{34.5}$ erg s^{-1}. The bright sources are clearly associated with accreting neutron stars in binary systems. The nature of the dim sources, however, remains in doubt. We review recent observations of globular-cluster X-ray sources with the ROSAT satellite. ROSAT detected bright sources in M31 globular clusters and greatly increased the number of dim sources known in galactic globular clusters. We discuss what these new observations have taught us about the distribution and nature of such sources, their spectral properties, and their underlying luminosity function.

1. Introduction

The discovery of X-ray sources in globular clusters was made with the UHURU satellite during the first survey of the X-ray sky (Forman *et al.* (1978) which found six sources associated with globular clusters (Katz 1975). A decade later, it was found with the more sensitive EINSTEIN satellite that cluster X-ray sources fell into two distinct categories: sources with X-ray luminosities $L_X \sim 10^{36}$–10^{38} erg s^{-1} (the "bright" sources), and sources with $L_X \lesssim 10^{34.5}$ erg s^{-1} (the "dim" sources; Hertz & Grindlay 1983; Hertz & Wood 1985). The bright sources all show X-ray bursts, and hence may fairly safely be associated with accreting neutron stars in a binary system.

The nature of the dim sources, on the other hand, is much less clear. Prior to ROSAT, there were eight such sources known in the cores of clusters, none of which had convincing counterparts. Hertz & Grindlay (1983) suggested they were cataclysmic variables; however, Verbunt et al. (1984) pointed out that the X-ray luminosities of some dim sources are up to 300 times higher than those of cataclysmic variables in the galactic disk, and argued that these are more likely to be quiescent X-ray transients. Other possible sources for the X-ray emission include the combined luminosity of many RS CVn stars (Belloni et al. 1993), millisecond pulsars, or even unrelated fore- or background objects.

The launch of ROSAT promised new developments in our understanding of the dim sources. ROSAT has three modes which are important for the investigation of cluster X-ray sources: *(a)* the ROSAT PSPC All-Sky Survey, which covered the entire sky in 6 months; *(b)* pointed PSPC observations, and *(c)* pointed HRI observations. The Survey gives us a measurement of the X-ray flux for every globular cluster, the PSPC gives a spectral resolution of $\sim 45\%$ at 1 keV , and the HRI gives a spatial resolution of $\sim 5''$. Together with the ability of the Hubble Space Telescope to resolve stars even in the dense cores of globular clusters, real progress is now being made on identifying and studying these sources. In Section 2 we discuss ROSAT observations of bright globular-cluster sources in the Galaxy and in M31; in Section 3 we discuss new ROSAT observations of dim sources, particularly their spectra and multiplicity. In Section 4 we discuss how we can combine these observations to derive the underlying luminosity function of the cluster X-ray sources.

2. Bright Globular-Cluster X-ray Sources

2.1. GLOBULAR CLUSTERS IN THE GALAXY: THE ROSAT ALL-SKY SURVEY

The ROSAT All-Sky Survey is the first complete survey of the X-ray sky since HEAO-1. The Survey took place using the Position-Sensitive Proportional Counter (PSPC) starting in August 1990, and lasted for six months. The positions of 141 globular clusters were searched for emission, and a flux measurement or upper limit obtained for every cluster (Verbunt et al. 1994). Since the average integration time per cluster was only ~ 300 s, the only cluster in which a dim source was seen during the Survey was 47 Tuc.

The bright sources, however, were easily detectable.

Prior to ROSAT, ten bright sources were known in globular clusters. That they were variable was already known; three of the six sources detected in the original UHURU survey sank below the detection limits of satellites observing in the following decade (Bradt & McClintock 1983). The ROSAT detections followed the same pattern: ten bright sources were detected, only eight of which had been seen before. ROSAT failed to detect the sources in NGC 6440 (which was a transient source seen only in December 1971–January 1972: Bradt & McClintock 1983) and Liller 1, not surprisingly since this source, the "rapid burster" (MXB 1730–335) has periods of burst activity, lasting $\sim$2–6 weeks, approximately every six months (see, e.g., Lewin & Joss 1981). ROSAT did, however, detect sources in Terzan 6 and NGC 6652 (Predehl et al. 1991). Thus approximately a quarter of the bright sources are variable on time scales of about 15 years, by a factor of $\gtrsim 10^2$–10^3 in luminosity.

The spectra of the detected sources are shown in Fig. 1.

2.2. A SUPERSOFT SOURCE IN M3: A NEW KIND OF OBJECT?

Another new source was detected in the All-Sky Survey, which represents a new type of source in globular clusters. An extremely soft source, RX 1342.1+2822, was discovered in the globular cluster M3 = NGC 5272 during the Survey, and was also seen in pointed observations using the HRI. The spectrum of the source is extremely soft, with $kT \sim 30$–$80\,\mathrm{eV}$ and $L_X \sim 1.3\ 10^{35}\,\mathrm{erg\,s^{-1}}$ (Verbunt et al. 1994). Two separate HRI observations showed the source to vary by a factor of at least 40 in six months (Hertz et al. 1993). This cluster contained a dim source in the EINSTEIN survey, but because EINSTEIN had no spectral information, we cannot say whether this is the same source or not.

If the source is a white dwarf, however, it should be noted that the fitting of a model atmosphere spectrum may give rather different luminosities than the fitting of a black body (see the contribution by A. van Teeseling et al. to these proceedings).

The temperatures found by this spectral fitting are very similar to those of the new class of supersoft sources found in the Galaxy, in the Magellanic Clouds and in M31 with ROSAT (Hasinger 1994; P. Kahabka, these proceedings). However, supersoft sources have X-ray luminosities approaching the Eddington luminosity, $L_X \simeq 10^{37}$–$10^{38}\,\mathrm{erg\,s^{-1}}$, while the source in M3 is a factor of 10^2–10^3 fainter. Its relationship to the class of supersoft sources is unclear.

The extremely soft spectrum of RX 1342.1+2822 meant that it was detected only in channels 7–50 of the PSPC (see Fig. 1). Its detection in

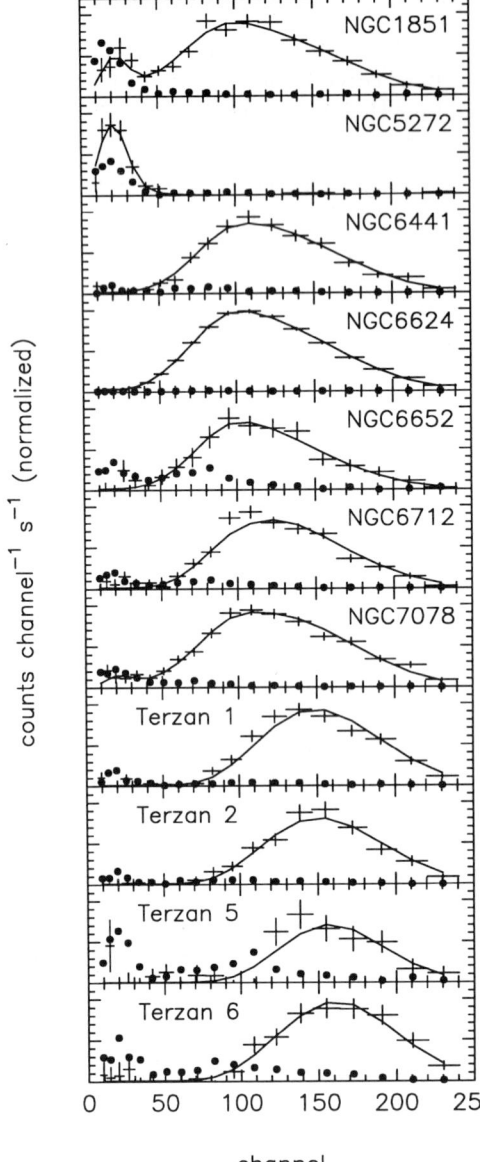

Figure 1. Spectra obtained with the XRT during the ROSAT All Sky Survey for eleven globular-cluster cores. The data have been binned in 19 energy intervals, where channel n corresponds roughly to $n \times 0.01\,\mathrm{keV}$. The source spectra are shown with crosses indicating horizontally the width of the energy bin and vertically the error, together with a spectral fit. The background spectrum is indicated with •. (From Verbunt et al. 1994).

M3 was only possible because of the extremely low reddening towards this cluster ($N_\mathrm{H} \sim 5.4\,10^{19}\,\mathrm{cm}^{-2}$). This raises the question of how many similar sources may be hiding in other clusters. In only 16 other clusters would a 40 eV blackbody spectrum with bolometric luminosity $10^{35}\,\mathrm{erg\,s^{-1}}$ source have been detectable in the Survey; we can rule out the presence of such a source in 15 of these (Verbunt et al. 1994). Thus the total population of such sources is probably limited. However, several clusters could contain similarly soft sources at a lower luminosity; see Section 3.2.

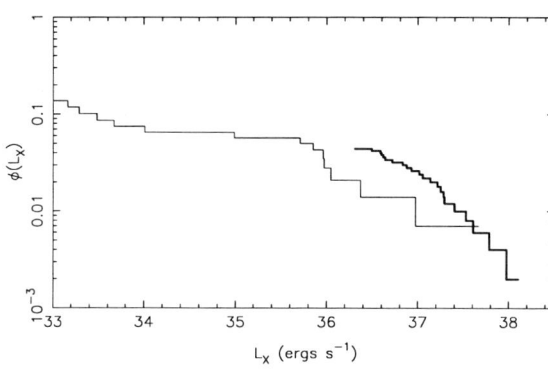

Figure 2. Cumulative luminosity function of globular-cluster X-ray sources in M31 (from Supper et al. 1995; thick line), and the Galaxy, (as described in Section 4 and Fig. 6; thin line). The curves represent the Kaplan-Meier estimator of the distribution. Increasing the number of upper limits shifts the curve vertically downwards. (After Supper 1994).

2.3. GLOBULAR CLUSTERS IN M31

ROSAT observed M31 in a 200 ks pointed PSPC observation in July 1991 (Supper 1994; Supper et al. 1995) with a sensitivity limit of $\sim 10^{36}$ erg s^{-1}, which, while a factor of ten deeper than that reached by EINSTEIN allows detection of only "bright" cluster sources, above the $10^{34.5}$–10^{36} erg s^{-1} gap. Eighteen sources had positions consistent with the position of a globular cluster (Supper et al. 1995).

Previous investigations had found significant differences between the luminosity function of globular-cluster X-ray sources in our Galaxy and in M31, suggesting that the sources in M31 were either more luminous or more numerous per cluster than in our Galaxy (see, e.g., figure 4 of Primini et al. 1993). The new ROSAT observations, combined with a new CCD program of detecting M31 globular clusters (Magnier 1994) reveal that at least part of the earlier discrepancy resulted from incompleteness in the catalogues of globular clusters. Fig. 2 shows the luminosity function derived from the M31 globular clusters compared with that of the Galaxy from ROSAT data (see Fig. 6). The Galactic luminosity function is still lower than that of M31.

3. Dim Globular-Cluster X-ray Sources

The study of the dim sources, with $L_X \lesssim 10^{34.5}$ erg s^{-1}, requires the use of pointed ROSAT observations. Only one dim core source was detected in the All-Sky Survey, the source in 47 Tuc, with $L_X = 5\ 10^{33}$ erg s^{-1}. The sky density of EINSTEIN dim sources that are not located in the cores of clusters indicates that they are probably not associated with the clusters (Verbunt et al. 1994); this has been confirmed by follow-up optical observations, which have identified several of the non-core sources as foreground dMe stars, quasars etc. (Grindlay 1994).

Pointed ROSAT observations, on the other hand, offer two important

TABLE 1. Properties of dim X-ray sources in the cores of globular clusters. L_X in the band 0.5–2.5 keV is calculated assuming a blackbody spectrum using the best-fit temperature of the source (shown in column 5), where available, or a 3 keV thermal bremsstrahlung spectrum otherwise. The distance and reddening to the cluster are assumed, from Djorgovski (1993). The bolometric luminosity L_{bol}, assuming a 0.04 keV blackbody spectrum, is shown in column 4 for sources which are compatible with a spectrum this soft (including sources for which we have no spectral information). Luminosities of clusters containing multiple sources are those of individual sources.

Cluster	No. of sources	log L_X (erg s^{-1})	log L_{bol}[a] (erg s^{-1})	kT_{bb} (keV)	Reference[b]
NGC 104 = 47 Tuc	5	32.6–33.1	-	0.6–1	H94, V94
Pal 2	1	33.9	38.2	-	R94
NGC 1904 = M79	1	33.7	34.0	-	HG83
NGC 5139 = ω Cen	2	32.0	-	0.6 ± 0.2	J94
NGC 5272 = M3	1	-	35.3	0.05	V94
NGC 5824	1	34.6	36.6	-	HG83
NGC 6304	1	33.0	36.0	-	R94
NGC 6341 = M92	1	32.5	32.6	< 0.3	J94
NGC 6397	5	31.2–31.8	33.2–33.8	< 0.1	C93, J94
NGC 6541	1	33.2	35.1	-	HG83
NGC 6626 = M28	1	32.8	36.1	< 0.3	J94
NGC 6656 = M22	1	30.9	-	$\gtrsim 0.3$[c]	J94
NGC 6752	3	31.6–32.1	-	0.35 ± 0.05	G93, J94
NGC 7099 = M30	1	32.7	33.5	< 0.2	J94

[a] For a 0.04 keV spectrum: see caption
[b] References: C93: Cool et al. 1993; G93: Grindlay 1993; H94: Hasinger et al. 1994; HG83: Hertz & Grindlay 1983; J94: Johnston et al. 1994; M94: Margon 1994; R94: Rappaport et al. 1994; V94: Verbunt et al. 1994.
[c] Spectral colours are not well represented by a blackbody.

elements for understanding the nature of the dim sources (not, unfortunately, both at the same time!): spatial resolution, particularly useful for determining the multiplicity of sources, and spectral resolution, which we can use to determine source temperatures. We now know of at least 25 dim sources in 14 clusters; Table 1 lists the known sources and their properties.

3.1. MULTIPLICITY OF SOURCES

The spatial resolution of ROSAT reveals that many of the sources in the cores of clusters are multiple. Sometimes this multiplicity is visible even in PSPC observations, which have spatial resolution of $\sim 25''$. PSPC observations of ω Cen and NGC 6752 showed the former to be double, with

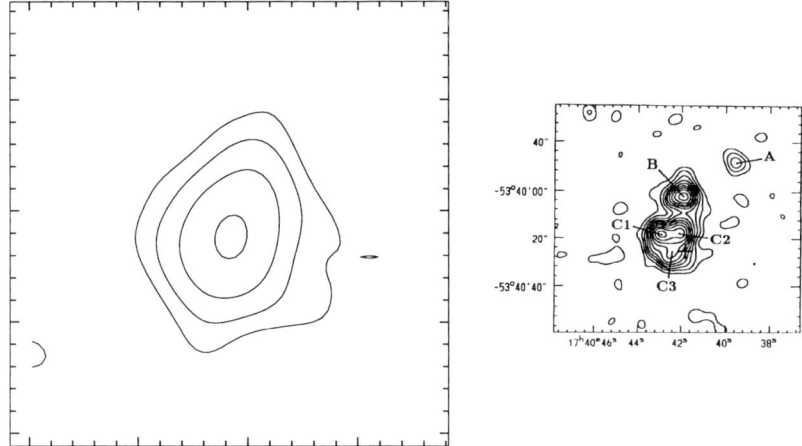

Figure 3. Comparison between a PSPC and an HRI observation of the globular cluster NGC 6397. Both figures are on the same scale, showing the inner ~$3' \times 3'$ and ~$1'.5 \times 1'.5$ respectively. The former is from Johnston et al. 1994, and shows a hint of asymmetry in the contours; the latter is from Cool et al. 1993 and clearly shows the source resolved into at least three, probably five separate components.

the fluxes of the two components approximately equal, while the latter is asymmetrical, suggesting a flux ratio of ~2:1 (Johnston et al. 1994).

However, the multiplicity of sources is better seen by far in HRI observations, with a spatial resolution of $5''$. Fig. 3 shows a comparison between a PSPC observation and an HRI observation of NGC 6397, showing the enormously superior ability of the latter instrument to resolve sources.

So far, HRI observations of four clusters have revealed the X-ray emitting sources to be multiple: in NGC 6397 the core source is resolved into 3-5 sources (Cool et al. 1993); in 47 Tuc five separate sources have been seen in the core, plus four more close to the core (Hasinger et al. 1994); in NGC 6752 at least three sources are resolved (Grindlay 1993), and in NGC 6304 the core source appears to be extended (Rappaport et al. 1994). The same observations show evidence of significant variability: of four core sources in 47 Tuc seen in 1992, only two were visible in 1993, plus one new one, indicating variability of a factor of 5 or more (Hasinger et al. 1994).

3.2. SPECTRAL PROPERTIES

The ROSAT PSPC can give information on the spectra of the dim X-ray sources. The number of photons in each observation is low – typically less than 50 counts per source in an observation of several thousand seconds – but we can use the observed colours to constrain their spectra. The high column density towards globular clusters ($N_\mathrm{H} \sim 10^{21}\,\mathrm{cm}^{-2}$)

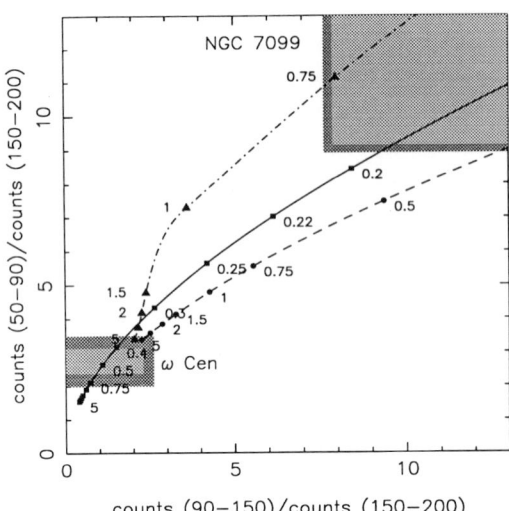

Figure 4. X-ray colour-colour diagram showing predicted colours for various spectral models. An absorption of $N_H = 5\ 10^{20}\ \text{cm}^{-2}$ was used. The observed colours for ω Cen ($N_H = 9\ 10^{20}\ \text{cm}^{-2}$) and NGC 7099 ($N_H = 3.6\ 10^{20}\ \text{cm}^{-2}$) are shown as shaded boxes (1σ error box with light shading, 2σ box with dark; where the box touches the edge of the graph it is a lower/upper limit to the true colour). Also plotted are three theoretical spectral models: blackbody (solid line), thermal bremsstrahlung (dashed line) and optically thin (dot-dashed line). The numbers next to points indicate the temperature in keV. (From Johnston et al. 1994).

means there are usually no counts in the lowest energy band. Thus we define three bands: a soft one, channels 50–90 (corresponding roughly to 0.5–0.9 keV), and two hard ones, channels 90–150 (0.9–1.5 keV) and 150–200 (1.5–2.0 keV). Comparison between the observed colours and various model predictions enables us to constrain the models. Fig. 4 shows the comparison between the observed colours and the colours of various spectral models as a function of temperature. It can be seen that the spectrum for NGC 7099 must be soft, with $kT \lesssim 0.2\,\text{keV}$ for a blackbody spectrum, $\lesssim 0.75\,\text{keV}$ for a bremsstrahlung or optically thin spectrum. For ω Cen, however, soft spectra are excluded: the observed colours predict temperatures of $kT = 0.6 \pm 0.2\,\text{keV}$ for a blackbody spectrum, $kT > 5\,\text{keV}$ for the other two spectra.

Performing this comparison for all sources observed shows that the spectra of the dim cluster sources are not all identical. Some, like NGC 7099, are soft, with $kT \lesssim 0.3\,\text{keV}$. Observations of 47 Tuc show the core sources there to fall into this category (Verbunt et al. 1994; Margon 1994). As shown in Table 1, four sources have only upper limits to their temperature. Assuming a blackbody spectrum of 40 eV for these sources yields bolometric luminosities between $4\ 10^{32}\,\text{erg s}^{-1}$ and $10^{36}\,\text{erg s}^{-1}$. Thus several of these sources could be similar to the source in NGC 5272 (Section 2.2). Soft spectra can be excluded for three clusters: ω Cen, M22 and NGC 6752 (Johnston et al. 1994). For several sources (e.g., Pal 2), we have no spectral information; if these sources have spectra as soft as 40 eV, their bolometric luminosities could be as high as $2\ 10^{38}\,\text{erg s}^{-1}$.

3.3. OPTICAL COUNTERPARTS OF DIM SOURCES

Having identified multiple sources in several clusters, it is natural to ask whether we can identify their counterparts at other wavelengths. However, here the limited positional accuracy of ROSAT hinders us; for the PSPC positions are accurate to $\sim 10''$, and for the HRI to $\sim 5''$. Since these sources are in the very crowded cores of globular clusters, the X-ray error circle can contain hundreds of stars in (for instance) an HST image. In cases where there is only one truly unusual object in a cluster, the identification can be regarded as fairly definite (e.g., the identification of an extremely bright UV source as 4U 1820-30 in NGC 6624 using HST observations; King et al. 1993). In cases where a proposed counterpart is less unusual, however, such as being a blue straggler or emission line object, the identification must remain more tentative (e.g., Paresce et al. 1992). Thus, positional coincidence of an unusual object with an X-ray error circle is not enough to prove it is a counterpart, especially when the error circle encompasses the whole core, as is the case for many of the clusters of interest. Since most of the unusual objects in a cluster are to be found in the core – pulsars, X-ray binaries, blue stragglers, etc. – we really require correlated variability or some such unambiguous signature at two wavelengths to prove association.

3.4. THE NATURE OF THE DIM SOURCES

The dim X-ray sources in clusters have luminosities down to a few times 10^{31} erg s^{-1} in the ROSAT band (Table 1). We can now begin to answer the question: what is the nature of these sources? Their spectral colours indicate we may be dealing with more than one type of source (see Section 3.2). Their luminosities are an important clue. In Fig. 5 we plot the X-ray luminosities of the observed sources, together with the luminosities of some of the proposed counterparts for these objects. The X-ray luminosities of the faintest sources are now compatible with the luminosities of disk cataclysmic variables. The brightest ones are probably not. Nearly all disk cataclysmic variables have $L_X < 10^{32}$ erg s^{-1} (though some, e.g., GK Per, can be much brighter in outburst: Watson et al. 1985); thus if the brightest sources, with $L_X \sim 10^{33.5}$–10^{34} erg s^{-1}, are cataclysmic variables, there should be *large* numbers visible at other wavelengths, of order ten or more for every X-ray object. This is particularly true because the brightest X-ray source may very well not be the brightest optical or ultraviolet source; Van Teeseling & Verbunt (1994) showed that these quantities are essentially uncorrelated for a sample of disk cataclysmic variables. Any large number of X-ray quiet cataclysmic variables also compounds the optical ID problem in the large X-ray error circles.

The observed luminosities in the range $L_X \sim 10^{32}$–10^{33} erg s^{-1} (Fig. 5)

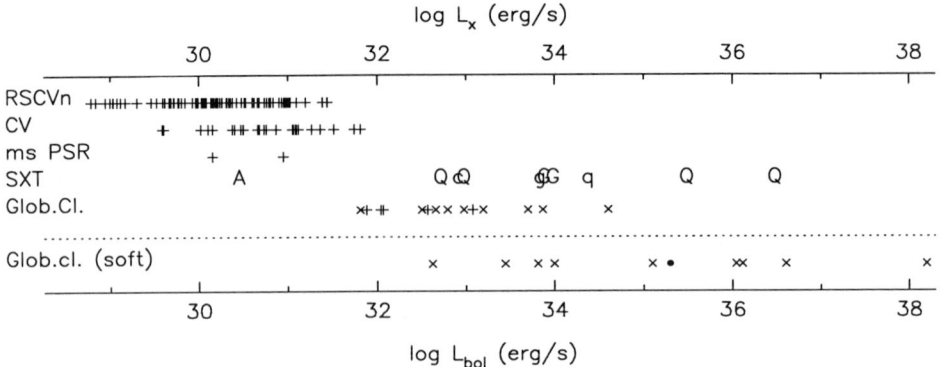

Figure 5. Comparison of the luminosities of dim globular cluster X-ray sources with some of their proposed counterparts in the disk. In the upper part of the figure the X-ray luminosities L_X in the band 0.5–2.5 keV are plotted for RS CVn systems (from Dempsey et al. 1993), cataclysmic variables (from the ROSAT Survey), millisecond pulsars (Kulkarni et al. 1992, Fruchter et al. 1992), soft X-ray transients (F. Verbunt, these proceedings), and globular-cluster dim sources (Table 1). For the SXTs, different letters indicate indicate different sources, with upper case letters being ROSAT observations: A = A 0620−00, G = GS 2023+338, C = Cen X-4, Q = Aql X-1. For the cluster sources, the sources whose spectra could be extremely soft (Table 1) are indicated with a ×. For these sources, the bolometric luminosity assuming a spectrum of 0.04 keV is plotted in the lower part of the figure, with the source in NGC 5272 indicated with a •. Note that the luminosities are much higher for a spectrum this soft.

and the soft colours, $kT \sim 0.3$ keV, which are measured for at least some of the core sources (Table 1), look very much like those of soft X-ray transients. X-radiation from four soft X-ray transients, including two neutron star systems and two black-hole systems, has now been detected in quiescence (F. Verbunt, and H. Inoue, these proceedings).

4. The X-ray Luminosity Function

Combining pointed observations and ROSAT All-Sky Survey observations, both detections and upper limits, we can derive the underlying luminosity function of cluster X-ray cores. The Survey improved the upper limits on the X-ray luminosities of the cluster centers by one or two orders of magnitude for many clusters. ROSAT also discovered more and fainter dim sources, so we can improve the determination of the luminosity function significantly.

The result of a maximum-likelihood determination of the luminosity function, using both the flux measurements of detected sources and flux upper limits (Avni et al. 1980) is shown in Fig. 6. The double-peaked nature, with a distinct gap between bright and dim sources first seen by EINSTEIN and HEAO-1, is still visible, but is less significant. This is partly due to the different bandpass used by ROSAT.

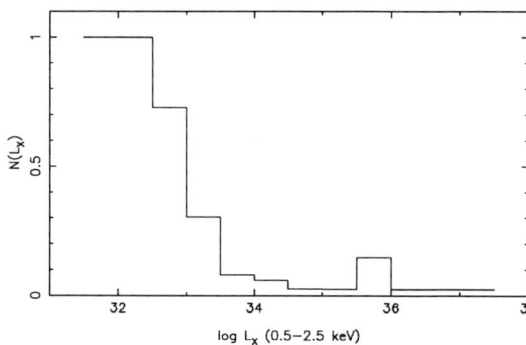

Figure 6. Maximum-likelihood estimate of the luminosity function of X-ray sources in globular clusters, using the method of Avni et al. 1980. The observational data used were detections and upper limits from the ROSAT All-Sky Survey as well as EINSTEIN and ROSAT pointed observations. (From Verbunt et al. 1994).

The luminosity function at low luminosities, $L_X \lesssim 10^{32.5}$ erg s^{-1}, is now much better constrained. In particular, the slope of the luminosity function is not rising as steeply at low luminosities as in previous determinations. Hertz & Grindlay (1983) using EINSTEIN data, and Hertz & Wood (1985), using HEAO-1 data combined with EINSTEIN data, had observed the luminosity function to rise steeply below $10^{34.5}$ erg s^{-1}. Hertz & Wood measured the slope below $10^{34.5}$ erg s^{-1} to be -1.4 ± 0.7; we measure -0.67 ± 0.2. This implies that no more than two-thirds of all globular clusters contain X-ray sources with $L_X \gtrsim 10^{32}$ erg s^{-1}.

5. Conclusions

ROSAT has added greatly to our understanding of the population of X-ray sources in globular clusters, particularly as regards the dim X-ray sources. It has increased greatly the number of such sources known, so that we now know of nearly as many dim X-ray sources in globular clusters as we do millisecond pulsars (Lyne 1992). Several clusters contain multiple sources, some of which are variable. The spectral characteristics show that they cannot all be described by the same spectrum; some sources are very soft, with $kT \lesssim 0.3$ keV, while others are much harder. The luminosity function suggests that no more than two-thirds of all globular clusters contain X-ray sources with $L_X \gtrsim 10^{32}$ erg s^{-1}.

Acknowledgements. We would like to thank Rodrigo Supper and Adrienne Cool for providing us with some of their data. FV and HMJ are supported by the Netherlands Organization for Scientific Research NWO under grant PGS 78-277.

References

Avni, Y. et al. 1980, ApJ 238, 800
Belloni, T., Verbunt, F. & Schmitt, J. 1993, A&A, 269 175
Bradt, H. & McClintock, J. 1983, ARA&A 21, 13
Cool, A. et al. 1993, ApJ, 410 L103
Dempsey, R. et al. 1993, ApJS 86, 599
Djorgovski, S. 1993, in *Structure and Dynamics of Globular Clusters*, S. Djorgovski & G. Meylan (Eds.), ASP Conf. Proc. Vol. 50, p. 373
Forman, W. et al. 1978, ApJS 38, 357
Fruchter, A. et al. 1992, Nat 359, 303
Grindlay, J. 1993, in *Structure and Dynamics of Globular Clusters*, S. Djorgovski & G. Meylan (Eds.), ASP Conf. Proc. Vol. 50, p. 285
Grindlay, J. 1994, in *The Evolution of X-ray Binaries*, S.S. Holt & C.S. Day (Eds.), AIP Conf. Proc. Vol. 308, p. 339
Hasinger, G. 1994, in *The Evolution of X-ray Binaries*, S.S. Holt & C.S. Day (Eds.), AIP Conf. Proc. Vol. 308, p. 611
Hasinger, G., Johnston, H. & Verbunt, F. 1994, A&A 288, 466
Hertz, P. & Grindlay, J. 1983, ApJ 275, 105
Hertz, P. & Wood, K. 1985, ApJ 290, 171
Hertz, P., Grindlay, J. & Bailyn, C. 1993, ApJ 410, L87
Johnston, H., Verbunt, F. & Hasinger, G. 1994, A&A 289, 763
Katz, J. 1975, Nat 253, 698
King, I. et al. 1993, ApJ 413, L117
Kulkarni, S. et al. 1992, Nat 359, 300
Lewin, W. & Joss, P. 1981, Space Sci. Rev. 28, 3
Lyne, A. 1992, in *X-ray binaries and the formation of binary and millisecond pulsars*, E.P.J. van den Heuvel & S. Rappaport (Eds.), Kluwer, Dordrecht, p. 79
Magnier, E. 1994, in *The Evolution of X-ray Binaries*, S.S. Holt & C.S. Day (Eds.), AIP Conf. Proc. Vol. 308, p. 640
Margon, B. 1994, in *New Horizon of X-ray Astronomy*, F. Makino & T. Ohashi (Eds.), Universal Academy Press, p. 395
Paresce, F., de Marchi, G. & Ferraro, F. 1992, Nat 360, 46
Predehl, P., Hasinger, G. & Verbunt, F. 1991, A&A 246, L21
Primini, F., Forman, W. & Jones, C. 1993, ApJ 410, 615
Rappaport, S. et al. 1994, ApJ 423, 633
Supper, R. 1994, in *The Evolution of X-ray Binaries*, S.S. Holt & C.S. Day (Eds.), AIP Conf. Proc. Vol. 308, p. 631
Supper, R. et al. 1995, (in preparation)
Van Teeseling, A. & Verbunt, F. 1994, A&A (in press)
Verbunt, F., Van Paradijs, J. & Elson, E. 1984, MNRAS 210, 899
Verbunt, F. et al. 1994, A&A (submitted)
Watson, M., King, A. & Osborne, J. 1985, MNRAS 212, 917

Discussion

R.-D. Scholz: What is the angular distance of the ROSAT source in M3 from the cluster center?

H. Johnston: It is in the core.

7

Cataclysmic Variables

AM HERCULIS BINARIES

K. BEUERMANN

Universitäts-Sternwarte Göttingen
Geismarlandstr. 11, D-37083 Göttingen, Germany

Abstract. AM Herculis binaries contain mass accreting magnetic white dwarfs which appear as bright X-ray sources in the ROSAT All Sky Survey. About 52 systems are presently known which allow detailed studies of the evolution of magnetic close binaries and of fundamental plasma-physical processes in the accretion region on the white dwarf.

1. Introduction

AM Herculis binaries are a subgroup of cataclysmic variables (CVs) which contain a mass losing late-type main-sequence star and an accreting magnetic white dwarf. Orbital periods range from 78 min to 4.6 hours with a further probable member of the class at 8.0 hours. The magnetic nature of the accreting white dwarf is manifested in a variety of ways, including the regular modulation of the optical and X-ray fluxes with superimposed flaring, the strongly polarized cyclotron emission, the occurrence of cyclotron lines in the optical/IR, and the Zeeman splitting of the Balmer absorption lines by typically several 100 Å. Both Zeeman splitting and the determination of the electron gyrofrequency from the spacing of the cyclotron lines, allow a determination of the field strength and to some extent also of the field structure. The latter determines the accretion flow, the location of the accretion spot(s), and, thereby, the overall appearance of the systems. Accreting matter of moderate density is shock-heated above the surface of the white dwarf and cools by hard X-ray bremsstrahlung and optical/IR cyclotron radiation. Dense filaments or blobs of matter, on the other hand, may penetrate into the photosphere. Their energy seems to be reverberated into the intense soft X-ray emission which is characteristic of AM Her stars.

2. Evolutionary Aspects

The accretion geometry depends on whether the white dwarf rotates freely (DQ Her stars) or synchronously with the orbital period (AM Her stars). Among the physical processes which may be responsible for the synchronization, a particularly simple mechanism is the magnetostatic interaction of the magnetic moments μ_1 and μ_2 of primary and secondary, respectively (Campbell 1985). Synchronization takes place when the corresponding torque G_{sync} exceeds the accretion torque G_{acc} with $G_{\text{sync}} \sim \mu_1 \mu_2/a^3 \propto P_{\text{orb}}^{-2}$ and $G_{\text{acc}} \simeq \dot{M}\Omega r_L^2 \propto P_{\text{orb}}^{1/3}$, where a is the binary separation, r_L the distance between the inner Lagrange point and the white dwarf, $\Omega = 2\pi/P_{\text{orb}}$, and, for simplicity, the accretion rate $\dot{M}$ and the masses of the primary and the secondary were taken as constant (i.e., $r_L \propto a \propto P_{\text{orb}}^{2/3}$). Equating the torques suggests as an order-of-magnitude estimate that synchronization takes place near $P_{\text{orb}} \sim 4\,\text{hours}$ for $\mu_1 \sim 10^{34}\,\text{G cm}^3$ ($B \sim 30\,\text{MG}$), an accretion rate $\dot{M} \sim 10^{16}\,\text{g s}^{-1}$, and an assumed magnetic moment of the secondary of $\mu_2 \sim 10^{33}\,\text{G cm}^3$ ($B \sim 100\,\text{G}$). Given the marked variations which $\dot{M}$ undergoes in individual systems, it is possible that a system which is synchronous at one time may break loose if $\dot{M}$ increases for a prolonged time. An estimate of the time scale for such variability is given by the period of harmonic oscillations of the magnetic axis about the line connecting the two stars. This period is of the order of 50 yrs (Campbell 1985; Wickramasinghe & Wu 1991; King & Whitehurst 1991). Observational evidence for a secular change in the orientation of the white dwarf is scarce so far, with the possible exception of DP Leo (Beuermann & Schwope 1994; Robinson & Córdova 1994). The long-period magnetic nova V1500 Cyg ($P_{\text{orb}} = 201\,\text{min}$) lost synchronism in the common-envelope phase of the eruption and is currently re-establishing it on a time scale of 170 yrs (Schmidt et al. 1994). Studies of this system can provide important information about the mechanism of synchronization. Unfortunately, V1500 Cyg is faint and the field strength of the white dwarf is uncertain. BY Cam with $P_{\text{orb}} = 202\,\text{min}$ is another system which is seemingly slightly out of synchronism (Silber et al. 1992; Piirola et al. 1994). It will also be worthwhile to search for the degree of synchronism in the other 9 systems with still longer periods of which 7 are newly discovered with ROSAT. Presently, the systems with the longest periods are RX J1313−32 ($P_{\text{orb}} = 255\,\text{min}$), RX J0203+29 ($P_{\text{orb}} = 275\,\text{min}$) and RX J0515+01 ($P_{\text{orb}} = 480\,\text{min}$) (Beuermann & Schwope 1994; Garnavich et al. 1995; Shafter et al. 1995; Walter et al. 1995).

Since DQ Her stars dominate among the known longer-period magnetic CVs and AM Her stars among the short-period ones, King et al. (1985) suggested that DQ Her stars evolve into AM Her stars once they become synchronized. This appears to be a viable hypothesis, save for the fact that

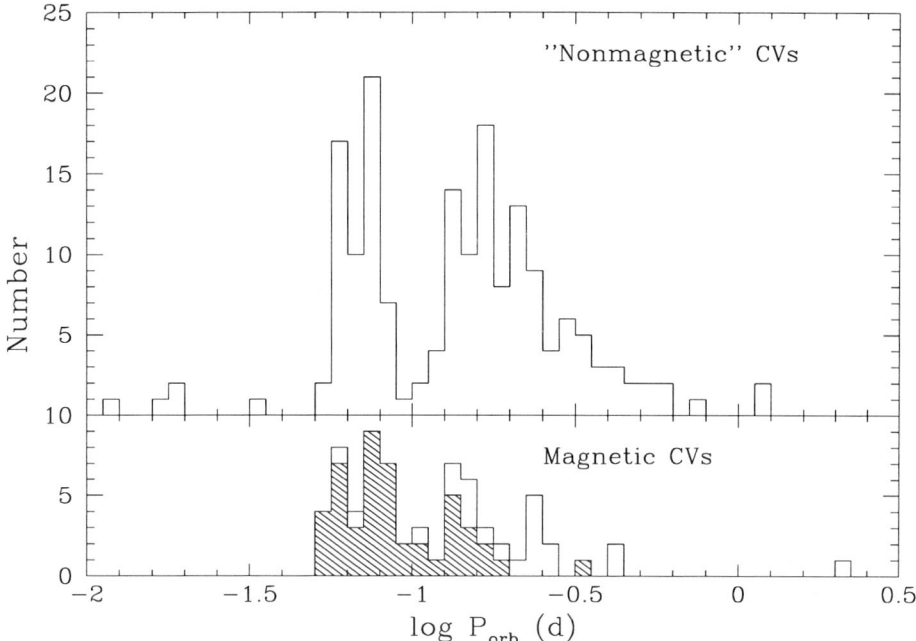

Figure 1. Orbital-period distribution of non-magnetic and magnetic CVs. The shaded histogram in case of the magnetic CVs denotes the AM Herculis binaries, the open part the DQ Her stars. Both diagrams indicate a paucity of systems with periods between 2 and 3 hours around $\log P_{orb}(\text{days}) = -1$.

most DQ Her stars seem to have lower polar-field strengths than AM Her stars. Nevertheless, the space density of DQ Her stars is comparable to that of long-period AM Hers ($n \sim 7\,10^{-8}\,\text{pc}^{-3}$) and at least some DQ Her stars may be expected to ultimately become synchronized. Several new ROSAT-discovered DQ Her stars with extremely soft X-ray spectra similar to those of AM Her stars (Mason et al. 1992; Haberl et al. 1994) may be progenitors of AM Her stars. The field strength of RX J0751+14 = RE 0751+14, $B \simeq$ 8–18 MG (Piirola et al. 1993), is comparable to those of white dwarfs in some AM Her stars.

The principal impact of the ROSAT mission on the study of AM Herculis binaries was to substantially increase the number of known systems to beyond 50. This allows to proceed from the study of individuals to the study of the properties of the class.

Fig. 1 shows the distribution of orbital periods P_{orb} of magnetic CVs along with that of non-magnetic CVs (Ritter & Kolb 1993). A CV appears on the long-period side of the distribution when the secondary first gets in contact with its Roche surface and evolves towards lower orbital periods by loss of angular momentum until the secondary becomes degenerate at $P_{\rm orb} \simeq 80$ min. A pronounced feature is the famous 'period gap' between 2 and 3 hours. At $P_{\rm orb} > 3$ hours, a CV loses angular momentum presumably by 'magnetic braking', i.e., by a stellar wind driven from the magnetosphere of the secondary (Verbunt & Zwaan 1981), in addition to gravitational radiation which acts on all systems. The conventional explanation of the period gap involves cessation of magnetic braking at $P_{\rm orb} \simeq 3$ hours when the secondary becomes fully convective and loses its magnetic activity. The associated reduction in $\dot{M}$ causes the secondary to relax to thermal equilibrium and recede from its Roche surface. Mass transfer resumes when gravitational radiation has re-established contact at $P_{\rm orb} \simeq 2$ hours.

It has been suggested that the magnetic configuration of AM Her stars prevents the wind loss, causing them to evolve merely by gravitational radiation (Wickramasinghe & Wu 1994) in which case there would be no gap. The observational situation seems to refute that suggestion. If we define the gap as extending from $\log P_{\rm orb}({\rm days}) = -0.90$ to -1.05 (2.1 to 3.0 hours) then the numbers below the gap and in the gap, and the corresponding ratios are 57, 7, 0.12 ± 0.05 for the non-magnetic systems and 30, 5, 0.17 ± 0.08 for the AM Her stars, respectively. While these two ratios are not significantly different, there are clearly much fewer magnetic systems above the gap than non-magnetic ones. This holds for AM Her stars alone and for all magnetic systems. One reason is that a typical CV will appear at $P_{\rm orb} \sim 8$ hours while a typical magnetic CV synchronizes only when $P_{\rm orb} \sim 4$ hours. Nevertheless, the presence of a gap indicates that the same process which causes the gap in non-magnetic CVs is acting also in magnetic ones.

The distributions in Fig. 1 are not free from selection effects in the sense that they do not represent the true space densities at a given period. Longer-period systems are generally drawn from a larger volume but not as large as may be expected from their luminosities because absorption of the soft X-ray emission causes a systematic loss of more distant systems. In summary, magnetic and non-magnetic CVs seem to evolve under the same angular-momentum loss processes but differences in the relative efficiencies of these processes for the two subclasses can not be excluded.

3. Magnetic Field of the White Dwarf

The strength and structure of the white-dwarf magnetic field determine not only the evolution of a system but also the accretion geometry and the radiative properties. Part of the luminosity is released as cyclotron radiation which displays the higher harmonics of the electron gyrofrequency as severely broadened emission lines from the $\sim 10^8$ K post-shock plasma.

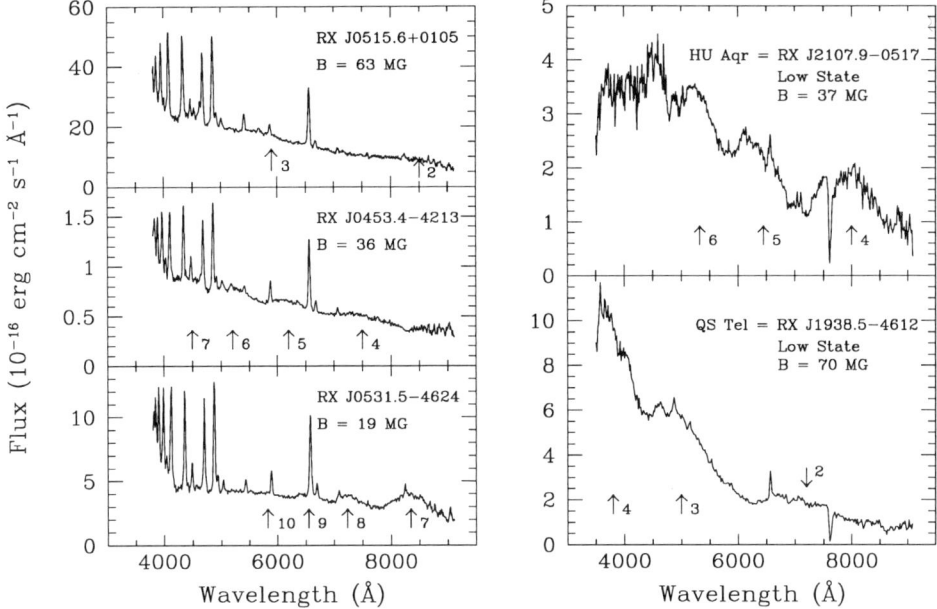

Figure 2. Examples of cyclotron lines in five AM Herculis binaries newly discovered with ROSAT. Line positions, harmonic numbers, and field strengths derived from their separations are indicated.

After minor special-relativistic corrections, the separation of these lines directly measures the field strength B_{cyc} in the accretion spot. Fig. 2 shows examples of cyclotron lines in five of the AM Herculis binaries newly discovered with ROSAT (Burwitz *et al.* 1994; Reinsch *et al.* 1994; Schwope *et al.* 1993a, 1994; Shafter *et al.* 1994).

Cool matter surrounding the accretion spot may lead to Zeeman lines in the cyclotron quasi-continuum. This effect measures the field strength B_{h} in a gaseous halo around the accretion spot and possibly at some height above the photosphere. Finally, photospheric Zeeman lines may be detected if accretion ceases or the spot is hidden behind the white dwarf. In this case, a flux-weighted mean field $\bar{B}_{\text{phot}}$ over the visible hemisphere is measured. In lucky circumstances, a combination of such measurements provides insight into the field structure of the white dwarf. We may expect $B_{\text{h}} \simeq B_{\text{cyc}} \simeq B_{\text{pole}}$ because emission and absorption occur near one of the poles and nearby in space; furthermore, in a dipole geometry B_{pole} would exceed $\bar{B}_{\text{phot}}$ by a factor of up to 2. Surprisingly, however, in some systems $\bar{B}_{\text{phot}}$ is similar to B_{cyc} (or B_{h}) or even significantly exceeds it, a result which indicates substantial deviations of the surface field from a dipole geometry.

Table 1 provides an overview of currently available field measurements of the white dwarfs in 21 AM Her stars. Field strengths for the main accretion region cluster around 30 MG with a standard deviation of 15 MG. There are no field strengths below ~ 10 MG which probably indicates that even short-period systems do not synchronize below this value. There are also no field strengths in the main accretion spot in excess of 61 MG. This is surprising because single white dwarfs show a much broader distribution which extends to beyond 500 MG (Chanmugam 1992). A viable hypothesis for the absence of high-field magnetic CVs involves a strong wind which couples onto the white-dwarf field lines and cause rapid evolution of such systems with a correspondingly small space density and chance of discovery (Hameury et al. 1988).

For some systems, two values are given for B_{cyc}. In these cases, two systems of cyclotron lines were observed which refer to two accretion regions. If these are located essentially opposite to each other, field strengths different by a factor of ~ 2 are incompatible with a dipole. In addition, there is evidence for non-dipolar fields also in other systems in which either no cyclotron lines are observed (e.g., because of excessive temperature broadening) or the second pole is hidden behind the white dwarf. E.g., BL Hyi displays pronounced Zeeman absorption lines in the cyclotron continuum which indicate a field strength of 12 MG in the accretion region. In low states, however, shallow Zeeman absorption lines appear which indicate a mean photospheric field of 22 MG with a substantial spread in field strength (Schwope et al. 1995). This is impossible in a dipole geometry. A somewhat less extreme example is MR Ser with a field of 24 MG in the main accretion spot and a mean photospheric field of 27 MG. In this case, it is noteworthy that all individual non-degenerate Zeeman components of Hα σ^- and Hα σ^+ are detected which imply a field spread of less than 1 MG over the visible hemisphere (Schwope et al. 1993b). Assuming a field structure of aligned dipole and quadrupole components (or a dipole offset along its axis) implies field strengths of ~ 60 MG at the second pole. Representing the observed spectra by synthetic Zeeman spectra for such configuration was quite successful for MR Ser but less so for BL Hyi, suggesting perhaps a more complex field structure in the latter (Schwope et al. 1993b, 1995).

Accepting direct and indirect evidence, there are 9 systems with significantly different field strengths at two 'poles'. In all cases, the main accretion spot is located at the low-field 'pole'. For neutron stars, Romani (1993) suggested that accretion may cause the horizontal component of the field to be submerged and carried away from the pole. Unfortunately, no quantitative treatment exists and the applicability to white dwarfs remains open because the total accreted mass is uncertain. Also, nova explosions may counteract submersion. Alternatively, the non-dipolar field structure could be a property of the internal field generating process in the progenitor star and the magnetic orientation of the white dwarf a consequence of the mechanism

TABLE 1. Field strengths of magnetic white dwarfs in cataclysmic variables. $B_{cyc,1}$ and $B_{cyc,2}$ refer to field strengths obtained from the spacing of cyclotron lines, B_h to the Zeeman effect in the cool halo of the accretion spot, and $\bar{B}_{phot}$ to the flux-weighted mean photospheric field from Zeeman lines.

System	P_{orb} (min)	$B_{cyc,1}$ (MG)	$B_{cyc,2}$ (MG)	B_h (MG)	$\bar{B}_{phot}$ (MG)	References
EF Eri	81			13		1
DP Leo	90	31	59			2,3,4
RX J1149+28	90:	43				5
RX J1957−57	99			14		6
VV Pup	100	31	54			3,7
V834 Cen	101	23		23	22	8,9
RX J0453−42	102:	36				10
MR Ser	113	24		24	27	11
BL Hyi	114			12	22	12
ST LMi	114	12			19:	13,14
EK UMa	114	35:/47:				15
AN UMa	115	29				3,16
RX J2107−05	125	35				17
EU Cnc	125	42:				18,19
UZ For	126	53	75(113)			20
RX J0531−46	133	19				21
RX J1938−46	140	47	70(80)			22
AM Her	186	14			13	23,24
BY Cam	202	28				3,16
QQ Vul	222	36:				3
RX J0515+01	480	61:				25

(1) Östreicher et al. 1990, (2) Cropper et al. 1990a, (3) Schwope 1991, (4) Cropper & Wickramasinghe 1993, (5) Schwope (in prep.), (6) Thomas et al. (in prep.), (7) Wickramasinghe et al. 1989, (8) Schwope & Beuermann 1990, (9) Ferrario et al. 1992, (10) Burwitz et al. 1995, (11) Schwope et al. 1993b, (12) Schwope et al. 1995, (13) Schmidt et al. 1983, (14) Ferrario et al. 1993, (15) Cropper et al. 1990b, (16) Cropper et al. 1988, (17) Schwope et al. 1993a, (18) Pasquini et al. 1994, (19) This work, (20) Schwope et al. 1990, (21) Reinsch et al. 1995, (22) Schwope et al. 1994, (23) Young et al. 1981, (24) Bailey et al. 1991, (25) Shafter et al. 1995.

of synchronisation (Wu & Wickramasinghe 1993).

In summary, our knowledge of the field structure in magnetic CVs is still utterly incomplete but improving. In the future, high-sensitivity spectropolarimetry of the cyclotron radiation in high accretion states and of Zeeman absorption spectra in low states promise to yield interesting results. Availability of the polarization spectra provides additional information which is needed to distinguish between different possible field geometries (Putney & Jordan 1994).

4. Accretion Scenarios

In AM Herculis binaries, the magnetic field of the white dwarf prevents the formation of an accretion disk. After leaving the inner Lagrange point L_1, the accretion stream penetrates into the field along a quasi-ballistic trajectory until equality of ram pressure and magnetic pressure is reached. We know very little about the plasma-dynamical processes in the magnetosphere which cause break-up of the stream and coupling onto near-polar field lines (Hameury et al. 1986). Observationally, emission line measurements delineate the hot and partially photoionized section of the stream close to the white dwarf, but not the distant cooler sections, including that between L_1 and the stagnation point. Break-up of the stream probably determines the density distribution of the matter and thereby the spatial and temporal distribution of the mass flow rate arriving at the white-dwarf surface. Additional information on the density and ionization structure of the stream in the outer magnetosphere can be obtained from X-ray and IR absorption measurements at the orbital phase when the stream passes in front of the hot spot on the white dwarf (e.g., Watson et al. 1989).

Standard theory (Lamb & Masters 1979; King & Lasota 1979) assumes a stationary flow of matter $\dot m$ (in $\mathrm{g\,cm^{-2}\,s^{-1}}$) which is shock-heated close to the surface of the star to $\sim 10^8$ K; subsequently, the plasma cools by bremsstrahlung and cyclotron radiation until is settles onto the photosphere. Since part of the emission will be intercepted by the white dwarf, a third component will be due to reprocessed radiation. Directly below the shock, the effective temperature of the heated photosphere may exceed $\sim 3\,10^5$ K if the shock height is small; the reprocessed flux then emerges at soft X-ray energies. If, on the other hand, the shock height in units of the white-dwarf radius is large, $h/R_{\mathrm{wd}} \sim 0.1$, as expected for low mass flow rates, the irradiated fraction of the white dwarf surface becomes large, too, $f \simeq h/2R_{\mathrm{wd}} \sim 0.05$. The temperature drops to $<10^5$ K, and the emission peak moves into the UV. Beaming of the cyclotron emission perpendicular to the field and, therefore, parallel to the surface adds to a wider spread of the irradiation. Gänsicke et al. (1995) have studied the energy balance in AM Herculis itself and found that bremsstrahlung and cyclotron irradiation can quantitatively account for the UV flux of the large ($f \sim 0.08$) heated spot, both in the low and in the high state. Hence, in AM Her this reprocessed component seems to emerge primarily in the UV and not in the soft X-ray regime. In other systems, however, this component may be irretrievably submerged in the accretion-produced soft X-ray emission.

Observationally, most AM Herculis stars are spectacular soft X-ray sources, but the soft X-ray luminosity is much too large to be due to reprocessing. For years, this so-called 'soft X-ray problem' has plagued theoreticians and

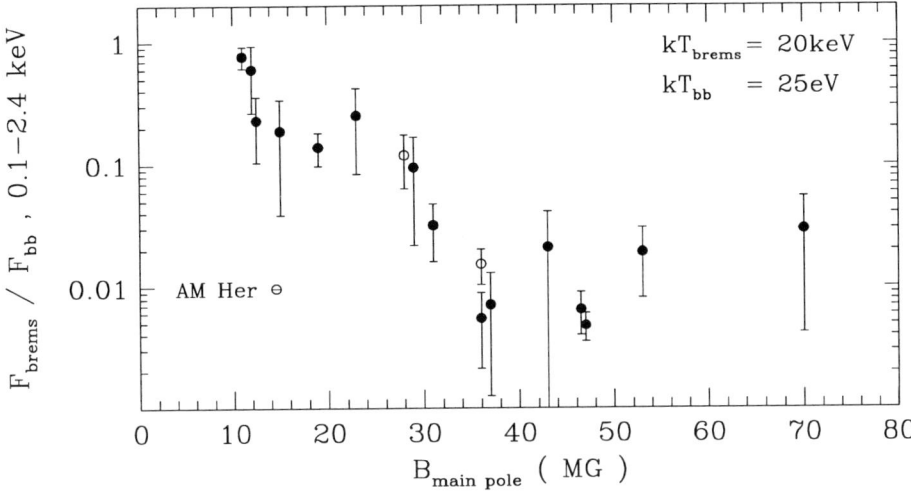

Figure 3. Ratio of the unabsorbed energy fluxes $F_{\rm brems}$ and $F_{\rm bb}$ in the ROSAT (0.1–2.4 keV) window as a function of field strength in the accretion spot. Bremsstrahlung is increasingly suppressed at high field strengths. Filled and open circles refer to short-period and long-period ($P_{\rm orb} > 3$ hours) systems, respectively.

observers alike until Kuijpers & Pringle (1983) and Morfill et al. (1984) suggested that sufficiently dense blobs of matter might transfer their kinetic energy to sub-photospheric layers and heat the photosphere from below. Short time-scale fluctuation studies of soft and hard X-rays indicate that these two components are not correlated (e.g., Beuermann et al. 1991) which argues against an origin of the soft X-rays from reprocessing of the *observed* hard X-ray component. On the other hand, Van Teeseling et al. (1994) showed that the EXOSAT grating spectrum of AM Her is not consistent with an undisturbed hot LTE atmosphere, but is much better fitted by an irradiated atmosphere or an atmosphere heated by more diffuse blobs which disperse their energy at small optical depths. Hence, the hydrodynamics and the radiative transfer are certainly more complex than suggested by the original Kuijpers & Pringle picture. The actual accretion region is probably structured with a wide range of co-existing mass flow rates. Cyclotron emission, hard X-ray bremsstrahlung and soft X-ray emission are probably due to different regimes in the mass flow rates of order $\sim 10^{-2}$, ~ 1, and $\sim 100 \, {\rm g \, cm^{-2} \, s^{-1}}$, respectively (e.g., Beuermann et al. 1987).

The large number of AM Her stars observed with ROSAT has, for the first time, allowed a systematic study of the emission properties as a function of system parameters. Fig. 3 shows the ratio of the bremsstrahlung to quasi-blackbody fluxes as a function of field strength in the accretion

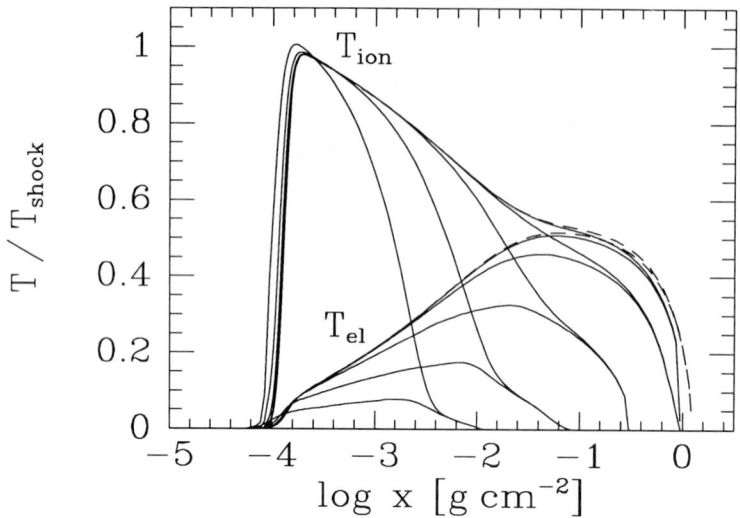

Figure 4. Two-fluid temperature structure of the shocked accretion flow for $B = 30$ MG and mass flow rates of $\dot{m} = 0.01$ to 100 g cm^{-2} s^{-1} (solid lines, from bottom to top). For comparison, the pure bremsstrahlung case for zero field and/or high mass flow rates is shown (dashed lines). Abscissa is column density of a pure hydrogen plasma, ordinate is temperature in units of $T = (3/16)(m/k)v_{\rm ff}^2$. The flow enters from the left and, for each set of curves, the drop to zero temperature indicates the location of the stellar photosphere.

spot. $F_{\rm brems}$ and $F_{\rm bb}$ are the fluxes in the ROSAT 0.1–2.4 keV band corrected for interstellar absorption (fluxes 'at the source'). The flux ratio is seen to decrease systematically with field strength, indicating a suppression of bremsstrahlung and/or a preference of soft X-ray emission at high field strengths. *Suppression* is an obvious choice because cyclotron emission increases with B and dominates over bremsstrahlung. *Preference of soft X-ray emission* may also occur because the mass flow rate in a magnetic funnel scales as $\dot{m} = \rho v \propto B$. Why AM Her, the supposed 'prototype of the class', avoids the general trend remains to be clarified.

Time-dependent radiation-hydrodynamical calculations are required to take full advantage of the plasma-diagnostic information content of the available observations. Such calculations which account for the fully angle-dependent and frequency-dependent radiative transfer of the cyclotron radiation are not yet available. As a step towards this goal, Woelk (1995) has solved the 1-D stationary two-fluid hydrodynamic equations together with the full radiative transfer for a constant field. His solutions are applicable for pill-box shaped accretion regions (i.e. not for tall columns which require 2-D radiative transfer). As an example of his results, Fig. 4 shows the electron and ion temperatures in the post-shock region for $B = 30$ MG and mass flow rates between $\dot{m} = 0.01$ and 100 g cm^{-2} s^{-1}, along with the pure bremsstrahlung solution which is the limiting case for zero field and also for very high $\dot{m}$. With decreasing mass flow rate (and thereby decreasing post-shock density), cyclotron cooling reduces the electron and the collisionally coupled ion temperatures far below the Hugoniot temperature. At the same time, the shock height x (in $\mathrm{g\,cm^{-2}\,s^{-1}}$) collapses to a small fraction of the bremsstrahlung case. The geometric shock height h (in cm), however, still increases with decreasing $\dot{m}$ but less so than $h \propto \dot{m}^{-1}$ which is the expected increase for cooling by bremsstrahlung alone. The calculated spectra yield a smooth cyclotron continuum for the higher $\dot{m}$ values, whereas cyclotron lines appear for small $\dot{m}$. Bremsstrahlung originates mainly from the high-$\dot{m}$ regions. At low $\dot{m}$, the solutions connect to the bombardment case studied earlier by Woelk & Beuermann (1992, 1993).

In summary, we are beginning to understand the relevant plasma physics of the accretion spot, although many questions relating, e.g., to the internal structure of the spot or to the stability of the shock solutions remain open. Our knowledge of the processes in the magnetosphere of the white dwarf are still fragmentary at best.

5. Conclusions

The main impact of the ROSAT mission on the field lies in the substantial increase in the number of recognized magnetic CVs. We can now proceed from the study of individuals to the study of the class, use these systems as plasma laboratories, investigate the magnetic-field structure of compact stars, and learn more about the evolution of cataclysmic variables.

Acknowledgements. I thank A.D. Schwope, Potsdam, for some of the data in Table 1, and V. Burwitz and U. Woelk for preparing Figs. 1–3 and Fig. 4, respectively. I enjoyed interesting discussions with these colleagues and with B. Gänsicke, K. Reinsch, and A. van Teeseling, and acknowledge useful comments on the manuscript. This work was supported in part by the DARA under project number 50OR9210.

References

Bailey, J., Ferrario, L. & Wickramasinghe, D.T. 1991, MNRAS 251, 37P
Beuermann, K. & Schwope. A.D. 1994, ASP Conf. Ser. 56, p. 119
Beuermann, K., Stella, L. & Patterson, J. 1987, ApJ 316, 360
Beuermann, K., Thomas, H.-C. & Pietsch, W. 1991, A&A 246, L36
Burwitz V. et al. 1995, A&A (in press)
Campbell, C.G. 1985, MNRAS 215, 509
Chanmugam, G. 1992, ARA&A 30, 143
Cropper, M. & Wickramasinghe, D.T. 1993, MNRAS 260, 696
Cropper, M. et al. 1988, MNRAS 236, 29P

Cropper, M., Mason, K.O. & Mukai, K. 1990a, MNRAS 243, 565
Cropper, M. et al. 1990b, MNRAS 245, 760
Ferrario, L. et al. 1992, MNRAS 256, 252
Ferrario, L., Bailey, J. & Wickramasinghe, D.T. 1993, MNRAS 262, 285
Gänsicke, B., Beuermann, K. & de Martino, D. 1995, A&A (submitted)
Garnavich, P.M. et al. 1995, ApJ (in press)
Haberl, F. et al. 1994, A&A 291, 171
Hameury, J.M., King, A.R. & Lasota, J.P. 1986, MNRAS 218, 695
Hameury, J.M., King, A.R. & Lasota, J.P. 1988, MNRAS 237, 48
King, A.R. & Lasota, J.P. 1979, MNRAS 188, 653
King, A.R. & Whitehurst, R. 1991, MNRAS 250, 152
King, A.R., Frank, J. & Ritter, H. 1985, MNRAS 213, 181
Kuijpers, J. & Pringle, J.E. 1982, A&A 114, L4
Lamb, D.Q. & Masters, R. 1979, ApJ 234, L117
Mason, K.O. et al. 1992, MNRAS 258, 749
Morfill, G.E. et al. 1984, A&A 137, 7
Östreicher, R. et al. 1990, ApJ 350, 324
Pasquini, L., Belloni, T. & Abbott, T.M.C. 1994, A&A 290, L17
Piirola, V., Hakala, P. & Coyne S.J., G.V. 1993, ApJ 410, L107
Piirola, V. et al. 1994, A&A 283, 163
Putney, A. & Jordan, S. 1994, ApJ (in press)
Reinsch, K. et al. 1994, A&A 291, L27
Ritter, H. & Kolb, U. 1993, in *X-ray Binaries*, W.H.G. Lewin, J. van Paradijs & E.P.J. van den Heuvel (Eds.), Cambridge Univ. Press, (in press)
Robinson, C.R. & Córdova, F. 1994, ApJ (in press)
Romani, R. 1993, Nat 347, 741
Schmidt, G.D., Stockman, H.S. & Grandi, S.A. 1983, ApJ 271, 735
Schmidt, G.D., Liebert, J. & Stockman, H.S. 1994, ApJ (in press)
Schwope, A.D. 1991, PhD thesis, Technical University of Berlin
Schwope, A.D. & Beuermann, K. 1990, A&A 238, 173
Schwope, A.D., Beuermann, K. & Jordan, S. 1995, A&A (in press)
Schwope, A.D., Beuermann, K. & Thomas, H.-C. 1990, A&A 230, 120
Schwope, A.D., Thomas, H.-C. & Beuermann, K. 1993a, A&A 271, L25
Schwope, A.D. et al. 1993b, A&A 278, 487
Schwope, A.D. et al. 1994, A&A (in press)
Shafter, A.W. et al. 1995, ApJ (in press)
Silber, A. et al. 1992, ApJ 389, 704
Van Teeseling, A., Heise, J. & Paerels, F. 1994, A&A 281, 119
Verbunt, F. & Zwaan, C. 1981, A&A 100, L7
Walter, F.M., Wolk, S.J. & Adams, N.R. 1995, ApJ (in press)
Watson, M.G. et al. 1989, MNRAS 237, 299
Wickramasinghe, D.T. & Wu, K. 1991, MNRAS 253, 11P
Wickramasinghe, D.T. & Wu, K. 1994, MNRAS 266, L1
Wickramasinghe, D.T., Ferrario, L. & Bailey, J. 1989, ApJ 342, L35
Woelk, U. 1995, (in preparation)
Woelk, U. & Beuermann, K. 1992, A&A 256, 498
Woelk, U. & Beuermann, K. 1993, A&A 280, 169
Wu, K. & Wickramasinghe, D.T. 1993, MNRAS 260, 141
Young, P., Schneider, D.P. & Shectman, S.A. 1981, ApJ 245, 1043

LUMINOUS SUPERSOFT X-RAY SOURCES

S. RAPPAPORT
Department of Physics and Center for Space Research
MIT, Cambridge, MA 02139, U.S.A.

AND

R. DI STEFANO
Harvard-Smithsonian Center for Astrophysics
60 Garden St., Cambridge, MA 02138, U.S.A.

1. Introduction

Supersoft X-ray sources exhibit spectra that are remarkably steep, in that the ratio of low-to-high energy X rays is much larger than is characteristic of the spectra associated with the previously known classes of luminous X-ray sources. The first supersoft sources were discovered during a survey of the Large Magellanic Cloud with the EINSTEIN Observatory (Long et al. 1981). The all-sky X-ray survey carried out with ROSAT has now established that luminous supersoft X-ray sources constitute a distinct astronomical class (see, e.g., Hasinger 1994). A number of the identified optical counterparts of the supersoft X-ray sources exhibit blue continua with emission lines of H and He II (Smale et al. 1988; Pakull et al. 1988; Cowley et al. 1990), which are characteristic of accretion disks. The X-ray emission of some sources is steady, while others exhibit significant time variability. Table 1 briefly summarizes what is known thus far about the numbers and characteristics of supersoft X-ray sources (see Hasinger 1994, and references therein).

In Section 2 we focus on a particular model for supersoft sources, one in which the observed luminosity is generated by the nuclear burning of matter that is accreting onto the surface of a white dwarf. We summarize the results of an investigation in which the total population of sources consistent with this model was calculated. Section 3 describes a nearly model-independent theoretical investigation in which we utilized the observed numbers and characteristics of supersoft sources to compute the

TABLE 1. Supersoft X-Ray Sources: Properties and Statistics

Range of Properties	Numbers	Identifications
$kT \sim 15$–55 eV	M31: ~ 15	Binaries ($0.5 < P_{\rm orb} < 3.5$ days): 4
$L \sim 10^{37}$–10^{38} ergs s^{-1}	Galaxy ~ 6	Symbiotic Novae: 3
	LMC ~ 6	Classical Nova: 1
	SMC ~ 4	Planetary Nebula*: 1
		X-Ray Pulsar: 1
		Globular Cluster Source (*low L*): 1

*We note that this source also has a hard X-ray component.

size of the underlying population of these sources. In Section 4 we continue along a model-independent line of inquiry and sketch several investigations of the ionization nebulae which may be associated with supersoft sources. Section 5 is a summary and a look forward.

2. Population Synthesis for the Steady Nuclear Burning Model

Proposed models for the supersoft X-ray sources include nuclear burning on the surface of a white dwarf (Van den Heuvel et al. 1992; Rappaport, Di Stefano & Smith 1994 [hereafter RDS]), and accreting black holes (Cowley et al. 1990) or neutron stars (Kylafis & Xilouris 1993; Hughes 1994). In this section, we focus almost exclusively on the nuclear burning model since (1) it is the most quantitative and leads to specific predictions, and (2) the preponderance of evidence points toward a majority of the systems thus far identified having a white dwarf as the X-ray emitting object. Qualitatively, the model we consider invokes steady nuclear burning of accreted matter on the surface of a ~ 1 $M_\odot$ white dwarf with accretion rates of between ~ 1 and $5 \ 10^{-7} M_\odot$ yr^{-1}; such rates are required to sustain the luminosity (Van den Heuvel et al. 1992; RDS and references therein). In the case of supersoft X-ray sources that are in binary systems with orbital periods in the range of 0.5–3 days, the companion star is expected to be a main-sequence or subgiant of mass ~ 1.3–$2.7 M_\odot$.

The high mass-transfer rates are a natural consequence of unstable mass transfer, on a thermal time scale, via Roche lobe overflow from the more massive donor star to a less massive accreting white dwarf. In the case of supersoft X-ray sources that are associated with symbiotic novae, the mass transfer may be driven by the expansion of a low-mass giant companion as it ascends the giant branch.

An evolutionary scenario for the formation of supersoft X-ray sources with orbital periods near a day is shown in Fig. 1 (which has been adapted from Van den Heuvel 1994).

We have carried out a population synthesis calculation of supersoft X-ray sources that are formed via the scenario described above (RDS). We used Monte Carlo techniques to generate the initial binary system, including the primary mass, the mass ratio, and the orbital period. A range of distributions for each of these system parameters was tested. The evolution of each binary was followed through a series of phases (several of which are depicted in Fig. 1) to determine which binaries evolved into systems consisting of a white dwarf accreting matter within the requisite range of rates from a low-mass companion. When a "successful" system was obtained, we recorded the properties of that system.

When properly normalized to the absolute birth rate of stars in the Galaxy, our results yield the number of supersoft X-ray sources, with orbital periods near a day, that should be active in the Galaxy at the present epoch. The results are shown in Fig. 2 in the form of normalized histograms of the predicted properties of (a subset of) the luminous supersoft X-ray sources.

The calculated distributions of properties of supersoft X-ray sources shown in Fig. 2 were for a set of "nominal" input model parameters and assumptions (see RDS). For these assumptions (the "standard" model of RDS), the total number of expected supersoft sources in the Galaxy at the current epoch is about 1000. We can also scale to the blue magnitudes of other galaxies in the local group to estimate their population of supersoft X-ray sources (see Di Stefano & Rappaport 1994). Such a procedure yields $\sim$2500, 100, and 30 sources for M31, the LMC, and the SMC, respectively. We have also repeated the above population synthesis calculations for a wide range of input parameters and assumptions (see RDS). The range of values for the predicted numbers of supersoft X-ray sources for the Milky Way, M31, the LMC, and the SMC are summarized in Table 2.

3. Estimating the Underlying Population of Supersoft Sources

Radiation at the wavelengths emitted by supersoft sources is readily absorbed by the interstellar medium. It is therefore clear that, whatever the physical nature of the sources (and it is unlikely that any single model will apply to all supersoft sources), ROSAT has not been able to detect a large fraction of the supersoft source population. The question we ask then is: how can we use the relatively small numbers of observed sources to derive an estimate of the true underlying population of presently active sources?

The approach that we have taken (Di Stefano & Rappaport 1994) is to use Monte Carlo techniques to "seed", with a population of sources, each of the galaxies in which supersoft sources have been detected. We then model the gas distribution of each galactic system in order to be able to compute

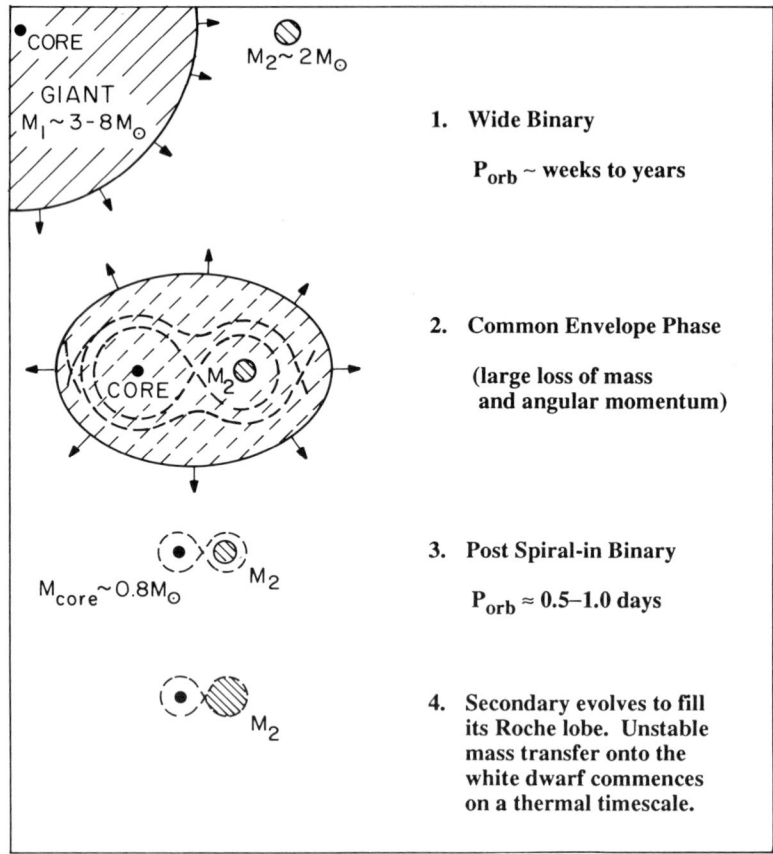

Figure 1. Evolutionary scenario for the formation of a particular class of supersoft X-ray sources. The initial (primordial) binary consists of two main-sequence stars where the primary has a mass in the range of $\sim$2–8 $M_\odot$, the secondary has a mass of $>$0.8 $M_\odot$, and the orbital period is weeks to years. If the primary overfills its Roche lobe while ascending the giant branch, the result will likely be a common-envelope stage (see, e.g., Sparks & Stecher 1974; Paczyński 1976; Taam, Bodenheimer & Ostriker 1978; Meyer & Meyer-Hofmeister 1979; Livio & Soker 1988; Webbink 1992), leading to the ejection of the giant's envelope. The resultant post-spiral-in binary may consist of a white dwarf (the core of the giant) and a low-mass main-sequence star with an orbital period of about a day. After some time has elapsed, either the orbit or the main-sequence star will evolve so that mass transfer onto the white dwarf will commence. If the donor star is more massive than the accreting white dwarf, the mass transfer will tend to be unstable. As long as the donor star is not sufficiently evolved, the mass transfer will be unstable on a thermal, rather than a dynamical timescale (Paczyński 1965, 1967; Kippenhahn, Kohl & Weigert 1967; Webbink 1979, 1985, 1992; De Kool 1992). It is important to understand whether the matter so accreted can undergo stable (rather than explosive) nuclear burning. This question has been studied by a number of authors (Paczyński 1970; Sion, Acierno & Tomcszyk 1979; Taam 1980; Nomoto 1982; Iben 1982; Sion & Starrfield 1986; Livio, Prialnik & Regev 1989; Prialnik & Kovetz 1994). Although the range of mass transfer rates consistent with steady nuclear burning depends on the mass of the white dwarf, typical values are in the range of $\sim$(1–5) 10^{-7} $M_\odot$ yr^{-1}.

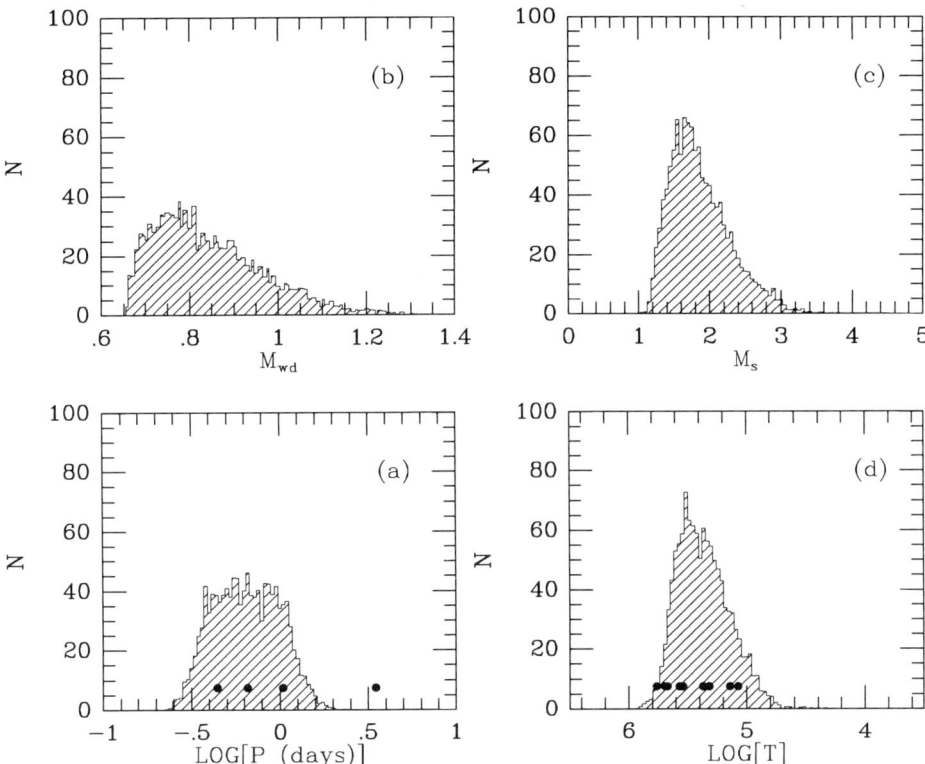

Figure 2. Panels (a) through (d) show the expected distributions of $P_{\rm orb}$, white dwarf and donor star mass, and effective temperature, $T_{\rm eff}$, respectively, as derived in the population synthesis study of RDS. The measured values of $T_{\rm eff}$ for 12 observed systems are superposed on the calculated distribution in (d), while the measured values of $P_{\rm orb}$ for 4 observed systems are superposed on the calculated distribution in (a). In each case the expected distribution is in good agreement with the observational results. The skew in the measured values of $T_{\rm eff}$ toward the high end of the distribution is almost certainly an observational selection effect. (See Section 3.) The distribution of white-dwarf and donor star masses indicate most probable values in the ranges of 0.65–1.1 $M_\odot$ and 1.5–2.5 $M_\odot$, respectively.

the column density encountered by the radiation from each source as it travels toward Earth. Using the PIMMS software (Mukai 1993), we then compute the number of counts that would be recorded by ROSAT and, depending on the ROSAT observation time for that portion of the sky, we determine whether or not ROSAT would have detected the seeded source.

TABLE 2. Derived Numbers of Supersoft X-Ray Sources

	Population Synthesis	Inferred from Observation
M31	400–6000	800–5000
Milky Way	100–1500	400–3000
LMC	20–300	15–60
SMC	5–60	10–40

In this way, we have been able to compute the fraction, f, of all supersoft sources that would have been detected by ROSAT, and hence to use the size of the observed population to infer the size of the total population of presently active sources. The results are summarized in the right-hand column of Table 2, above. In comparing the population estimates (Table 2) with the numbers of sources observed (Table 1), one can see that the fraction of sources that ROSAT detects is different for different host galaxies in the Local Group. The value of $1/f$ lies between ~ 2.5 and 10 for the Magellanic Clouds, but ranges from ~ 100 to 750 for the Milky Way, where the majority of sources would have to be detected through the large column densities associated with the disk. Distance plays a significant role in decreasing the number of sources detectable in M31. However, whatever the differences in the values of f from galaxy to galaxy within the Local Group, in all cases f is small.

The fact that there might be a large population of supersoft sources not detectable with ROSAT motivates a search for other ways to detect and study supersoft sources. A promising line of research is the study of the ionization nebulae which may be associated with at least some supersoft sources. These nebulae are described in the next section.

4. Ionization Nebulae Surrounding Supersoft X-Ray Sources

4.1. THEORY

Regardless of what the nature of luminous supersoft X-ray sources is, it is clear that they emit copious quantities of highly ionizing photons in the range 20 eV to 200 eV. Thus, we expect that there will be ionization nebulae surrounding these sources (Rappaport et al. 1994).

We have carried out calculations of the ionization and temperature structure of the nebulae expected to surround supersoft X-ray sources, as well as their optical line fluxes (Rappaport et al. 1994). The models for ionization nebulae were calculated with the XSTAR code (Kallman & Krolik 1993) which uses the computational methods described by Kallman & McCray (1982). The models consist of a spherical gas cloud with a point source of continuum radiation at the center. The input parameters include the source spectrum, the gas composition and density, the initial ionization parameter (which determines the initial radius), and the column density of the cloud (which determines the outer radius). Construction of a model consists of the simultaneous determination of the state of the gas and the radiation field as a function of distance from the source. The state of the gas at each radius follows from the assumption of a stationary local balance between heating and cooling and between ionization and recombination. The calculated characteristic radius of the ionized regions ($\sim 6\,\mathrm{pc}$ for an ISM density of $10\,\mathrm{cm}^{-3}$) agrees with what one obtains from a rough estimate based on the Strömgren-sphere formula.

An important question to answer is whether it is possible to distinguish the nebulae associated with supersoft sources (hereafter called supersoft nebulae) from the nebulae associated with other astrophysical objects. In this regard, geometrical considerations are useful. For example, the size of supersoft nebulae can be used to distinguish them from planetary nebulae which typically have radii that are an order of magnitude smaller. It may also be possible to use geometrical considerations to distinguish supersoft nebulae from H II regions. The reason for this is that it is expected that the boundary of supersoft nebulae will be less sharply defined than that of classical H II regions since there is substantial power in photons with energies (100–300 eV) that are able to reach moderate and even arbitrarily large distances from the source.

Moreover, it may also be possible to draw spectral distinctions between supersoft nebulae and other astrophysical nebulae. For example, Rappaport et al. (1994) showed that these ionization nebulae should be very bright in [O III] $\lambda 5007$, with the ratio of [O III]/Hβ in the range of ~ 12–26 for a solar composition. In the inner parts of these nebulae, this ratio may be larger than for any other known astrophysical object (see, e.g., Osterbrock 1989).

The detection of an ionization nebula around a supersoft X-ray source should lead to a better understanding of both the source luminosity (since, in a sense, the interstellar medium acts as a giant bolometer) and the properties of the surrounding interstellar medium itself. The development of efficient techniques for searching for such ionization nebulae would represent a new means of discovering supersoft X-ray sources using ground-based observations (i.e., at optical wavelengths). This could lead to the optical discovery of many of the supersoft X-ray sources that are undetectable in the UV and soft X-ray bands because of severe attenuation by interstellar gas (Di Stefano & Rappaport 1994).

4.2. OBSERVATIONS

Recently, a comprehensive search for ionized gaseous nebulae surrounding 9 known supersoft X-ray sources in the Large and Small Magellanic Clouds was carried out (Remillard, Rappaport & Macri 1994). (One source, CAL 83, was already known to have an associated nebula [Pakull & Motch 1989]). Deep images were made using narrow-band filters to isolate the emission lines of Hα and [O III] (λ5007). In the narrow-band images of Remillard et al. (1994) the CAL 83 nebula is detected out to distances as far as 20 pc from the central source, and the integrated luminosity in each line is of the order of 100 $L_\odot$. If we interpret the observed line luminosities in terms of the model nebula calculations (Rappaport et al. 1994), we conclude that the time-averaged X-ray luminosity of CAL 83 over the past $\sim 3 \, 10^4$ years is $\sim 3 \, 10^{37}$ ergs s^{-1}. The bright inner nebula contains $\sim 150 \, M_\odot$ within ~ 8 pc of the central source, which clearly indicates that the nebular material is part of the interstellar medium and has not been ejected from the binary system.

In contrast, Remillard et al. (1994) reported no detections of nebulae associated with the other 8 luminous supersoft X-ray sources in the Magellanic Clouds, with upper limits for the [O III] luminosity that are a factor of ~ 10 below that for the CAL 83 nebula. For these sources, either the time-averaged X-ray luminosity of the central source is substantially below that of CAL 83, or the local interstellar medium in the vicinity of these sources is much less dense.

4.3. IMPLICATIONS

The association of nebulae with even a subset of all active supersoft sources may have interesting implications. For example, typical supersoft nebula luminosities in the λ5007 line of [O III] can be on the order of several times 10^{36} erg s^{-1}. This is comparable to the value of the cut-off luminosity in this line for the planetary nebula luminosity function (PNLF) (see, e.g., Jacoby et al. 1992). We therefore note that these nebulae may possibly be confused with the most luminous planetary nebulae in distant galaxies (Di Stefano, Paerels & Rappaport 1994). Such confusion would have two potential consequences. First, it might mean that a subset of the nebulae identified through programs designed to study the PNLF may, in fact, be supersoft nebulae. If this subset can be identified, then this would give us a new store of supersoft nebulae to study. Second, such an effect would possibly have a small, but systematic influence on the results obtained by using the PNLF to compute intergalactic distances.

5. Conclusions and Future Directions

In Sections 2 and 3 we have described investigations which indicate that the underlying population of presently active supersoft sources may be large—on the order of several thousand sources for a galaxy such as M31. Although it will be difficult for observations at X-ray wavelengths to study this population, the ionization nebulae that are likely to be associated with at least a subset of them may provide another way for us to discover and study supersoft sources.

An important implication of the white dwarf accretor model for supersoft sources is the possibility that some of these systems will accrete enough mass to explode as type Ia supernovae. The computations in RDS assess the number of potential supernovae across a range of model parameters, including the uncertain retention of accreted matter by the white dwarf for different mass accretion rates. For at least one set of plausible model parameters, the computed rate of type Ia supernovae was comparable to the observed rate (Cappellaro et al. 1993). A more detailed investigation is underway.

In short, supersoft sources seem likely to be related to a number of interesting astrophysical phenomena. The study of these objects should provide continuing insights for a number of years to come.

Acknowledgements. This work was supported in part by the National Aeronautics and Space Administration under contract NAS 5-29298 and grant NAGW-1545.

References

Cappellaro, E. et al. 1993, A&A 273, 383
Cowley, A.P., Schmidtke, P.C., Crampton, D. & Hutchings, J.B. 1990, ApJ 350, 288
de Kool, M. 1992, A&A 261, 188
Di Stefano, R. & Rappaport, S. 1994, (in press)
Di Stefano, R., Paerels, F. & Rappaport, S. 1994, (in preparation)
Hasinger, G. 1994, in *Evolution of X-Ray Binaries*, S.S. Holt & C.S. Day (Eds.), AIP Conf. Proc. 308, Amer. Inst. Phys. (New York), p. 611
Hughes, J.P. 1994, ApJ 427, L25
Iben Jr., I. 1982, ApJ 259, 244
Jacoby, G.H. et al. 1992, PASP 104, 599
Kallman, T.R. & McCray, R.A. 1982, ApJS 50, 263
Kallman, T.R. & Krolik, J.H. 1993, (preprint)
Kippenhahn, R., Kohl, K. & Weigert, A. 1967, Zeits. für Astrophys. 66, 58
Kylafis, N.D. & Xilouris, E. 1993, A&A 278, L43

Livio, M. & Soker, N. 1988, ApJ 329, 764
Livio, M., Prialnik, D. & Regev, O. 1989, ApJ 341, 299
Long, K.S., Helfand, D.J. & Grabelsky, D.A. 1981, ApJ 248, 925
Meyer, F. & Meyer-Hofmeister, E. 1979, A&A 78, 167
Mukai, K. 1993, (private communication)
Nomoto, K. 1982, ApJ 253, 798
Osterbrock, D.E. 1989, *Astrophysics of Gaseous Nebulae* Univ. Sci. Books (Mill Valley)
Paczyński, B. 1965, Acta Astr. 15, 89
Paczyński, B. 1967, Acta Astr. 17, 193
Paczyński, B. 1970, Acta Astr. 20, 287
Paczyński, B. 1976, in *Structure and Evolution of Close Binary Systems*, IAU Symp. 73, P. Eggleton, S. Mitton & J. Whelan (Eds.), Reidel, p. 75
Pakull, M.W. & Motch, C. 1989, in *ESO Workshop on Extranuclear Activity in Galaxies*, E.J.A. Meurs & R.A.E. Fosbury (Eds.), p. 285
Pakull, M.W., Beuermann, K., Van der Klis, M. & Van Paradijs, J. 1988, A&A 203, L27
Prialnik, D. & Kovetz, A. 1994, (preprint)
Rappaport, S., Di Stefano, R. & Smith, J.D. 1994, ApJ 426, 492 (RDS)
Rappaport, S., Chiang, E., Kallman, T. & Malina, R. 1994, ApJ 431, 237
Remillard R., Rappaport, S. & Macri, L. 1994, ApJ (in press)
Sion, E.M. & Starrfield, S.G. 1986, ApJ 303, 130
Sion, E.M., Acierno, M.J. & Tomcszyk, S. 1979, ApJ 230, 832
Smale, A.P. et al. 1988, MNRAS 233, 51
Sparks, W.M. & Stecher, T.P. 1974, ApJ 188, 149
Taam, R.E. 1980, ApJ 242, 749
Taam, R.E., Bodenheimer, P. & Ostriker, J.P. 1978, ApJ 222, 269
Van den Heuvel, E.P.J. 1994, (private communication)
Van den Heuvel, E.P.J., Bhattacharya, D., Nomoto, K. & Rappaport, S.A. 1992, A&A 262, 97
Webbink, R. 1979, in *White Dwarfs and Variable Degenerate Stars*, IAU Colloquium 53, H. Van Horn & V. Weidemann (Eds.), University of Rochester Press, p. 426
Webbink, R.F. 1985, in *Interacting Binary Stars*, J.E. Pringle & R.A. Wade (Eds.), Cambridge Univ. Press, p. 39
Webbink, R. 1992, in *X-Ray Binaries and Recycled Pulsars*, E.P.J. Van den Heuvel & S. Rappaport (Eds.), Kluwer Academic Publishers, p. 269

SUPERSOFT ROSAT SOURCES IN THE GALAXIES

P. KAHABKA
Astronomical Institute 'Anton Pannekoek, University of Amsterdam, and Center for High-Energy Astrophysics, Kruislaan 403, 1098 SJ Amsterdam, The Netherlands

AND

J. TRÜMPER
Max-Planck-Institut für extraterrestrische Physik D-85740 Garching, FRG

1. Introduction

Before the X-ray surveys performed with EINSTEIN and ROSAT soft X-ray (or EUV) sources were claimed to exist (Iben 1982; Fujimoto 1982); they were looked for in the ultraviolet and indeed such sources were found in the symbiotic systems (which count as CV-like systems) and termed the *hot component* of symbiotics. Although the nature of these hot components has been subject to debate (in terms of either nuclear burning white dwarfs or accretion phenomena) observational facts were in many systems favouring the first scenario (Mikolajewska & Kenyon 1992). Symbiotic binaries require wide orbits in order to keep the big giant star within its Roche lobe. It is natural to look for the short-period counterpart, but it was much more difficult to detect them in the optical due to the faintness of the secondary, which is supposed to be an evolved main-sequence star or even smaller. What turns out in these systems to be predominant is the much brighter accretion disk. It was the unique chance of the satellite borne X-ray imaging instruments to discover these EUV and soft X-ray sources and with the EINSTEIN observatory the first firm candidates were found. However, complete coverage of the soft X-ray sky was needed in order to get them all and ROSAT was the instrument which mapped the whole sky.

Before describing the definition of *supersoft* sources applied to select the sample, we will outline the currently most successful model to explain the phenomenon of supersoft sources, i.e., steady nuclear burning on white dwarfs (Van den Heuvel et al. 1992). The question, whether supersoft systems comprise a homogenous class or not has not been finally answered: they include close and wide binaries and single stars as well. But two features may be common, a white dwarf (WD) as the hot and compact object, and steady nuclear burning occurring in the envelope of the WD. Stable burning of hydrogen occurs within a narrow range of accretion rates $3.4 \, 10^{-7} f \leq \dot{M} \, (M_\odot \, \text{yr}^{-1}) \leq 8.5 \, 10^{-7} f$, dependent on the mass of the WD by $f = M_{WD}[M_\odot] - 0.52$. The required accretion rates are supplied in short-period ($P \sim 70^m - 2^d$) binaries by Roche lobe overflow from a donor star (1.4–$2.2 \, M_\odot$) more massive than the WD (0.7–$1.2 \, M_\odot$) (Van den Heuvel et al. 1992) and in long-period ($P \sim 100$ days–few years) binaries from a low-mass giant (1–$3 \, M_\odot$) to a WD (~ 0.5–$0.7 M_\odot$) (De Kool et al. 1986; Sion & Starrfield 1994). From Fujimoto's (1982) theory of hydrogen shell flashes on accreting WDs the main observational features of such supersoft sources become obvious: there is a lower limit to the mass of the envelope of a WD for which the stability criterium is fulfilled, ranging from $\sim 10^{-4}$ to $\sim 10^{-7} \, M_\odot$ for M_{WD} between $\sim 0.4 \, M_\odot$ and $1.4 \, M_\odot$; the corresponding upper limits are $\sim 3 \, 10^{-4}$ to $\sim 10^{-6} \, M_\odot$. As hydrogen is burnt at a rate $(1$–$4) \, 10^{-7} \, M_\odot \, \text{yr}^{-1}$, this envelope will be burnt within ~ 0.25–1 years for a $1.4 \, M_\odot$ WD, and within ~ 100–2000 years for a $0.5 \, M_\odot$ WD. Such time scales define systems, in which the envelope is refreshed at an accretion rate below the stable accretion rate of hydrogen burning of ~ 1–$4 \, 10^{-7} \, M_\odot \, \text{yr}^{-1}$. If the accretion rate is within the critical range then the lifetime of the system will depend on the thermal time scale of the donor star (cf., Paczynski 1971; Van den Heuvel et al. 1992). At very low rates ($\leq 10^{-9} \, M_\odot \, \text{yr}^{-1}$) nova (like) systems form with ejection of part of the envelope during the nova explosion, drastically reducing the above estimated time scales. Systems accreting at rates in excess of the stability rate may undergo a limit cycle as described in Van den Heuvel et al. (1992) of duration ~ 100 years with visibility in X-rays for ~ 10 years.

2. Definition

When the EINSTEIN X-ray observatory performed a survey of the LMC in soft X-rays (Long et al. 1981) it became obvious that two sources had extremely soft spectra. These sources are numbers 83 and 87 in the Columbia Astrophysics Laboratory catalog, i.e., CAL 83 and CAL 87. A similar soft X-ray survey has been performed of the SMC by Seward & Mitchell (1981). They discovered two sources which were later on found to be supersoft, but at the time the nature of these objects was not yet recognized. In a later comprehensive analysis of the SMC observations by Wang & Wu (1992) these sources were found to be the softest in the sample. The softness was defined in terms of a hardness ratio which in the work of Wang & Wu is defined as $Q = (H - S)/(H + S)$ with S and H the countrates in the soft (0.16–0.8 keV) and hard (0.8–3.5 keV) energy band, respectively. Long et al. (1981) used somewhat different energy bands ($S = 0.15$–1.5 keV, $H = 1.5$–4.5 keV) but this does not affect the general trends. From the analysis of Long et al. (1981) it became clear, that different source types occupy different ranges in hardness ratios Q, -1.0 to 0.0 for supernova remnants, 0.0 to 1.0 for binaries, -0.3 to 0.3 for AGNs. As the LMC supersoft sources had $Q = -0.9$ and -0.7, respectively, it becomes obvious that they form a separate class of objects.

About 10 years after the EINSTEIN observations a complete all-sky-survey in soft (0.1–2.4 keV) X-rays was made with ROSAT; it included the fields of the LMC and SMC with a high sensitivity (Trümper et al. 1991; Pietsch & Kahabka 1993; Kahabka & Pietsch 1993). More than 500 X-ray sources were found in a $12° \times 12°$ field centered on the LMC and 72 sources in an $8° \times 8°$ field covering the SMC. Several additional pointed observations towards the Magellanic Clouds (MCs) with considerably deeper exposures complement the survey observations. A hardness ratio criterium has been applied to select candidates for supersoft sources: $HR1 \pm \sigma(HR1) \leq -0.8$, with $HR1 = (H - S)/(H + S)$ and $S = 0.1$–0.4 keV, $H = 0.5$–2.1 keV. Six sources have since been found in the LMC and four in the SMC including the 4 sources known from the EINSTEIN observations (cf., Fig. 1 and Table 1). It was recognized, that the extremely soft spectral appearance alone did not yet make a new class of objects as galactic WDs have similar soft X-ray spectra (Fleming et al. 1991), and a few AGNs look very soft as well. A second criterium was introduced to better constrain the characteristics of this class, the luminosity criterium. It was claimed, that supersoft sources radiate at or close to the Eddington limit of a 1 $M_\odot$ compact object ($\sim 1.3 \ 10^{38}$ erg s^{-1}). This excluded any weak galactic foreground object and luminous background sources. Initially, blackbody spectra were applied to the X-ray data; the inferred (bolometric) luminosities were very uncertain, because of the large bolometric corrections, and the substantial galactic absorption along the line of sight. Also, in most cases the distance to the source was unknown. In case of the MCs the distance is well known and the extent of the Clouds was assumed to be small compared to the distance. However, due to the considerable angle of the sky covered by the Clouds a few galactic WDs or CVs could have a chance superposition. A third criterium necessarily had to be fulfilled to uniquely classify a source as supersoft: an optical counterpart had to be identified which showed the signatures of high excitation (He II $\lambda 4686$, Hα, Hβ) lines, known from the optical spectra of LMXBs (Van Paradijs 1983) and understood to originate in an irradiated accretion disk.

3. The Magellanic Clouds

The *first light* observation of ROSAT was dedicated to discover X-ray emission from supernova 1987A. While the supernova was not seen, two supersoft sources CAL 83 and RX J0527.8−6954 were detected, the latter for the first time (Trümper et al. 1991). CAL 83 is the prototype of supersoft sources. It shows three periodicities, a 1.04 day orbital period, occasional flux variability at ~ 2 h (Smale et al. 1988) and a moving feature in the broad wings of the He II 4686 line with a recurrence time of ~ 69 d possibly associated with a precessing disk or a jet (Crampton et al. 1987). The optical spectrum shows a strong He II 4686 line with a flux of $\sim 4 \; 10^{33}$ erg s^{-1} (Smale et al. 1988), two orders of magnitude more luminous than in galactic LMXBs. An intrinsic X-ray luminosity of $5 \; 10^{39}$ erg s^{-1} was estimated, significantly larger than the Eddington limit for accretion onto a 1.4 M$_\odot$ compact object (Smale et al. 1988). Extended emission in O III is seen surrounding the source. This nebula has a shell-like structure and is most likely ionized by X-rays (Pakull & Angebaut 1986). Crampton et al. discuss the scenario of a massive post-AGB star collapsing into a WD which has lost the H-rich envelope in a close binary system. However, they argue that the observed fast variability indicates that the system is an interacting binary with mass transfer going on via an accretion disk, and they favour the idea that it is an unusual LMXB system in the LMC. The low X-ray luminosity observed in the EINSTEIN band (0.15–4.5 keV) of $\sim 3.2 \; 10^{36}$ erg s^{-1} and the low ratio of X-ray to optical luminosity $L_X/L_{opt} \sim 0.7$ (compared to values of $\sim 10^2$–10^3 for LMXBs) led Cowley et al. to classify CAL 83 as an ADC system similar to X 1822−371 and X 0921−630 (White & Holt 1982; Mason et al. 1987). The observed flux would then be due to X rays scattered by an extended disk corona. But it remained unclear how such a corona can generate the observed soft spectrum. The X-ray spectra of CAL 83 measured with ROSAT revealed the very soft ($kT_{bb} \sim 20$–40 eV) and luminous ($L_{bol} \sim L_{Edd}$ for a 1 M$_\odot$ compact object) nature of the emission (Greiner et al. 1991). Such a soft spectral distribution was hard to be explained by an ADC, and other scenarios like an optically thick cocoon engulfing a NS or near-Eddington accretion onto a NS were invoked (Greiner et al. 1991; Kylafis & Xilouris 1993). However, Van den Heuvel et al. (1992) demonstrated in a self-consistent way how the manifold observational facts could be reconciled with a ~ 0.7–1.2 M$_\odot$ WD accreting from an evolved main-sequence star transferring mass unstably on a thermal time scale (cf., Pylyser & Savonije 1988) with steady nuclear burning occurring on the WD surface. Heise et al. (1994) and Van Teeseling et al. (1994) have shown that WD atmosphere spectral fits have to be applied to the X-ray data. Consistency could be obtained with a steady-state burning WD for

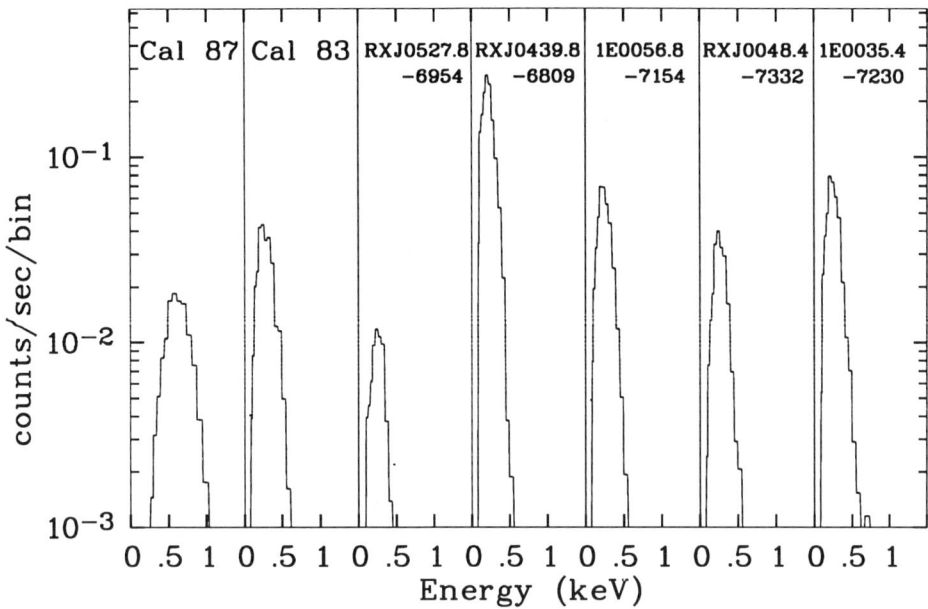

Figure 1. ROSAT PSPC spectra of the LMC & SMC supersoft sources.

CAL 83 (cf., Van Teeseling et al., these proceedings) in full agreement with the model of Van den Heuvel et al. CAL 83 was stated to be a prototype of the supersoft sources and the other members may be of a similar nature (as long as the supersofts form a homogenous class). We will further discuss the other MC systems, and concentrate on the X-ray observations as the optical observations are covered by the contribution of A. Cowley to these proceedings.

CAL 87 is another outstanding supersoft source in the LMC. A binary orbital period of 10.6 hr could be established (Callanan et al. 1989; Cowley et al. 1990). The system shows optical (Pakull et al. 1988; Cowley et al. 1990) and X-ray eclipses (Schmidtke et al. 1993; Kahabka et al. 1994) indicative of a high inclination (70°–90°). The primary eclipse shows asymmetry in ingress with variability from cycle to cycle similar to pre-eclipse variations known from LMXBs. The observed (0.5–2.4 keV) X-ray luminosity is low ($\sim 10^{36}$ erg s^{-1}) but has to be high ($\sim 10^{38}$ erg s^{-1}) in order to explain the X-ray heating of the secondary seen in the optical light curve (Van den Heuvel et al. 1992). An optically thick disk has been invoked to shield the X-ray source. The X-ray spectrum was poorly constrained by EINSTEIN observations (cf., Brown et al. 1994). ROSAT observations showed that the X-ray source has a very soft spectrum ($kT_{bb} \sim 30$–55 eV), is luminous ($\geq 5\ 10^{38}$ erg s^{-1}) and suffers a large absorption ($N_H \sim 10^{22}$ cm^{-2}

(Schmidtke et al. 1993; Kahabka et al. 1994) which probably cannot be accounted for by galactic foreground ($\sim$8 10^{20} cm^{-2}) and LMC intrinsic ($\sim$3 10^{21} cm^{-2}) absorption. However, WD atmosphere spectral fits are consistent with CAL 87 being located at the far side of the LMC. Then the absorption contribution by an accretion disk would be minor. This raises the question, whether the X-ray source is extended or not. The WD model atmosphere fits are consistent with the X-ray emission originating from a region of the size of the surface of a high-mass steady-state nuclear burning WD. On the basis of radial-velocity variations of the He II 4686 line it has been suggested that CAL 87 harbours a black hole (Cowley et al. 1990); however, the origin of the line more probably is connected with a strong wind from the optical star (cf., Van den Heuvel et al. 1992), arguing against the black-hole nature.

ROSAT observations added four further supersoft sources to the LMC sample. RX J0513.9–6951 was discovered during the ROSAT all-sky survey (Schaeidt et al. 1993) and was observed for 22 days. Within a few days the count rate increased by a factor of $\sim$20 and oscillated on a time scale of several days (cf., Fig. 2). The source was not detected in calibration data taken 4 months before, confirming the transient nature of the source. Monitoring of the source every three months revealed a second outburst about 1 year and 9 month after the first outburst. This makes RX J0513.9–6951 the first recurrent supersoft X-ray transient (Hasinger 1994). The X-ray spectrum is very soft ($kT_{bb} \sim 40$ eV) and the bolometric luminosity is high ($L_{bol} \sim 2.3\ 10^{38}$ erg s^{-1}) (Schaeidt et al. 1993). WD atmosphere spectral fits are consistent with a steady-state nuclear burning WD (Van Teeseling et al. 1994). Optical and IUE observations of RX J0513.9–6951 (Pakull et al. 1993) imply the presence of a luminous ($\sim$3 10^{37} erg s^{-1}) accretion disk indicating a very high mass transfer rate ($\dot{M} \gg 10^{-8}$ M$_\odot$ yr^{-1}). Such rates occur on a thermal time scale from a donor star that is more massive than the WD (Paczynski 1971; Van den Heuvel et al. 1992). The outburst was not accompanied by optical brightening of the companion star and therefore appears to be fundamentally different from that of X-ray novae. Pakull et al. proposed that the outburst corresponds to an episode of reduced mass transfer that lead to a considerable contraction of the WD envelope, decreasing the radius and increasing the temperature from <20 eV (most radiation unobservable) to 40 eV. The binary nature of the source is indicated by radial-velocity variations of the optical emission lines.

RX J0527.8–6954 was discovered during the *first light* observation of ROSAT of the LMC (Trümper et al. 1991). The source is soft ($kT_{bb} \sim 30$ eV), luminous ($L_{bol} \sim L_{Edd}$) (Greiner et al. 1991) and transient (Trümper et al. 1991), as it was not seen 10 years before with EINSTEIN. Orio & Ögelman (1993) found the source intensity to have decreased by a factor of

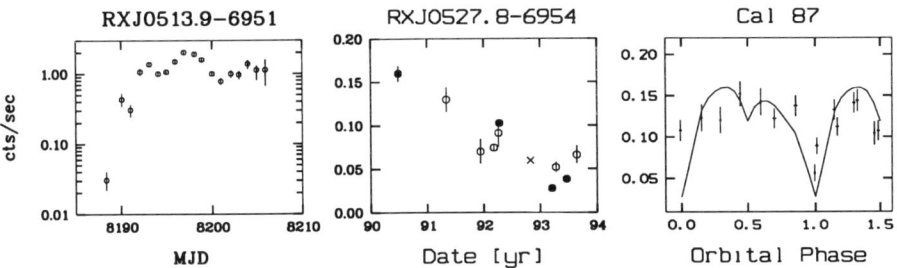

Figure 2. X-ray light curves of RX J0513.9−6951 (from Schaeidt et al. 1993), RXJ 0527.8−6954 (Hasinger 1994) and CAL 87 (Kahabka et al. 1994).

4 two years after the ROSAT observations. A complete record of the source intensity spanning a period of ∼3.5 years covering the all-sky survey and numerous dedicated and serendipitous observations (cf., Fig.2) revealed an almost linear decrease of the count rate by a factor of ∼4 (Hasinger 1994). This gives the source the appearance of a slow nova. Any optical counterpart has to be fainter than 17^m (Hasinger 1994).

In the SMC two supersoft sources were known from EINSTEIN observations. 1E 0056.8−7154 coincides within $4''$ with the bright $V = 16.6$ nebula N67 in the SMC (Seward & Mitchell 1981; Wang 1991). The extremely soft spectrum and strong emission were explained by the radiation from the surface of the central star of N67 which is evolving into a WD. In a typical evolutionary model for a planetary nebula the duration is less than 10^3 years. The deduced temperature ($T_{\text{eff}} \sim 3 \pm 1 \times 10^5$ K) and luminosity ($L_{bol} \sim 4\ 10^{38}$ erg s^{-1}), although quite uncertain, place the source at the location of a high-mass ($\sim 0.9\,M_\odot$) WD in the Hertzsprung-Russell diagram of planetary nebula nuclei (Wang 1991). IUE observations of the nebula show strong lines of NV and HeII indicating high excitation by a hot nucleus (Aller et al. 1987). ROSAT observations of 1E 0056.8−7154 confirm the presence of a hot ($T_{bb} = 1.6$–$4\ 10^5$ K) and luminous ($\sim 2\ 10^{38}$ erg s^{-1}) X-ray source (Kahabka et al. 1994). This shows that the source temperature and luminosity have remained constant for ∼10 years, in agreement with the track of a planetary nebula nucleus. Heise et al. (1994) have shown, that the luminosity of 1E 0056.8−7154 is overestimated by a factor of ∼10 if a blackbody description is used. They applied WD atmosphere model fits with log g=8 to the data and deduce a temperature of $4.5\ 10^5$ K and a luminosity of $2\ 10^{37}$ erg s^{-1}. Although a single hot nucleus explains the observations in a consistent way, a possible close binary system nature of the central source of N67 has been considered (cf., Iben & Tutukov 1993) but no decision on this can be drawn from the X-ray data alone. A search for a weak optical counterpart and/or an accretion disk in the hypothetical binary system should be initiated.

1E 0035.4–7230 is the second source in the SMC known from EINSTEIN observations (Seward & Mitchell 1981). The source appeared point like in the HRI image, an extremely soft spectrum with $T_{\rm eff} \leq 4\ 10^5$ K and a luminosity in the EINSTEIN band of $\sim 10^{37}$ erg s^{-1} were deduced. ROSAT observations confirm the source to be soft ($T_{\rm bb} \sim 4\ 10^5$ K) and luminous ($\sim 2\ 10^{37}$ erg s^{-1}) (Kahabka et al. 1994). The source appears not to have varied drastically since the EINSTEIN observations. From WD atmosphere model fits it follows that 1E 0035.4–7230 is consistent with a steady-state nuclear burning WD (Van Teeseling et al. 1994). A possible optical counterpart at $V \sim 21$ was claimed by Jones et al. (1985). Orio et al. (1993) report about the identification of 1E 0035.4–7230 with a variable star of magnitude $B = 19.9$–20.2. The star shows strong UV excess, a blue continuum and weak lines of high excitation. A variation of ~ 0.2 mag within ~ 2 h of observations was detected and a possible eclipse event was considered. This would indicate that the source is a binary with an orbital period of order hours to tens of hours. The existence of two possible eclipse events in this source has been claimed from X-ray data (Kahabka et al. 1994) and an upper limit of 30 d was estimated for an orbital period. 1E 0035.4–7230 has a similar optical brightness as CAL 87, and if eclipsing may be seen under a similar inclination. The high absorption seen in the X-ray data of CAL 87 is absent in 1E 0035.4–7230.

Two additional SMC sources were found with ROSAT. RX J0048.4–7332 was discovered in the ROSAT all-sky-survey data (Kahabka & Pietsch 1993) and monitored with several pointed SMC observations (Kahabka et al. 1994). The count rate remained essentially constant for ~ 2 years. The source may have been missed in the EINSTEIN survey of the SMC as it lies in an unexposed small region between two fields (Wang & Wu 1992). Blackbody fits yield $kT_{\rm bb} \sim 10$–40 eV and $L_{\rm bol} \geq 0.2\,L_{\rm Edd}$ (for a 1 M$_\odot$ compact object). The spectrum shows substantial absorption additional to the galactic foreground which may either be due to the SMC or related to the system itself (Kahabka et al. 1994). The X-ray source coincides with the symbiotic star SMC 3 which is located in, or in front of, the SMC cluster NGC 269. The optical star is of spectral type M0. The hot component had an outburst sometime between 1980 December and 1981 November, increasing by 3 mag in the U band. In the infrared band no outburst was detected. The intensity of the hot component has stayed nearly constant since this outburst (Morgan 1992). Recent IUE observations of SMC 3 (Vogel & Morgan 1994) show an enrichment of N similar to galactic novae, which was related to a (possible) recent thermonuclear event connected with the optical outburst. A pronounced Si overabundance deduced from the IUE spectrum was explained by collisional ionization due to a high-density ($N_e \geq 10^9$ cm^{-3}) symbiotic nebula in the system (Vogel & Morgan 1994; Nussbaumer & Stencel 1987). Such a nebula of size ~1 AU would result in a line of sight absorbing column in agreement with the column of ~2–3 10^{21} cm^{-2} deduced from the X-ray observations. RX J0048.4–7332 is most likely a symbiotic (slow) nova. This would imply a low-mass (~0.5–0.7 M$_\odot$) nuclear burning WD. RX J0048.4–7332 is the first and only symbiotic nova seen in X-rays and identified in the Magellanic Clouds. As the giant M star in such systems is easily detectable in the optical any similar system might already have been found.

RX J0058.6–7146 is another ROSAT discovery (Kahabka et al. 1994). The source was seen in outburst during one pointed observation. Several additional observations before and after the outburst (Kahabka et al. 1994; Hasinger 1994) did not reveal the source; it varies by at least a factor of 50. During the detection observation the source luminosity increased from below the detection limit to ~3 10^{36} erg s^{-1} within ~2 days (Kahabka et al. 1994). Unfortunately, a possible further increase of the source flux could not be followed as the observation ended. A very soft (~40 eV) and absorbed (~4 10^{20} cm^{-2}) spectrum was deduced. As no optical identification has yet been made SMC membership of the transient remains open. The source resembles the low-luminosity globular-cluster X-ray source 1E 1339.8+2837 in M3 (Hertz, Grindlay & Bailyn 1993). A soft ($kT_{bb} \sim 20$ eV) outburst with a luminosity $L_{bol} \sim 1.2\ 10^{36}$ erg s^{-1} has been detected with ROSAT in this source, which may be a CV with steady nuclear burning of accreted material on the WD surface. Both the SMC and M3 globular cluster source may be recurrent EUV sources (Fujimoto 1982).

4. The Galaxy

As the supersoft X-ray sources have been recognized as a new class of objects which have been found in the MCs, it was natural to look for them in other galaxies and the closest galaxy is our Galaxy itself. A clear definition of this source type has been set up (cf., Section 2), the whole sky has been mapped in soft X-rays with ROSAT and a first version of a point source catalog with ~50,000 entries has been worked out (Voges 1992). A selection according to the hardness ratio criterium and a count rate threshold has been applied to the catalog and this subset has been further considered as input for an optical identification program (Greiner et al. 1994; Motch et al. 1994). Motch et al. started a galactic-plane survey within $b^{II} = \pm 20°$. As supersoft sources will be significantly absorbed due to the galactic column concentrated towards the galactic plane, Motch et al. found by simulations a somewhat different hardness ratio criterium to be appropriate, i.e., $HR1 \geq -0.4$ (for $N_H \geq 5\ 10^{20}$ cm^{-2}) and $HR2 \leq -0.8$ (for $N_H \leq 4\ 10^{22}$ cm^{-2}). An

TABLE 1. System parameters of supersoft sources

Name	log ($T_{\rm eff}$ [K])*	log ($L_{\rm bol}$ [$L_\odot$])*	log g*	binary/ single	$P_{\rm orb}$ [d]	variab. class
LMC						
RX J0439.8−6809	5.4–5.5	3.2–3.6	7.0–8.3	?		persistent?
RX J0513.9−6951	5.6–5.8	3.5–4.1	7.3–9.3	b?		recurrent
RX J0527.8−6954	5.5–5.9	2.0–3.9	7.0–9.9	b?		variable
CAL 83	5.6–5.8	3.0–4.1	7.3–9.9	b	1.04	persistent
CAL 87	5.9–6.0	3.5–4.4	8.3–9.9	b	0.44	eclipsing
RX J0550.0−7151				?		variable
SMC						
1E 0035.4−7230	5.6–5.8	3.0–3.5	7.5–9.8	b	0.1–2?	eclipsing
RX J0048.4−7332°	5.0–5.2	3.1–3.5		b	> 100?	slow nova
RX J0058.6−7146				b?		transient
1E 0056.8−7154	5.5–5.7	3.6–4.0	7.0–9.0	s?	−	persistent
Galaxy						
GQ Mus	∼5.5	∼4.4		b	0.059	post-nova
RX J0925.7−4758	5.7–5.8			b	3.5	
RX J0019+21	5.4–5.5	4.1–3.4		b	0.658	recurrent?
1E 1339.8+2837 (M3)	∼5.4	∼2.5		b?		recurrent?
RR Tel	5.2	3.5	≥6.5	b		slow nova
AG Dra				b	554	slow nova
RX J2117.1+3412	5.2	4.0	5.6–6.3	s	−	persistent

*) values for LMC & SMC sources from Van Teeseling et al. (1994), °) parameters determined from IUE observations (Vogel & Morgan 1994).

additional count rate criterium (≥ 0.1 cts s^{-1}) was applied and 98 candidates were found, from which all but 6 were rejected due to an association with a late-type star.

The brightest remaining candidate source RX J0925.7−4758 was selected for follow-up X-ray and optical studies. A ROSAT pointed observation showed the source to be soft ($kT_{\rm bb} \sim 30$–55 eV) and heavily absorbed ($N_{\rm H} \sim 1.3\ 10^{22}$ cm^{-2}). The large column indicates that RX J0925.7−4758 may be located behind the nearby Vela sheet molecular cloud. A maximum source distance of ∼2 kpc is deduced for a source luminosity of 2×10^{38} erg s^{-1}. Optical photometric monitoring of the 17.1 mag counterpart suggest a possible (orbital) period of 3.5 days. This source appears to be quite similar to

CAL 87 in the LMC (Motch et al. 1994). From the absence of further bright ($>0.27\,\mathrm{cts\,s^{-1}}$) candidates in the galactic plane Motch et al. conclude, that sources may be more concentrated to the galactic plane than the LMXBs which may show a wider distribution due to a kick obtained during the NS formation.

RX J0019+21 is the second close binary ($P_\mathrm{orb} = 15.8\,\mathrm{h}$) supersoft X-ray source discovered in the Galaxy (Reinsch et al. 1993). The optical/UV continuum is reminiscent of a bright accretion disk with strong He II and Balmer emission lines.

RR Tel is a symbiotic nova (Jordan et al. 1994) that started an outburst in 1944. Symbiotic novae are wide binaries (orbital periods 10^{2-3} days) with an evolved late-type star and a very hot companion, a WD undergoing nuclear burning (of hydrogen-rich matter) during outburst. They may be embedded in a dense nebula due to the cool giant's wind. The outbursts last of the order of decades, i.e., much longer than classical novae. The giant in RR Tel is of spectral type M5 and a Mira variable with heavy mass loss ($\dot{M} \approx 5\,10^{-6}\,M_\odot\,\mathrm{yr}^{-1}$). The UV spectrum indicates $T_\mathrm{eff} \geq 140,000\,\mathrm{K}$ and $\log g \geq 6.5$ for the WD; the X-ray data are consistent with $T_\mathrm{eff} = 142,000\,\mathrm{K}$ and $L = 1.3\,10^{37}\,\mathrm{erg\,s^{-1}}$. An additional hard X-ray component of low luminosity ($\sim 10^{33}\,\mathrm{erg\,s^{-1}}$) may be due to hot plasma from a colliding wind in the interacting binary (Jordan et al. 1994).

GQ Muscae is the first classical nova detected in X-rays (with EXOSAT) during outburst (Ögelman et al. 1987). ROSAT observations 9 years after the outburst revealed a soft X-ray source (Ögelman et al. 1993). Assuming Eddington luminosity for a $1\,M_\odot$ WD a temperature of $\sim 3.5\,10^5\,\mathrm{K}$ was deduced. It was concluded, that GQ Mus is burning hydrogen-rich matter. As the other 25 recently detected novae were not detected in X rays GQ Mus either has not ejected its envelope during outburst or is burning recently accreted material (Ögelman et al. 1993).

The probably recurrent EUV source 1E 1339.8+2837 in M3 has been discussed together with RX J0058.6–7146 in the SMC which may be of a similar nature.

5. M31 and other galaxies

A soft X-ray (0.1–2.4 keV) survey of M31 has been performed with the ROSAT PSPC in July 1991 (Supper et al. 1994). A total of 396 X-ray sources have been detected with luminosities from 10^{35} to $3\,10^{38}\,\mathrm{erg\,s^{-1}}$ (690 kpc). 15 sources were considered to belong to the class of supersoft sources. For the brightest source (# 309) a blackbody fit was applied, and a temperature $kT \sim 30\,\mathrm{eV}$ and a luminosity $L_\mathrm{bol} \sim 10^{38}\,\mathrm{erg\,s^{-1}}$ deduced. From evolutionary calculations (Rappaport et al. 1994; Rappaport, these proceedings) ~1000 supersoft sources are expected to exist in M31, but obviously most of them will be highly absorbed.

A systematic search for supersoft sources in other nearby galaxies has not yet been completed. There may be detectable candidates in M101, NGC 253 and probably M33.

6. Conclusions

Of the 17 supersoft systems observed in the MCs and in the Galaxy 4 have been identified as close binaries with orbital periods of 0.4–3.5 days (cf., Van den Heuvel et al. 1992), 2 further sources (RX J0513.9–6951 and 1E 0035.4–7230) may turn out to belong to the same class, 3 supersoft systems are symbiotic (wide) binaries, and 2 are hot central stars of PNe or PG1159 stars. Among the 6 remaining sources are post-novae (1) and recurrent EUV sources (3). The two LMC sources RX J0439.8–6809 and RX J0550.0–7151 can presently not yet be classified. This gives a ratio of close to wide binaries of 1–2 and shows the importance of the close binary systems contributing to the supersoft systems. The ratio of binary to single stars is 4–7 and may reflect the longer lifetime of binaries compared to planetaries.

The question arises whether there are (predominantly) helium accreting systems among the supersoft sources as discussed in Iben & Tutukov (1993, 1994). Such systems are expected to have orbital periods above 10 min (cf., Savonije et al. 1986) and may be characterized by strongly suppressed hydrogen Balmer lines (Hα and Hβ) in the optical spectra of the accretion disks in these systems. In a few systems (CAL 87 and RX J0925.7–4758) Hβ is missing (Pakull et al. 1988; Cowley et al. 1990; Motch et al. 1994).

Supersoft sources have been discussed as progenitors of type Ia supernovae and as systems that eventually may undergo accretion induced collapse (AIC). Livio (1994) concludes that supersoft systems with CO degenerates of mass 0.7–1.2 M$_\odot$ will, if they grow beyond the Chandrasekhar limit due to accretion, experience carbon deflagration and hence a type Ia supernova event, but they do not undergo AIC (initial white-dwarf masses in excess of 1.2 M$_\odot$ are required). As type Ia supernovae comprise a rather inhomogeneous class (cf., Della Valle & Livio 1994), their progenitors may be found among the supersofts, and recurrent novae but also among WD mergers. In late-type galaxies CV-type supersofts and recurrent novae are favoured, and in early-type galaxies double degenerates (Della Valle & Livio 1994). It is interesting to note, that supersofts in which the WD does not grow beyond the Chandrasekhar limit, will become double degenerates (cf., Iben & Tutukov 1984).

Acknowledgements. P. Kahabka is a Human Capital and Mobility fellow.

References

Aller, L.H. & Keyes, C.D. 1987, ApJ 320, 159
Brown, T., Cordova, F., Ciardullo, R. et al. 1994, ApJ 422, 118
Callanan, P.J., Machin, G., Naylor, T. et al. 1989, MNRAS 241, 37P
Cowley, A.P., Schmidtke, P.C., Crampton, D. et al. 1990, ApJ 350, 288
Crampton, D., Cowley, A.P., Hutchings, J.B. et al. 1987, ApJ 321, 745
De Kool, M., Van den Heuvel, E.P.J. & Rappaport, S.A. 1986, A&A 164, 73
Della Valle, M. & Livio, M. 1994, ApJ 423, L31
Fleming, T.A., Schmitt, J.H.M.M., Barstow, M.A. & Mittas, J.P.D. 1991, A&A 246, L47
Fujimoto, M.Y. 1982, ApJ 257, 767
Greiner, J., Hasinger, G. & Kahabka, P. 1991, A&A 246, L17
Greiner, J., Hasinger, G. & Thomas, H.-C. 1994, A&A 281, L61
Hasinger, G. 1994, in: *The Evolution of X-Ray Binaries*, AIP Conference Proceedings 308, S.S. Holt & C.S. Day (Eds.), p. 611
Heise, J., Van Teeseling, A. & Kahabka, P. 1994, A&A 288, L45
Hertz, P., Grindlay, J.E. & Bailyn, C.D. 1993, ApJ 410, L87
Iben, I. 1982, ApJ 259, 244
Iben, I. & Tutukov, A.V. 1984, ApJS 54, 335
Iben, I. & Tutukov, A.V. 1993, ApJ 418, 343
Iben, I. & Tutukov, A.V. 1994, ApJ 431, 264
Jones, L.R., Pye, J.P., McHardy, I.M. et al. 1985, Sp. Sci. Rev. 40, 693
Jordan, S., Mürset, U. & Werner, K. 1994, A&A 283, 475
Kahabka, P. & Pietsch, W. 1993, in: *New Aspects of Magellanic Cloud Research*, Lecture Notes in Physics, Vol. 416, B. Baschek, G. Klare & J. Lequeux (Eds.), p. 71
Kahabka, P., Pietsch, W. & Hasinger, G. 1994, A&A 288, 538
Kylafis, N.D. & Xilouris, E.M. 1993, A&A 278, L43
Livio, M. 1994, in: *Proceedings of the Aspen Conference on Millisecond Pulsars: A Decade of Surprise*, (in press)
Long, K.S., Helfand, D.J. & Grabelsky, D.A. 1981, ApJ 248, 925
Mason, K.O., Branduardi-Raymont, G., Cordova, F.A. et al. 1987, MNRAS 226, 423
Mikolajewska, J. & Kenyon, S.J. 1992, MNRAS 256, 177
Morgan, D.H. 1992, MNRAS 258, 639
Motch, C., Werner, K. & Pakull, M.W. 1993, A&A 268, 561
Motch, C., Hasinger, G. & Pietsch, W. 1994, A&A 284, 827
Nussbaumer, H. & Stencel, R.E. 1987, in: *Exploring the Universe with the IUE Satellite*, Y. Kondo (Ed.), Reidel (Dordrecht), p. 203
Ögelman, H., Krautter, J. & Beuermann, K. 1987, A&A 177, 110
Ögelman, H., Orio, M. & Krautter, J. 1993, Nat 361, 331
Orio, M. & Ögelman, H. 1993, A&A 273, L56
Orio, M., Della Valle, M., Massone, G. et al. 1994, A&A 289, L11
Paczynski, B. 1971, ARA&A 9, 183
Pakull, M.W. & Angebaut, L.P. 1986, Nat 322, 511
Pakull, M.W., Beuermann, K., Van der Klis M. et al. 1988, A&A 203, L27
Pakull, M.W., Motch, C., Bianchi, L. et al. 1993, A&A 278, L39
Pietsch, W. & Kahabka, P. 1993, in: *New Aspects of Magellanic Cloud Research*, Lecture Notes in Physics, Vol. 416, B. Baschek, G. Klare & J. Lequeux (Eds.), p. 59
Pylyser, E. & Savonije, G.J. 1988, A&A 208, 52
Reinsch, K., Beuermann, K. & Thomas, H.-C. 1993, Astronomische Gesellschaft Abstract Series No. 9, 41
Rappaport, S., Di Stefano, R. & Smith, J.D. 1994, ApJ 426, 692
Rappaport, S., Chiang, E., Kallman, T. et al. 1994, ApJ 431, 237

Savonije, G.J., De Kool, M. & Van den Heuvel, E.P.J. 1986, A&A 155, 51
Schaeidt, S., Hasinger, G. & Trümper, J. 1993, A&A 270, L9
Schmidtke, P.C., McGrath, T.K., Cowley, A.P. et al. 1993, PASP 105, 863
Seward, F.D. & Mitchell, M. 1981, ApJ 243, 736
Sion, E.M. & Starrfield, S.G. 1994, ApJ 421, 261
Smale, A.P., Corbet, R.H.D. & Charles, P.A. 1988, MNRAS 233, 51
Supper, R., Hasinger, G., Pietsch, W. et al. 1994, (preprint)
Trümper, J., Hasinger, G., Aschenbach, B. et al. 1991, Nat 349, 579
Van den Heuvel, E.P.J., Bhattacharya, D., Nomoto, K. et al. 1992, A&A 262,97
Van Paradijs, J. 1983, in: *Accretion-driven Stellar X-ray Sources*, W.H.G. Lewin & E.P.J. van den Heuvel (Eds.), Cambridge University Press, p. 189
Van Teeseling, A., Heise, J. & Kahabka, P. 1994, (preprint)
Vogel, M. & Morgan, D.H. 1994, A&A 288, 842
Voges, W. 1992, in: *Space Sciences with particular emphasis on High Energy Astrophysics*, Proc. European ISY Symposium, p. 223
Wang, Q. 1991, MNRAS 252, 47
Wang, Q. & Wu, X. 1992 ApJS 78, 391
White, N.E. & Holt, S.S. 1982, ApJ 257, 318

Discussion

V. M. Lipunov: Some supersoft superluminous sources (about several percent as determined from our calculations) must be related with super-accreting neutron stars. This is confirmed by the discovery of the supersoft source in the X-ray pulsar in the SMC by Hughes (1994, ApJ 427, L25).

R. Sunyaev: It is obvious that the spectrum of a WD with a steady burning hydrogen envelope must be very different from a black body. Therefore the luminosity might strongly differ from that estimated using the black body approximation. UV or even optical measurements might be useful for the comparison of the more realistic WD atmosphere radiation spectrum with the observed one.

SUPERSOFT X-RAY SOURCES IN THE LMC

A.P. COWLEY AND P.C. SCHMIDTKE
Dept. of Physics & Astronomy
Arizona State University
Tempe, AZ 85287, U.S.A.

AND

D. CRAMPTON AND J.B. HUTCHINGS
Dominion Astrophysical Obs.
5071 W. Saanich Rd.
Victoria, B.C., V8X 4M6, Canada

1. Introduction

Data obtained with ROSAT have established the "supersoft" X-ray sources (SSS) as a separate class of objects. Their unique features include temperatures of ~ 30 eV and luminosities of $\sim 10^{38}$ erg s^{-1}. Hasinger (1994) has recently summarized the properties of the SSS presently known in the Galaxy and the Magellanic Clouds (cf., Kahabka & Trümper 1995). The six which are thought to be LMC members are listed in Table 1 and discussed individually, based primarily on optical data we have obtained at CTIO.

2. Individual Systems

2.1. CAL 83

The prototype of the SSS class, CAL 83, was known as an extremely soft X-ray source from EINSTEIN data (Long *et al.* 1981), a decade before SSS were recognized as a class. The optical spectrum of CAL 83 shows emission lines (primarily He II and H) indicating the presence of a bright accretion disk. The system is a close binary with an orbital period of 1.04 days (Crampton *et al.* 1987; Smale *et al.* 1988; Cowley *et al.* 1991). Mass and luminosity determinations indicate the system is composed of a low-mass dwarf and a degenerate star, but neither star is seen in the optical spectrum.

TABLE 1. Supersoft X-ray Sources in the LMC

Source	V	Comments
CAL 83	~17.3	emission-line binary; P=1.04 days
CAL 87	~18.9	emission-line binary; eclipsing; P=0.44 days
RX J0439.8−6809	no id.; $B>19$	see Greiner et al. 1994
RX J0513.9−6951	~16.8	HV 5682; emission-line star
		1990 X-ray outburst, but no opt. increase
RX J0527.8−6954	no id.; $B>18$	probably not HV 2554; steadily declining X-rays
RX J0550.0−7151	no id.	possibly very blue, ~18th mag. star?

Crampton et al. concluded that the system probably contains a neutron star, however Van den Heuvel et al. (1992) have suggested that the compact star is an accreting white dwarf undergoing steady nuclear burning. Both Greiner et al. (1991) and Kylafis & Xilouris (1993) explain the SSS by near-Eddington accretion onto neutron stars.

Several spectral characteristics of CAL 83 are unusual. Evidence of very high ionization in the disk is revealed by O VI emission lines at $\lambda\lambda 3881$ & 3834. Even more unusual is the very broad emission seen near He II $\lambda 4686$, implying radial velocities up to ~$2300\,\mathrm{km\,s^{-1}}$. Crampton et al. (1987) have shown that this feature shifts from one side of the He II line to the other, with a velocity range of several thousand kilometers per second, in a time scale of months. They suggest an origin in a collimated outflow seen over a range of angles as the disk precesses, analogous to the jets in SS 433 but with lower velocities.

Additionally, CAL 83 is the only known SSS to show a resolved Hα and [O III] nebula, extending ~25 pc from the star (Pakull & Motch 1989; Remillard et al. 1994). This may indicate a previous period of extensive mass loss from the binary or interaction with the ISM.

2.2. CAL 87

CAL 87 is the only SSS presently known to eclipse (Pakull et al. 1988; Naylor et al. 1989; Cowley et al. 1990). The orbital period is 10.6 hours, with the primary optical eclipse lasting almost half the cycle. This indicates the object being occulted (the accretion disk) is very large. Cowley et al. concluded the system may contain a black hole, but Van den Heuvel et al. interpret the degenerate star as an accreting white dwarf. A brief, shallow X-ray eclipse suggests the X-ray emitting region is an accretion-disk corona which is only partially occulted (Schmidtke et al. 1993).

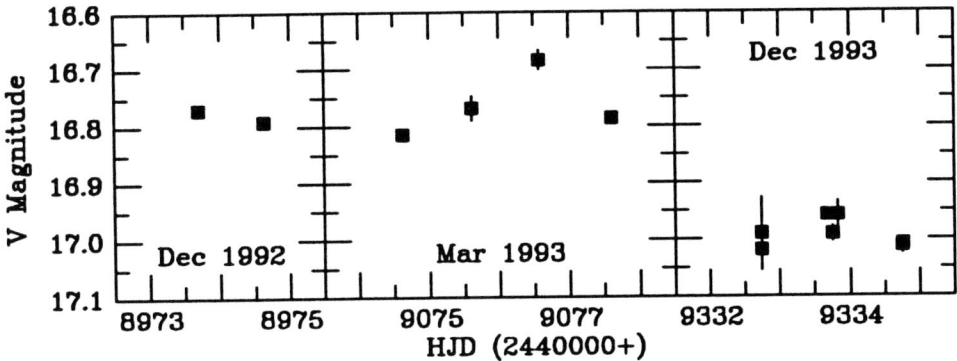

Figure 1. Optical light curve of RX J0513.9–6951 for three epochs.

CAL 87 is very faint (outside eclipse, $V \sim 19$), so that the optical spectra we obtained are low quality. They show a disk-dominated system with He II strongly in emission. In spite of the fact that the system eclipses (indicating a high inclination angle for the orbit), the emission lines exhibit only a low velocity amplitude. This was one factor which lead Cowley et al. to infer a large mass for the compact star.

2.3. RX J0439.8–6809

This source was discovered by Greiner et al. (1994) using ROSAT All-Sky-Survey data. X-ray spectral fitting gives $kT_{bb} = 20\,\mathrm{eV}$ for $N_H = 0.43\,10^{21}\,\mathrm{cm}^{-1}$. Within the error circle ($\pm 15''$) they found no suitable optical candidates brighter than 19th magnitude. Our only optical spectrum in this field is for star 'D' (see Greiner et al.) which lies well outside their error circle. It is a late B star in the LMC.

2.4. RX J0513.9–6951

This source was not detected with EINSTEIN but had a remarkable X-ray outburst in 1990 which was observed with ROSAT (Schaeidt et al. 1993). Although the X-ray count rate increased by >200 times, optical photometry shows the B magnitude did not brighten during the same period (Pakull et al. 1993). The star appears to be HV 5682 which historically has shown variations of about a magnitude (Leavitt 1908). Our photometry of RX J0513.9–6951 shows small changes (± 0.1 mag) on a night-to-night basis and ± 0.3 magnitude variations on a time scale of months (see Fig. 1). This is broadly in agreement with the behavior observed by Pakull et al. (1993).

Like CAL 83 and CAL 87, the optical spectrum of RX J0513.9–6951 is dominated by a luminous accretion disk with strong emission lines of He II

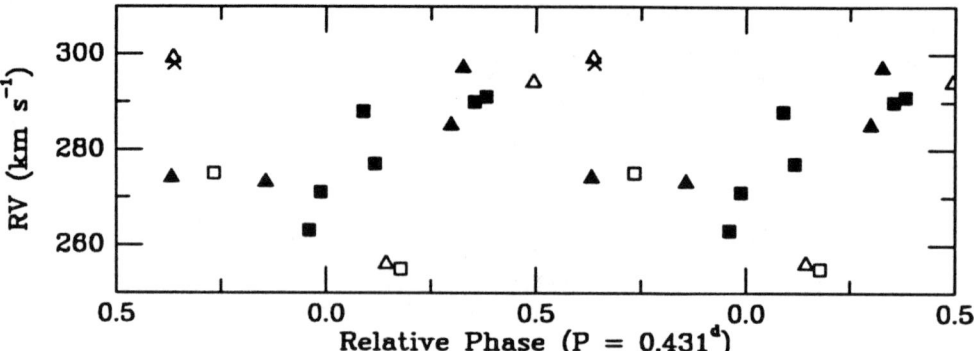

Figure 2. Radial velocities of RX J0513.9−6951 folded on a 0.431-day period. Different symbols represent data from five consecutive nights in Dec. 1993.

and H. Its spectrum bears many similarities to that of CAL 83, including relative line strengths, a broad emission near 4686 Å, and presence of O VI lines in the near ultraviolet (Pakull *et al.* 1993). A comparison of spectra of these two stars is shown in figure 2 of Cowley *et al.* (1993).

A series of 15 spectra obtained at CTIO in Dec. 1993 reveal ~ 40 km s^{-1} nightly velocity variations in the emission lines. Our data are insufficient to define the period uniquely, but $P = 0.43$ days gives a reasonable orbital fit. Fig. 2 shows the velocities folded on this period. More data are needed to better define the period, to rule out alias periods (e.g., 0.32 or 0.77 days), and to determine other system parameters.

Numerous weak, unidentified lines are present in most spectra. Some can be identified assuming they are high velocity components of He II and H, shifted by $\sim +4200$ and -3700 km s^{-1}. This implies a bipolar, collimated outflow may be present. Pakull (1994) has reached a similar conclusion based on entirely independent data. Recall that such a high velocity outflow may also be responsible for the changing broad He II wing in CAL 83. Thus, it is possible that such flows may be an additional characteristic of SSS.

2.5. RX J0527.8−6954

Trümper *et al.* (1991) discovered this supersoft source in a PSPC survey of the LMC. The X-ray luminosity is variable. It was not detected with EINSTEIN but was a prominent source when first observed with ROSAT. Its X-ray count rate has steadily declined since 1990 (Hasinger 1994). Although its X-ray position was shown on a finding chart (Cowley *et al.* 1993), this needs correction to take into account a small field rotation (0.4°) announced by the ROSAT group (Kuerster 1993) after publication of our paper. Our revised position is shown on the finding chart in Fig. 3 (marked

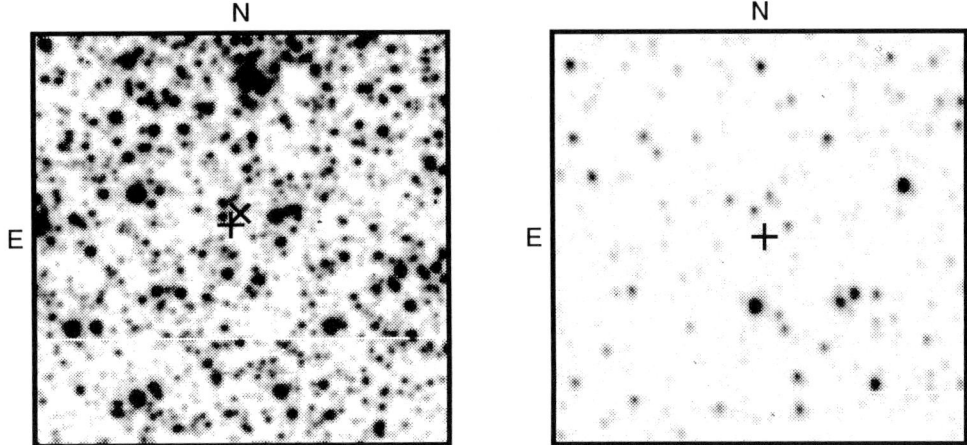

Figure 3. Finding charts for two SSS ($1.5' \times 1.5'$). Left: RX J0527.8–6954, '+' marks the position derived from a single HRI pointing and '×' marks the average position given by Hasinger (1994). Right: RX J0550.0–7151, '+' is the X-ray position from an off-axis PSPC observation.

by a '+'). Also shown is Hasinger's (1994) position (marked with a '×') which is based on an average of one HRI and four PSPC pointings. It is unlikely that HV 2554 (the easternmost star of the pair $\sim 10''$ west of the X-ray position) is the optical counterpart since it is too far from the new position. Furthermore, the spectrum of HV 2554 shows it is an ordinary A5 star with nothing to indicate it might be an X-ray source. Spectra of several other blue stars brighter than 19th magnitude near the X-ray position reveal normal B or A stars in the LMC and hence unlikely candidates. Thus, this source is still optically unidentified.

2.6. RX J0550.0–7151

The X-ray spectrum of RX J0550.0–7151 is very soft and similar to that of RX J0439.8–6809. A black-body fit to the ROSAT data gives $kT_{bb} \sim 32\,\text{eV}$. The position is based on an off-axis PSPC image, so it is probably only accurate to $\sim \pm 1'$ (larger than the field shown in the finding chart in Fig. 3!). Spectra were obtained for seven candidate blue stars in Dec. 1993. Most appear to be normal LMC early-type stars. One star in the field shows very broad lines and may be a galactic white dwarf. However, Beuermann (1994) reports that RX J0550.0–7151 is variable and was not detected in an HRI observation his group obtained. This makes it unlikely that the source could be a white dwarf. If the source brightens again, an HRI observation is essential to obtain an accurate X-ray position for making a more concentrated search for the optical counterpart.

3. Concluding Remarks

Three of the six supersoft X-ray sources in LMC have been identified with disk-dominated, close binaries whose absolute magnitudes range between $M_V \sim -1.6$ and $+0.5$. We note that the orbital periods for these systems are longer than the ones for typical low-mass X-ray binaries. Two of the systems (CAL 83 and RX J0513.9–6951) show evidence of high-velocity outflows or possibly collimated, precessing jets. optical counterparts for three systems have not been found, but they must be considerably fainter than those already identified. Not only is there a range in absolute magnitudes, but some of the X-ray spectra differ (e.g., X-ray emission in CAL 87 extends to much higher energies than in other SSS). The nature of the component stars in the supersoft sources is not yet resolved, but models include white dwarfs, neutron stars, or black holes as the collapsed star. It seems likely that the SSS include several types of systems, and more than one model may be needed to understand them.

References

Beuermann, K. 1994, private communication at IAU Symposium 165
Cowley, A.P., Schmidtke, P.C., Crampton, D. & Hutchings, J.B. 1990, ApJ 350, 288
Cowley, A.P. et al. 1991, ApJ 373, 228
Cowley, A.P. et al. 1993, ApJ 418, L63
Crampton, D. et al. 1987, ApJ 321, 745
Greiner, J., Hasinger, G. & Kahabka, P. 1991, A&A 246, L17
Greiner, J., Hasinger, G. & Thomas, H.-C. 1994, A&A 281, L61
Hasinger, G. 1994, MPE Preprint #286 for *Reviews in Modern Astronomy*
Kahabka, P. & Trümper, J. 1995, these Proceedings
Kuerster, M. 1993, ROSAT Status Report #67
Kylafis, N.D. & Xilouris, E.M. 1993, A&A 278, L43
Leavitt, H.S. 1908, Harvard Ann. 60, 87
Long, K.S., Helfand, D.J. & Grabelsky, D.A. 1981, ApJ 248, 925
Naylor, T., Callanan, P., Machin, G. & Charles, P.A. 1989, IAU Circ. 4747
Pakull, M.W. 1994, private communication at IAU Symposium 165
Pakull, M.W. & Motch, C. 1989, ESO Workshop: *Extranuclear Activity in Galaxies*, E.J.A. Meurs & R.A.E. Fosbury (Eds.), p. 285
Pakull, M.W., Beuermann, K., Van der Klis, M. & Van Paradijs, J. 1988, A&A 203, L27
Pakull, M.W. et al. 1993, A&A 278, L39
Remillard, R.A., Rappaport, S. & Macri, L.M. 1994, (preprint)
Schaeidt, S., Hasinger, G. & Trümper, J. 1993, A&A 270, L9
Schmidtke, P.C., McGrath, T.K., Cowley, A.P. & Frattare, L.M. 1993, PASP 105, 863
Smale, A.P. et al. 1988, MNRAS 233, 51
Trümper, J. et al. 1991, Nat 349, 579
Van den Heuvel, E.P.J., Bhattacharya, D., Nomoto, K. & Rappaport, S.A. 1992, A&A 262, 97

ARE SUPERSOFT X-RAY SOURCES CONSISTENT WITH WHITE DWARFS?

A. VAN TEESELING
Universitäts-Sternwarte Göttingen
Geismarlandstr. 11, D-37083 Göttingen

J. HEISE
SRON Laboratory for Space Research
Sorbonnelaan 2, NL-3584 CA Utrecht

AND

P. KAHABKA
Astronomical Institute "Anton Pannekoek"
Kruislaan 403, NL-1098 SJ Amsterdam

1. Introduction

Van den Heuvel *et al.* (1992) argued that the emission from supersoft X-ray sources can be explained by steady nuclear burning of hydrogen accreted onto massive white dwarfs. An argument that has been used against this model is that the bolometric luminosities inferred from blackbody fits are sometimes above the Eddington luminosity of a $\sim 1\,M_\odot$ star. However, hot white dwarfs don't radiate blackbody spectra (Heise et al. 1994). Therefore, we have calculated model spectra for hot (accreting) white dwarfs, and investigated whether the emission from supersoft X-ray sources is consistent with an origin from white dwarfs. Here we summarize some preliminary results. For a review of supersoft X-ray sources we refer to the contribution of Kahabka & Trümper to these proceedings.

2. LTE White-Dwarf Model Atmospheres

Fig. 1 shows a selection of the calculated spectra for LMC abundances. The spectra are almost flat up to a strong absorption edge of hydrogen-like or helium-like C, N, or O. Our model atmospheres emit relatively more

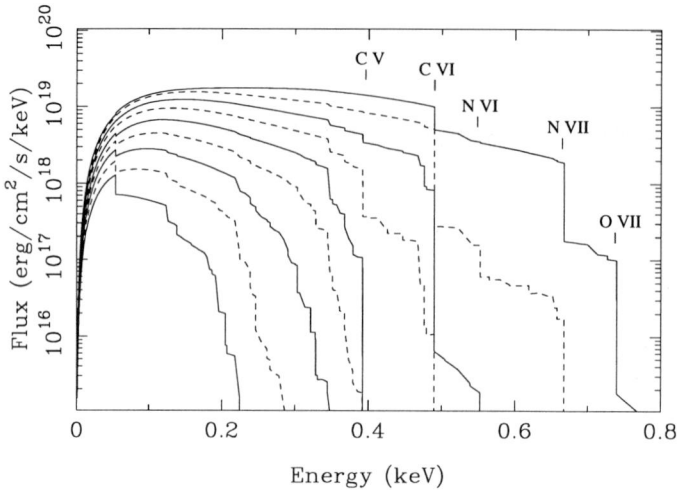

Figure 1. Emergent fluxes of LTE white-dwarf model atmospheres for 0.25× cosmic abundances, $\log g = 8$, and $T_{\text{eff}} = 2\,10^5$ K to $6\,10^5$ K in steps of $5\,10^4$ K from bottom to top.

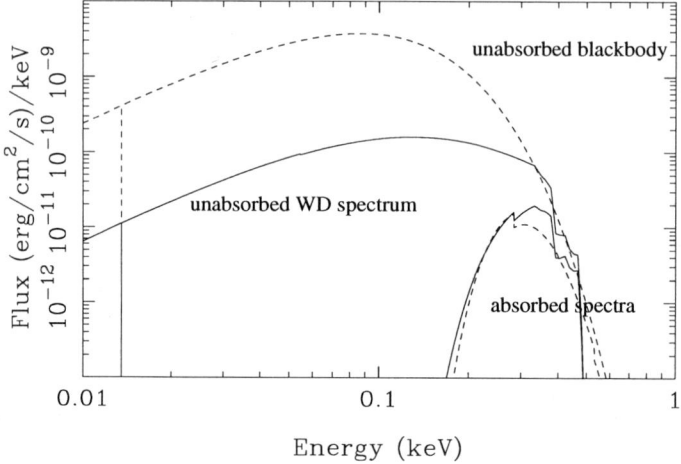

Figure 2. The best-fit blackbody spectrum and the best-fit $\log g = 8$ white-dwarf model atmosphere spectrum of the supersoft X-ray source 1E 0056.8−7154.

soft X-rays than blackbodies. This shows that very hot white-dwarfs are predominantly soft X-ray sources.

To illustrate the effect of fitting white-dwarf model atmosphere spectra instead of blackbody spectra to the observed spectra of supersoft X-ray sources, we first consider 1E 0056.8−7154. This source has been identified with the planetary nebula N67 in the SMC (Wang 1991). Fig. 2 shows the

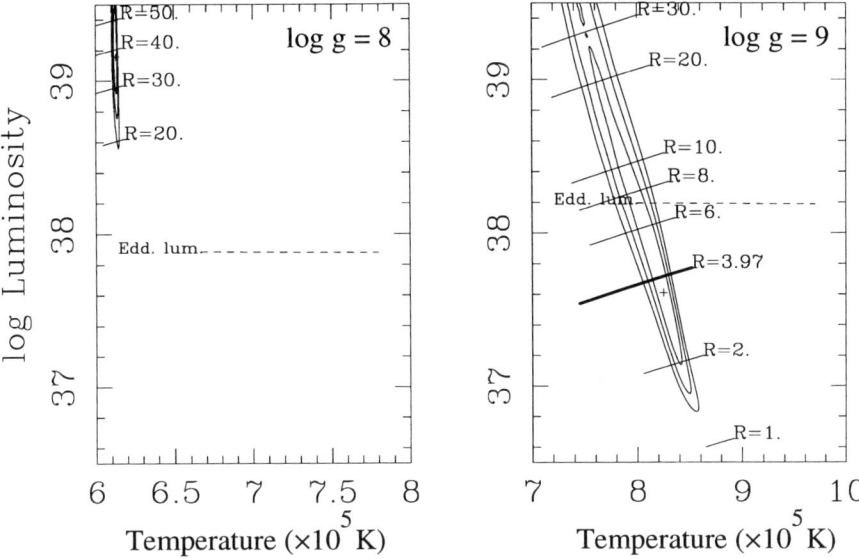

Figure 3. χ^2 contours at 1, 2 and 3-σ levels for fits of model atmospheres to the spectrum of CAL 87. The bolometric luminosity is given in erg s^{-1}. Contours of equal normalization constant are labeled with the radius ($\times 10^8$ cm) of the source at 52 kpc. The thick line is for the normalization constant corresponding to the radius of a non-expanded white dwarf with the log g of the models.

best-fit blackbody spectrum and the best-fit $\log g = 8$ model atmosphere spectrum for the ROSAT spectrum of 1E 0056.8−7154. The unabsorbed bolometric luminosity of the model atmosphere (2 10^{37} erg s^{-1}) is well below the Eddington luminosity of a $\sim$1 M$_\odot$ star and more than a factor of ten lower than the luminosity of the blackbody.

3. Inferred Luminosities and Radii

The spectral resolution of the ROSAT PSPC is not sufficient to discriminate between blackbody spectra and model atmosphere spectra. Every spectrum that has a steep cut off at $\sim$0.5 keV will fit the observed ROSAT spectra of supersoft X-ray sources. However, for a particular type of star (e.g., a white dwarf) the inferred radius has to be consistent with the gravity used to calculate the model atmospheres. For 1E 0056.8−7154 the inferred radius is indeed consistent with a massive $\log g = 8$ white dwarf.

To investigate whether other supersoft X-ray sources, for which it is not certain that they contain a white dwarf, are also consistent with white dwarfs, we analysed ROSAT PSPC observations of seven other supersoft X-ray sources in the Magellanic Clouds.

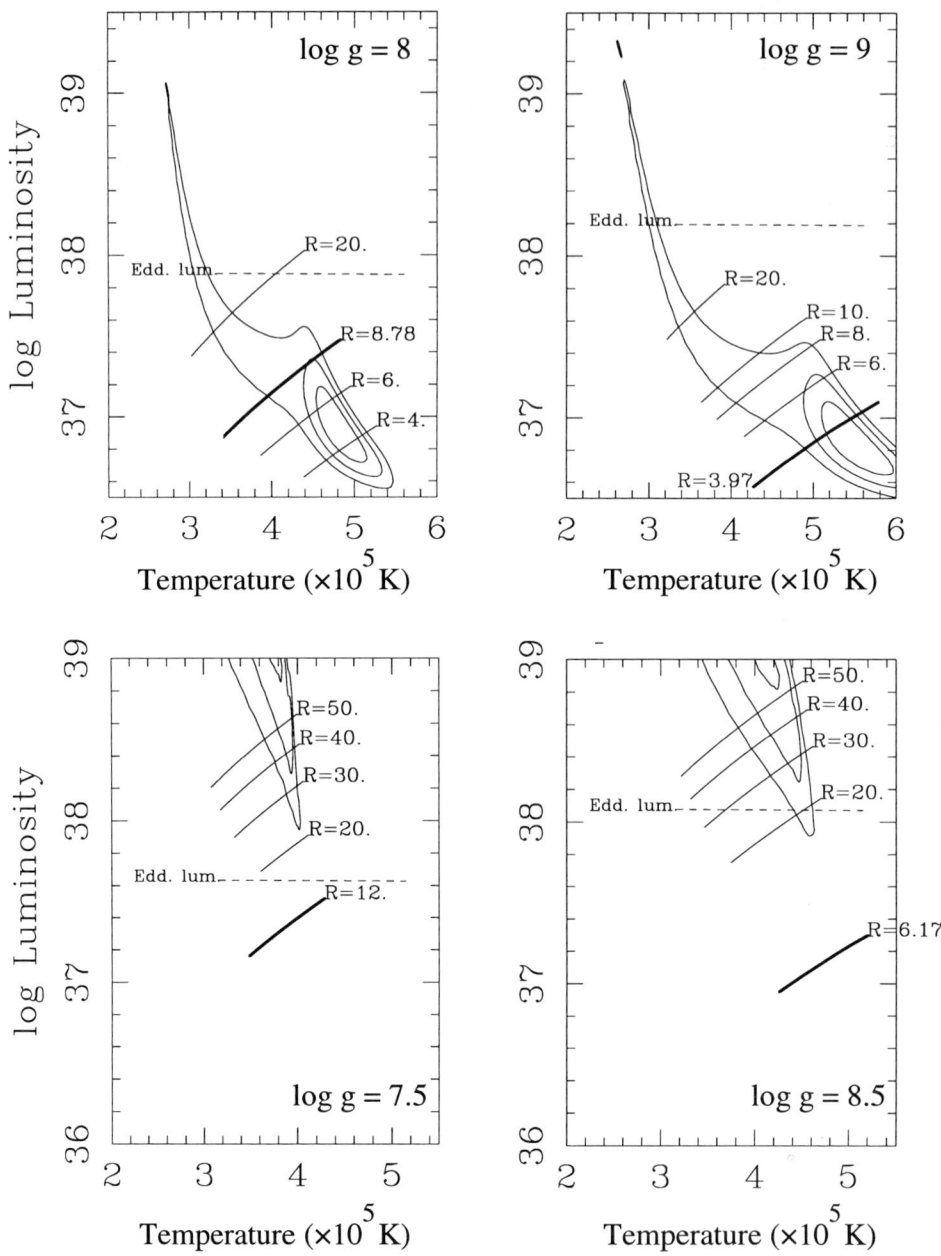

Figure 4. 1, 2, and 3-σ contours for 1E 0035.4−7230 (*top panels*), and RX J0048.4−7332 (*bottom panels*). The bolometric luminosities are given in erg s^{-1}. Lines of equal inferred radii are given in units of 10^8 cm. The Eddington luminosities and the thick lines correspond to the radii of non-expanded white dwarfs with the appropriate log g.

Fig. 3 shows the confidence levels for CAL 87. It is not possible to get a consistent fit with $\log g = 8$, because the inferred radii are too large for a white dwarf. However, CAL 87 has a harder spectrum than most other supersoft X-ray sources, indicating a high effective temperature. This is possible with larger gravities. Indeed, the spectrum and count rate of CAL 87 are consistent with a $\log g = 8.5$ and $\log g = 9$ white dwarf.

Fig. 4 shows the χ^2 contours for 1E 0035.4–7230 in the SMC. The observations are consistent with both $\log g = 8$ and $\log g = 9$ LTE atmospheres. For seven of the analysed sources we find radii which are consistent with white dwarfs. The only exception is RX J0048.4–7332 in the SMC. For all gravities the inferred radii are significantly larger than the radii of non-expanded white dwarfs (Fig. 4). RX J0048.4–7332 has been identified with the symbiotic system SMC3 (Kahabka et al. 1994), which probably had a significant increase in luminosity in 1981 (Vogel & Morgan 1994).

4. Steady Nuclear Burning of Accreted Hydrogen

In Fig. 5 we have plotted for the five analysed LMC sources the 1-σ confidence contours. The observed spectra are consistent with LTE white-dwarf spectra within these contours. Also, within these contours the inferred radii are consistent with a not significantly expanded white dwarf. The elongated shape of the contours is due to the fact that a spectrum for higher gravity and temperature is similar to a spectrum for lower gravity and lower temperature. CAL 87 is consistent with a high-mass white dwarf with a high effective temperature. RX J0439.8–6809 is consistent with an intermediate-mass white dwarf with a lower effective temperature.

5. Conclusions

LTE model atmosphere spectra may give bolometric luminosities of a factor of ten lower than blackbody spectra. For seven of the eight analysed supersoft X-ray sources the inferred best-fit luminosities are below the Eddington luminosity of a solar mass star. The derived radii of these sources are consistent with white dwarfs, and the bolometric luminosities and effective temperatures are consistent with white dwarfs with steady nuclear burning of accreted hydrogen. We did not prove that supersoft X-ray sources are white dwarfs, but our analysis indicates that the X-ray observations of these sources can be explained with emission from hot white dwarfs.

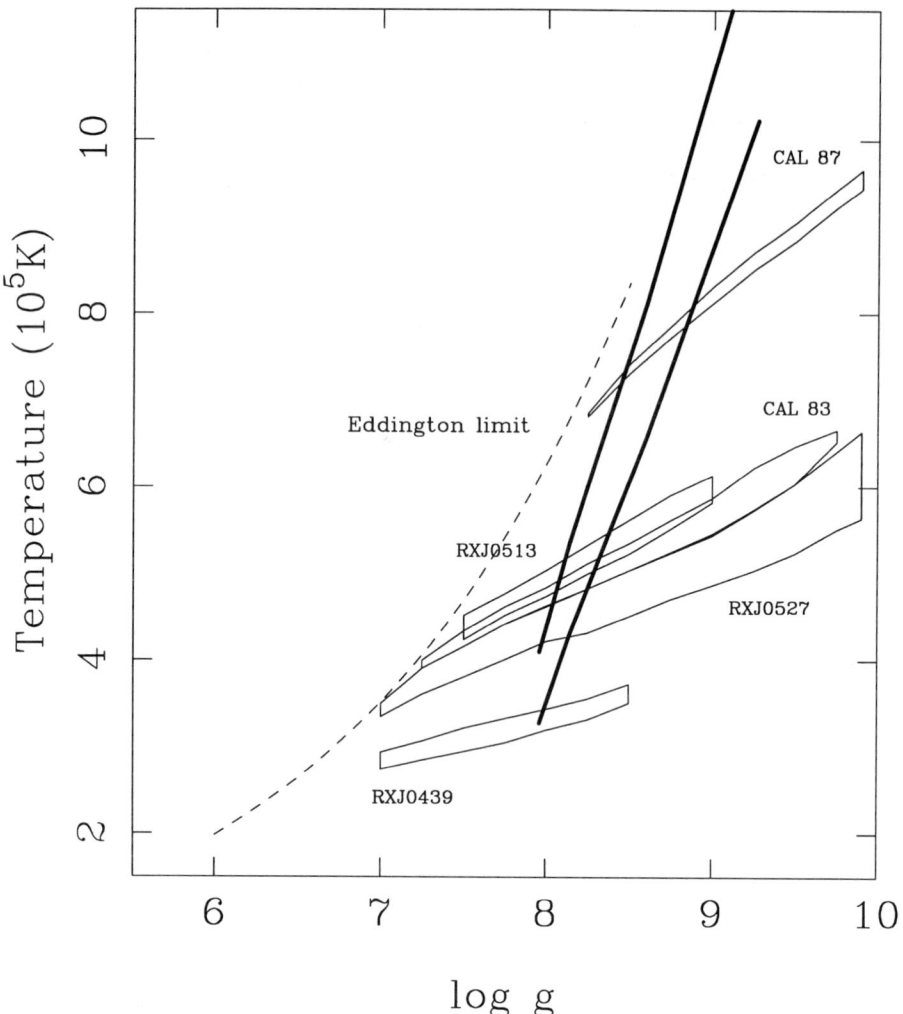

Figure 5. Results of fitting LTE model atmospheres to supersoft X-ray sources in the LMC. The 1-σ range of effective temperatures and $\log g$ for which self-consistent model fits can be obtained is plotted. The solid thick lines bound the region where steady nuclear burning of hydrogen on non-expanded white dwarfs with masses larger than $0.6\,M_\odot$ is stable (Van den Heuvel *et al.* 1992).

References

Heise, J., Van Teeseling, A. & Kahabka, P. 1994, A&A 288, L45
Kahabka, P., Pietsch, W. & Hasinger, G. 1994, A&A 288, 538
Van den Heuvel, E.P.J. *et al.* 1992, A&A 262, 97
Vogel, M. & Morgan, D.H. 1994, A&A 288, 842
Wang, Q. 1991, MNRAS 252, 47P

PROPELLERS – A NEW CLASS OF INTERACTING BINARIES

M. MIKOŁAJEWSKI
Institute of Astronomy, N. Copernicus University
Chopina 12/18, PL-87100 Toruń, Poland

J. MIKOŁAJEWSKA
Copernicus Astronomical Center
Bartycka 18, PL-00716 Warsaw, Poland

AND

T. TOMOV
National Astronomical Observatory Rozhen
4700 Smoljan, PB 136, Bulgaria

Abstract. CH Cyg and MWC 560 are very peculiar symbiotic binaries consisting of an M giant and a white-dwarf companion. The systems have many features in common. In particular, both show occasional eruptions with sub-Eddington luminosity accompanied by flickering activity, and appearance of high-velocity jets. We present arguments that objects like CH Cyg and MWC 560 form a new subclass of interacting binaries distinguished by the presence of wind accreting magnetic white dwarfs.

1. CH Cygni and MWC 560 as Binary Systems

CH Cyg is a long-period, $P_{\rm orb} = 15.6\,\rm yr$, eclipsing binary which contains an SRa variable (M6–7 III), and a white-dwarf companion (Mikołajewski et al. 1987; Mikołajewski et al. 1988). The length of the orbital period suggests the M giant should not fill its Roche lobe. This picture is supported by available optical and IUE data. An M6–7 giant filling its Roche lobe will transfer matter at such a high rate, that the accreting material will burn steadily on the white-dwarf surface giving rise to a large luminosity, $L_{\rm h} \sim 1000\text{--}50\,000\,L_\odot$, and a very high effective temperature,

$T_{\rm eff} \sim 10^5$ K (Mikołajewska & Kenyon 1992), entirely inconsistent with observations (Mikołajewski et al. 1988; Mikołajewski & Mikołajewska 1994). Thus, in the following we assume that the giant never fills its tidal lobe, and the white-dwarf secondary accretes material from a stellar wind.

The case of MWC 560 is more complicated. The optical spectrum is dominated by narrow emission lines of singly ionized metals (Kolev & Tomov 1993). Radial velocities of these lines measured in our spectra obtained in 1990–94, do not show any changes related to orbital motion, while they do follow the orbital motion in CH Cyg (Skopal et al. 1989). The lack of evidence for orbital motion, and the jets expanding practically along the line of sight (see below) suggest a very low orbit inclination, $i \sim 0°$. On the other hand, analysis of the photographic light curve derived from Sonneberg plates collected in the period 1930–1990 (Luthardt 1991) revealed sharp 1 mag maxima occurring at $\sim$1900 d intervals. A similar periodicity was recently found by Doroshenko et al. (1993). We attribute this periodicity to the orbital motion of MWC 560. The maxima of the light curve may be due to increased activity of the accreting component while passing the periastron in an elliptical orbit. As in the case of CH Cyg, the adopted orbital period is too long for the giant component – classified as M4–5 III (Thakar & Wing 1992) – to fill its tidal lobe.

Basic parameters of both systems are collected in Table 1.

TABLE 1. Binary characteristics

	CH Cygni	MWC 560
Orbital period:	5700 d	$\sim$ 1900 d
Inclination:	$\sim$90° (eclipses)	$\sim$0° (no orbital motion)
Eccentricity:	$\sim$0.5	high?
Cool component:	M6–7 III; SRa	M4–5 III
	$\dot{M}_{\rm cool} \sim (1{-}4)\ 10^{-7}\ M_\odot\ {\rm yr}^{-1}$	???
Hot component:	wind accreting magnetic white dwarf	wind accreting magnetic white dwarf

2. Activity and Jets

The hot components in both systems show very spectacular activity. In CH Cyg, we deal with irregular outbursts, during which the luminosity of the hot component varies by a factor of 10^4, while in MWC 560 the hot component is always relatively luminous, and its brightness changes by a factor of $\sim$3 to 4. Nevertheless, the main characteristics of their activity show striking similarities (Table 2). These are: (1) presence of a blue A–B type continuum veiling the M giant absorption features in the optical; (2) appearance of H I Balmer emission lines and numerous emission lines of ionized metals; (3) rapid variability (flickering) on time scales from minutes to hours; (4) appearance of rapid jets and bipolar outflows.

TABLE 2. Basic Properties of Activity

	CH Cygni	MWC 560
Luminosity:	sub-Eddington $L_{max} \sim 300\,L_\odot$ A–B type continuum + "iron curtain" with variable τ; H I line & continuum emission	sub-Eddington $L_{max} \sim 300–1000\,L_\odot$ A–B type continuum + "iron curtain" with variable τ; H I line & continuum emission
Light curve:	low & high activity stages, long inactive periods exist	high activity stages appear periodically with $P \sim 1900^d$
Flickering:	only during active phase; coherent periodic component $A \sim 5\%$, $P \sim 8$ min	always present; coherent periodic component $A \sim 5\%$, $P \sim 1$ hour
Jets:	accompanying *high → low* state transition: $v \sim 800\,\mathrm{km\,s^{-1}} \sim v_{esc}(R_m)$ $\perp$ orbital plane	slow continuous: $v \sim 200–2000\,\mathrm{km\,s^{-1}} \sim v_{esc}(R_m)$ rapid occasional at max: $v \sim 2000–7000\,\mathrm{km\,s^{-1}} \sim v_{esc}(R_{WD})$ $\perp$ orbital plane

MWC 560 became very fashionable in 1990, when Tomov et al. (1990) discovered variable high-velocity ($\sim$2000 to 6000 km s^{-1}) blueshifted absorption components in H I Balmer lines. A similar effect was found in the "iron curtain" features observed in the UV (Michalitsianos et al. 1991). This phenomenon was interpreted as due to expansion of highly collimated, unstable jets ejected along the line of sight (Tomov et al. 1992; Shore et al. 1994). Since 1990 MWC 560 has shown various symptoms of activity. Moreover, all – rather sporadic – spectroscopic and photometric observations made before 1990 also revealed some kind of activity, although none of these observations was done during the sharp maxima of the photographic light curve.

A detailed analysis of the photometric behavior of CH Cyg over a century (1885–1988) showed that, except for the period 1963–1994 when the hot component experienced a series of outbursts, the system remained inactive (Mikołajewski et al. 1990a). The light curve for the largest 1977–86 outburst suggests that CH Cyg oscillates between a *high* and a *low* accretion stage. This resembles the light curve of MWC 560. The *high* state of CH Cyg, observed as a flat maximum of the light curve during 1981–1984, corresponds to the sharp maxima of MWC 560. However contrary to CH Cyg, the hot component of MWC 560 is always active, although it spends most time in the *low* state.

The active phases in both systems are characterized by flickering. The flickering may have low-amplitude periodic components (Mikołajewski et al. 1990b; Michalitsianos et al. 1993).

The most spectacular phenomenon is, however, the occurrence of jets. In CH Cyg the jet features were first detected in VLA radio images (Taylor et al. 1985, 1988), and later also in optical emission lines (Solf 1987; Leedjärv et al. 1994). In MWC 560 the jet which is approaching practically along the line of sight, can be monitored by observations of the blueshifted absorption components – well separated from the emission components – of H I Balmer lines, and in the "iron curtain" features in the UV (Tomov et al. 1994). In both systems, the jets seem to be related to transitions between the *high* and the *low* stage of accretion. In CH Cyg jets developed just after a sudden drop from the high to the low state in July 1984. MWC 560 showed quasi-stationary outflow with moderate velocities, ~ 200 to $2000\,\mathrm{km\,s^{-1}}$, in the low state, before and after the 1990 maximum (Wachter et al. 1994; Tomov et al. 1994), while in the high state in 1990 occasional jets with high velocity, ~ 2000 to $7000\,\mathrm{km\,s^{-1}}$, and time scales of order of a few hours were observed. In both systems the jets seem to be highly collimated and moving nearly perpendicularly to the orbital plane.

3. Accretor–Propeller Model

The active phases (outbursts) in CH Cyg and MWC 560 have many features in common, which suggests similar nature of the active components and mechanisms of their activity. The presence of flickering, and the appearance of highly collimated jets in the sub-Eddington accretion regime, indicate the presence of a strong magnetic field in the accreting component. We attribute the periodic flickering component in each system to the spin period of the magnetic white dwarf. The presence of a rotating disk-like envelope and an extended pseudo-photosphere, points to the presence of a centrifugal barrier (propeller action). Finally, the occurrence of low- and high-activity stages, the appearance of jets during the transition to the low state in CH Cyg, and episodic ejection of very high-velocity jets during high states in MWC 560, indicates *propeller* ↔ *accretor* transitions.

In our study we assume, following Lipunov (1987), that the rotational luminosity: $L_{rot} \propto R_t^3$, where the distance scale R_t should be close to the corotation radius, R_c, in the *accretor* state, and to a magnetosphere radius, R_m, in the *propeller* state. Unfortunately, the magnetosphere radius so far is, but poorly, known in the propeller state, and can highly deviate from the standard expression for the Alfven radius adopted in most studies (Lipunov 1987). Including accretion and assuming that for some critical values, $R_m = R_{cat}$ ($\sim R_c$) and $\dot{M} = \dot{M}_{cat}$, the rotator is in *catastrophic* equilibrium, and a small change in the accretion rate makes the rotator pass from *propeller* (low) into *accretor* (high) state and vice versa. The luminosity of the rotator will interchange between two values:

$$L_{prop} = \frac{GM_{WD}\dot{M}_{cat}}{R_{cat}} + \kappa_t \mu^2 \frac{\omega}{R_{cat}^3} \iff L_{accr} = \frac{GM_{WD}\dot{M}_{cat}}{R_{WD}} + \kappa_t \mu^2 \frac{\omega}{R_{cat}^3}$$

where $\kappa_t = 2/3 \sin^2 \beta$, μ is the magnetic dipole moment, β the angle between the magnetic and rotational axes, and ω the angular velocity of the rotator. If L_{prop}, L_{accr}, R_{WD}, M_{WD}, and ω are known, practically all parameters of the rotator can be estimated (Mikołajewski *et al.* 1990b; Mikołajewski & Mikołajewska 1994). The inclination of the rotator, β, remains the only free parameter of our model (Table 3).

TABLE 3. Parameters of Oblique Rotator

	CH Cygni	MWC 560
B	$\sim 10^7$–10^8 G	$\sim 10^7$–10^8 G
P_{rot}	~ 8 min	~ 1 hour
$\dot{M}_{cat}$	a few 10^{-8} M$_\odot$ yr^{-1}	a few 10^{-8} M$_\odot$ yr^{-1}
R_{cat}	$\sim$ a few R_{WD}	$\sim$ a few R_{WD}
β	large? ($\sim 45°$)	small? ($< 5°$)

In CH Cyg, the jets appeared only after the *accretor* → *propeller* transition, and almost all matter accumulated around the rotator was expelled. It suggests very strong propeller action along the rotational axis, i.e., in the direction of maximum gradient of magnetic pressure changes, and thus $\beta \sim 45°$ (Illarionov & Sunyaev 1975). On the other hand, sporadic jets which appear during *high* (accretor) stages of MWC 560, have very high velocities comparable to the escape velocity from the white-dwarf surface. Thus the jets appear to be driven by super-Eddington accretion in the accretion columns. Simultaneously, the observed high collimation of the jets requires a magnetic axis nearly parallel to the rotational axis ($\beta < 5°$).

CH Cyg and MWC 560 seem to form a new subclass of interacting binaries distinguished by a wind accretion onto a magnetic white dwarf. The salient characteristic of the class are: (1) sub–Eddington brightness at maximum; (2) presence of flickering; (3) high and low stages of activity; (4) ocassional appearance of bipolar outflows or jets. Because these objects spend most time in the propeller (low) state, and the ejection of jets is due to specific propeller action, we propose to call them *propeller stars* or *propellers*. Other possible candidates are R Aqr, o Cet, and the symbiotic recurrent novae T CrB and RS Oph. R Aqr is a well known symbiotic Mira with jets, while the remaining candidates are the only known symbiotic stars with flickering.

Acknowledgements. This study was partially supported by Polish KBN research grant No. 2 P304 007 06, and Bulgarian NFSR grant No. F–35/1991.

References

Doroshenko, V.T. et al. 1993, Inf. Bull. Var. Stars 4980
Illarionov, A.F. & Sunyaev, R.A. 1975, A&A 39, 185
Kolev, D. & Tomov, T. 1993, A&AS 100, 1
Leedjärv, L., Mikołajewski, M. & Tomov, T. 1994, A&A 287, 543
Lipunov, V.M. 1987, Ap&SS 132, 1
Luthardt, R. 1991, Inf. Bull. Var. Stars 3563
Michalitsianos, A.G. et al. 1991, ApJ 371, 761
Michalitsianos, A.G. et al. 1993, ApJ 409, L53
Mikołajewska, J. & Kenyon, S.J. 1992, MNRAS 256, 177
Mikołajewska, J., Selvelli, P.L. & Hack, M. 1988, A&A 198, 150
Mikołajewski, M. & Mikołajewska, J. 1994, MNRAS (submitted)
Mikołajewski, M., Mikołajewska, J. & Khudyakova, T.N. 1990a, A&A 235, 219
Mikołajewski, M. et al. 1990b, Acta Astr. 40, 129
Mikołajewski, M., Szczerba, R. & Tomov, T. 1988, in *The Symbiotic Phenomenon*, J. Mikołajewska et al. (Eds.), Kluwer, p. 221
Mikołajewski, M., Tomov, T. & Mikołajewska, J. 1987, Ap&SS 131, 733
Shore, S.N., Aufdenberg, J.P. & Michalitsianos, A.G. 1994, AJ 108, 671
Skopal, A., Mikołajewski, M. & Biernikowicz, R. 1989, Bull. Astr. Inst. Czechosl. 40, 433
Solf, J. 1987, A&A 180, 207
Taylor, A.R., Seaquist, E.R. & Kenyon, S.J. 1988, in *The Symbiotic Phenomenon*, J. Mikołajewska et al. (Eds.), Kluwer, p. 231
Taylor, A.R., Seaquist, E.R. & Mattei, J. 1985, Nat 359, 38
Thakar, A. & Wing, R.F. 1992, BAAS 24, 801
Tomov, T. et al. 1994, in *Evolutionary links in the Zoo of interacting binaries*, F. d'Antona (Ed.), (in press)
Tomov, T. et al. 1990, Nat 346, 637
Tomov, T. et al. 1992, MNRAS 258, 23
Wachter, S. et al. 1994, in *Interacting Binary Stars*, A.W. Shafter (Ed.), ASP Conference Series Vol. 56, p. 401

ANALYSIS OF LONG-TERM AAVSO OBSERVATIONS OF RS OPHIUCHI

B.D. OPPENHEIMER AND J.A. MATTEI
American Association of Variable Star Observers
25 Birch Street
Cambridge, Massachusetts 02138
U.S.A.

Abstract. We present here a study of the five reported outbursts of the recurrent nova RS Ophiuchi, together with an analysis of its light variation between outbursts. We find that the shapes and decline rates of the outbursts are strikingly similar to each other. We note the possibility of an additional, unreported outburst in 1945, supported by the similarity of its decline to that of other outbursts. Each interval between outbursts shows periodicity within itself, however, there is no periodicity common to all, and the possible periods range between 892 and 2283 days.

1. Introduction

RS Ophiuchi is a cataclysmic variable of the recurrent nova type. It is an interacting binary system consisting of an M red giant and a blue compact companion (Barbon, Mammano & Rosino 1968; Sanduleak & Stephenson 1973). Due to its composite spectrum it has also been classified as a symbiotic star (Barbon et al. 1968). Its orbital period has been derived from radial-velocity variations (Dobrzycka & Kenyon 1994) to be 460 days, and its distance has been estimated to be 1.6 kpc (Hjellming et al. 1986).

It has had five reported outbursts, in 1898, 1933, 1958, 1967, and 1985. All of the outbursts show very similar light curves.

RS Ophiuchi is the only known recurrent nova that shows significant light fluctuations between outbursts. In this study we analyzed the long-term behavior of RS Ophiuchi both during outbursts and between outbursts, using mostly AAVSO observations spanning 104 years.

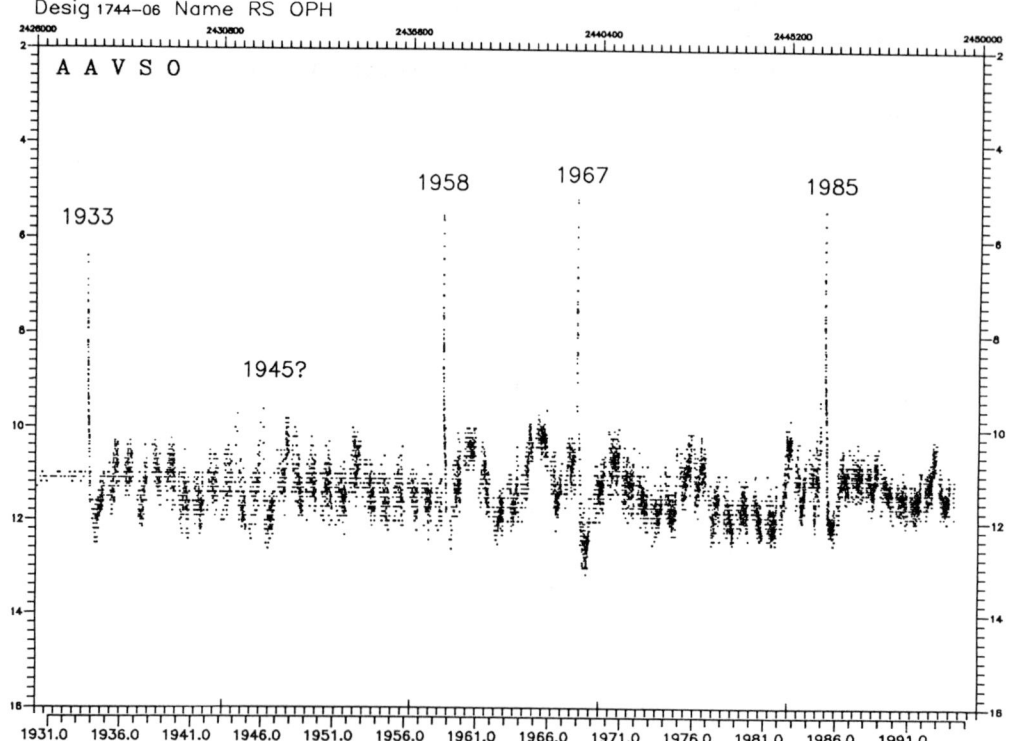

Figure 1. Light curve of AAVSO observations of RS Ophiuchi from 1930 to mid-1993, plotted date (top - Julian Day, bottom - year) vs. visual magnitude.

2. Observations and Analysis

RS Ophiuchi has been in the observing program of AAVSO and well-monitored by hundreds of observers since its 1933 outburst (Fig. 1). For this study we have utilized 26 966 visual observations from the AAVSO International Database and Archives between 1920 and May 1993, and 565 photographic observations from Harvard plates from 1890 to 1916 (Campbell 1920). We have studied the outbursts and the intervals between them separately. We used 0.5-day means of the observations for the outbursts and 1-day means for the non-outburst intervals.

2.1. OUTBURSTS

An outburst is defined as the sudden rise in brightness from quiescent state to about 5th magnitude within 24 hours. Each outburst was studied individually, and then superimposed onto each other. Three stages of decline were found from maximum to minimum. These stages were isolated and decline rates obtained for each of them. Particular attention was given to the interval following an observing gap, beginning in December 1945, when the star was observed to be bright at magnitude 9.6. This interval was studied thoroughly and the decline rate of the light curve was compared with the rest of the outbursts to check whether this may have been an outburst caught at the tail-end.

2.2. INTERVALS BETWEEN OUTBURSTS

Six intervals between outbursts from 1898 to 1993 were identified and each interval was analyzed using the Date-Compensated Discrete Fourier (DCDF) Analysis technique (Ferraz-Mello 1981) developed for MS-DOS at AAVSO. For each interval, a primary Fourier transform was taken, followed by a residual and a second residual, to find the possible strong periods. Particular attention was given in search of both the 230-day period and the 460-day spectroscopic orbital periods reported by Garcia (1986), then later revised by Dobrzycka & Kenyon (1994).

3. Discussion

3.1. OUTBURSTS

Each of the five reported outbursts is strikingly similar to each other. In each outburst, the rise to maximum is fast, as the star brightens by about 6 magnitudes within 24 hours, reaching maximum visual magnitude of about 5.0; the star then immediately starts to fade. The decline from an outburst is in three stages. The first stage is the fastest, averaging $0.09\,\mathrm{mag\,day}^{-1}$

and lasting 43 days, followed by a decline rate of only 0.02 mag day^{-1} until 85 days after outburst. The average rate of decline of the final stage is 0.05 mag day^{-1} lasting until 122 days after outburst (see Table 1).

TABLE 1. Outbursts of RS Ophiuchi

Outburst Year	JD Max (2400000+)	Max mag	$\Delta t(1)$ (days)	Rate of $\Delta t(1)$ (mag d^{-1})	$\Delta t(2)$ (days)	Rate of $\Delta t(2)$ (mag d^{-1})	$\Delta t(3)$ (days)	Rate of $\Delta t(3)$ (mag d^{-1})
1898	14442?	—	51	.06?	112	.01?	128	.06?
1933	27297	6.4?	48	.06?	81	.02	118	.06?
1945?	31786?	—	—	—	76	—	118	.05?
1958	36399	5.4	41	.10	89	.02	117	.04
1967	39791	4.9	39	.12?	89	.02?	140	.04
1985	46094	5.4	38	.10	65	.02	111	.05
Average		5.5	43	.09	85	.02	122	.05

During the secondary decline phase, the star brightens slightly and the decline slows down. This interval coincides with the appearance of the coronal lines of [Fe X], [Fe XIV], and [A X], forbidden lines of [O III] and [N II], and permitted lines of N III, He I, and He II in the spectrum of each outburst (Barbon, Mammano & Rosino 1968; Hack & Selvelli 1993). The appearance of some of the coronal lines in the visual spectra, such as [Fe X] and [Fe XIV], may cause the overall visual brightness to increase. As they weaken the star fades and continues to decline at the third, faster rate.

The star reaches minimum at about magnitude 12.5, on the average 216 days after outburst. It then brightens, reaching a mean maximum of magnitude 10.3 in 501 days, or 717 days after outburst maximum.

3.1.1. Outburst in 1945

There is strong evidence in the AAVSO observations that RS Ophiuchi had another outburst in December 1945 (JD 2431786) which has not been reported previously. An observation made at magnitude 9.9 on JD 2431784, 2 days before the possible maximum outburst, may have been the beginning of the rise to maximum. For the next 68 days, the star was in its seasonal observing gap. On JD 2431854, 65 days after the possible maximum, it was observed at magnitude 9.6. Over the next 69 days, the brightness fell 2.9 magnitudes, a rate of change never before observed in this star for such an extended amount of time outside outbursts. The light curve and the rate of decline is similar to the third stage of other outbursts. After reaching minimum the star brightened by 2.5 magnitudes in 521 days, a behavior also seen in other outbursts. Thus, we strongly believe that RS Ophiuchi indeed had another outburst in 1945 which has not been reported in the literature until now. The occurrence of this outburst may have bearing on the modeling of the system in that with the 5 known outbursts, the average interval between outbursts is 22 years, whereas with an outburst in 1945, the average reduces to 13 years between the 1933 and 1985 outbursts. With so few observations between 1898 and 1933, one or more outbursts could have been missed, thus this interval has not been included in the averages.

3.2. INTERVALS BETWEEN OUTBURSTS

The light curve outside of the outbursts resembles the light curves of other symbiotic stars such as Z Andromedae and CH Cygni. These stars show significant oscillations with multiple periods.

The intervals between outbursts of RS Ophiuchi vary from each other both in the amplitude of variation, ranging between one and three magnitudes, and in the complexity of the modulations. The AAVSO light curve from 1930 to May 1993 (Fig. 1) shows the longterm behavior and the activity between outbursts clearly. The intervals between the 1958–1967 and the 1967–1985 outbursts appear to be more active, showing the largest range of variation.

After the decline of each outburst, there is a secondary brightening, reaching a minor-maximum in the average 717 days after the outburst. This is usually followed by two or three similar modulations until the next outburst.

The Date-Compensated Discrete Fourier Transform (DCDFT) applied to each of the intervals indicates that there is no common period between these intervals. Modulations between outbursts are multiperiodic, with the possible periods ranging from 892 to 2283 days. The whole data set, excluding the outbursts, was also studied, and the DCDFT applied to this set indicates a period of 2016 days.

A likely period of 471 days was found for the time interval 1985 to 1993. This period is within the errors of the 460-day orbital period derived by Dobrzycka & Kenyon (1994). No period at or close to 260 days (Garcia 1986) was found by the DCDFT.

Further analysis of the AAVSO light curves is being carried out in search of a 50 to 70-day period reported by Tempesti (1975) in the photoelectric data between 1972 and 1973.

4. Conclusions

The outburst light curves are very similar to each other and show three stages of decline. The first stage is the fastest, the second is slower, and the third is intermediate in rate of decline. There is minor brightening during the second stage that may coincide with the appearance of coronal lines in the spectra.

The outbursts may be phase-dependent, occurring near the minimum of a modulation and followed by rise to secondary minor-maximum after each outburst. The mechanism causing the modulations, particularly the longer ones, may be connected to the occurrence of the outbursts.

There is strong evidence in the AAVSO data of an unreported outburst in 1945.

RS Ophiuchi shows significant fluctuation between outbursts, varying between magnitude 9.6 and and 12.8. The light variation shows a wide variety of behavior from one interval to the next, with no period of variation common to all intervals. The most prominent periods of the intervals range from 892 to 2283 days. The only prominent period for the entire data set was found to be 2016 days.

Acknowledgments. We thank hundreds of AAVSO observers worldwide who made about 27 000 observations on RS Ophiuchi that made this study possible. B.D.O. gratefully thanks NASA Astrophysics Grant Supplements for Education, NAGW-3228 to the AAVSO, and the AAVSO Margaret Mayall Assistantship, which provided the funds for his high school summer internship in the AAVSO. We thank the National Science Foundation for Education Grant ESI-9154091, which supported the development by E.G. Foster, AAVSO technical staff, of the data analysis software used in this project.

References

Barbon, R., Mammano, A. & Rosino, L. 1969, in *Non-Periodic Phenomena in Variable Stars*, L. Detre (Ed.), Academic Press (Budapest), p. 527
Campbell, L. 1920, Harvard Ann. 81, No. 5
Dobrzycka, D. & Kenyon, S.J. 1994, AJ (in press)
Ferraz-Mello, S. 1981, AJ 86, 619
Garcia, M.R. 1986, AJ 91, 1400
Hack, M. & Selvelli, P.L. 1993, in *Cataclysmic Variables and Related Objects*, M. Hack & C. la Dous (Eds.), NASA SP-507 (Washington D.C.), p. 511.
Hjellming, R.M. et al. 1986, ApJ 305, L71
Pottasch, S.R. 1967, Bull. Astron. Inst. Netherlands 19, 227
Sanduleak, N. & Stephenson, C.B. 1973, ApJ 185, 899
Tempesti, P. 1975, I.A.U. Inf. Bull. Variable Stars 974

Discussion

K. Beuermann: Is the coverage sufficient to exclude an outburst between 1967 and 1985?

J. Mattei: Yes, there was no outburst.

F. Meyer: Accretion disks of the size of the disk in RS Ophiuchi should have regions of midplane temperatures around 10 000 K. One would expect therefore "dwarf nova outbursts", as found by Wolfgang Duschl (1986, A&A 163, 56) for symbiotic systems.

8

Gamma-ray Bursts

GAMMA-RAY BURSTS: OBSERVATIONAL OVERVIEW

GERALD J. FISHMAN
Space Sciences Laboratory
NASA/Marshall Space Flight Center
Huntsville, AL 35812 USA

Abstract. Gamma-ray bursts remain one of the greatest mysteries in astrophysics in spite of new and more detailed observations made with the BATSE experiment on the Compton Observatory. The new observation with the greatest impact has been the observed isotropic distribution of bursts along with a deficiency of the weak bursts which would be expected from a homogeneous burst distribution. This is not compatible with any known Galactic population of objects. Other recent important observations include an enormous variety of burst morphologies and gamma-ray burst photons extending to GeV energies. A time dilation effect has also been reported to be observed in gamma-ray bursts.

1. Introduction

Gamma-ray bursts are a phenomenon without precedent in astronomy, having no quiescent counterpart in any other wavelength region, no observations that would provide a direct measure of their distance and no comprehensive model that can explain their origin. Furthermore, the bursts have an extremely wide variety of durations, temporal profiles and spectral variations, which makes modeling them all the more difficult.

It is now over 25 years since the discovery of gamma-ray bursts, and their origin appears as elusive as ever. The field of gamma-ray bursts has undergone a rapid, dramatic, and to many, a surprising change over the past three years as a result of new, more sensitive observations of the gamma-ray sky distribution. The observed isotropy and inhomogeneity of these objects represent a distribution unlike any other known galactic objects. Over a hundred theories of their origin have now been catalogued (Nemiroff 1994).

These models cover distance scales from the local Oort cloud to cosmological distances. Whatever the distance scale, it will most likely represent a new class of objects, processes and/or emission mechanisms. This paper describes some of the observed properties of gamma-ray bursts, primarily their temporal and spectral characteristics and their distribution. Models based upon bursts at cosmological distances are described in another paper in these proceedings (Piran 1995).

Considerable observational progress has been made in the past few years as more sensitive space-borne detectors have become available. Most of the observations in this paper were made with the Burst and Transient Source Experiment (BATSE) on the Compton Gamma-Ray Observatory. While many of the observational results are relatively straight-forward, some of the properties and interpretations of ensembles of bursts have become subject to debate. Details of some of the more recent observational results can be found in conference proceedings that have been published in the past three years (Paciesas & Fishman 1992; Friedlander, Gehrels & Macomb 1993; Fishman, Brainerd & Hurley 1994).

2. Time Profiles

Perhaps the most striking features of the time profiles of gamma-ray bursts are their morphological diversity and the large range of burst durations. Coupled with this diversity is the general inability to place many gamma-ray bursts into well-defined classifications. Several attempts have been made in the past to categorize gamma-ray burst morphologies. This difficult task is always hampered by bursts with multiple characteristics, bursts that are too weak to classify, and the rather arbitrary and subjective (non-quantitative) ways that classes are defined. Examples of extreme differences in burst morphologies and durations are shown in a sample page (Fig. 1) from the First BATSE Burst Catalog (Fishman et al. 1994).

Weaker bursts have been shown to have the same temporal diversity as the stronger bursts even though the temporal variations are of lower statistical significance (Lestrade 1994). A cursory examination of burst profiles indicates that some are chaotic and spiky with large fluctuations on all timescales, while others show rather simple structures with few peaks. However, some bursts are seen with both characteristics present within the same burst. No periodic structures have been seen from gamma-ray bursts.

The durations of gamma-ray bursts range from about 10 ms to over 1000 s in the energy range in which most bursts are observed (below 1 MeV). Sub-millisecond structure has been detected in at least one burst (Bhat et al. 1992). Recent EGRET observations show high-energy (>100 MeV) emission lasting 1.6 hours after the burst trigger (Hurley et al. 1994) from a burst

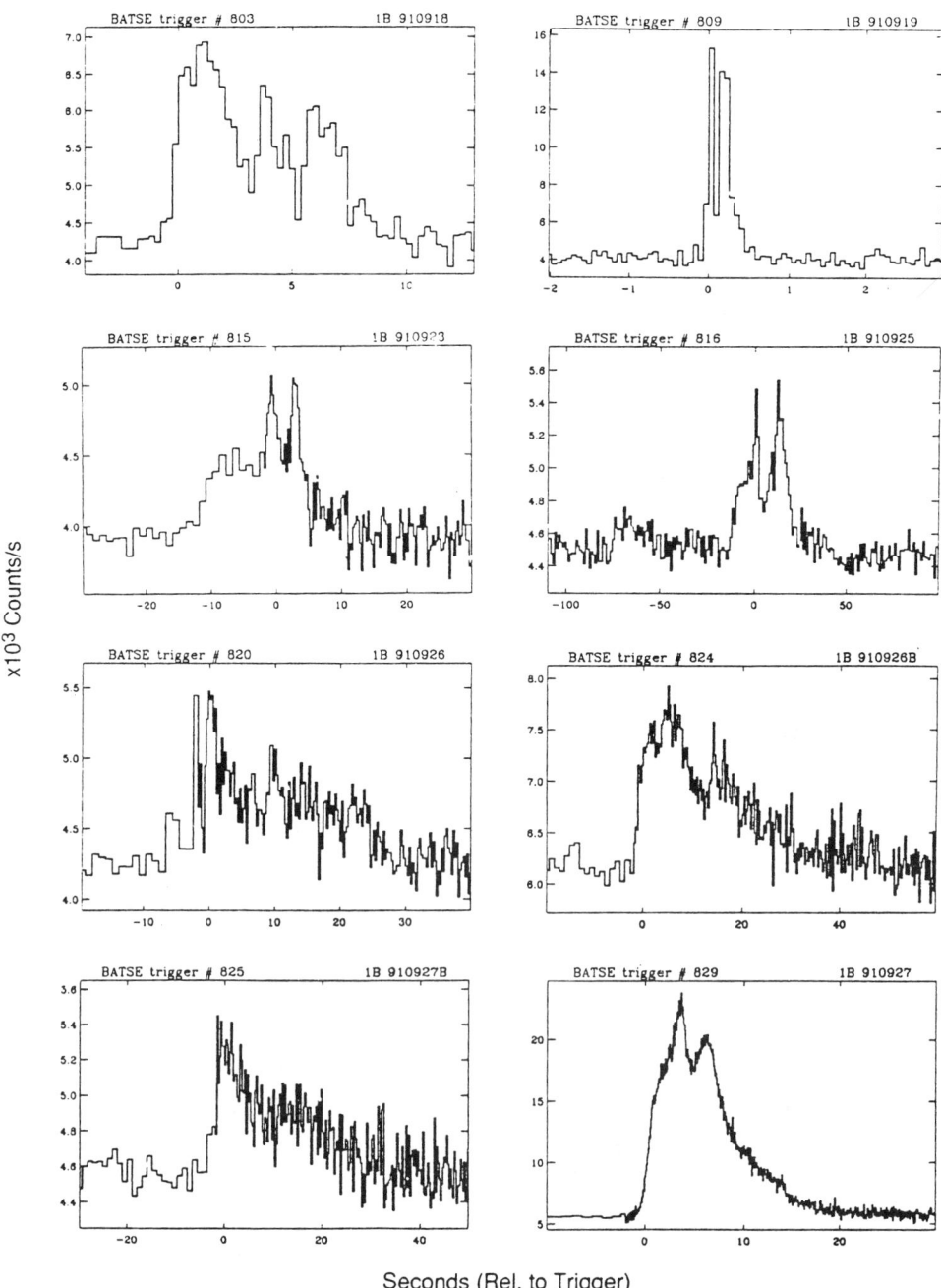

Figure 1. A sample of eight gamma-ray bursts from the first BATSE Catalog (Fishman et al. 1994), showing the extreme range of burst time profiles and durations.

that occurred on 17 February 1994. At the lower photon energies, characteristic of that observable with the BATSE and Ulysses detectors, this gamma-ray burst lasted only 180 s. During this initial time, EGRET observed about a dozen high-energy photons, with energies as high as 4 GeV. EGRET high energy photons are seen coming from the burst direction as late as 1.6 hours after the initial outburst. Fig. 2 (from Hurley et al. 1994) shows the composite time profiles of this burst, as seen with the EGRET, BATSE, and Ulysses experiments.

A bimodality is seen in the logarithmic distribution of gamma-ray burst durations, with broad, unresolved peaks at about 0.3 s and 20 s and a minimum at around 2 s. The shorter bursts are also seen to have harder spectra, as measured by a hardness ratio (Kouveliotou et al. 1994a). Another general property of the gamma-ray burst time profiles is that they tend to have shorter rise-times and fall-times (sharper spikes) at higher energies. Most bursts also show an asymmetry, with shorter leading edges than trailing edges. This has been quantified by Link, Epstein & Priedhorsky (1993) and by Nemiroff et al. (1994). Other analyses have used a variety of temporal parameters and constructs to quantify and characterize gamma-ray burst temporal properties.

A recent analysis of time profiles by Norris et al. (1994) shows a systematic widening or stretching of gamma-ray burst time profiles as bursts become weaker. This analysis was performed by artificially weakening the stronger gamma-ray bursts and introducing the appropriate background so that all bursts could be analyzed in a consistent manner. The quantitative analysis of the time profiles was made through the use of wavelets. The observed stretching of the profiles of bursts is consistent with that expected from the effects of time-dilation from bursts at cosmological distances. However, the time dilation observation and its interpretation are still somewhat controversial.

3. Spectral Characteristics

A distinctive feature of gamma-ray bursts is their high-energy emission: almost all of the observed power is above 50 keV. Some bursts show emission as low as 1 keV, but this power is less than 1 or 2 percent of the total power. Most bursts show rather simple continuum spectra which appear similar in shape when integrated over the entire burst and when sampled on various time scales within a burst. Fig. 3 shows a typical burst spectrum from 0.1 to 10 MeV, with the peak power at ~ 600 keV (Share et al. 1994).

Spectral shapes which have been fit to burst spectra include broken power laws (Schaefer et al. 1992), log-normal distributions (Pendleton et al. 1994), and exponential spectra with power-law high energy tails (Band

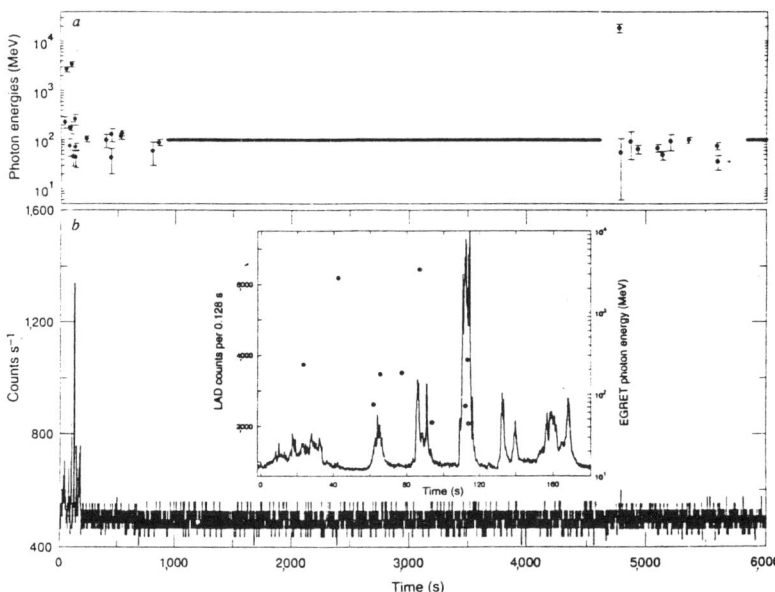

Figure 2. The extremely long high photon energy gamma-ray burst of 17 February 1994, as seen with the EGRET, BATSE and Ulysses experiments. Only the EGRET experiment shows photons above 100 MeV, up to 1.6 hours after the initial outburst (from Hurley *et al.* 1994).

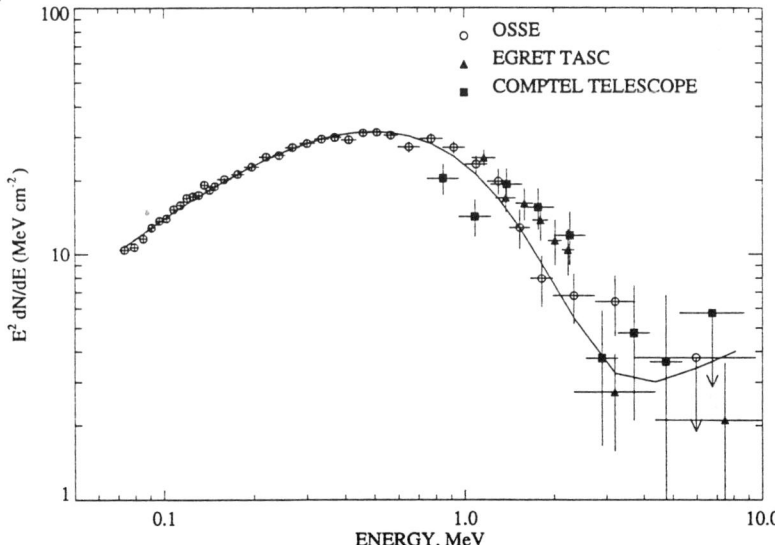

Figure 3. The high-energy spectrum of GRB 910601, as measured by three of the experiments on the Compton Observatory (Share *et al.* 1994), integrated over a large portion of the burst. A characteristic broad spectral shape, with peak power at about 0.6 MeV is seen. (The spectral up-turn at high energies is not real.)

et al. 1993). Although the spectral shapes of many bursts are similar, the energy at which peak power is emitted changes greatly from burst to burst and it is seen to change rapidly within a burst. Some significant changes on time scales as short as tens of milliseconds have been observed (cf., Ford *et al.* 1995). Earlier observations with the gamma-ray spectrometer on the Solar Maximum Mission showed that in many bursts, the high energy emission follows the same power law to over 80 MeV (Share *et al.* 1992). The EGRET observation of the long-duration burst (Fig. 2) shows a single photon from the burst direction with an energy of about 20 GeV, occurring late in the burst. Within most (but not all) bursts, there is a hard-to-soft spectral evolution, resulting in the lower energies peaking earlier (Pendleton *et al.* 1994; Ford *et al.* 1995).

A search for unambiguous gamma-ray line features with BATSE/GRO has thus far been unable to confirm the earlier reports of spectral line features from gamma-ray bursts (Palmer *et al.* 1994). Several recent papers from the proceedings of the last Huntsville Gamma-ray Burst Workshop (Fishman, Brainerd & Hurley 1994) have also discussed the preliminary BATSE line search analyses and their results.

4. Burst Counterparts

There is no doubt that a great advance in our understanding of gamma-ray bursts can be attained through successful correlated observations of gamma-ray bursts at other wavelengths. This fact was demonstrated recently by the combined gamma-ray, X-ray, optical and radio observations of Soft Gamma-ray Repeaters (SGR's) (Kouveliotou *et al.* 1994b; Murakami *et al.* 1994; Kulkarni *et al.* 1994). Within the past four years, there have been major, renewed efforts to find a counterpart to a gamma-ray burst in other wavelength regions as evidenced by either simultaneous emission or afterglow emission. Some of the world's most powerful ground-based facilities are involved with these attempts for correlated burst observations. A sensitive, wide-field transient optical camera has been operating for over three years at Kitt Peak (Vanderspek *et al.* 1994). Space-borne correlated observations of well-located gamma-ray bursts have also been attempted in the UV, EUV, and X-ray regions. Comprehensive studies of archival plates also have been made. There have been several suggestions for counterparts although the results are inconclusive and problematic. In view of the importance of the implied results, further observational evidence is needed before these results are accepted. A recent review of the present status of correlated gamma-ray burst observations is given by Schaefer (1994).

A new near-realtime BATSE burst location system called BACODINE (BAtse COordinates DIstribution NEtwork) (Barthelmy *et al.* 1994) is now operational. This system, when linked to a rapid slewing optical telescope, opens the exciting possibility of obtaining optical images of burst regions while the burst is in progress. Although the present BATSE location accuracies are coarse (~4 deg.), plans are being made for new, powerful wide-field CCD camera systems dedicated for such burst counterpart searches.

A joint BATSE-COMPTEL capability also exists that is able to provide more accurate (~1 deg) locations within several hours for those gamma-ray bursts which also happen to be within the COMPTEL field-of-view. This capability has been demonstrated for the intense gamma-ray burst of 31 January 1993, when an extraordinary effort involving over 30 instruments observed the burst region within hours and days of its occurrence (Schaefer et al. 1994).

5. Repetition and Burst Distributions

There have been reports of burst repetition in the BATSE data but the evidence is not statistically compelling and additional data have not supported these claims of burst repetition. Recent papers by the BATSE team have detailed the observational and statistical arguments concerning burst repetition (Meegan et al. 1995; Hartmann et al. 1995) and an analysis of time-dependent repetition has also been made, with negative results (Brainerd et al. 1995). Typical upper limits of classical gamma-ray burst repeaters on time scales of years are ~20 percent. From BATSE data alone, the coarse error locations cannot provide greater constraints on burst repetition.

The isotropy of the BATSE gamma-ray burst distribution, coupled with its inhomogeneity (as measured by the deficiency of weak gamma-ray bursts) continues to be the most surprising recent observation of gamma-ray bursts, and the one that has eliminated most of the usual Galactic distribution models (Meegan et al. 1992; Briggs et al. 1995). Fig. 4 shows the distribution of 921 BATSE gamma-ray bursts, plotted in Galactic coordinates. The BATSE sky exposure used in the derivation of this map is uniform to within $\pm 20\%$ (Fishman et al. 1994). When corrected for sky exposure, no significant dipole exists with respect to the Galactic center and there is no significant quadrupole moment with respect to the Galactic plane. The inhomogeneity for the measurable bursts in this distribution, as measured by $V/V_{\max}$ (cf., Schmidt 1968), is $V/V_{\max} = 0.32 \pm 0.01$. A value of 0.5 for $V/V_{\max}$ is expected for a homogeneous distribution. Recently, the BATSE intensity distribution has been combined with the PVO intensity distribution to yield a combined data set over almost four decades in intensity (Fenimore et al. 1993). The composite intensity distribution matches well in the overlap region, showing a smooth transition to the $-3/2$ power law expected at the higher intensities.

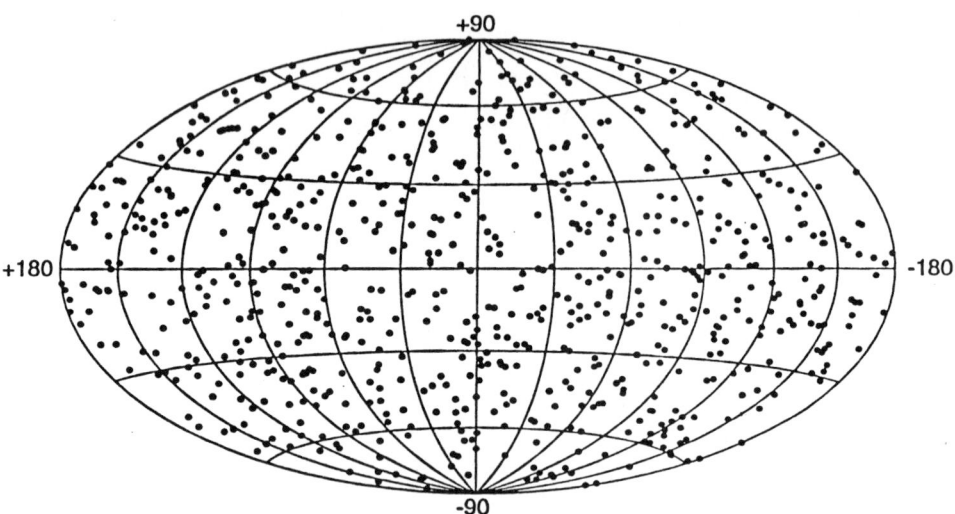

Figure 4. The sky distribution of 921 gamma-ray bursts observed with BATSE on the Compton Gamma-Ray Observatory. The isotropy of the bursts is apparent.

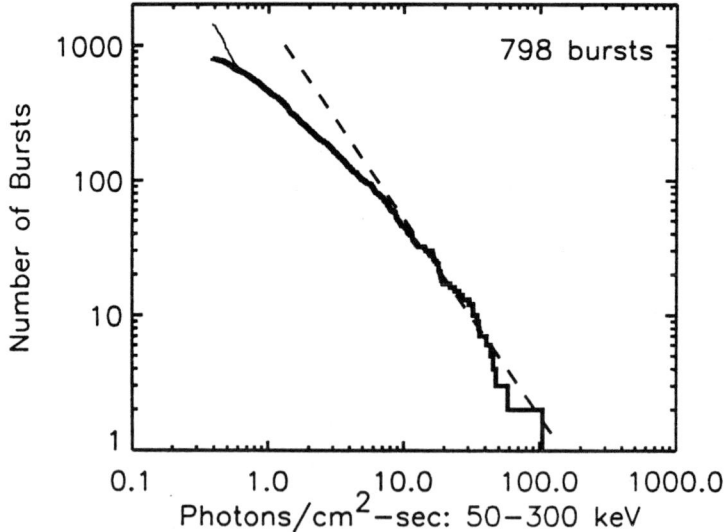

Figure 5. The intensity distribution (peak flux measured on a 256 ms time scale) of gamma-ray bursts, measured with BATSE (from Pendleton *et al.* 1994). There is a clear deviation from a homogeneous distribution (dashed line).

6. Future Observations

Two US spacecraft containing gamma-ray burst instruments are scheduled to begin operation soon: TRGS/WIND in 1994 and HETE in 1995. The TRGS (Transient Gamma-Ray Spectrometer) is an experiment on the US WIND spacecraft (Owens et al. 1991). The detector is a high-resolution, passively cooled germanium detector that operates between 20 keV and 8 MeV. It has a nearly hemispherical field-of-view and a typical energy resolution of about 2 keV.

The HETE (High Energy Transient Explorer) satellite is a small satellite mission dedicated to the study of gamma-ray bursts (Ricker et al. 1992). The prime objective is the precise localization and rapid follow-up observation of gamma-ray burst locations by on-board UV detectors and observatories on the ground. HETE consists of an array of wide-field scintillation detectors which can operate from 6 keV to greater than 1 MeV with good energy resolution, a set of two coded-mask X-ray proportional counters, and an array of sensitive UV CCD detectors. Burst localization to 0.1 degree can be achieved with the X-ray detectors and to 3 arcsec with the CCD, if there is concurrent, detectable UV emission. Data can be distributed in near real-time to a large number of primary and secondary receiving sites for rapid follow-up observations.

There has not been a successful interplanetary probe launched with a gamma-ray burst detector since Ulysses in 1990 (Hurley 1992). The Russian Mars-96 spacecraft will carry several burst detectors. It will become an important component of the Interplanetary Network (IPN) of gamma-ray burst detectors.

7. Summary

The gamma-ray burst enigma appears to be as difficult now as when it was described over 20 years ago (Ruderman 1975). A wealth of new data on time profiles, spectral characteristics and burst distributions has thus far failed to provide conclusive evidence on the distance scale, central object(s) or emission mechanism(s) for the classical gamma-ray bursts. The isotropy and inhomogeneity of the bursts only shows that we are at the center of the apparent burst distribution. Many feel that the identification of a burst with an object in another wavelength region may be the key to understanding these objects. The recent EGRET-Compton Observatory discovery of delayed GeV emission from a burst is yet another severe constraint for many burst models. The field continues to be exciting and frustrating.

Acknowledgements. This paper is based upon lectures given by the author at the Les Houches School of Theoretical Physics/ NATO Advanced

Studies Institute "The Gamma Ray Sky with Compton GRO and SIGMA," January 1994.

References

Band, D. et al. 1993, ApJ 413, 281
Barthelmy, S. et al. 1994, in *Proc. Gamma-Ray Burst Workshop* (1993, Huntsville), G.J. Fishman, J.J. Brainerd & K. Hurley (Eds.), AIP Conf Proc. Vol. 307, p. 643
Brainerd, J.J. et al. 1995, ApJ (in press)
Bhat, N. et al. 1992, Nat 359, 217
Briggs, M. et al. 1995, ApJ (in press)
Fenimore, E. et al. 1993, Nat 366, 40
Fishman, G. et al. 1994, ApJS 92, 229
Fishman, G.J., Brainerd, J.J. & Hurley, K. (Eds.) 1994, *Proceedings Gamma-Ray Burst Workshop* (1993, Huntsville), AIP Conf. Proc. Vol. 307
Ford, L. et al. 1995, ApJ 439, 307
Friedlander, M., Gehrels, N. & Macomb, D. (Eds.) 1993, *Proceedings Compton Gamma-Ray Observatory Symposium* (1992, St. Louis), AIP Conf. Proc. Vol. 280
Hartmann, D. et al. 1995, ApJ (in press)
Hurley, K. 1992, in *Gamma-ray Bursts*, C. Ho, R. Epstein & E. Fenimore (Eds.), Proc. Los Alamos Workshop (1990, Taos NM), Cambridge University Press, p. 183
Hurley, K. et al. 1994, Nat 372, 652
Kouveliotou, C. et al. 1994a, ApJ 422, L59
Kouveliotou, C. et al. 1994b, Nat 368, 125
Kulkarni, S. et al. 1994, Nat 368, 129
Lestrade, J.P. 1994, ApJ 429, L5
Link, B., Epstein, R. & Priedhorsky, W. 1993, ApJ 408, L81
Meegan, C. et al. 1992, Nat 355, 143
Meegan, C. et al. 1995, ApJ (submitted)
Murakami, T. et al. 1994, Nat 368, 127
Nemiroff, R.J. 1994, Comm. in Astrophys. 17, 189
Nemiroff, R. et al. 1994, ApJ 423, 432
Norris, J.P. et al. 1994, ApJ 424, 540
Owens, A. et al. 1991, IEEE Trans. Nuc. Sci. 38(2), 559
Paciesas, W.S. & Fishman, G.J. (Eds.) 1992, *Proceedings Gamma-Ray Burst Workshop* (1991, Huntsville), AIP Conf. Proc. Vol. 265
Palmer, D.M. et al. 1994, ApJ 433, L77
Pendleton, G. et al. 1994, ApJ 431, 416
Piran, T. 1995, these Proceedings
Ricker, G. et al. 1992, in *Gamma-ray Bursts*, C. Ho, R. Epstein & E. Fenimore (Eds.), Proc. Los Alamos Workshop (1990, Taos NM), Cambridge University Press, p. 288
Ruderman, M. 1975, Ann. N.Y. Acad. Sci. 262, 164
Schaefer, B. 1994, in *Proc. Gamma-Ray Burst Workshop 1993*, G.J. Fishman, J.J. Brainerd & K. Hurley (Eds.), AIP Conf. Proc. Vol. 307, p. 382
Schaefer, B. et al. 1992, ApJ 393, L51
Schaefer, B. et al. 1994, ApJ 422, L71
Schmidt, M. 1968, ApJ 151, 393
Share, G. et al. 1992, in *Gamma-ray Bursts*, C. Ho, R. Epstein & E. Fenimore (Eds.), Proc. Los Alamos Workshop (1990, Taos NM), Cambridge University Press, p. 249
Share, G. et al. 1994, in *Proc. Gamma-Ray Burst Workshop* (1993, Huntsville), G.J. Fishman, J.J. Brainerd & K. Hurley (Eds.), AIP Conf. Proc. Vol. 307, p. 283
Vanderspek, R., Krimm, H. & Ricker, G. 1994, in *Proc. Gamma-Ray Burst Workshop* (1993, Huntsville), G.J. Fishman, J.J. Brainerd & K. Hurley (Eds.), AIP Conf. Proc. Vol. 307, p. 438

SOFT GAMMA REPEATERS REVISITED WITH BATSE

C. KOUVELIOTOU
USRA, NASA/MSFC,
ES-84, Huntsville, AL 35812, USA

Abstract. After the first 4 years of its operation, the Burst and Transient Source Experiment (BATSE) onboard the Compton Gamma Ray Observatory (CGRO), detected recurrent emission from two of the three known Soft Gamma Repeater (SGR) sources, SGR 1900+14 and SGR 1806-20. The reactivation of the latter prompted a quick international campaign resulting in the identification of the X-ray counterpart of the source, which also coincides with a compact radio source. The absence of detection of new sources in the 4 years of BATSE operation and the reactivation of two of the three already known SGRs, indicates that these objects are rare. We give here a short review of the previously detected SGR emissions and present the recent results obtained with BATSE.

1. Introduction

Of the over one thousand detected cosmic gamma-ray sources of transient emission, only three have clearly exhibited recurrent activity (Hurley 1986; Atteia *et al.* 1987; Laros *et al.* 1987; Kouveliotou *et al.* 1987). Apart from their recurrent nature, these events differ from the majority of gamma-ray bursts in the following characteristics:

- their temporal profiles often consist of single, simple pulses, usually with a triangular or trapezoidal shape;
- their durations are very short, typically of the order of 0.1 s,
- their spectra are relatively soft, with characteristic temperatures in thermal-bremsstrahlung fits of the order of 20–30 keV.

The objects that emit these events can thus be considered as members of a distinct class – the Soft Gamma Repeaters (SGRs).

The total number of events observed before BATSE from SGR 0526-66, SGR 1900+14, and SGR 1806-20, were 16, 3, and over 100, respectively, with intervals between events that varied stochastically from hours to years (Norris et al. 1991). The embryonic SGR sky distribution suggested that they are galactic objects: the first source is located in the Large Magellanic Cloud and the remaining two are located in the Galactic plane (Norris et al. 1991). Two of the sources (SGR 1900+14 and SGR 1806-20) have been recently detected with BATSE (Kouveliotou et al. 1993, 1994); in the following sections we review these observations together with the previous activity of the sources.

2. SGR 1900+14

2.1. FIRST DETECTED ACTIVATION

On 1979 March 24, 25 and 27, the KONUS experiments onboard the Venera 11-13 spacecraft recorded three soft transient events with very similar properties (Mazets et al. 1979); all three outbursts came from the same direction on the sky, centered at a rather large (almost 0.5 square degree) error box at R.A. = 286.47° and dec = 10.43° (epoch 2000.0). In galactic coordinates the center of that region was located at $b^{II} = 1.6°$ and $l^{II} = 43.8°$, very close to the Galactic plane.

TABLE 1. KONUS events from SGR 1900+14

Date Mar 1979 day (UT)	Dur. ms	Rise time ms	Fluence 10^{-6} erg cm^{-2}	Peak flux 10^{-6} erg cm^{-2} s^{-1}	Peak rate$^\alpha$ cts (15.6 ms)$^{-1}$	ΔT^β hours	kT^γ keV
24.6714	120	<8	1.0	8.3	130	—	35±2
25.0822	190	<8	1.5	7.9	80	9.9	36±2
27.4379	50	<8	0.35	7.0	30	56.5	30±10

$^\alpha$ estimated from Fig. 1a
$^\beta$ time elapsed since previous event
$^\gamma$ from a thermal bremsstrahlung fit

The temporal profiles of the KONUS events are shown in Fig. 1a with a time resolution of 0.0156 s. A feature that strikes the eye is the monotonic decrease of their peak count intensities: the first event is ~1.5 times more intense than the second, which in turn is ~3.0 times more intense than the third. The time separations of the events are ~10 and ~56.5 hours, respectively. Fig. 1b shows their spectra accumulated between 50-150 keV and fitted with a thermal bremsstrahlung function; they are identical within

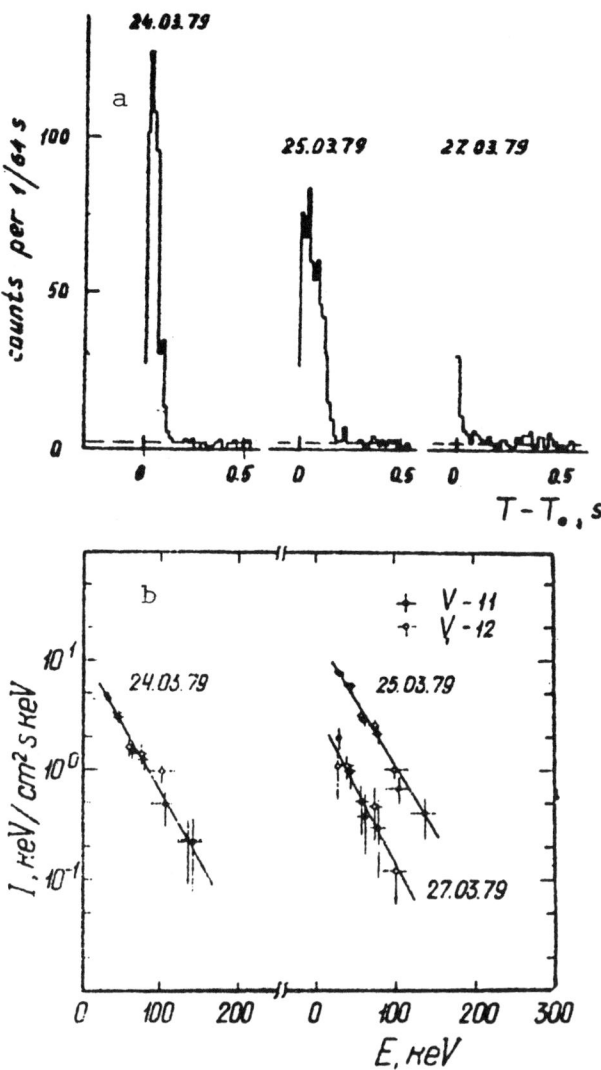

Figure 1. (a) Temporal profiles of the 3 outbursts of SGR 1900+14 detected with the KONUS instrument in 1979, (b) Spectra from the same events (Mazets *et al.* 1979).

errors with an average $kT \sim 34$ keV. Table 1 contains information of these early SGR 1900+14 outbursts, such as durations, temperatures, fluxes, fluences, etc. From Table 1 we see that the average peak flux detected with KONUS was $7.7\,10^{-6}$ erg cm^{-2} s^{-1}. The rise times of <8 ms are upper limits which reflect the instrument time resolution.

According to Mazets et al. (1979), the KONUS instruments did not detect any other emission from this source; their post-factum search extended before the first trigger for a total period of 9 months (1978 September – 1979 May). All in all the source burst for three times in March 1979.

2.2. SECOND ACTIVATION

On June 19, 1992 BATSE was triggered by a short, soft event clearly reminiscent of an SGR emission. The event was located near SGR 1900+14, albeit with a large error box. The center of the error circle was at R.A. = 288° and dec = 11° (epoch 2000.0), with a ~5° 1σ radius. As repetition is a defining characteristic of an SGR, we decided to wait for the next outburst before we would alert the community. We had another reason to "wait it out": a new X-ray transient source, GRS 1915+105, had just erupted in the vicinity (Castro-Tirado et al. 1992) and the BATSE error circle could not exclude it as the origin of the SGR event. In the next two months, BATSE recorded two more SGR-like outbursts coming from the part of the sky including the transient and SGR 1900+14. Fig. 2 shows the three location circles for these events, the BATSE contour map of the Aquila transient and the 2σ "old" error box obtained from the KONUS events (epoch 2000.0). Although the BATSE data do not exclude GRS 1915+105 as the source of the triggered emission, the KONUS error box (which is known with arcmin accuracy) is 1°.5 away from the transient and cannot be related to it.

TABLE 2. BATSE events from SGR 1900+14

Date 1992 DOY (UT)	Dur. ms	Rise time ms	Fluence 10^{-8} erg cm^{-2}	Peak flux 10^{-6} erg cm^{-2} s^{-1}	Peak rate$^\alpha$ kcts s^{-1}	ΔT^β days	kT^γ keV
171.7484A	40	<0.5	4.5	1.1	40	—	—
171.7484B	40	15	4.3	1.0	100	—	39±3
190.2194	<64	<64	3.0	0.5	16	18.5	—
232.2067	80	<5	6.6	0.8	60	42.0	—

$^\alpha$ from Kouveliotou et al. 1993
$^\beta$ time elapsed since previous event
$^\gamma$ from a thermal bremsstrahlung fit

The overall properties of the bursts were very similar to the previous emissions from SGR 1900+14 (see Table 2). Figs. 3a and 3b show the temporal structure of the first event recorded with 0.512 ms time resolution. This event was the longest of the three: it comprised two pulses separated by ~0.5 seconds with a total duration of ~0.545 s. The profiles of the pulses

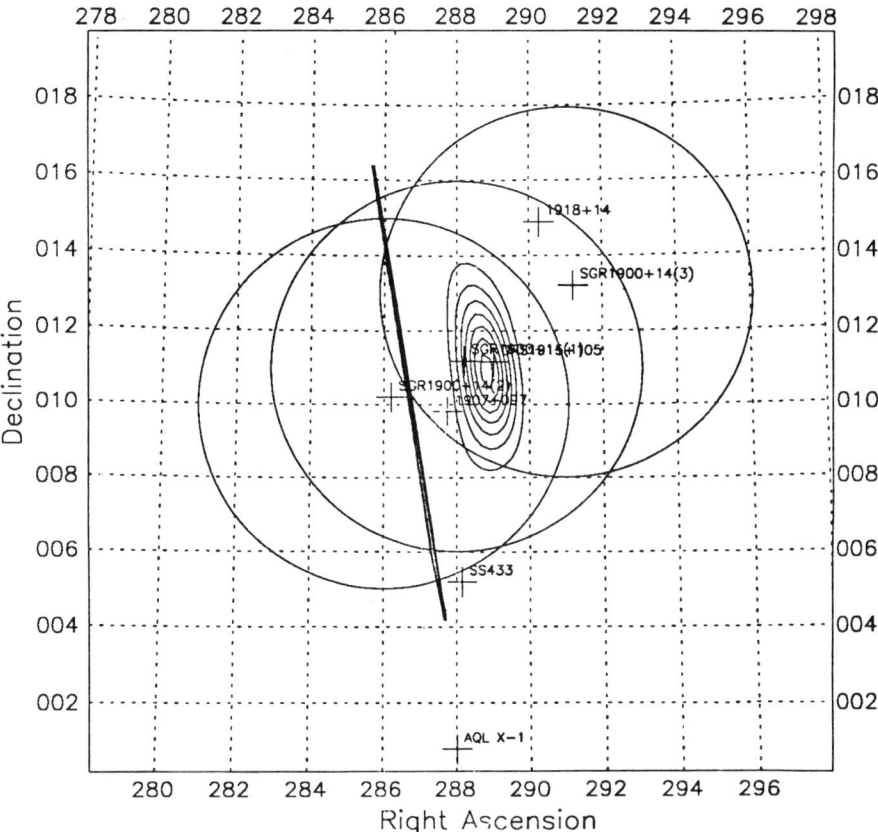

Figure 2. Sky-map of the area around SGR 1900+14. The large circles are the three BATSE detections. The thin vertical box is the KONUS error box of 1979. The intensity contours are centered on the Aquila transient as detected with BATSE.

are trapezoidal and triangular, respectively. Each pulse is described with separate entries in Table 2, indicated with A and B. We notice here that the rise time of the first pulse is well resolved and extremely small: 0.5 ms; it is the shortest *resolved* rise time of an SGR event so far.

Fig. 3c shows the spectrum of the second pulse (the only one for which we had spectral data). The two lines designated by (1) and (2) in Fig. 3c correspond to best fits of an OTTB function to consequtive 16-ms time bins, with $kT = 41 \pm 5$ and 36 ± 3 keV, respectively; the spectra are identical within errors and thus show no spectral evolution during the pulse.

Conclusions: The combined KONUS-BATSE results indicate that activity from SGR 1900+14 was detected again ∼13 years after its discovery.

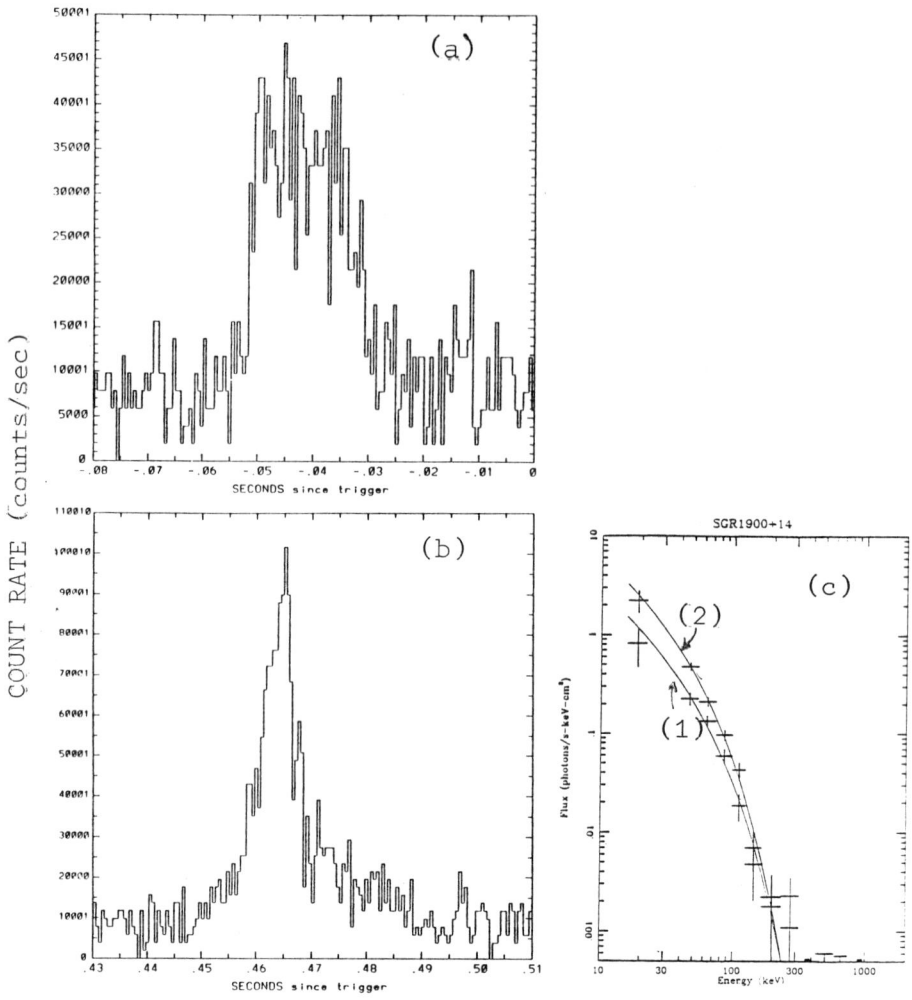

Figure 3. (a) Light curve of the first pulse of the 19 June 1992 trigger, (b) Second pulse of the same trigger. The counts are integrated between 20 and 100 keV. Time resolution is 0.512 ms. (c) Photon spectrum for two consequtive 16-ms bins of the pulse in Fig. 3b. (1) the first 16-ms (2) the second 16-ms. The curves indicate the best-fit OTTB function, of the form $A \exp(-E/kT)/E$, with $kT = 41 \pm 5$ and 36 ± 3 keV, respectively.

The rise times were extremely short (<0.5 ms) and there was no spectral evolution within an event (at the 16-ms resolution). BATSE detected events with peak fluxes ∼10 weaker than the previous detections. If we assume a galactocentric distance for the source, these fluxes would correspond to super-Eddington luminosities of $\sim 10^{41}$ ergs s^{-1}.

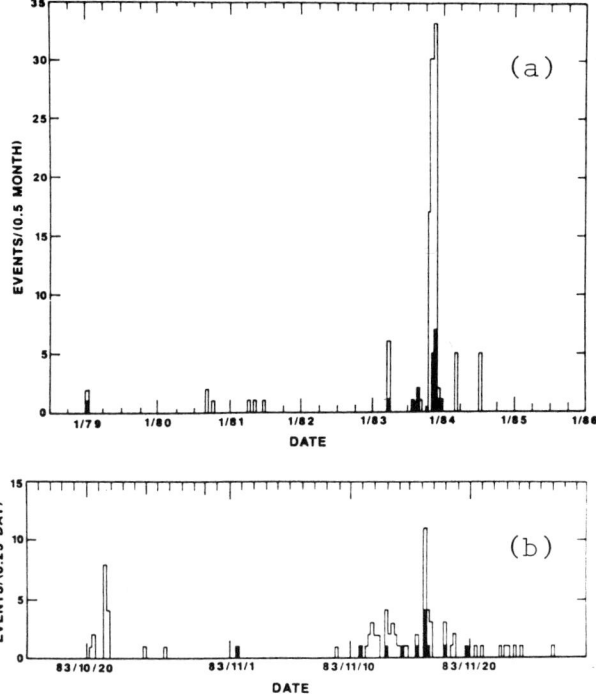

Figure 4. ICE events *vs* time. The filled-in part of each histogram is the subset of events also observed with other spacecraft. In (a) 8 years of data are shown at 0.5 month time resolution; (b) presents 1.5 months of data at 0.25 day resolution. The temporal coverage is ~75% in 1978–1983, and ~20% thereafter (from Laros *et al.* 1987).

3. SGR 1806−20

3.1. EARLIER DETECTIONS

SGR 1806−20 was the last discovered repeater. Although the first emission of this source was immediately recognized as unusual due to the softness of its spectrum (Laros *et al.* 1986), it was later through the efforts of K. Hurley that subsequent emissions were associated with this source. In 1986, he discovered, while investigating the Prognoz satellite data, that several short, soft transient events appeared to come from the same general direction of the sky. A wide search effort (Atteia *et al.* 1987; Laros *et al.* 1987; Kouveliotou *et al.* 1987) to confirm this result, established 18, well-localized emissions with a common intersection at $(\alpha, \delta)_{1950.0} = (18^h06^m, -20°)$. In addition, between 1978 August 13 and 1986 June 27, 111 events were assigned with high probability to the repeater (see Fig. 4) based on their SGR-like properties, although most were observed only with one instrument, the International Cometary Explorer (ICE) (Laros *et al.* 1987).

The statistical properties of these 111 emissions have been studied extensively by Laros et al. (1987). They find little or no correlation between event arrival times and intensities and that clustering persists on shorter time scales (intervals range from 1 s to >1 year). In fact, for the very weak bursts (log $I < 1.4$), 12 of 15 events are associated with intervals shorter than the median interval of 1400 s, while the most intense bursts apparently have no preferred separations (Laros et al. 1987). The event spectra are best fit with an OTTB function with temperatures between 24 and 40 keV; there are strong indications for a spectral turnover below 10 keV. The range in the burst intensities spans a factor of 30; the log N − log P function can be approximated by a power law that flattens at low luminosities. The average peak flux of these bursts was $\sim 3.2 \ 10^{-6}$ ergs cm^{-2} s^{-1} (Norris et al. 1991).

3.2. BATSE DETECTIONS OF SGR 1806−20

On 29 September 1993, BATSE detected three bursts from SGR 1806−20 within 14 hours; the source appeared again on 5 Oct, on 9 Oct, and on 10 Nov 1993, respectively (Kouveliotou et al. 1994). Four of the six bursts have simple, triangular profiles. The third event on 29 September consisted of two brief pulses separated by one second, closely resembling a previous emission from the source in 1983 (Kouveliotou et al. 1987). Table 1 from Kouveliotou et al. (1994) has all the pertinent information on the properties of these outbursts. We note here that their peak fluxes range between 2 and 9 10^{-6} ergs cm^{-2} s, values which are 4 to 15 times smaller than those of the brightest events detected from this source (Norris et al. 1991) and comparable to the ones detected with the Solar Maximum Mission (SMM) (Kouveliotou et al. 1987). The intervals between these 6 bursts range from 0.1 to 31.5 days.

We have derived a spectrum for the third burst of 29 September 1993, over 4.096 s from background-subtracted 16 energy channel data. A fit with an optically-thin thermal bremsstrahlung (OTTB) function yields a value of $kT = 28 \pm 5$ keV, and a normalization of 1.8 ± 0.6 photons (s^{-1} keV^{-1} cm^{-2}) at 1 keV. The kT value is similar to the ones reported before (Atteia et al. 1987; Kouveliotou et al. 1987; Laros et al. 1986; Fenimore et al. 1993), indicating very little change in the emission mechanism over 10 years. The fluences cluster around an average value of $\sim 10^{-7}$ erg cm^{-2}.

Fig. 5 shows the overlapping error circles of the 6 bursts, drawn with the 1σ radius of 8°. The very narrow diamond indicated by I is the updated (K. Hurley, personal communication) SGR 1806−20 location (epoch 2000.0) computed with data from the 2nd Inter-Planetary Network (IPN). Clearly, the BATSE circles overlap the IPN location.

Conclusions: SGR 1806−20 has been found active again $\sim$14 years after

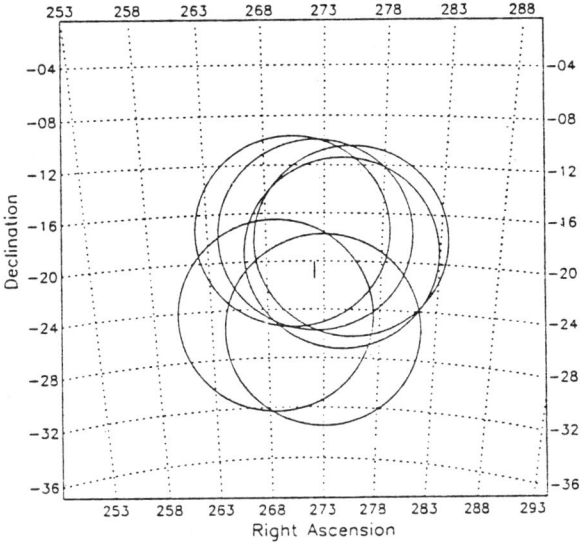

Figure 5. Sky map of the region near $\alpha = 271°$, $\delta = -20°$. The six large circles represent the locations of the BATSE events. The size of the circles (8°) reflects mainly the statistical errors of the measured positions. Burst locations are obtained from the relative strength of their signals in the relevant subset of the 8 identical BATSE Large Area Detectors (LADs). The position determination includes the detailed spectral dependence of the angular sensitivity of the detectors, and corrections for photon back-scattering by the Earth's atmosphere. The narrow diamond (I) represents the 2σ error box of SGR 1806–20 (K. Hurley, private communication).

its discovery. BATSE has detected outbursts 4 to 15 times weaker than the brightest KONUS events. For distances in the range 8.5 to 17 kpc (Kulkarni & Frail 1993) the peak luminosity for the brightest BATSE events is between $7\,10^{40}$ and $3\,10^{41}$ ergs s^{-1}, i.e., the events are highly super-Eddington if they are unbeamed and originate from neutron stars.

4. Counterpart searches

Based on the identification of all three emissions on 29 September with SGR 1806–20, the BATSE team initiated an international multi-wavelength campaign to search for counterparts in the IPN SGR error box. As a result, on 9 October 1993 the Japanese X-ray satellite ASCA detected (Murakami et al. 1994) a burst from the area of the source simultaneous with BATSE. ASCA also discovered (Murakami et al. 1994) the steady X-ray counterpart of SGR 1806–20.

Earlier searches (Kulkarni & Frail 1993) had suggested the identification of SGR 1806–20 with G 10.0–0.3, a SNR which was later identified to be of a rare type, known as a plerion (also known as "filled-center"or "Crab-like"

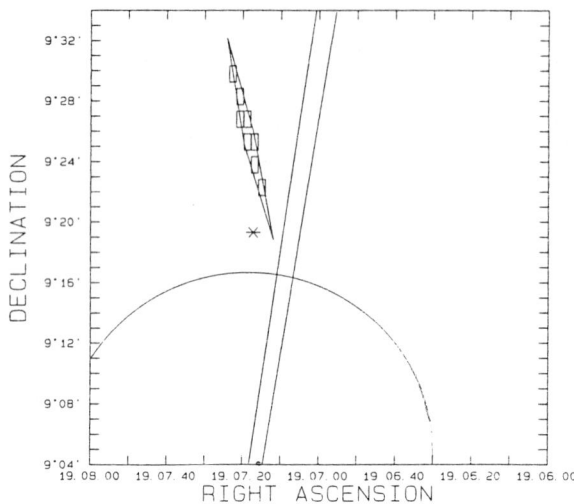

Figure 6. Map of the region around SGR 1900+14. The two vertical lines are the boundaries of the KONUS error box (see also Fig. 2), the diamond with the rectangles is the *virtual* network synthesis error box, the circle encompasses the radio contours of G 42.8+0.6, and the asterisk is at the position of the ROSAT source mentioned in the text (from Hurley et al. 1994).

SNRs). Plerions are young SNRs which have centrally brightened morphologies, and are powered by a central object (known to be a pulsar in the case of the Crab nebula). SGR 0526-66 has been (controversially) associated with a SNR (N49) in the Large Magellanic Cloud (Cline 1981). Although this remnant is not a plerion, it is relatively young (Vancura et al. 1992) and could contain a pulsar.

Given the apparent radio and optical associations found for the other two SGRs, we searched for currently catalogued SNRs (Green 1991; Reich et al. 1990) in the error box of SGR 1900+14, and found three that are overlapped by this error box: G 43.9+1.6, G 42.8+0.6 and G 39.7-2.0 (= W50). Recently, Hurley et al. (1994) have developed a virtual network synthesis method which enabled them to reduce the 430 arcmin^2 error box of SGR 1900+14 by a factor of 70. Fig. 6 shows this smaller error box for the SGR together with the "old" KONUS error box and the outer contours of G 42.8+0.6. Searching in that region Vasisht et al. (1994) have identified a ROSAT source which is very close to both objects (SGR and SNR); it is thus a prime candidate for the counterpart and several planned future observations will shed light on the nature of this object.

5. Discussion

From the fact that reactivation of two of the three known SGRs has been observed with BATSE, but that no new SGRs have been detected, it follows that the total number, N_{SGR}, of SGR sources in the Galaxy with properties similar to those already known, (including repetition time scales), is probably very limited. A precise limit is difficult to derive in view of gaps in monitoring as well as the differences in sensitivity, sky exposure and trigger criteria of the various experiments capable of observing SGRs. We have estimated a rough upper limit on N_{SGR} by assuming approximately equal sky exposures of BATSE and the experiments with which the SGR sources were detected ($\sim$12 and $\sim$13 ster yrs, respectively). For a series of assumed values of N_{SGR} we calculated the probability that at least two of the known sources are detected again with BATSE, but no new SGR sources are detected, and from this we infer that $N_{SGR} < 7$ (95% level). In view of the above mentioned uncertainties we suggest that it is unlikely that our Galaxy contains at any given time more than a few active SGR sources.

Our detections show that SGRs keep their ability to be active for many years. The extended duration of SGR activity strengthens the argument that these sources are related to galactic (likely Population I) objects, quite plausibly neutron stars (Kouveliotou et al. 1987; Norris et al. 1991), as recurrent SGR emissions do not signify a unique (catastrophic) event in the life cycle of the source. The association of SGR 1806−20 with a plerion suggests that the events originate from young neutron stars, with ages less than $\sim 10^4$ years. With neutron star birth rates of order 1 per 100 years (Strom 1994; Lorimer et al. 1993) this would imply that either the active SGR phase lasts less than $\sim 10^3$ years, or that only a small fraction (~ 10 percent) of these young neutron stars become SGR sources. It is currently unknown what special property would turn a young neutron star into an SGR source.

References

Atteia, J.-L. et al. 1987, ApJ 320, L105
Castro-Tirado, A.J. et al. 1992, IAU Circular 5590
Cline, T.L. 1981, Ann. NY Acad. Sci. 375, 314
Green, D.A. 1991, PASP 103, 209
Fenimore, E.E., Laros, J.G. & Ulmer, A. 1993, ApJ 418, 395
Hurley, K. 1986, talk presented at *Taos Gamma Ray Stars Conference*
Hurley, K. et al. 1994, ApJ 431, L31
Kouveliotou, C. et al. 1987, ApJ 322, L21
Kouveliotou, C. et al. 1993, Nat 362, 728
Kouveliotou, C. et al. 1994, Nat 368, 125
Kulkarni, S. & Frail D.A. 1993, Nat 365, 33

Laros, J.G. et al. 1986, Nat 322, 152
Laros, J. et al. 1987, ApJ 320, L111
Lorimer, D.R. et al. 1993, MNRAS 263, 403
Mazets, E.P., Golenetskii, S.V., Guryan, Yu.A. 1979, SvA Lett. 5 (6), 343
Murakami, T. et al. 1994, Nat 368, 127
Norris, J.P. et al. 1991, ApJ 366, 240
Reich, W. et al. 1990, A&AS 85, 633
Strom, R. 1994, in *The Lives of the Neutron Stars*, M.A. Alpar, Ü. Kızıloğlu & J. van Paradijs (Eds.), NATO ASI Vol. C450, p. 23
Vancura, O. et al. 1992, ApJ, 394, 158
Vasisht, G. et al. 1994, ApJ 431, L35

GAMMA-RAY BURSTS AND BINARY NEUTRON STAR MERGERS

TSVI PIRAN
Racah Institute for Physics
The Hebrew University, Jerusalem, Israel 91904

Abstract. Neutron star binaries, such as the one observed in the famous binary pulsar PSR 1913+16, end their life in a catastrophic merger event (denoted here NS^{2}M). The merger releases $\sim 5 \; 10^{53}$ ergs, mostly as neutrinos and gravitational radiation. A small fraction of this energy suffices to power γ-ray bursts (GRBs) at cosmological distances. Cosmological GRBs must pass, however, an optically thick fireball phase and the observed γ rays emerge only at the end of this phase. Hence, it is difficult to determine the nature of the source from present observations (the agreement between the rates of GRBs and NS^{2}Ms providing only indirect evidence for this model). In the future a coinciding detection of a GRB and a gravitational-radiation signal could confirm this model.

1. Introduction

A binary neutron star merger (NS^{2}M) is the last event in the complex evolution of some massive binary systems. A massive binary becomes a massive X-ray binary after one of the stars undergoes a core collapse and a supernova explosion. The second supernova disrupts most systems. A small fraction survives and forms a neutron star binary. There is a chance to detect a neutron star binary if one of the neutron stars is a pulsar. So far three such binaries have been detected. General Relativity predicts that the binary will emit gravitational radiation and its orbit will shrink. Taylor & Weisberg (1982) measured the orbit of PSR 1913+16 and confirmed this prediction with an amazing precision. The gravitational radiation emission from the known binary pulsars is too low for a direct detection. The emission increases as the distance between the two neutron star decreases and gravitational radiation detectors such as LIGO (Abramovici *et al.* 1992) and VIRGO (Giazotto et al. 1988), that are being built now, will be able to detect gravitational signals from the last three minutes (Cutler et al. 1994) of binary neutron stars at cosmological distances.

The outcome of the merger is most likely a rotating black hole of $\sim$2.2–2.8 $M_\odot$ (a massive rapidly rotating neutron star, very near the upper mass limit for a rotating neutron star, cannot be ruled out yet (Davies et al. 1994). In either case the released binding energy is $\sim 5 \; 10^{53}$ ergs. Shortly after the discovery of the first binary pulsar, Clark & Eardley (1977) estimated that most of this energy is emitted as neutrinos, and a smaller fraction as gravitational radiation. The neutrino signal resembles a supernova neutrino burst. At present there is no way to detect such bursts from cosmological distances. Even if such a detector would have existed it would have been impossible to distinguish the rare NS^2M neutrinos bursts from the much more frequent SN neutrino bursts.

NS^2Ms conspire to release their energy in the two channels that are hardest to detect: gravitational radiation and neutrinos. However, if a small fraction of the total energy is channeled into electromagnetic energy it would be detectable. Several years ago Eichler et al. (1989) suggested that $\nu\bar{\nu}$ annihilation ($\nu\bar{\nu} \to e^+e^-$) can convert about 10^{-3} of the total neutrino energy to electromagnetic energy which, in turn, would produce a GRB, a possibility mentioned without a detailed discussion earlier by Goodman (1986), Paczyński (1986), and Goodman et al. (1987). According to this model GRBs signal the final stage in the complicated evolution of what was initially a massive binary system: the merger of a neutron star binary into a black hole (or a rapidly rotating maximal neutron star).

This idea met with skepticism. At the time it was generally believed that GRBs originate from neutron star in the Galaxy (see, e.g., the review by Higdon & Lingenfelter (1990)). However, BATSE on the COMPTON-GRO observatory has revolutionized our understanding of GRBs (Meegan et al. 1992; see also Fishman 1994). The BATSE observations have shown that the angular distribution of GRBs is isotropic and the count distribution is incompatible with a homogeneous one. The isotropy and inhomogeneity rule out a Galactic disk population, which would be either anisotropic (if the sources are at typical distances of more than several hundred pc) or homogeneous (if the sources are at typical distances of less than several hundred pc). This leaves us with the possibility of a galactic halo population (with a very large core radius) or a cosmological population. While the controversy between a galactic (mostly galactic halo) and cosmological populations is still going on, there is a growing consensus (see the contribution of Fishman to this Volume) that GRBs are cosmological. With this emerging consensus the NS^2M model has made the full transition from one of the least likely amongst more than a hundred GRB models to the most conservative one. It is the only model based on an independently observed phenomenon which, as we discuss shortly, takes place at a comparable rate and can easily account for the energy required.

In this lecture I discuss three issues. I examine first the GRB distribution and I show that the count distribution is compatible with a cosmological distribution (an arbitrary inhomogeneous distribution will not necessarily be compatible with a cosmological one). The analysis of the count distribution enables us to estimate the rate of GRBs, which I compare with the estimated rate of NS^2Ms. The agreement between the two rates clearly supports the NS^2M model.

I turn, next, to the issue of fireballs: the sudden release of copious γ-ray photons into a compact region, as required for a cosmological GRB, creates an opaque photon-lepton fireball due to the prolific production of electron-positron pairs. Once this was an argument against any cosmological source (Schmidt 1978; Cavallo & Rees 1978). However, Goodman (1986) and Paczyński (1986) have shown that the fireball expands relativistically, and releases its energy at a later stage when it is optically thin. This optically thick phase between the energy injection and the final emission stage makes the task of deciphering the GRB enigma much more difficult than what was otherwise expected.

Finally, I return to NS^2Ms. I discuss some recent numerical simulations of the hydrodynamics of the merger and I examine how this model satisfies the constraints introduced earlier. I conclude with some open questions and predictions.

2. The Distribution of GRBs

2.1. ANGULAR DISTRIBUTION

The angular distribution of the bursts is isotropic to within the statistical errors of the sample (Fishman 1994). This is clearly in agreement with a cosmological model. The isotropy sets some severe constraints on Galactic halo models (Hartmann 1994), requiring a large homogeneous core (significantly larger than the homogeneous core expected in the dark-matter distribution) and pushing the GRBs to almost inter-galactic distances. The lack of observations of GRBs from M31 sets an upper limit to the possible distance to GRBs and the combined observations begin to rule out Galactic halo models (see however Podsiadlowski *et al.* 1994, for a recent discussion of Galactic models).

2.2. COUNT DISTRIBUTION

A homogeneous population of sources in an Euclidian space-time will produce a cumulative source count distribution $N(>C) \propto C^{-3/2}$. Several relativistic effects influence a cosmological distribution of sources. First, space is not Euclidian, and the relation $F \propto r^{-2}$ which is the basis for $N(>C) \propto C^{-3/2}$ is not valid. Second, the redshift factor, $1 + z$, changes the observed spectrum and introduces a K correction. Third, the number of counts is diluted by a $(1 + z)$ factor if the detector operates within a fixed interval ΔT that is shorter than the total duration of the burst. These effects combine to yield:

$$C(\tilde{L}, z) = \frac{\tilde{L}(1+z)^{2-\alpha}}{4\pi d_l^2(z)}, \qquad (1)$$

for a detector with a fixed energy range, ΔE, that operates for a fixed time interval, Δt, and sources with a photon spectrum: $N_{\rm ph}(E) \propto E^{-\alpha}$. $\tilde{L}$ depends on the luminosity of the source, L, in the relevant energy range, ΔE, on the average energy $\bar{E}$ and on the observation time, Δt: $\tilde{L} = L(\Delta E)\Delta t/\bar{E}$. $d_l(z)$ is the luminosity distance (Piran 1992). The number, $N(>C)$, of events with an observed count rate larger than C is:

$$N(>C) = 4\pi \int_0^\infty n(L)\, dL \int_0^{z(C,L)} \frac{d_l^2}{(1+z)^3} n(z) \frac{dr_{\rm p}(z)}{dz} dz \qquad (2)$$

where $r_{\rm p}(z)$ is the proper distance to a redshift z and $z(C, L)$ is obtained by inverting Eq. 1. $n(L)$ is the luminosity function and $n(z)$ is the intrinsic rate: the number of events per unit proper volume and unit proper time at redshift, z. $N(C)$ depends on the cosmological parameters: Ω and Λ, and on the source parameters: α, $n(z)$ and $n(L)$.

Cohen & Piran (1994) calculated the likelihood function that the BATSE 2B (the 2B catalogue) data results from a cosmological distribution of the form given by Eq. 2 for a variety of cosmological models and source parameters. Following Kouveliotou et al. (1993) and Mao, Narayan & Piran (1993), Cohen & Piran divided the GRB population to two sub-populations, short ($\delta t_{90} \leq 2\,{\rm s}$) and long ($\delta t_{90} \geq 2\,{\rm s}$) bursts [BATSE is more sensitive to the long bursts (Mao et al. 1994), hence it is meaningless to perform the analysis on the whole population as one group].

The maximal red shift up to which BATSE detects long bursts (for standard candles with no source evolution and spectral index $\alpha = 1.5$ (Schaefer et al. 1992) is $z_{\rm max}({\rm long}) = 2.1^{+1.1}_{-0.7}$. With an estimated BATSE detection efficiency of ~ 0.3 this corresponds to $(2.3^{-0.7}_{+1.1}) \, 10^{-6}$ events per galaxy per year (for a galaxy density of $10^{-2}h^3\,{\rm Mpc}^{-3}$, see Kirshner et al. 1983); the rate per galaxy is independent of H_0 and it is only weakly dependent on Ω. For $\Omega = 1$ and $\Lambda = 0$ the typical energy of a burst whose observed fluence is F_7 (in units of 10^{-7} ergs cm^{-2}) equals $(7^{+11}_{-4})\, 10^{50} F_{-7}$ ergs. These numbers vary slightly if the bursts have a wide luminosity function. The distance to the sources decreases and correspondingly the rate increases and the energy decreases if the spectral index is 2 and not 1.5. The rate increases and the luminosity decreases if there is a positive evolution of the rate of bursts with cosmological time.

Short bursts are detected only up to much smaller distances: again assuming standard candles and no source evolution we find $z_{\max}(\text{short}) = 0.4^{+1.1}$. There is no significant lower limit on $z_{\max}$ for this sub-class. The estimate of $z_{\max}(\text{short})$ corresponds to a comparable rate of $(6.3^{-5.6})\, 10^{-6}$ events per year per galaxy and a typical energy of $(3^{+39})\, 10^{49} F_{-7}$ ergs (note that there is no lower limit on the energy or upper limit on the rate since there is no lower limit on $z_{\max}(\text{short})$.

Luminosity functions with an effective width of up to factor of ten fit the data. This width is comparable to, and even wider than, the width of some observed luminosity distributions such as the luminosity functions of different types of supernovae.

Norris et al. (1994) found that the dimmest bursts are longer by a factor of ~ 2.3 compared to the brightest ones. With our canonical value of $z_{\max}(\text{long}) = 2.1$ the bright bursts originate at $z_{\text{bright}}(\text{long}) \sim 0.2$. The corresponding expected ratio due to cosmological time dilation, 2.6, agrees with this measurement. If the interpretation is correct then this result confirms the cosmological model. It implies that intrinsic source evolution is insignificant and it rules out the low $z_{\max}$ obtained form the combined PVO and BATSE data (Fenimore et al. 1993).

3. Fireballs

A cosmological GRB releases 10^{51} ergs (if the energy emission is isotropic) in a very small volume. The rapid rise time observed in some bursts implies that the sources are compact with sizes, R_i, as small as 100 km. This results in what we call a "fireball": an optically thick radiation - electron - positron plasma whose initial energy is larger than its rest mass. The initial optical depth in cosmological GRBs for $\gamma\gamma \to e^+ e^-$ is:

$$\tau_{\gamma\gamma} = f_\gamma E \sigma_T / R^2 m_e c^2 \approx 10^{19} f_\gamma E_{i,51} R_{i,7}^{-2}, \qquad (3)$$

where $E_{i,51}$ is the initial energy of the burst in units of 10^{51} ergs, $R_{i,7}$ is the initial radius in units of 10^7 cm and f_γ is the fraction of primary photons with energy larger than $2m_e c^2$. Since $\tau_{\gamma\gamma} \gg 1$ the system rapidly reaches thermal equilibrium with a temperature: $T = 6.4 E_{i,51}^{1/4} R_{i,7}^{-3/4}$ MeV. At this temperature there is a copious number of e^+e^- pairs that contribute to the opacity via Compton scattering. It is interesting that Eq. 3 yields $\tau_{\gamma\gamma} \gg 1$ even for Galactic halo objects (Piran & Shemi 1993).

The huge initial optical depth prevents us from observing directly the radiation released by the source. The observed γ rays emerge only after the fireball has expanded significantly and became optically thin. The fireball phase determines, therefore, the observational features of GRBs and it screens the specific nature of the energy source. A comparable situation is familiar in stars where the energy generated in the core leaks out through an optically thick envelope and the observed spectra are independent of the details of the energy generation mechanism at the core. Another analogous situation occurs in SNRs where the observed radiation depends just on the total energy of the ejected material and is independent of any other specific features of the source. I review here some essential physics of fireballs [see Piran et al. (1994) for details].

3.1. FIREBALL EVOLUTION

Consider, first, a pure radiation fireball. Initially, the local temperature $T \gg m_e c^2$ and the opacity due to e^+e^- pairs, $\tau_p \gg 1$ (Goodman 1986). The fireball expands and cools until at $T_p \sim 20\,\text{keV}$, $\tau_p \sim 1$ and the photons escape freely as the fireball becomes transparent.

Astrophysical fireballs include baryonic matter in addition to radiation and e^+e^- pairs. The baryons affect the fireball in two ways. The electrons associated with the baryons increase the opacity, $\tau = \tau_p + \tau_b$, delaying the escape of radiation. The baryons are also dragged by the accelerated leptons and this requires a conversion of the radiation energy into a kinetic energy. Thus, two important transitions take place as a loaded fireball evolves: (i) the transition from an optically thin to an optically thick fireball, which takes place at:

$$R_\tau = \left(\frac{\sigma_T E}{\eta m_p c^2}\right)^{1/2} = (6\ 10^{12}\ \text{cm})\ E_{i,51}^{1/2} \eta_4^{-1/2}, \tag{4}$$

and (ii) the transition from a radiation dominated phase to a matter dominated phase which takes place at:

$$R_\eta = 2 R_i \eta = (2\ 10^{11} \text{cm})\ R_{i,7} \eta_4. \tag{5}$$

$\eta \equiv E_i/Mc^2$, the ratio of the initial radiation energy E to the rest energy Mc^2 controls these transitions.

The overall outcome of the fireball depends critically on whether $R_\eta > R_\tau$ or vice versa. If $R_\eta > R_\tau$, most of the energy comes out as high-energy radiation, while if $R_\tau > R_\eta$, the fireball results in a relativistic expanding shell of baryons. Energy conservation dictates that in this case $M\gamma_F \sim E$ and $\gamma_F \sim \eta$. We can classify four situations (Shemi & Piran 1990; Piran et al. 1993): (i) $\eta > \eta_{\text{pair}} = (3\sigma_T^2 E_i \sigma T_p^4/4\pi m_p^2 c^4 R_i)^{1/2} \sim 10^{10} E_{i,51}^{1/2} R_{i,7}^{-1/2}$; the effect of the baryons is negligible. The pair opacity τ_p drops to 1 and $\tau_b \ll 1$ at T_p while the fireball is still radiation dominated and the radiation escapes carrying all the energy. (ii) $\eta_{\text{pair}} > \eta > \eta_b = (3\sigma_T E_i/8\pi m_p c^2 R_i^2)^{1/3} \sim 10^5 E_{i,51}^{1/3} R_{i,7}^{-2/3}$; the optical depth becomes dominated by τ_b. The comoving temperature decreases far below T_p before τ reaches unity. The fireball is, however, still radiation dominated when $\tau_b = 1$ and the escaping radiation carries most of the energy. (iii) $\eta_b > \eta > 1$; the fireball becomes matter dominated before it becomes optically thin. The total energy is the bulk kinetic energy of extreme relativistic baryons. The final Lorentz factor is $\gamma_F \sim \eta$. (iv) $\eta < 1$: this is the Newtonian regime. The rest energy exceeds the radiation energy and the expansion is not relativistic. This is the situation, for example in supernova explosions in which the energy is deposited into a massive envelope.

A quick glance at the corresponding mass limits (the transition from case (ii) to case (iii) for $E_i \sim 10^{51}$ ergs is for $M < 10^{-8} M_\odot$) reveals that case (iii) is the most likely one and even this requires a rather "clean" fireball (the transition to the Newtonian regime is at $M = 3 \; 10^{-3} M_\odot$). Initially such a fireball is radiation dominated and it accelerates with $\gamma \propto r$. The fireball is roughly homogeneous in its local rest frame but due to the Lorentz contraction its width in the observer frame is $\Delta r \sim R_i$, the initial size of the fireball. From R_η onwards the baryons coasts asymptotically with $\gamma_F \sim \eta$.

3.2. ENERGY CONVERSION MECHANISMS

The kinetic energy of the bulk motion of the relativistic baryons can be recovered if shock waves form. The shocks could be either internal (Rees & Meszaros 1994; Narayan et al. 1992; Waxman & Piran 1994) or due to interaction with the ISM (Meszaros & Rees 1992, 1993a,b). Variations of η (and hence in γ_F) as a function of radius become important at R_w:

$$R_w \approx \eta^2 R_i \approx (10^{15} \text{ cm}) \; \eta_4^2 R_{i,7}. \tag{6}$$

If γ_F decreases outward, then inner shells take over outer shells and internal shocks form (Rees & Meszaros 1994). Quite generally this is preceded by an unstable phase (Waxman & Piran 1994). The shocks (Rees & Meszaros 1994; Narayan et al. 1992) or the instability that preceded them (Waxman & Piran 1994) could convert a significant fraction of the kinetic energy of the baryons back to thermal energy. Since $R_w > R_\tau$ (for most reasonable parameters) this takes place in an optically thin region and the emitted photons can escape freely, producing the observed GRB.

If γ_F increases monotonically outward then the observed pulse width increases linearly with the radius: $\Delta r \sim R_i/\gamma_F^2$ and there are no internal shocks. In this case the baryons can still interact with the ISM (Meszaros & Rees 1992, 1993a,b). A similar situation occurs in SNRs where the interaction of the ejecta with the ISM produces a shock in which the kinetic energy of the ejecta is converted into radio emission. The mean free path of a relativistic baryon in the ISM is $\sim 10^{26}$ cm, hence the interaction between the baryons and the ISM cannot be collisional. However, from the existence of SNRs we can infer that a collisionless shock can form (possibly via magnetic interaction). The interaction becomes significant at R_γ where the fireball sweeps an external mass of $M_0/\gamma_F = (E_i/\eta c^2)/\gamma_F$ and loses half of its initial momentum:

$$R_\gamma = \left[\frac{M_0}{(4\pi/3)n\gamma_F}\right]^{1/3} = (1.3\ 10^{15} \text{cm})\ E_{i,51}^{1/3}\eta_4^{-2/3}n^{-1/3} \quad (7)$$

R_w increases with η while R_γ decreases with η. Quite generally $R_\gamma < R_w$ for $\eta > 10^4$ and vice versa for $\eta < 10^4$. We should expect, therefore, that internal shocks will be important for heavily loaded "slow" fireballs while interaction with the ISM will be dominant for lightly loaded "fast" fireballs.

3.3. BEAMING AND TIMING

The fireball reaches relativistic velocities and the emitting source moves relativistically towards the observer. Thus, each observer detects radiation only from a narrow angle $\sim 1/\gamma$. This does not mean, however, that the overall GRB is beamed in a narrow angle. The angular spread of the GRB emission depends on the width of the emitting region $\Delta\theta$ which, depending on the source model, could be as large as 4π (but not less than $1/\gamma$).

The duration of the burst depends on several factors. The fireball appears as a narrow shell whose width could be as short at the original duration of the pulse at the source or significantly longer due to spreading of the pulse:

$$\Delta T_\parallel \approx \begin{cases} (1.5\ 10^{-3}\ \text{sec})\ E_{i,51}^{1/3}\eta_4^{-8/3}n^{-1/3} & \text{for } R > R_w \\ (10^{-3}\ \text{sec})\ R_{i,7} & \text{otherwise} \end{cases} \quad (8)$$

A given observer will detect radiation from an angular scale $1/\gamma$ around his line of sight. This will lead to a typical duration of (Katz 1994a):

$$\Delta T_\perp \approx R_\gamma/\gamma_F c = (5\ \text{sec})\ E_{i,51}^{1/3}\eta_4^{-5/3}n^{-1/3}. \quad (9)$$

The overall duration, $\Delta T = \max(\Delta T_\parallel, \Delta T_\perp)$. The strong dependence of ΔT on η is an advantage, as it provides a possible explanation to the large variability in durations of GRBs.

4. Binary Neutron Star Mergers

Using the three binary pulsars observed in the Galaxy, Narayan, Piran & Shemi (1991) and Phinney (1991) estimated the rate of NS²Ms to be $10^{-5.5\pm.5}$ per year per galaxy. This estimate is probably too high, mostly due to the fact that the current estimates of the distances to pulsars are larger than the one used in those calculations (see, e.g., Bailes 1995). Still, the rate is within the range of rates quoted earlier for GRBs (recall that the rates of long and short bursts are comparable). This is a crucial, albeit indirect, evidence for the NS²M model for GRBs. A small fraction of the total energy released in a NS²M suffices to power a GRB at cosmological distances. Thus the NS²Ms satisfies the two essential requirements for a viable model.

There are still many open questions. The two basic ones are: (i) What is the energy transfer mechanism into the electromagnetic channel? (ii) Is the resulting fireball sufficiently free of baryons to reach the ultra-relativistic velocities ($\gamma \gtrsim 10^2$) needed to produce a GRB?

To address these issues Davies et al. (1994) developed a numerical code that follows neutron star binary mergers and calculates the thermodynamic conditions of the coalesced binary. The process of coalescence, from initial contact to the formation of an axially symmetric object, takes only a few orbital periods. Some material from the two neutron stars is shed, forming a thick disk around the central, coalesced object. The mass of this disk depends on the initial neutron star spins; higher spin rates resulting in greater mass loss, and thus more massive disks. For spin rates that are most likely to be applicable to real systems, the central coalesced object has a mass of $2.4\,M_\odot$, tantalizingly close to the maximum mass allowed by any neutron star equation of state for an object supported in part by rotation. Using a realistic nuclear equation of state we estimate the temperatures after the coalescence: the central object is at a temperature of $\sim 10\,\text{MeV}$, while shocks heat the disk to a temperature of 2–4 MeV.

Fig. 1 depicts a typical density cut perpendicular to the equatorial plan. The disk is thick, almost toroidal, the material having expanded on heating through shocks. This disk surrounds a central object that is somewhat flattened due to its rapid rotation. An almost empty centrifugal funnel forms around the rotating axis and there is practically no material above the polar caps. This funnel provides a region in which a baryon free radiation-electron-position plasma could form (Mochkovich et al. 1994). Neutrinos

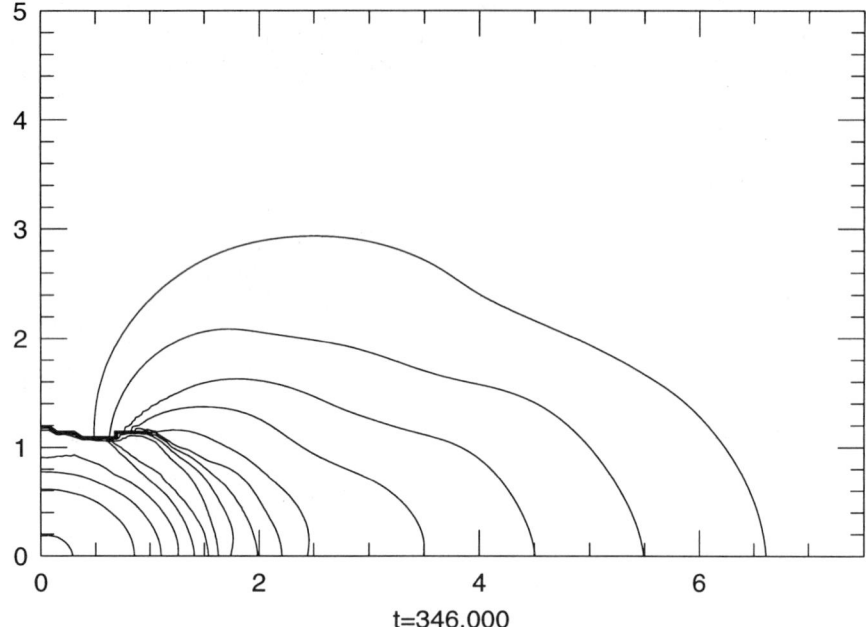

Figure 1. Logarithmic density contour lines at the end of the computation of the merger. The contours are logarithmic, at intervals of 0.25 dex.

and antineutrinos from the disk and from the polar caps would collide and annihilate preferentially in the funnel (the energy in the comoving frame is larger when the colliding ν and $\bar{\nu}$ approach at obtuse angle, a condition that easily holds in the funnel). The numerical computations do not show any baryons in the funnels. The resolution of our computation is insufficient, however, to check whether the baryonic load in the funnel is low enough. The neutrino radiation pressure on polar-cap baryons can generate a baryonic wind that will load the flow. This effect depends strongly on the temperature of the polar caps (Duncan et al. 1986; Woosley & Baron 1992). The estimated temperature from our computations, $\sim 2\,\mathrm{MeV}$, is marginal. Our estimate is, however, least certain in low temperature regions like this.

If the core does not collapse directly to a black hole, it will emit its thermal energy as neutrinos; then the reaction $\nu\bar{\nu} \to e^+e^-$ could convert about 10^{-2} to 10^{-3} of the neutrino flux to electron-positron pairs and produce a GRB. The time scale for the neutrino burst is short enough to accommodate even the shortest rise times observed. Accretion of the disk onto the central object and magnetic field reconnection around the disk (Narayan et al. 1992) are two additional energy sources that could power a GRB. This energy source can operate on a longer time scale independently of the dynamics of the central object.

The numerical calculations support earlier suggestions (Piran et al. 1994) that the energy release is anisotropic and that an empty funnel forms around the rotating axis of the binary system. The fireball is highly non-spherical and it expands along the polar axis and forms a jet. This poses an immediate constraint on the model. If the width of the jet is θ than we observe GRBs only from a fraction $2\theta^{-2}$ of NS^{2}Ms. The rates of GRBs and NS^{2}Ms agree only if $\theta \gtrsim 0.2$ unless the rate of NS^{2}Ms is much higher than the current estimates. In fact, Tutukov & Yungelson (1994) suggested recently that most neutron star binaries are born with a short orbital period and their life time is too short to detect them as binary pulsars. Such systems would escape the binary pulsar statistics and could increase by one or two orders of magnitude the rate of NS^{2}M.

The duration and spectra of GRBs vary from one burst to another. The fireball phase determines both the duration and the spectra of the bursts. The source might add variability by producing fireballs with different Lorentz factors and different initial durations. Within the funnel the baryonic load varies as a function of the angular position leading to varying final Lorentz factors that, in turn, produce bursts with different durations and spectra. Another source of variability could arise from the interplay between the two energy sources in NS^{2}Ms: neutrino annihilation and accretion energy of the disk. These mechanisms would operate on different time scales and produce bursts that look different. An additional source of diversity (Davies et al. 1994) is the distinction between systems that collapse directly to a black hole and those that undergo a longer rotating core phase. Finally, black hole-neutron star binaries are predicted to be as common as neutron star binaries (Narayan et al. 1991). Black hole - neutron star mergers (Paczynski 1991) would produce GRBs with different characteristics than NS^{2}Ms.

NS^{2}M events can take place in a variety of host systems including dwarf galaxies, or even in the intergalactic space if in the process of formation the neutron star binary is ejected from the host galaxy (Narayan et al. 1992). Hence, unlike in other cosmological models it is not essential that optical counterparts will be observed in the error boxes of GRBs (Schaefer et al. 1994). A unique prediction of the NS^{2}M model is that gravitational radiation signals from the final stages of the merger should accompany GRBs (Piran 1992; Kochaneck & Piran 1993) and vice versa (the latter is true only up to the anisotropic emission factor discussed earlier). This coincidence, which could serve to increase the sensitivity of the gravitational radiation detectors (Kochanek & Piran 1993), would prove or disprove this model.

Acknowledgements. I thank Ehud Cohen, Tsafrir Kollat, Ramesh Narayan and Eli Waxman for many helpful discussions. This research was supported by a BRF grant to the Hebrew University and by a NASA grant NAG5-1904.

References

Abramovici, A. et al. 1992, Science 256, 325.
Bailes, M. 1994, these Proceedings
Cavallo, G. & Rees, M.J. 1978, MNRAS 183, 359
Clark, J.P.A. & Eardley, D. 1977, ApJ 215, 311
Cohen, E. & Piran, T. 1994, (preprint)
Cutler, C. et al. 1994, Phys. Rev. Lett. 70, 2984
Davies, M.B. et al. 1994, ApJ (in press)
Duncan, R., Shapiro, S.L. & Wasserman, I. 1986, ApJ 309, 141
Eichler, D. et al. 1989, Nat 340, 126
Fenimore, E.E. et al. 1993, Nat 366, 40.
Fishman, J. 1994, these Proceedings
Giazotto, A. et al. 1988, Nucl. Instrum. Meth. A289, 518
Goodman, J. 1986, ApJ 308, L47
Goodman, J., Dar, A. & Nussinov, S. 1987, ApJ 314, L7
Hartmann, D. 1994, in *Gamma-Ray Bursts, Second Workshop*, G.J. Fishman, J.J. Brainerd & K. Hurley (Eds.), AIP Conf. Proc. Vol. 307, p. 562
Higdon, J.C. & Lingenfelter, R.E. 1990, ARA&A 28, 401
Katz, J.I. 1994a, ApJ 422, 248
Katz, J.I. 1994b, preprint Astro-ph 9405033
Kirshner, R.P. et al. 1983, AJ 88, 1285
Kochaneck, C. & Piran, T. 1993, ApJ 417, L17
Kouveliotou, C. et al. 1993, ApJ 413, L101
Mao, S., Narayan, R. & Piran, T. 1994, ApJ 420, 171
Meegan, C.A. et al. 1992, Nat 355, 143
Mészaros, P. & Rees, M.J. 1992, MNRAS 258, 41P
Mészaros, P. & Rees, M.J. 1993a, ApJ 405, 278
Mészaros, P. & Rees, M.J. 1993b, ApJ 418, L59
Mészaros, P. & Rees, M.J. 1994, preprint, Astro-ph 9404056.
Mochkovich, R. et al. 1994, in *Gamma-Ray Bursts, Second Workshop*, G.J. Fishman, J.J. Brainerd & K. Hurley (Eds.), AIP Conf. Proc. Vol. 307, p. 537
Narayan, R., Paczyński, B. & Piran, T. 1992, ApJ 395, L83
Narayan, R., Piran, T. & Shemi, A. 1991, ApJ 379, L1
Norris, J.P. et al. 1994, ApJ 423, 432
Paczyński, B. 1986, ApJ 308, L51
Paczyński, B. 1991, Acta Astron. 41, 257
Phinney, E.S. 1991, ApJ 380, L17
Piran, T. 1990, in *Supernovae*, J.C. Wheeler, T. Piran & S. Weinberg (Eds.), World Scientific (Singapore)
Piran, T. 1992, ApJ 389, L45
Piran, T. 1994, in *Gamma-Ray Bursts, Second Workshop*, G.J. Fishman, J.J. Brainerd & K. Hurley (Eds.), AIP Conf. Proc. Vol. 307, p. 495
Piran, T. & Shemi, A. 1993, ApJ 403, L67
Piran, T., Narayan, R. & Shemi, A. 1992, in *Gamma-Ray Burst, Huntsville*, W.S. Paciesas & G.J. Fishman (Eds.), AIP Conf. Proc. Vol. 265, p. 149
Piran, T., Shemi, A. & Narayan, R. 1993, MNRAS 263, 861
Podsiadlowski, P., Rees, M.R. & Ruderman, M. 1994, (preprint)

Rees, M.J. & Mészaros, P. 1994, ApJ 430 L93
Schaefer, B.E. et al. 1992, in *Gamma-Ray Burst, Huntsville*, W.S. Paciesas & G.J. Fishman (Eds.), AIP Conf. Proc. Vol. 265, p. 180
Schaefer, B. et al. 1994, *Gamma-Ray Bursts, Second Workshop*, G.J. Fishman, J.J. Brainerd & K. Hurley (Eds.), AIP Conf. Proc. Vol. 307, p. 382
Schmidt, W.K.H. 1978, Nat 271, 525
Shemi, A. & Piran, T. 1990, ApJ 365, L55
Taylor, J.H. & Weisberg J.M. 1982, ApJ 253, 908
Tutukov, A.V. & Yungelson, L.R. 1994, MNRAS 268, 871
Waxman, E. & Piran, T. 1994, ApJ 433, L85
Woosley, S.E. & Baron, E. 1992, ApJ 391, 228

Discussion

P.C. Joss: I have a comment that is pertinent both to your talk and to that of the previous speaker. You have concentrated your attention on an essentially cosmological model for γ-ray bursts, and Gerry Fishman argued that there is no reason to invoke more than one mechanism for producing γ-ray bursts and that, in particular, Galactic neutron stars are now excluded as a source of γ-ray bursts. In fact, as shown by Gerry in his talk, γ-ray bursts encompasses a very wide range of phenomenology in terms of overall bursts duration, temporal structure, spectral shape, and mean photon energy. I'd like to suggest that it would, if anything, be somewhat surprising if this broad range of phenomenology was due to a single underlying physical mechanism and that, in particular, it may be premature to exclude Galactic neutron star as the source of a modest fraction (perhaps 10%) of observed γ-ray bursts. The current situation is reminiscent of that surrounding cosmic X-ray bursts twenty years ago. There, with a much narrower range of phenomenology, Occam's razor was frequently invoked to argue that all X-ray burst have a single physical source. This argument led theorists to speculate about a wide variety of exotic burst mechanisms, such as massive accreting black holes in the cores of Globular clusters, etc. When the work of Walter Lewin and his collaborators demonstrated that there were, in fact, two distinct types of X-ray bursts with different physical origins the theoretical situation was greatly clarified. Perhaps the same sort of clarification lies ahead in the future of our understanding of γ-ray bursts.

T. Piran: I completely agree that the variability of GRBs suggests that there may be two distinct populations of sources. I tried to stress in my talk that the NS^2M model allows for a lot of variability, but this might not be enough and one could easily hide a sub-class of about 10% of Galactic bursts amongst the observed GRBs. In fact, I was led to the NS^2M model while considering NS^2Ms as gravitational radiation sources. I realized that NS^2M release their binding energy in practically invisible channels and I was looking for a way to convert a fraction of this energy to electromagnetic energy. I was certain that if such a mechanism exists then NS^2Ms would be observed via this channel. This has led me to the NS^2M-GRB model (Eichler et al. 1989, Nat 340, 126; Piran 1990, in *Supernovae*, J.C. Wheeler, T. Piran & S. Weinberg (Eds.), World Scientific [Singapore]) in spite of the consensus that existed at the time that GRBs originate from Galactic neutron stars. In 1990 I was asked, in a seminar on this model at the IOA, if I really believed that GRBs are not Galactic. My reply was: "GRBs are such a diverse phenomenon that I could easily imagine that a small sub-class, say of 10%, are cosmological". I am glad that now after the BATSE observations the situations has reversed and in reply to your question I would say that "GRBs are such a diverse phenomenon that I could easily imagine that a small sub-class, say of 10%, are Galactic".

W. Kundt: You mentioned the difficulty for most models to account for the delayed super-high-energy detections ($\geq 10\,\mathrm{GeV}$) from a few GRBs. In our contribution to the Huntsville workshop, we explained them as the transient switch of a Geminga-like behaviour [see Kundt & Chang 1993, A&SS 200, 151].

T. Piran: Several mechanisms, within the context of the fireball model, could produce the delayed super-high-energy detections ($\geq 10\,\mathrm{GeV}$): (i) ultra-high-energy photons produced at the shock of the fireball with the ISM (the delay is between the emission from the internal shocks and the ISM shock) (Meszaros & Rees 1994). (ii) Production of the ultra-high energy photons from interaction of extremely relativistic protons with a dense cloud of interstellar matter in the host galaxy (the delay arises from the fact that the trajectory of the protons is not directly towards us; see Katz 1994b). (iii) Production of the ultra-high-energy photons from interaction of extreme relativistic protons with intergalactic protons or CMBR photons. The time delay arises from a time of flight delay of the protons.

Author Index

Abramovici, A., 18, 154–5, 157, 160, 163, 489
Abramowicz, M.A., 52
Acierno, M.J., 418
Aizu, K., 322–3
Aldering, G., 116, 147
Allakhverdiyev, A.O., 258
Allen, B., 207
Aller, L.H., 431
Alpar, M.A., 226, 236, 303, 251, 314
Anderson, B., 213–4, 217, 219,
Anderson, S.B., 201, 279, 281–2
Angebault, L.P., 428
Angelini, L. 303
Antia, H.M., 32
Ao, C.O., 276
Apostolatos, T.A., 164–7
Appel, A.W., 5
Applegate, J.H., 252, 257
Arnett, W.D., 133, 143
Arzoumanian, Z., 201, 203, 274, 382
Asai, K., 293
Aschenbach, B., 260
Ashtekar, A., 166
Atteia, J.-L., 477, 483–4
Augusteijn, T., 53
Avni, Y., 398

Backer, D.C., 197ff, 207–8, 225, 229, 274, 382
Bahcall, J.N., 379, 382

Bailes, M., 38, 89, 161, 195, 214, 218–220, 227–8, 230, 232, 236, 245, 257, 260, 497
Bailey, J., 409
Bailyn, C.D., 25, 203, 236, 279, 348, 352, 357, 363–4, 369, 434
Band, D., 470
Barbon, R., 111, 457, 460
Barkat, Z., 146
Barnes, J.E., 5
Baron, E., 75, 240, 498
Barthelmy, S., 472
Bartolini, C., 353–4
Bartunov, O.S., 130
Baym, G., 23
Becker, R.H., 257
Becker, W., 228
Beckwith, S.V.W., 187
Begelman, M.L., 38
Bell, J.F., 228–9
Belloni, T., 304, 390
Bender, P., 175, 178, 181
Bennett, D.P., 207
Benz, W., 26
Bessell, M.S., 228
Bethe, H.A., 170
Beuermann, K., 404, 409, 411, 413, 443
Bhat, N., 468
Bhattacharya, D., 12, 33, 35, 61, 99, 121, 193, 218–9, 243, 245–6, 283, 289, 294

Biehle, G.T., 12, 30–1, 34
Bildsten, L., 22, 24, 162, 168, 309
Binney, J., 336
Bisnovatyi-Kogan, G.S., 35, 82, 84, 91, 245–6
Bjorkman, J.E., 60–1
Blanchet, L., 166
Blandford, R.D., 91, 200, 236–7, 257
Blaskiewicz, M., 198
Blinnikov, S.I., 130
Blondin, J.M., 246
Bodenheimer, P., 5, 9–11, 32, 418
Boesgaard, A.M., 32
Bonnell. I.A., 172
Bonnet–Bidaud, J.M., 292–3
Bonsema, P.F.J., 252
Boothroyd, A.I., 32
Boriakoff, V., 225, 229
Bouchet, F.R., 207
Bradaschia, C., 18, 157
Bradt, H., 391
Brainerd, J.J., 468, 472–3
Branch, D., 120, 135
Brandt, W.N., 34, 38
Briggs, M., 473
Brown, G.E., 96, 249
Brown, T., 429
Brumberg, V.A., 91
Buccheri, R., 229
Bunk, W., 389
Burderi, L., 73, 78
Burkert, A., 6
Burrows, A., 170, 283
Burwitz, V., 407, 409

Caldwell, R.R., 207
Callanan, P.J., 53, 351, 353, 357–9, 429
Cameron, A.G.W., 31

Camilo, F., 199, 201, 218, 227, 229, 232–3, 238, 250
Campbell, C.G., 404
Campbell, L., 459
Campbell-Wilson, D., 364, 366, 369–370, 373
Canal, R., 26
Cannizzo, J.K., 44–6, 49–50
Cannon, R.C., 12, 30ff
Cappellaro, E., 111ff, 133, 423
Caraveo, P.A., 259
Casares, J., 39 , 342ff
Cassinelli, J.P., 60–1
Castro-Tirado, A.J., 480
Caswell, J.L., 258, 261
Cavallo, G., 491
Centrella, J.M., 24, 172
Chandrasekhar, S., 18, 20, 174
Chang, H., 502
Chanmugam, G., 408
Charbonneau, P., 40
Charles, P.A., 33, 39–40, 344ff, 352, 357–8
Chen, K., 25
Chen, W., 49, 51, 53
Chen, X., 52
Cheng, F.H., 51
Chernoff, D., 18–9, 165, 167
Chevalier, C., 53, 348, 356ff
Chevalier, R.A., 13, 35, 96–7, 139, 249, 257
Christodoulou, D., 169
Clark, D.H., 258
Clark, J.P.A., 21, 490
Clayton, D.D., 31
Cline, T.L., 486
Clocchiatti, A., 130
Cognard, I., 200–1
Cohen, E., 492

Colgate, S.A., 25
Conti, P.S., 39, 136
Cook, G.B., 23
Cool, A., 394, 395
Corbet, R.H.D., 232, 291, 293
Cordes, J.M., 75, 89, 198–201, 213–8, 229, 238, 259, 271, 274
Cordova, F., 404
Cornwell, T.J., 257
Coté, J., 238, 256
Counselman, C.C., 4
Cowley, A.P., 99, 415–6, 429–30, 436, 439–42
Crampton, D., 428, 439–40
Crary, D., 365
Cropper, M., 409
Crotts, A.P.S., 146
Cumming, R.J., 147–8
Cutler, C., 18–9, 22–4, 162, 164–9, 490
Czerny, M., 334

D'Amico, N., 258
D'Antona, F., 65
Damour, T., 167
Davies, M.B., 19, 24, 168, 385, 490, 497, 499
Davies, R.E., 84
De Kool, M., 5, 62, 93, 96, 418, 426
De Loore, C., 13, 143, 146
Deeter, J.E., 290, 293–4
Della Valle, M., 363, 436
Demiański, M., 195
Dempsey, R., 336–7, 398
Dermer, C.D., 261
Detweiler, S., 207
Dewey, R.J., 33, 75, 89, 195, 201, 217–8, 229
Di Stefano, R., 76, 415–7, 419, 422
Djorgowski, S., 229, 394

Dobrzycka, D., 457ff
Doroshenko, V.T., 452
Dougherty, S.M., 267
Downs, G.S., 274
Duncan, R.C., 261, 493
Dwarakananth, K.S., 260

Eardley, D.M., 21, 490
Ebisawa, K., 297, 304
Eggleton, P.P., 97, 142
Eichler, D., 19, 490, 502
Elias, J.H., 108, 135
Ensman, L.M., 129
Epstein, R., 470
Ergma, E., 65, 73ff
Evans, C.R., 25

Fabian, A.C., 146, 226, 327–8
Falk, S.W., 143
Fauci, F., 229
Faulkner, J., 46
Fenimore, E.E., 473, 484, 492
Ferrario, L., 409
Ferras-Mello, S., 459
Filippenko, A.V., 108ff, 116, 120–1, 127–30, 135–6, 138, 147–8
Finger, M.H., 303, 313–5
Finn, L.S., 18–9, 165, 167, 170, 180
Fishman, G.J., 314, 467ff, 490–1
Flanagan, E.E., 19, 23, 165–6, 168
Flannery, B.P., 13, 283
Fleming, T.A., 427
Fomalont, E.B., 214
Ford, L., 472
Forman, W., 389
Fortner, B., 303
Foster, R.S., 199, 208, 229, 274, 382
Frail, D.A., 187, 215, 218, 227–9, 257ff, 485

Frank, J., 50, 65, 323
Fransson, C., 116, 127, 130, 132
Freese, K., 246
Friedlander, M., 468
Friedmann, J.L., 174
Fruchter, A.S., 203, 228–9, 252, 398
Fujimoto, M.Y., 11, 425–6, 434
Fujimoto, R., 323
Fukushima, T., 205

Gänsicke, B., 410
Gamow, G., 30
Garcia, M.R., 359, 459, 461
Garmany, C.D., 39
Garnavich, P.M., 404
Gehrels, N., 49, 53, 468
Geppert, U., 246
Ghosh, P., 46, 49, 57, 59ff, 85, 245, 314, 317–8
Giazotto, A., 490
Gingold, R.A., 5
Giovannelli, F., 313
Goldstein, H., 20
Gontikakis, C., 53, 65
Goodman, J., 378, 490–1, 494
Goss, W.M., 257, 260
Gott, J.R., 216
Graziati, L.S., 313
Grebenev, S.A., 304, 309
Green, D.A., 486
Greengard, L., 5
Greiner, J., 428, 430, 433, 440–1
Grindlay, J.E., 76, 236, 279, 390, 393–395, 399, 434
Groth, E.J., 274
Grove, E., 57, 318
Guinot, B., 204–5
Gunn, J.E., 38, 213, 216, 237

Haberl, F., 405
Habets, G.H.M.J., 13, 93–4, 121, 129, 285
Hachisu, I., 127
Hack, M., 460
Hameury, J.M., 51–3, 65, 408, 410
Hamuy, M., 112ff, 136
Hansen, B.M.S., 193
Harkness, R.P., 107, 109, 121, 135
Harlaftis, E., 349, 359
Harmon, B.A., 313, 327, 364, 367, 373
Harpaz, A., 65, 250
Harrison, E.R., 219
Harrison, P.A., 214, 258
Hartmann, D., 473, 491
Hashimoto, M., 121, 125, 129, 132
Hasinger, G., 302–4, 389, 391, 394–5, 415, 430–1, 439, 442–3
Haswell, C.A., 342, 351ff
Hatchett, S., 296–7
Heathcote, S.R., 146
Heggie, D.C., 379–80, 382, 387
Heise, J., 82, 431, 445
Helfand, D.J., 198, 257, 259
Hellier, C., 292–3
Hellings, P., 143
Hellings, R.W., 198, 207
Henrichs, H., 59
Hernquist, L., 5, 257
Hertz, P., 76, 390–1, 394, 399, 434
Hester, J.J., 229
Higdon, J.C., 490
Hillebrandt, W., 146
Hills, J.G., 33, 78, 379–80, 382
Hils, D., 178, 181
Hirose, M., 47, 352
Hjellming, R.M., 364, 370, 372, 457
Ho, L.C., 138, 147
Holt, S.S., 428

Horne, K., 344
Houser, J.L., 172
Hsu, J.J.L., 34, 141ff
Hu, W., 199
Huang, M., 49–50
Hughes, J.P., 416
Hughes, S.A., 168
Hulse, R., 159
Humphreys, R.M., 35, 147
Hunstead, R.W., 364, 366, 369–70, 372
Hurley, K., 468ff, 475, 477, 484–6
Hut, P., 5, 377ff
Hutchings, J.B., 439

Iben Jr, I., 5, 25, 89, 141, 253, 254, 418, 425, 431, 436, 418
Ichikawa, S., 47
Illarionov, A.F., 59–60, 82, 87, 455
Ilovaisky, S.A., 53, 348, 356ff
Inoue, H., 307, 321, 326, 335–6, 398
Isern, P., 26
Ishida, M., 322
Ishizaka C., 46
Iwamoto, K., 125

Jacoby, G.H., 422
Jahan Miri, M., 246
Jaranowski, P., 21, 165
Jauncey, D., 364, 370
Jeffery, D.J., 109ff
Jeffrey, L.C., 252
Johnston, H.M., 52, 356, 389, 394–6
Johnston, K.J., 372
Johnston, S., 57ff, 84, 89, 91, 227–30, 236, 257–8, 263ff, 272
Jordan, S., 409, 435
Joss, P.C., 34, 64, 141ff, 232–3, 236, 250, 391
Junker, W., 19

Kahabka, P., 391, 427, 430–3, 439, 445, 449
Kallman, T.R., 296, 421
Kalogera, V., 100
Kaluzienski, J.L., 52
Karitskaya, E.A., 353
Kasim, N.E., 257ff, 502
Kaspi, V.M., 58, 84, 91, 198ff, 205–7, 258, 271ff
Kato, T., 148, 348, 359
Katz, J.I., 389, 496
Katz, N., 5
Kelley, R.L., 289–90, 293–4
Kembhavi, A.K., 32
Kennefick, D., 169
Kennicutt, R.C., 136
Kenyon, S.J., 425, 452, 457ff
Kesteven, M., 364, 370
Kidder, L.E., 19, 164–5
Kim, S.W., 52
King, A.R., 47, 50–3, 73, 78, 343, 404, 410
King, I., 397
Kippenhahn, R., 143, 418
Kirshner, R.P., 109, 116, 125–6, 132, 135, 492
Kitamoto, S., 292–3, 295, 298–9, 304
Kley, W., 6
Kochanek, C.S., 22, 24, 61, 162, 168, 499
Kohl, K., 418
Kolb, U., 49, 65, 69–71, 89, 405
Kolev, D., 452
Komberg, B.V., 82, 91, 245–6
Konar, S., 246
Kondo, Y., 26, 96, 284
Kopal, Z., 4
Kornilov, V.G., 81–2, 88–9
Kotani, K., 329

Kouveliotou, C., 470, 472, 477ff, 492
Kovetz, A., 418
Kroeger, R.A., 366
Krolak, A., 21, 165, 167
Krolik, J.H., 37, 421
Kudritzki, R.P., 35
Kuerster, M., 442
Kuijpers, J., 411
Kulkarni, S.R., 62, 198, 218, 229, 236, 238, 250, 254, 257–8, 261, 325, 334, 398, 472, 458
Kumar, P., 275
Kundt, W., 502
Kuulkers, E., 53, 309
Kylafis, N.D., 416, 428, 440

Lai, D., 17ff, 172
Laidler, V.G., 108
Lamb, D.Q., 410
Lamb, F.K., 303, 314, 317–8
Lamzin, S.A., 35
Landau, L.D., 30
Langmeier, A., 306
Lapshov, I., 327
Laros, J.G., 477, 483–4
Lasota, J.P., 43, 51–3, 100, 335, 410
Leavitt, H.S., 441
Leedjärv, L., 454
Leibundgut, B., 107ff, 135, 142
Lemonick, M.D., 147
Leonard, P.J.T., 25, 33–4
Lestrade, J.-F., 200
Lestrade, J.P., 468
Levine, A., 78, 291, 293–4
Levreault, R., 109, 135
Levy, E.H., 193
Lewin, W.H.G., 333, 341, 391
Lewis, J.R., 108ff
Lewis, W., 232–3, 250

Li, H., 261
Lin, D.N.C., 46, 49
Lincoln, C.W., 19, 164
Lindblom, L., 175
Lingenfelter, R.E., 490
Link, B., 470
Lipunov, V.M., 81ff, 100, 102, 161–2, 276, 454–5
Livio, M., 5, 25–6, 46–7, 49, 53, 77, 418, 436
Lombardi, J., 384
Long, K.S., 334, 415, 426–7, 439
Lorimer, D.R., 34, 75, 98, 213–5, 219, 221, 230–3, 235ff, 250, 259, 261, 487
Lubow, S.H., 47
Lucy, L., 5, 121
Ludwig, K., 46
Lund, N., 49
Lundgren, S.C., 218, 229
Luthardt, R., 452
Lyne, A.G., 34, 73–4, 78, 98, 195, 202, 213–5, 217, 219, 226–7, 240, 259, 261, 399

Mönchmeyer, R., 170
Macomb, D., 468
Macri, L., 422
Maeder, A., 136
Magnier, E., 393
Makino, J., 380
Makino, Y., 304, 306–7
Malhotra, R., 188, 192
Mammano, A., 457, 460
Manchester, R.N., 61, 91, 215, 231, 258–9, 265, 269
Mao, S., 492
Marcaide, J.M., 132
Margon, B., 394, 396

Marinus, M., 379
Markovic, D., 19, 167
Marsh, T.R., 344, 347, 352, 356
Martin, E.L., 39, 346–7
Mason, K.O., 405, 428
Massey, P., 39, 138
Masters, R., 410
Matheson, T., 147
Matsuoka, M., 323
Mazets, E.P., 478–9
McAdam, W.B., 258
McClintock, J.E., 323, 335–6, 341–4, 352ff, 391
McClure, R.D., 111, 114, 116
McConnell, D., 271
McCray, R.A., 296–7, 421
McKay, D.J., 364, 370
McMillan, S.L.W., 24, 380, 382
Mead, R., 192
Meegan, C.A., 19, 473, 490
Meers, B.J., 19, 169
Meier, D.L., 38
Melatos, A., 269
Mendell, G., 175
Meszaros, P., 495, 497, 502
Meyer, F., 46–7, 146, 418
Meyer-Hofmeister, E., 46–7, 143, 418
Meylan, G., 76
Michalitsianos, A.G., 453–4
Michaud, G., 40
Michel, F.C., 236, 252
Middleditch, J., 279
Mihalas, D., 336
Mikkola, S., 382
Mikołajewska, J., 425, 451–2, 455
Mikołajewski, M., 451ff
Miller, G.S., 303
Milne, D.K., 258, 261
Milsom, J.A., 52

Mineshige, S., 46, 49ff, 59, 100, 334, 352, 357
Mirabel, I.F., 366
Mitchell, M., 426, 431–2
Miyamoto, S., 304–5
Mochkovich, R., 25–6, 497
Moffett, D.A., 257
Molteni, D., 180
Monaghan, J.J., 5
Morfill, G.E., 411
Morgan, D.H., 433–4, 449
Motch, C., 304, 314, 422, 433, 435–6, 440
Mukai, K., 420
Muller, R.A., 112ff, 125
Murakami, T., 289–90, 293, 472, 485
Muslimov, A.G., 78

Nagase, F., 57, 290–1, 293, 298
Nakamura, N., 323
Nakamura, T., 24, 166, 168
Narayan, R., 18–9, 91, 96, 100, 102, 161–2, 178, 220, 236–7, 257, 492, 495, 497–99
Nauenberg, M., 69
Naylor, T., 38–9, 440
Nelder, J.A., 192
Nemiroff, R.J., 19, 467, 470
Nicastro, L., 230
Nice, D.J., 199, 218, 229
Nomoto, K., 25–6, 96, 108–9, 116, 119ff, 135, 142, 147, 284, 418
Nordveth, K., 167
Norris, J.P., 470, 478, 484, 487, 492
Novikov, I.D., 75
Nussbaumer, H., 433

Oda, M., 304
Ögelman, H., 430, 435

Oohara, K., 24
Oosterbroek, T., 304, 306
Orio, M., 430, 432
Orosz, J.A., 343–4, 356
Osaki, Y., 46–7, 49, 352
Osborne, J.L., 258
Osminkin, E.Yu., 85
Osterbrock, D.E., 421
Östreicher, R., 409
Ostriker, J.P., 3, 32, 38, 213, 216, 237, 418
Owens, A., 475

Paciesas, W.S., 313, 366, 369, 370, 373, 468
Paczyński, B., 3, 19, 25, 74ff, 82, 143, 261, 418, 426, 430, 490–1, 499
Paerels, F., 422
Pakull, M.W., 415, 422, 428ff, 436, 440–442
Palmer, D.M., 472
Panagia, N., 108–9, 133
Papaloizou, J., 46
Paresce, F., 397
Parkinson, M.L., 258
Parmar, A.N., 293
Pasquini, L., 409
Patat, F., 108, 132
Patterson, J., 352–3
Pavlenko, E.P., 358
Peale, S.J., 188, 191
Pendleton, G., 470, 472, 474
Perlmutter, S., 112
Petit, G., 204–5
Phillips, J.A., 199
Phinney, E.S., 18, 161–2, 178, 193, 198, 202, 220, 233, 237, 250, 279, 380, 497
Pietsch, W., 427, 432

Piirola, V., 404–5
Piran, T., 18–9, 96, 161, 220, 468, 489ff
Podsiadlowski, Ph., 25, 29, 32, 34, 38, 65, 97–8, 127–8, 141ff, 161, 193, 248, 250, 253, 261, 491
Poisson, E., 165–6
Pols, O., 121, 123–4, 379
Portegies-Zwart, S.F., 385
Porter, A.C., 136
Postnov, K.A., 81, 84–5, 88–9, 276
Pounds, K.A., 49
Prószyński, M., 195
Predehl, P., 391
Prialnik, D., 418
Priedhorsky, W., 339, 470
Primini, F., 393
Prince, T.A., 280
Pringle, J.E., 25, 46–7, 84, 172, 226, 411
Prins, S., 303
Prokhorov, M.E., 84–5, 90–1, 276
Pryor, C., 76
Putney, A., 252, 409
Pylyser, E.H.P., 64, 238, 251, 256, 428

Quataert, E.J., 276
Quinlan, G., 163

Radhakrishnan, V., 217–8, 236
Raine, D.J., 50
Rajagopal, M., 207–8
Rappaport, S.A., 13, 65, 76, 89, 143, 146, 232–3, 236, 250, 252, 394–395, 415ff, 435
Rasio, F.A., 17ff, 168, 188, 202, 229, 380, 382
Rathnasree, N., 128, 146
Ray, A., 32, 127–8, 147
Rees, M.J., 25, 32, 146, 226, 261, 491, 495, 502

Regev, O., 418
Reich, W., 486
Reimers, D., 35
Reinsch, K., 407, 409, 435
Remillard, R.E., 323, 333, 343–4, 352ff, 422, 440
Reynolds, A.P., 294
Reynolds, J., 364, 370
Richmond, M., 147
Richter, O.-G., 115
Ricker, G., 475
Ricketts, M.J., 49
Ritter, H., 46, 65, 68ff, 250, 405
Robertson, J.G., 369
Robinson, C.R., 404
Robinson, E.L., 358
Rodriguez, L.F., 366
Rokhlin, V., 5
Romani, R.W., 39, 93ff, 207–8, 218, 246, 408
Rosa, M., 115
Rosenberg, F.D., 313
Rosino, L., 457, 460
Rozyczka, M., 9–10
Ruden, S.P., 193
Ruderman, M.A., 228, 246, 252, 261, 475
Rupen, M., 364, 370
Ryan, F., 167, 180
Ryba, M.F., 199, 252

Sackmann, I.-J., 32
Saffer, R.A., 25, 384
Sallmen, S., 274, 382
Salter, M.J., 213
Samimi, J., 306
Sargent, A.I., 187
Sargent, W.L.W., 109
Sarna, M.J., 73, 78

Sasaki, M., 166
Sathyaprakash, B.S., 167
Savonije, G.J., 64, 82, 95, 250–1, 256, 428, 436
Schäfer, G., 19
Schaefer, B.E., 470, 472–3, 492, 499
Schaeidt, S., 430–1, 441
Schaudt, K.J. 257
Schmidt, B., 120, 125–6
Schmidt, G.D., 404, 409
Schmidt, M., 473
Schmidt, W.K.H., 491
Schmidtke, P.C., 430, 439–40
Schoembs, R., 352
Schrijver, C., 337
Schutz, B.F., 19, 164, 167, 172, 174
Schwope, A.D., 404, 407–9
Seidelmann, P.K., 205
Selvelli, P.L., 460
Seward, F.D., 426, 431–2
Shafter, A.W., 404, 407, 409
Shaham, J., 53, 228, 252, 303, 314
Shahbaz, T., 336, 344, 347, 356–7
Shakura, N.I., 44, 82, 339
Shankar, A., 6
Shapiro, I.I., 200
Shapiro, S.L., 17ff, 173, 168, 170–1, 229
Shara, M.M., 76
Share, G., 470, 472
Shemi, A., 18, 16, 96, 220, 494–5, 497
Shibata, M., 24, 166, 168
Shields, G.A., 94
Shigeyama, T., 121, 129–32
Shitov, Y.P., 198
Shklovskii, I.S., 218, 232
Shore, S.N., 453
Shrinivasan, G., 217
Shukre, C.S., 218
Shull, J.M., 258, 261

Shvartsman, V.F., 2
Sigurdsson, S., 202, 279, 380
Silber, A., 404
Singh, K.P., 147
Sion, E.M., 418, 426
Skillman, D.R., 352-3
Skopal, A., 452
Slettebak, A., 266
Smak, J.I., 44ff
Smale, A.P., 415, 428, 439
Smarr, L., 26, 236
Smith, J.D., 416-7, 419
Smith, S.C., 172
Snow, T.P., 276
Soker, N., 5, 418
Solf, J., 454
Sparks, W.M., 418
Sramek, R.A., 129, 135, 138
Srinivasan, G., 246
Standish, E.M. Jr., 205-6
Starrfield, S.G., 418, 426
Stebbins, R.T., 25
Stecher, T.P., 418
Steigman, G., 32
Stella, L., 337
Stencel, R.E., 433
Stinebring, D.R., 228-9
Stokes, G.H., 225, 229
Stollman, G.M., 261
Stoltz, B., 352
Strain, K.A., 19
Strom, R., 111, 487
Struve, O., 263
Suleymanova, S.A., 198
Sunyaev, R.A., 44, 59-60, 82, 87, 339, 366, 455
Supper, R., 393, 435
Sutaria, F.K., 147
Suzuki, T., 130

Swank, J.H., 52
Swartz, D.A., 109, 117, 120-1, 130, 133
Szedenits, E., 173

Taam, R.E., 5, 9-11, 32, 34-5, 49, 52, 96, 245, 254, 418
Tademaru, E., 219, 259
Tagoshi, H., 166
Takizawa, H. 307
Tammann, G.A., 109, 114, 136
Tanaka, Y., 39, 321, 327-9, 333
Tananbaum, H., 304
Tassoul, J.-L., 20-1
Tatarinzeva, V.S., 81
Tauris, T.M., 89, 100
Tavani, M., 229, 250
Taylor, A.R., 259, 267, 454
Taylor, J.H., 22, 91, 159, 164-5, 198-9, 203, 215, 218, 228-9, 231, 238, 252, 271ff, 382, 489
Tempesti, P., 461
Tennant, A.F., 304, 306
Terman, J.L., 5-6, 10
Teukolsky, S.A., 20, 91, 173, 229
Thakar, A., 452
Thieleman, F.-K., 133
Thomas, H.-C., 409
Thompson, C., 261
Thorne, K.S., 12, 30, 35, 153ff, 181
Thorsett, S.E., 22, 192, 195, 202-3, 238, 247, 274, 342, 382
Timmes, F.X., 240
Tingay, S.J., 364, 370-2
Tomcszyk, S., 418
Tomov, T., 451-4
Toor, A., 306
Torres-Dodgen, A.V., 138
Tout, C.A., 142
Trümper, J., 228, 427ff, 439, 442, 445

Treffers, R.R., 114
Truran, J.W., 76
Tsunemi, H., 323
Tuohy, I.R., 259
Turner, M.J., 49
Tutukov, A.V., 19, 25, 78, 82, 89, 100, 141, 161–2, 253–4, 431, 436, 499

Uomoto, A., 109, 135
Urpin, V.A., 246
Usov, V.V., 25

Vacca, W.D., 108, 110
Van Dyk, S.D., 108–9, 131, 135ff
Van Kerkwijk, M.H., 84, 289, 292, 295
Van Paradijs, J., 48–9, 62, 306, 333–4, 343, 427
Van Teeseling, A., 391, 397, 409, 428ff, 445ff
Van den Bergh, S.A., 109ff, 125, 136
Van den Heuvel, E.P.J., 12–13, 33, 35, 61–3, 82, 84, 93–5, 109, 117, 121, 161, 193, 218, 220, 245, 252–4, 283, 289, 294, 343, 416–417, 426ff, 436, 440, 445
Van der Klis, M., 292–3, 301ff, 314
Van der Laan, H., 372
Van der Woerd, H., 48
Vanbeveren, D., 142–3, 146
Vancura, O., 486
Vanderspek, R., 472
Vasisht, G., 486
Verbunt, F., 49, 51, 62, 64, 75–6, 97, 252, 325, 333–5, 337, 343, 389ff, 406
Vikhlinin, A., 304
Vilenkin, A., 207
Vivekanand, M., 237

Vogel, M., 433–4, 449
Voges, W., 433
Vogt, N., 48

Wachter, S., 454
Wade, R.A., 344
Wagner, R.M., 335–6, 358
Wallace, B.J., 260
Walter, F.M., 404
Wampler, E.J., 147
Wang, Q., 426, 431–2, 446
Warner, B., 352
Waters, L.B.F.M., 60–1, 266–9
Watson, M.G., 397, 410
Waxman, E., 495
Weaver, T.A., 143
Webbink, R.F., 4, 25, 62, 94, 178, 250, 418
Weber, J., 181
Weigert, A., 418
Weiler, K.W., 138–9, 257ff
Weisberg, J.M., 22, 198, 218, 272, 276, 489
Wellington, K.J., 258, 261
Wheeler, J.C., 46, 49–50, 52, 94, 107ff, 116, 120–1, 127, 130, 135, 146
Whelan, J.A.J., 82
White, N.E., 39, 52, 333, 342, 428
Whitehurst, R., 47–8, 100, 352, 404
Whiteoak, J.B.Z., 257
Wickramasinghe, D.T., 82, 404, 406, 409
Wijers, R.A.M.J., 74ff, 253
Will, C.M., 19, 164–6, 169
Wilson, C.A., 363
Wilson, R.B., 313–4
Wing, R.F., 452
Wiseman, A.G., 19, 164, 169
Woelk, U., 412–3

Wolszczan, A., 22, 25, 187–8, 192, 199, 203, 227–9, 247, 279
Wood, J., 48
Wood, K., 390, 399
Woosley, S.E., 75, 95, 98, 108, 116, 121, 126–7, 129, 132, 143, 147, 192, 240, 283, 498
Wu, K., 404, 406, 409
Wu, X., 426, 432

Xilouris, E.M., 416, 428, 440

Yamaoka, H., 98, 121, 161
Yamasaki, T., 46
Ye, T., 369
Yoshida, K., 303, 306
Young, P., 409
Yungelson, L.R., 19, 25, 82, 89, 100, 141, 161–2, 499

Zeldovich, Ya., 75
Zepka, A.F., 229
Zhang, S.N., 363–4, 369
Zhang, Z., 65
Zhao, P., 350
Zimmermann, H.-U., 132
Zimmermann, M., 173
Zughe, X., 24, 168
Zwaan, C., 337, 406
Zwitter, T., 73–5
Żytkow, A.N., 12, 30, 35, 161

Subject Index

AAT, 345, 347
AAVSO, 457ff
Abrikosov fluxoid, 246
Accretion disk
    beat-frequency model, 303, 313–4, 317
    characteristics, 415
    disk corona (ADC), 428, 440
    instability, 44, 46–7, 49, 51
        outburst types, 45
    orientation, 60
    precession, 352, 355, 440
    quiescence, 45, 47
    radius, 356
    relativistic disk, 329
    viscosity, 44ff
Accretion induced collapse (AIC), 26, 75–8, 96–7, 236, 284, 436
Accretion rate
    Eddington, 33, 245
    near-Eddington, 251, 440
    super-Eddington, 455
Accretion torque, 61–2
    Ghosh-Lamb model, 313–4
Accretor, 82, 86, 455
Accretor-propellor model, 454–5
AGB phase, 253
AGN, 207
AM Her systems, 403ff
    accretion flow, 410, 412
    cyclotron lines, 407
    magnetic fields, 409
    orbital periods, 405, 409
    relation to DQ Her systems, 404
    'soft X-ray' problem, 410
    space density, 405
    synchronization, 404
    X-ray pre-heating, 324
    Zeeman lines, 407–8
Annihilation line, 347
Apsidal motion, 275
Arecibo, 188, 208–9, 227, 229, 280
ASCA, 130, 289ff, 321ff, 336, 485
Asiago supernova catalog, 111
Atoll sources, 302–3, 306
Australia Telescope Compact Array (ATCA), 370

B-P diagram, 57–8, 244–5
    death line, 244
    Hubble line, 244
    spin-up line, 61, 64, 244
Barycentric Coordinate Time (BCT), 204
BATSE, 19, 313–4, 363ff, 369–71, 467ff, 477ff, 490, 492
    burst catalog, 468–9
BATSE Coordinates Distributions Network (BACODINE), 472
Beat-frequency model, 303, 313–4, 317
Berkeley SN search, 112–3

Be stars, 263ff
    circumstellar disk, 263, 267
    equatorial disk, 59
    magnetic field, 267
    models, 268
    radio observations, 266
    stellar wind, 263, 275
    with millisecond pulsar, 59
    with X-ray pulsar, 313
Be/X-ray binaries, 13, 99
Binary stars
    accretion scenarios, 143
    binary evolution, 3ff, 82
        coalescence, 13, 17ff, 25, 32
        common-envelope evolution, 3ff, 89, 144
        common-envelope phase, 146
        common-envelope spiral in, 62, 95–6
        disruption, 32, 57
        dynamical time scale, 418
        He star core collapse, 95–6, 98
        hydrodynamical instability, 11
        irradiation of companion, 250, 252
        mass-transfer instability, 44ff
        scenario machine, 81ff
        tidal instability, 47
    binary neutron stars, 17ff, 236
        coalescence, 17ff, 32, 159ff, 167, 219, 489ff
        global instability, 18–24
        gravitational radiation, 18–23, 157ff
        paucity, 219
        progenitor, 249
        synchronization, 24
    binary radio pulsars, see Radio pulsars — binary pulsars

binary white dwarfs, 25
    coalescence, 25
    type Ia supernovae, 25
black-hole binaries, see Black-hole binaries
black hole – neutron star binaries, 95–6, 499
high-mass binary pulsars, see High-mass binary pulsars
high-mass X-ray binaries, see High-mass X-ray binaries
hypercritical accretion, 97, 249
intermediate-mass X-ray binaries, see Intermediate-mass X-ray binaries
low-mass binary pulsars, see Low-mass binary pulsars
low-mass X-ray binaries, see Low-mass X-ray binaries
soft X-ray transients, see Soft X-ray transients
X-ray binaries, see X-ray binaries
X-ray pulsars, see X-ray pulsars
Birth rate
    low-mass binary pulsars, 235–6, 238, 241, 250
    low-mass X-ray binaries, 34, 250
    Thorne-Żytkow objects, 34, 250
Birth rate problem, 236, 241
Blaauw mechanism, 98
Black-hole binaries, 93ff
    coalescence, 167
    evolution, 93ff
        triple scenarios, 97–8
    formation, 93ff
    galactic population, 99
    population, 100
    progenitor, 95
    relativistic accretion disk, 329

X-ray PHA spectra
   reflection component, 327
X-ray transients, 99
   recurrence time, 51
Black-hole candidates, 37–9, 93, 102, 304ff, 326–9, 333, 341ff, 351ff, 366, 430
   fluorescent iron line, 327
   similarities to neutron stars, 306
   source states, 304, 326
   X-ray PHA spectra, 326–9
Blue stragglers, 384
   formation in globular clusters, 384ff

Calan-Tololo SN search, 112
Cambridge, 227
Cataclysmic variables, 457
   companions
      irradiation, 65ff
      superadiabatic convection zone, 66
   disk sources, 397
   dwarf novae, see Dwarf novae
   evolution, 65ff, 76
      irradiation, 65ff
      irradiation instability, 70–1
   formation, 3ff
   globular clusters, 76
   intermediate polars, see Intermediate polars
   irradiation, 65ff
   irradiation-induced mass transfer, 67–70
   limit cycle, 45, 70–1
   orbital period, 405, 409
   period gap, 70
   polars, see AM Her systems
   progenitors, 10
   recurrent novae, 457ff

   superhumps, 352
Cerro Tololo Interamerican Observatory (CTIO), 439, 442
CfA archive, 109
Christodoulou memory, 169
Circumpulsar disk, 193
Circumstellar matter
   around supernovae, 130ff
Classical novae, 435
Coalescence
   binary neutron stars, 17ff, 32, 159ff, 167, 219, 489ff
      neutrino emission, 497–8
      disk formation, 497
   binary white dwarfs, 25
   black-hole binaries, 167
   massive black holes, 179
   rates, 161ff, 219, 497–9
Common-envelope evolution, 3ff, 89, 144
   coalescence, 13
COMPTEL, 472
Compton Gamma-Ray Observatory (CGRO), 313, 467ff, 477, 490
Corbet diagram, 232
Coronal lines, 460
Corotation radius, 83
Cosmic strings, 207
CTIO, 136
Cyclotron lines, 407
C+O stars
   as SN progenitors, 121ff

Death line, 244
Dim X-ray sources, 394
   counterparts, 390, 397
   X-ray PHA spectra, 395
Dispersion measure, 90, 263ff, 271, 275
Doppler tomography, 349

DQ Her systems
    comparison to AM Her systems, 404
Dwarf novae, 43ff
    coronal flow, 47
    irradiation of companion, 48
    mass transfer instability model, 44ff
    outbursts, 43ff
        recurrence behaviour, 48
    superoutbursts, see Superoutbursts
    UV lag, 46

Eclipsing systems
    binary pulsars, 73, 203
    super-soft sources, 440
    symbiotic binaries, 451
Eddington
    accretion rate, 33, 245
        near-Eddington, 251, 440
        super-Eddington, 455
    limit, 13, 35, 427, 445–8
    sub-Eddington luminosity, 454
Effelsberg, 208
EGRET, 468, 470–2, 475
EINSTEIN, 321, 334, 389, 391, 393, 398–9, 415, 425ff, 439, 441–2
Ejector, 82, 86
Ellipsoidal variations, 343–4, 351ff
Equation of state (EOS), 17, 20–3
EXOSAT, 292, 334, 411
$e^{\pm}$ annihilation line, 347

Faraday rotation, 90
Fireballs, 493ff
    baryon loading, 494
    beaming, 496, 499
    evolution, 494
    gamma-ray bursts, 495
    interaction with ISM, 495

Galaxies
    distribution, 207
    mass distribution, 199
Gamma-ray bursts, 19, 25, 168, 261, 467ff, 477, 490ff
    counterparts, 472
    cosmological models, 491–2
    distribution, 467, 473–4
    duration distribution, 470
    event rate, 492–3
    fireball model, 493ff
    galactic models, 491
    gamma-ray line features, 472
    homogeneity, 490–1
    isotropy, 490, 492
    location, 475
    luminosity function, 493
    morphologies, 467–8
    repetition, 473
    soft gamma-ray repeaters, see Soft gamma-ray repeaters
    spectral characteristics, 470, 472
    time dilation effect, 470
    time profiles, 468
General relativity (tests), 157, 203
Georotator, 82
GEO600, 159
GINGA, 51–2, 57, 290–1, 293, 334
Globular clusters, 32, 73ff, 199, 201–2, 377ff, 389ff
    binary radio pulsar, 73
    blue stragglers, 384ff
    cataclysmic variables, 76
    collisions, 32
    core sources, 396–7
    evolution, 76, 199
    foreground stars, 393

'heating' mechanisms, 378
luminosity function, 393
millisecond pulsars, 279ff, 382
primordial binaries, 378–9
radio pulsars
  proper motions, 197, 199
  stability, 202
stellar dynamics, 377ff
stellar evolution, 379
triple systems, 380, 382ff
X-ray bursters, 390
X-ray colour-colour diagram, 396
X-ray luminosity function, 398
X-ray sources, 433
  optical counterparts, 397
  population, 392
GMRT, 209
GPS, 198, 205
GRANAT, 327
Gravimagnetic parameter, 83
Gravitational capture radius, 83
Gravitational radiation, 153ff, 203, 207, 250
  background, 197, 207–8
    spectrum, 208
  chirp, 21, 163
  detectors, 153ff
  from binary neutron stars, 18–23, 157ff
  radiation efficiency, 23
  relativistic effects, 164ff
  signal estimates, 160, 171, 177
  sources
    binary stars, 178
    coalescing compact binaries, 159ff, 499
    coalescing massive black holes, 179
    cosmic strings, 207

spinning neutron stars, 173ff
stellar collapse, 169ff
waveforms, 19, 162ff, 167ff
Green Bank, 208–9, 227
  NRAO, 208–9
  Telescope (GBT), 209

H II regions, 421
Hα nebulae around pulsars, 229
Harvard plate archive, 459
Hierarchical tree algorithm, 5
High Energy Transient Explorer (HETE), 475
High-mass binary pulsars (HMBP), 247
  orbital period, 249
  origin, 248
High-mass X-ray binaries (HMXB), 32–33, 38, 62, 247, 289ff
  companions
    blue supergiant, 96
    evolution, 3ff, 93–7, 102, 248
    spiral in, 248–9
    supernova, 38, 62
  orbital period changes, 290ff
  progenitor, 34, 36
  Wolf-Rayet star companions, 289, 292
  X-ray spectroscopy, 295ff
He stars, 13–4, 34, 142, 248–9
  core collapse, 95–6, 98
  white dwarfs, 250, 253
HEAO-1, 390, 398–9
Hubble constant, 19, 167
Hubble line, 244
Hubble Space Telescope (HST), 51, 132, 137–8, 203, 390, 397
Hydrodynamical instability, 11

ICE, 483

IMF, 101
Instability
    disk instability, 44ff, 100
    global instability, 18–24
    hydrodynamical instability, 11
    irradiation instability, 70–1
    mass-transfer instability, 44ff
    thermal instability, see disk instability
    tidal instability, 47
    tidal-thermal instability, 47–8
Intermediate-mass X-ray binaries (IMXB), 100–2, 247, 253
    descendants, 254
Intermediate polars, 85, 322–3
    cooling flow, 322
    ionization equilibrium, 322
    multi-temperature plasma, 322
International Atomic Time (TAI), 204
International Ultraviolet Explorer (IUE), 451
Interplanetary Network (IPN), 475, 483
Ionization nebulae, 420–1
Iron line, 327
Irradiation-induced mass transfer, 67–70
Irradiation instability, 70–1
ISAS, 321
Isolated pulsars, 243, 245, 248

Jets, 329, 372, 440, 451–5
    precession, 329, 444
    radiative cooling, 330
Jodrell Bank, 208, 227, 230

Kelvin time, 143
Keplerian-frequency model, see Beat-frequency model
Kick velocity, 14, 78, 217, 240

Kitt Peak, 472
KONUS, 478, 480, 486
KPNO, 136

LAGOS, 25–6
Leuschner SN search, 114
Lick Observatory, 136
Light cylinder radius, 83
LIGO, 17–8, 157ff, 490
Limit cycle, 45, 70–1
LISA, 175ff
Lithium
    abundances, 32, 39
    production, 31
    soft X-ray transients, 346
    lithium stars, 32
Low-mass binary pulsars (LMBP), 62, 78, 235ff, 247, 250, 252
    birth rate, 235–6, 238, 241, 250
    galactic distribution, 235, 237–8
    kick velocities, 78, 240
    orbital period distribution, 250–1
Low-mass X-ray binaries (LMXB), 34, 38, 43, 52, 62, 65, 70, 247, 250, 324, 333, 444
    angular momentum loss, 99
    atoll sources, 302–3, 306
    binary disruption, 252
    birth rate, 34, 250
    black-hole transients, 100
    evolution, 3ff, 97, 100–2, 250
        irradiation of companion, 52–3
        near-Eddington accretion, 251, 440
    formation, 96
    gravitational radiation, 250
    inclination effects, 308–10
    magnetic braking, 250
    orbital period changes, 292–3

orbital period distribution, 251
X-ray bursters, 324
X-ray colour-colour diagrams, 302
X-ray PHA spectra, 324
X-ray power spectra, 302ff, 365
   noise components, 302
   QPO, 302, 304
   Z sources, 302, 306
Lowell Observatory, 136

Magnetic braking, 250–1
Magnetic field
   AM Her systems, 409
   Be stars, 267
   binary pulsars, 62, 78, 245–6
      X-ray transients, 318
   decay, 62, 78
   neutron stars, 303
      Ohmic decay, 245–6
   soft X-ray transients
      companion, 336
   symbiotic stars, 454
   versus spin period diagram, see B-P diagram
   white dwarfs, 409
Magnetor, 82
Mariner, 206
Mars-96, 475
Mass-transfer (enhancement) instability model, 44ff
McDonald Observatory, 358
Millisecond pulsars, 37, 62, 187–8, 193, 197, 201, 243, 247, 251, 337, 390
   age, 201
   companions, 201–3
      evaporating companion, 252
   evolution, 247
   formation, 243ff, 253

   glitches, 201
   in globular clusters, 279ff, 382
   near-millisecond pulsars, 253
   optical counterparts, 202
   planets, 187ff, 203, 253
   population, 199, 252
   progenitor, 57ff, 326
   pulsar monitoring telescope, 210
   rotation parameter, 197, 201
   searches, 225ff
   Shklovskii effect, 232
   single, 57ff, 252
   spin evolution, 57ff
   tidal dissipation of companion, 252
   timing, 197ff, 207
   timing noise, 201
Molonglo Observatory Synthesis Telescope (MOST), 364, 369, 370–1
Monte Carlo techniques, 142–4, 417

Nançay, 208–9
Neutron stars, 321, 333, 416
   Abrikosov fluxoid, 246
   conductivity, 64
   crustal plate tectonics, 246
   Eddington limit, 35
   equilibrium spin period, 245
   equation of state, 17, 20–3
   formation by AIC, 26, 96–7, 236, 184
   halo population, 261
   kick velocity, 14, 217, 240
   magnetic field, 303
      Ohmic decay, 245–6
   mass (limit), 279ff, 342
   radius, 23
   similarity to black-hole candidates, 306
   spin evolution, 57ff, 81ff, 337, 363ff

superfluid vortices, 246
Novae
    dwarf novae, see Dwarf novae
    recurrent novae, 416, 456ff
    supernovae, see Supernovae
    symbiotic novae, 416, 456
    X-ray novae, 342
NRAO Green Bank, 208–9

OB-association, 147
Oblique rotator, 455
Optical counterparts
    dim X-ray sources, 397
    millisecond pulsars, 202
    supersoft sources, 415, 443
Orbital periods 290ff, 405, 409
    AM Her systems, 405, 409
    distribution
        binary radio pulsars, 233
        high-mass binary pulsars, 249
        low-mass binary pulsars, 250–1
        low-mass X-ray binaries, 251
        magnetic cataclysmic variables, 405
        non-magnetic cataclysmic variables, 405
    orbital period changes, 290ff
        measurements, 293
        mechanisms, 294ff
    period gap, 70, 250–1
OSSE, 57, 130, 318
Outbursts
    dwarf novae, 43ff
        recurrence behaviour, 48
    inside-out, 45, 47, 50
    outside-in, 45, 50
    radio outbursts, 369ff
    superoutbursts, see Superoutbursts
    symbiotic binaries, 451–2

Parkes, 208, 227, 230, 264, 272
Period gap
    cataclysmic variables, 70
    low-mass binary pulsars, 250–1
Planetary nebulae, 422, 431, 446
    luminosity function (PNLF), 422
Planets
    around black holes, 37
    around radio pulsars, 25, 38, 187ff, 193, 203, 247, 253
        circumpulsar disk, 193
        formation, 193, 253
        gravitational perturbation, 188
    around Sun-like stars, 195
    around white dwarfs, 25
    Solar system, 192–3, 195, 206
        dynamics, 205
    stellar spectroscopy, 195
Plasma torque, 59
Power spectra, 302ff, 313-5, 365
    power-law noise, 315
    quasi-periodic oscillations (QPO), see QPO
Prognoz, 483
Propellors, 82, 86, 454–6
    superpropellors, 82
    symbiotic binaries, 454–6
    torque
        subsonic, 59
        supersonic, 60
Pulsars, see Radio pulsars and X-ray pulsars
PVO, 473

Quasi-periodic oscillations (QPO), 302, 304, 314–5, 317
    beat-frequency model, 303, 313–4, 317

Radio outburst, 369ff
Radio pulsars, 85
    acceleration, 282
    ages, 259–60
    B-P diagram, see B-P diagram
    binary pulsars, 18, 22, 89, 201, 235ff, 243, 247, 279ff
        accretion disk orientation, 60
        apsidal motion, 275
        B/Be star companion, 263ff, 271ff, 313
        black-hole companion, 276–7
        companion star, 247
        disk spin-up/spin-down, 60
        eccentricities, 233, 382
        eclipse, 73, 203
        equatorial disk, 59
        evolution, 62, 73ff, 247
        formation, 102, 243ff
        globular cluster, 73
        magnetic field strength, 245
        magnetic field decay, 62, 78
        magnetic field distribution, 246
        magnetic field evolution, 245–6
        orbital periods, 233
        progenitors, 57ff
        recycling scenario, 236
        Reverse Mass Transfer scenario, 96, 100
        spin evolution, 57ff
        tidal capture, 74, 77
        tidal circularization, 75
        tidal heating, 75
        tidal interaction, 275
    black hole-neutron star pulsar formation, 96
    companions
        population, 243
    dispersion measure, 263ff, 271, 275
    distances, 259–60
    distribution, 213, 215
    evolution, 38
        pulsar ejection, 38
    Hα nebulae, 229
    isolated pulsars, 243, 245
        recycled, 248
    high-mass binary pulsars, see High-mass binary pulsars
    low-mass binary pulsars, see Low-mass binary pulsars
    Magellanic Clouds, 271
    millisecond pulsars, see Millisecond pulsars
    near-millisecond pulsars, 253
    planetary companions, 25, 38, 188, 193, 195, 228, 247
        circumpulsar disk, 193
        formation, 193, 253
    polarization, 265
    progenitor 247
    proper motions, 197, 199, 214, 259
    recycled pulsars, see Recycled pulsars
    rotation measure, 263ff
    runaway, 62
    scintillation, 214
    searches, 225ff, 257
    Shapiro delay, 200, 203
    SNR associations, 257ff
    spin evolution, 57ff
        accretion torque, 61–2
        electro-magnetic braking torque, 59
    spin period, 243
    spin-up/spin-down, 244
    timing, see Timing
    timing noise, 201, 274
    velocities, 213ff, 258

correlation with magnetic moment, 218
X-ray transients
quiescence, 325
Rapid proton (rp) process, 31, 39
Recurrent novae, 416, 457ff
Recycled pulsars, 37, 57, 61–2, 96, 100, 243, 245, 248, 333
binary disruption, 57
isolated pulsars, 248
magnetic field distribution, 62
Recycling scenario, 236
Reverse Mass Transfer scenario, 96, 100
Roche lobe, 343
ROSAT, 51, 130, 132, 292, 321, 325, 333ff, 389ff, 405, 411, 415, 417, 420, 425ff, 439ff, 447, 486
All-Sky Survey, 390
Rotational broadening, 343–4
RS CVn systems, 344

S-wave, 349
SAS-3, 290
Shapiro delay, 200, 203
SHEVE VLBI array, 370
Shklovskii effect, 239
Smooth particle hydrodynamics
Soft Gamma Repeaters (SGR), 472, 477ff
counterparts, 485ff
positions, 481
profiles, 479, 482
properties, 478, 480, 482
recurrence, 483
Soft X-ray transients, 39–40, 43ff, 93, 102, 333ff, 342, 351ff, 363ff, 398
accretion disk, 352
accretion rate, 334

black-hole candidates, 93, 102, 341ff, 351ff, 366
boundary layer, 336
companion star, 336
irradiation, 52–3, 335
magnetic field, 336
rotation period, 336
transient mass transfer, 335
coronal activity, 325
disk instability, 100
ellipsoidal variations, 351ff
formation, 39
infrared photometry, 356–8
inside-out/outside-in outbursts, 50
lithium
enhancement, 346
abundances, 32, 39
mass functions, 342
mass transfer instability model, 44ff, 67–70
neutron star
spin evolution, 337
progenitor, 38
quiescence, 51–2, 325–6, 333ff
radio observations, 364–5
rapid proton (rp) process, 39
recurrence time, 51
spots on companion, 352
superhumps, 352ff
superluminal motion, 364
transient mass transfer, 335
X-ray burstser, 324
X-ray pulsar, 313
Solar Maximum Mission (SMM), 472
Solar System
planets, 192–3, 195, 205–6
Source states, 304, 308–9
Space Telescope Science Institute (STScI) GASP system, 136

Stellar wind, 35, 59, 95, 142, 144, 147, 248, 263, 275, 452
Stopping radius, 83, 85
Superaccretor, 82
Superejector, 82
Superhumps
    in cataclysmic variables, 47, 352
    in soft X-ray transients, 352ff
    period, 348, 352, 356–7
    tidal effects, 352
Superluminal motion, 364
Supernovae
    accretion scenarios, 143
    discoveries, 111
    evolution
        merger scenario, 143
    hybrid, 109
    hydrodynamics, 144
    kicks, 32, 34
    multiple events in galaxies, 115
    precursor
        blue supergiant, 143–4, 146–7
    progenitors, 15–6, 108
        OB-associations, 147
        stellar wind, 144, 147
    radio observations, 109
    rates, 108, 111, 115–6
    searches, 112
    type I, 147
    type Ia, 25, 423, 436
        peak luminosity, 113
        rate, 423
    type Ib/c, 98, 107ff, 119ff, 135ff, 142, 144
        absence in E galaxies, 109
        association with star formation, 108
        association with WR stars, 98, 109
        binary evolution, 98, 115–6, 121ff
        environment, 135ff
        formation rate, 124
        hydrogen lines, 110
        light curves, 109, 111, 116, 125ff
        peak luminosity, 113, 15–6
        progenitors, 108, 116–7, 121ff, 135ff, 142
        radio observations, 109, 138–9
        spectra, 107–9
        statistics, 110ff
    type II, 141ff
        binary merger, 142
        binary progenitors, 142
        hydrodynamic calculations, 144
        light curves, 145–8
        peak luminosity, 116
        precursor, 143–4, 146–7
        progenitors, 142–4, 146–8
        rates, 116
    type IIb/L, 119ff, 129
    type IIn, 120, 129
    type IIp, 120
Supernova remnants
    ages, 259–60
    associations, 257ff
    distances, 259–60
    interaction composites, 261
    pulsar associations, 257ff
    radio imaging, 257
    X-ray observations, 261
Superoutbursts, 47–8
    superhumps, 47
    tidal instability, 47
    tidal-thermal instability, 47–8
Superpropellors, 82
Supersoft (X-ray) sources, 391, 415ff, 425ff, 439, 445ff
    accretion disk

characteristics, 415
  disk corona, 440
  precession, 440
 bipolar outflow, 442
 black-hole candidates, 416
 classical nova, 435
 collimated outflow, 440, 442
 companion star
  giant branch, 416, 418
 distribution, 417
 eclipsing systems, 440
 Eddington limit, 445–8
 evolution, 416, 418, 435
 finding charts, 443
 H II regions, 421
 ionized gaseous nebulae, 420–1, 422
 jets, 440
  precession, 444
 models, 425ff, 445
 nuclear burning, 416, 418
 optical
  counterparts, 415, 443
  light curves, 441
  observations, 430ff
 parameters, 434
 planetary nebula, 422, 431, 446
 population, 417, 420
 recurrence, 430
 recurrent novae, 416
 supersoft nebulae, 421–2
 symbiotic star, 432–5, 447
 thermal time scale, 418
 UV observations, 430–2
 white-dwarf (atmospheres), 416, 428, 432, 445–6
 X-ray PHA spectra, 445
SU UMa systems, 47–8, 352
Symbiotic binaries, 451ff
 accretor-propellor model, 454–5
  accretor state, 455
  companion star
   giant branch, 416, 418
  eclipsing systems, 451
  flickering, 454
  high and low state, 454
  jets, 451–5
  oblique rotator, 455
  outbursts, 451–2
  propellors, 454–6
  stellar wind, 452
  sub-Eddington luminosity, 454
  super-Eddington accretion, 455
  wind accretion, 451, 456
Symbiotic Mira, 456
Symbiotic novae, 416
  recurrent, 456
Symbiotic star, 457
Synchrotron
  radiation, 364
  self absorption, 372

TENMA, 290
Terrestrial Time (TT), 204–5
Thermal (disk) instability, 47
Thorne-Żytkow objects (TŻOs), 12, 29ff, 62, 97, 248, 253
  binary merger, 34
  birth rate, 34, 250
  envelope, 36–7
  evolution, 29ff
   common-envelope phase, 33, 62, 97–8
   neutrino-dominated regime, 35
  formation
   supernova kick, 32, 34
  globular clusters
   collissions, 32
  gravitational energy, 30

lithium abundances, 32, 39
neutrino loss, 35
neutron stars
  spin evolution, 36–7
nuclear burning, 30
rapid proton (rp) process, 31, 39
red-supergiant appearance, 30
similarities to lithium stars, 32
stellar wind, 35
structure, 29ff
time scales, 37

Tidal
  capture, 74, 77
  circularization, 75
  dissipation of companion, 252
  heating, 75
  interaction, 18–20, 275, 352
  instability, 47
  torque, 61–2

Tidal-thermal instability model, 47–8
Time dilation effect, 470
Time scale
  atomic time scale, 205
  Barycentric Coordinate Time (BCT), 204
  cooling time scale, 37
  dynamical time scale, 418
  International Atomic Time (TAI), 198, 204
  Kelvin time, 143
  Kelvin-Helmholtz time scale, 37
  Terrestrial Time (TT), 204–5
  thermal time scale, 418
  viscous time scale, 37

Timing, 197ff, 207
  analysis, 314
  (millisecond pulsar) array, 205, 208–9
  arrival time transformation, 206
  atomic time scale, 205
  Barycentric Coordinate Time (BCT), 204
  Doppler shifts, 207
  ephemeris (dipole) perturbations, 207
  galactic motion, 201
  International Atomic Time (TAI), 198, 204
  interstellar plasma, 199
  lenses, 200
  parallax, 197, 199
  planet perturbation, 188, 192
  power-law noise, 201
  power spectra, see Power spectra
  pulse arrival times, 192, 197
  residuals, 191, 207, 209
  Solar system barycenter, 199
  space-time metric, 200
  Terrestrial Time (TT), 204–5
  time (monopole) perturbations, 207
  times-of-arrivals (TOA), 188, 201
  timing noise, 201, 274

Torque
  accretion torque, 61–2
    Ghosh-Lamb model, 313–4
  electromagnetic braking, 59
  subsonic propellor, 59
  supersonic propellor, 60
  tidal, 61–2

Transient Gamma-ray Spectrometer (TGRS), 475
Triple scenarios, 97–8
Triple systems, 380, 382ff

UHURU, 290, 389, 391
UKIRT, 345
Ulysses, 470–11, 475

Usada, 208
UV lag, 46

Venera 11–13, 478
VIRGO, 18, 157ff, 490
Viscosity, 44ff
VLA, 370, 454
VLBA, 370
VLBI, 200, 369, 370–1
Voyager, 20

WATCH, 327
Westerbork, 209
White dwarfs
    binary white dwarfs, 25
    C-O white dwarfs, 25, 253
    He white dwarfs, 250, 253
    magnetic fields, 409
    model atmospheres, 428, 432, 445–446
    O-Ne-Mg white dwarfs, 25, 75
    planetary companions, 25
    radius, 447
WHT 344, 347
WIND, 475
Wolf-Rayet (WR) stars, 83, 87, 289, 292
    evolution, 98, 100
    SN Ib/c progenitors, 136–8
    winds, 139

X-ray binaries, 93ff
    adiabatic flow, 330
    bipolar jet, 329
    black-hole binaries, see Black-hole binaries
    black-hole candidates, 326–9, 341ff
      source states, 326
    evolution, 93
      Wolf-Rayet, 98, 100
    jets, 329
    kicks, 99
    line spectroscopy, 321
    neutron stars
      equilibrium spin period, 245
      orbital period changes, 290ff
      Roche lobe overflow, 248
      stellar wind, 248
X-ray binary pulsars, 57ff
    spin evolution, 57ff
    X-ray transient, 313ff
X-ray bursters
    low-mass X-ray binaries, 324
    globular clusters, 390
X-ray bursts, 40, 306, 390
X-ray colour-colour diagram
    globular cluster sources, 396
    low-mass X-ray binaries, 302
X-ray ionized nebula, 428
X-ray novae, 342
X-ray PHA spectra, 321ff
    black-hole binaries, 327
    black-hole candidates, 326–9
    dim X-ray sources, 395
    high-mass X-ray binaries, 295ff
    low-mass X-ray binaries, 324
    reflection component, 327
    supersoft sources, 445
X-ray pulsars, 57ff
    accretion powered phase, 61–2
    accretion torque model, 313–4
    Be-star companions, 313
    binary pulsars, 57ff
    cyclotron line, 57
    spin evolution, 57ff
    spin-up rate, 313, 317
    X-ray power spectra, 313–5
    X-ray transient, 313ff
X-ray transients, 363ff, 369ff
    Be/X-ray binaries, 99

binary pulsar
   magnetic field, 318
   spin-up rate, 313, 317
black-hole binaries, 99
expanding synchroton bubble model, 372
jets, 372
opacity effect, 369, 371
quiescence, 51–2, 325–6, 333ff
   radio pulsar activity, 325
radio
   counterparts, 369
   spectrum, 372
rapid proton process (rp-process), 39
recurrence time, 51
recurrent transients, 324
soft X-ray transients, see Soft X-ray transients
spots on companion, 352
superhumps, 352ff
superluminal motion, 364

Z sources, 302, 306
Zeeman lines, 407–8

Object Index

A 0535+262, 57, 84, 313ff
A 0620−00, 38−9, 51, 93, 97−8, 100, 333−6, 341ff, 353−4, 359, 398
Z And, 461
Aql X-1, 325, 333−9, 398
R Aqr, 456

44 i Boo, 178

Cal 83, 422, 426ff, 439ff, 434, 447, 449
Cal 87, 426, 429ff, 440−1, 444
BY Cam, 409
Z Cha, 48−9
Cen X-3, 84, 289ff
Cen X-4, 40, 324, 331, 333−9, 343−8, 398
ω Cen, 394, 396
V822 Cen, 40
V834 Cen, 409
Ceres, 195
o Cet, 456
Cir X-1, 304, 306−7
EU Cnc, 409
Crab Nebula, 259
Crab Pulsar, 257, 260
T CrB, 456
RS CVn, 75, 331, 336−7, 390, 398
Cyg X-1, 93, 95, 84, 289, 305, 307, 326−327, 341
Cyg X-3, 82, 84, 289ff
CH Cyg, 451−6, 461
N Cyg 1975, 79

V404 Cyg, 32−3, 38−9, 51, 100, 309, 334−6, 339, 341−8, 358
AG Dra, 434

E 2259+586, 58
Earth, 195
EF Eri, 409
EXO 0748−676, 292−3

UZ For, 409

G 5.4−1.2, 258ff
G 10.0−0.3, 485
G 39.7−2.0, 486
G 42.8+0.6, 486
G 43.9+1.6, 486
G 57.1+1.7, 259, 261
G 114.3+0.3, 259, 261
G 308.8−0.1, 258ff
G 341.2+0.9, 258−9
G 343.1−2.3, 258−9
G 354.1+0.1, 258−9
Galaxy, 94, 100−2, 391, 393, 416, 439
U Gem, 44, 46, 48
GS
    1124−68, 305, 307
    2000+25, 52, 334−5, 348, 352, 357−358
    2023+338, 309, 334, 341−8, 358, 398
GX 339−4, 309

GRB 910601, 471
GRO J0422+32, 53, 348–9, 359
GRO J1655–40, 363ff, 369ff
GRS 1009–45, 327–9
GRS 1915+105, 366, 480

H 1608–522, 336
H 1705–25, 335
HD 22403, 337
HD 133640, 178
HDE 245770, 313
Her X-1, 58, 84, 100, 247, 290, 293–4
AM Her, 74, 79, 82, 323, 409
HR 8857, 345
HV 2554, 440, 443
HV 5682, 440–1
EX Hya, 322–4, 330
Hydra A, 264
BL Hyi, 409
VW Hyi, 46, 48

Jupiter, 195, 206

Kes 32, 259

DP Leo, 409
Liller 1, 391
LMC, 415–7, 420, 422, 427ff, 439ff, 449, 478, 486
LMC X-3, 99, 341
LMC X-4, 289ff
ST LMi, 409
Local Group, 420
LSI +61°303, 82

M 3, 391–2, 394, 400, 433ff
M 4, 202, 377, 382ff
M 15, 280
M 15C, 279ff
M 22, 394, 396

M 28, 394
M 30, 394
M 31, 389–91, 393, 416–7, 420, 422, 435, 491
M 33, 436
M 51, 109, 116, 120, 135, 138
M 79, 394
M 81, 108, 120, 147
M 83, 137
M 92, 394
M 101, 436
Magellanic Cloud, 94, 391, 439
Mars, 195, 206
Mercury, 195, 206
Milky Way, 417, 420
V616 Mon, 32–3, 38–9, 341, 343
Moon, 195
MSH 15–52, 259, 261
GQ Mus, 434ff
N Mus 1991, 51, 100, 305, 307, 343–4, 347–8, 352, 356
MWC 560, 451–6
MXB 1730–335, 391

N 49, 486
N 67, 431, 446
Napoleon's Hat, 147
Neptune, 195, 206
NGC
    104, 394
    253, 436
    269, 432
    1851, 392
    1904, 394
    3310, 137–8
    4568, 139
    5139, 394
    5272, 391ff
    5824, 394

6304, 394–5
6341, 394
6342, 73, 76
6397, 394–5
6440, 391
6441, 392
6541, 394
6624, 392, 397
6626, 394
6652, 391–2
6656, 394
6712, 392
6752, 394–6
7078, 392
7099, 394, 396
Nova
    Cyg 1975, 79
    Mus 1991, 51, 100, 305, 307, 343–4, 347–8, 352, 356
    Per 1992, 348–9
    Sco 1994, 363ff, 369

OAO 1657–415, 364–5
RS Oph, 456, 457ff

Pal 2, 394, 396
GK Per, 397
N Per 1992, 348–9
ψ Per, 267
PKS 0915–118, 264
PSR
    J 0034–0534, 209, 230, 241
    B0042–73, 84, 91
    J 0045–7319, 57–8, 271ff
    J 0218+4232, 230
    J 0437–4715, 209, 230, 241
    0531+21, 259
    0540–69, 259
    J 0613–0200, 209, 230, 241
    0655+54, 97, 62
    B0655+64, 248, 253–4
    J 0712–68, 230
    J 0751+18, 228–9, 241
    B0820+02, 241, 253
    0833–45, 259
    J 1012+5307, 230
    J 1023+10, 249, 253–4
    J 1025+10, 229
    J 1025–07, 230
    J 1045–4509, 230
    1046–58, 259
    B1257+12, 25, 187ff, 203, 209, 227–229, 241, 247
    B1259–63, 57–9, 61, 263ff, 426
    1269–63, 84, 89
    1338–62, 258–9
    J 1455–3330, 230, 241
    1509–58, 259
    B1534+12, 13, 22, 219–21, 248–9
    1541–52, 217
    J 1604–72, 230
    1610–50, 259
    B1620–26, 202–3, 209, 247, 377, 382
    J 1643–1224, 230
    1643–43, 258–9
    1706–44, 258–9
    J 1713+0747, 209, 229, 241
    1718–19, 63, 73ff
    1727–33, 258–9
    J 1730–2304, 209, 230, 241
    1737–30, 259
    1744–24A, 78
    J 1745–11, 230
    1757–24, 258–9
    1758–23, 259
    1800–21, 258–9
    1804–08, 217

J 1804−27, 230
1820−11, 219
B1821−24, 201−2, 209
1823−13, 259
B1831−00, 63, 241
1853+01, 259
B1855+09, 209, 225, 229, 236, 241
B1913+16, 13, 22, 58, 62, 84, 159, 217−21, 236, 248−9, 283, 489
1930+22, 259
B1937+21, 58, 197, 199−201, 209, 227, 229
B1953+29, 58, 62, 225, 228−9
B1957+20, 58, 203, 228−9, 252
J 2019+2425, 229, 241
J 2052−08, 228, 230
J 2124−3358, 230
2127+11A, 280, 282
2127+11C, 279ff
J 2129−57, 230
J 2145−0750, 209, 230, 241, 249, 253−4
B2303+46, 22, 219, 249
J 2317+1439, 209, 229, 241
J 2322+2057, 229, 239, 241
2334+61, 259
VV Pup, 409

Rapid Burster, 391
RX
J 0019+21, 434ff
J 0048.4−7332, 432ff, 448−9
J 0058.6−7146, 433ff
J 0429.8−6809, 449
J 0439.8−6809, 434, 436, 440−3
J 0453−42, 409
J 0513.9−6951, 430ff, 440ff
J 0515+01, 409
J 0527.8−6954, 428ff, 434, 440, 442−3
J 0531−46, 409
J 0550.0−7151, 434, 436, 440, 443
J 0925,7−4758, 434, 436
J 1149+28, 409
1342.1+2822, 391
J 1938−46, 409
J 1957−57, 409
J 2107−05, 409
J 2117+3412, 434

Saturn, 195, 206
Sco X-1, 339
N Sco 1994, 363ff, 369
R Ser, 409
WZ Sge, 44, 48−50
SGR
0526−66, 478, 486
1806−20, 477ff
1900+14, 477ff
Sk −69°202, 146−7
SMC, 91, 416−7, 420, 422, 425ff, 432, 449
SMC 3, 432, 449
SMC X-1, 289, 291, 293−4
SNR 0540−693, 259
SN
1983N, 109−10, 137ff
1984L, 109, 134, 139
1985F, 109
1987A, 116, 133, 141−2, 146−8, 428
1987K, 109, 111, 129
1987M, 109−10, 121
1988T, 113
1988U, 113
1988Z, 130
1990B, 138
1990K, 133

1990W, 109
1991N, 137ff
1992ar, 113
1992bi, 113
1993J, 108ff, 119–20, 127ff, 141–2, 147–8
1994F, 113
1994G, 113
1994H, 113
1994I, 109ff, 116, 119–20, 125ff, 133, 135, 138ff
SS 433, 14, 82, 84, 289, 329, 440
SS 2883, 59, 61, 263ff

RR Tel, 434ff
Terzan 1, 392
Terzan 2, 392
Terzan 5, 391, 392
Terzan 6, 392
47 Tuc, 390, 393–6

AN UMa, 409
EK UMa, 409
SU UMa, 44, 47, 49
Uranus, 195, 206

Vela molecular cloud, 434
Vela pulsar, 257, 260
Vela X-1, 58, 88, 289ff
Vela XYZ, 259–60
Venus, 195, 206
Virgo cluster, 157
QQ Vul, 409

W 28, 259, 261
W 30, 258–9
W 44, 259
W 50, 486

X 0921−630, 428

X 1627−67, 78
X 1820−30, 78
X 1822−371, 292–3, 428
X 1916−05, 78

1E
    0035.4−7230, 432ff, 448–9
    0056.8−7154, 431, 434, 446–7
    1339.8+2837, 433ff
    2259+586, 25

4U
    1538−52, 291, 293
    1608−52, 306–7, 324, 331
    1700−37, 289
    1820−30, 292–3, 308, 397

Listing of Poster Papers

Neutron Star Theory
Chou, C.-K., et al. – *The Modified Kompaneets Equation with Astrophysical Applications*
Fujii, H., Maruyama, T., Muto, T. & Tatsumi, T. – *Kaon Condensation in Neutron Stars in Relativistic Mean-Field Theory*
Geppert, U. & Urpin, V.A. – *Magnetic Field Evolution in the Accretion influenced Crust of Neutron Stars*
Goldman, I. – *Limits on Long Range Fields from Binary Pulsars Timing*
Gusev, A. – *Elastic Deformations in the Neutron Star Core*
Hüseyinov, O.H. & Alpar, A. – *Birth Frequencies of Neutron Stars and Black Holes in Binaries*
Martemyanov, B.V. – *Conversion of Neutron Stars to Strange Stars in Binary Systems*
Muto, T. & Tatsumi, T. – *Dissipation Mechanism of Vibrations of Neutron Stars with Kaon Condensate*
Urpin, V.A. – *Magnetic Field Decay in Neutron Stars Entering Close Binaries*
Wiebicke, H.-J. & Geppert, U. – *Amplification of Large-scale Neutron Star Magnetic Fields by Thermoelectric Effects*

Accretion Processes
Beskin, G. & Minarini, R. – *Influence of Red Dwarf Activity on Accretion in Close Binaries*
Bisikalo, D.V., Boyarchuk, A.A., Kuznetzov, O.A. & Chechetkin, V.M. – *The Mass Transfer in Symbiotic Stars enforced by Stellar Wind and by Roche Lobe Overfilling*
Gvaramadze, V.V. – *Jet Formation near Accreting Stars with Strong Magnetic Field*
Horne, K. – *Emission Line Signatures of Anisotropic Turbulence in Accretion Disks*

Hoyng, P., van Niekerk, E.C.M., Schramkowski, G.P. & Achterberg, A. – *Distribution and Flow of Magnetic Energy in an Accretion Disk*
Karetnikov, V.G. & Nazarenko, V.V. – *Mass Transfer in the Region near the Inner Lagrangian Point in different Types of Contact and Semi-detached Eclipsing Binaries*
Meyer, F., Meyer-Hofmeister, E. & Liu, F.-K. – *Evaporation of Accretion Disks in Cataclysmic Binaries*
Mitronova, S., Beskin, G., Neizvestny, S., Plokhotnichenko, V., Popova, M. & Zhuravkov, A. – *Investigations of Optical Variability of Relativistic Objects with High Time Resolution*
Schramkowski, G.P. & Achterberg, A. – *Slender Fluxtubes in Accretion Disks*
Zampieri, L., Turolla, R., Zane, S. & Treves, A. – *Spherical Accretion onto Unmagnetized Neutron Stars*

Pulsars

Allahkverdiyev, A.O., Kasumov, F.K. & Rustamov, Y.S. – *Galactic Distribution and real Ages of Pulsars with known Proper Motions*
Arshakian, D.G. – *The Distribution of Space Velocities of Pulsars*
Bhatia, V.B., Misra, S. & Panchapakesan, N. – *Millisecond Pulsars as Sources of the Galactic Gamma-ray Background*
Björnsson, C.I. – *The Distribution of Radio Pulsars in the B versus P Plane*
Dermer, C.D. & Sturner, S.J. – *Gamma Ray Emission from Millisecond Pulsars*
Deshpande, M.R., Vats, H.O., Chandra, H., Janardhan, P., Dobra, A.D. & Vyas, G.D. – *Bursts from Pulsar 0950+08*
Foster, R.S., Edelstein, J. & Bowyer, S. – *Detection of the Binary Millisecond Pulsar J0437−4715 with the Extreme Ultraviolet Explorer*
Gil, J. – *Structure of Pulsar Beams and the Spectra of Millisecond Pulsars*
Gil, J. – *Microlensing of Pulsar Radiation in the Galactic Centre*
Gök, F., Alpar, A. & Hüseyinov, O. – *Evolutions of Single and Binary PSRs on the $\log P - \log \dot P$ Diagram*
Hartman, J.W. – *The Spatial Distribution of Radio Pulsars*
Hartman, J.W. & Verbunt, F. – *Distances of Neutron stars to the Galactic Plane*
Kaspi, V., Tavani, M., Nagase, F., Kawai, N. & Hoshino, M. – *Periastron X-Ray Observations of PSR B1259−63*
Li, X.-D. & Wang, Z. – *Populations and Evolutions of Radio Pulsars*
Ohnishi, K., Hosokawa, M., Fukushima, T. & Takeuti, M. – *Gravitational Time Delay of Pulsar Timing and Mass Measurement of Stars and MACHOs*
Spreeuw, J.N., van den Heuvel, E.P.J. – *An Explanation for the Low Num-*

ber of Observed Double Neutron Star Systems
Stringfellow, G.S., Pavlov, G. & Cordova, F. – *UV-Optical Observations of PSR 0656+14 with Post-COSTAR HST*
Tauris, T.M. – *Monte Carlo Studies of Binary Millisecond Pulsar Formation*
Thielheim, K.O. – *High Energy Particles from Pulsars*
Wielebinski, R. – *Studies of Pulsars at the Highest Radio Frequencies*

Supernovae
Asvarov, A.I., Kasumow, F.K. & Novruzova, H.I. – *The Role of Diffuse Shock Acceleration in the Radio Emission of Shell-type Supernova Remnants*
Duorah K. & Duorah, H.L. – *Neutrino-Nucleosynthesis of Long-lived Beta-Active Nuclei in Astrophysics*
Iwamoto, K. – *Hydrodynamics of SN 1993J and a Binary Model for its Progenitor*
Kryvdyk, V.G. – *Electromagnetic Radiation from Collapsing Stars*
Kumagai, S. – *X-rays from New Born Neutron Stars in SN 1993J and Type Ib/Ic Supernovae*
Li, Z. & Li, W. – *Studies of Multiple Supernova in Spiral Galaxies*
Pols O.R., Nomoto, K. & van den Heuvel, E.P.J. – *A C+O Star Model for the Type Ic Supernova 1994I*
Seidov, Z.F. – *Supernova Events as a Two-frequency Poisson Process*
Smit, J.M. – *Neutrino-electron Scattering and Supernova Collapse*
Suzuki, T. – *X-Ray Emission of SN 1993J and its Binary Nature*
Walton, N.A., Unger, S.W., Meikle, W.P.S., Martin, R. & Lewis, J.R. – *Optical Observations of SN 1994d and SN 1993j from La Palma*
Wanas, M.I., Melek, M., & Kahil, M.E.– *Is it true that SN 1987A Observations confirm WEP?*

High-Mass X-ray Binaries and X-ray Pulsars
Baykal, A. – *A Statistical Study of the 164 Day Clock Noise of The Relativistic Beams in SS 433*
Berger, M. & van der Klis M. – *HTR Observations of Cyg X-3 with EXOSAT*
Borisov, N., Beskin G. & Pulstil'nik L. – *On Estimates of Lower and Upper Limits for the Masses of Compact Components in Close Binaries*
Burderi, L., et al. – *A Model for the Emitting Region of the X-ray Pulsar 4U 0352+30*
Cherepashchuk, A., et al. – *Parameters of Wolf-Rayet Star in X-ray binary Cyg X-3*

Dolan, J.F., Wolinski, K.G., Boyd, P.T., Bless, R.C., Elliot, J.L., Nelson, M.J., Percival, J.W., Robinson, E.L., Taylor, M.J., Townsley, L.C. & van Citters, G.W. – *The UV polarization of 4U 1700−37, Vela XR-1 and Cyg XR-1*

Fender, R.P. & Bell Burnell, S.J. – *The Hot and the Wind of Cygnus X-3*

Greenhill, J., Watson, R.D., Clarke M., Pritchard, J.D. & Tobin, W. – *H-alpha Photometry of an X-ray Binary*

Jowett, F.H. & Spencer, R.E. – *MERLIN Observations of SS 433 at 5 GHz*

Kaper, L., Lamers, H.J.G.L.M., Ruymaekers, E., van den Heuvel, E.P.J. & Zuiderwijk, E.J. – *On the Nature of Wray 977: the Optical Counterpart of GX 301−2*

Li, X.-D. & Wang, Z. – *X-ray Pulsars with Disk in the Wind-fed Case*

Magnier, E., Prins, S., Augusteijn, T. & Supper, R. – *A Variability Search for MXRBs in M31*

Maisack, M. – *X-ray Pulsars: Pulse versus Orbit and the High-Energy Continuum*

Mereghetti, S., Israel, G.L. & Stella, L. – *The Discovery of 8.7 s Pulsations from the Ultrasoft X-ray Source 4U 0142+614*

Özdemir, S., Hüseyinov, O. Demircan, O. – *On the Progenitors of X-ray Binaries and Binary Pulsars*

Schulz, N.S. – *ROSAT Observations of the Transient X-ray Pulsar Cep X-4*

Trunkovsky, E.M. – *On the Short-time Optical Variability of the Be Star HDE 245770: Optical Counterpart of the Transient X-ray Pulsar A 0535+26*

Trushkin, S.A. – *Radio Flares from SS 433 in the RATAN-600 Multi-frequency Observations*

van der Klis, M., Finger, M., Vaughan, B., Lewin, W., Wilson, R.B., Kouveliotou, C. & Van Paradijs, J. – *BATSE Pulse Timing of Vela X-1 and Cen X-3*

Wilson, R.B., Finger, M.H., Harmon, B.A., Preece, R., Pendleton, G. & Fishman, G.J. – *BATSE Observations of a Giant Outburst from A 0535+26*

Low-Mass X-ray Binaries

Angelini L., et al. – *The LMXB pulsar 4U 1626−67*

Beskin, G., Neizvestny, S., Plothoknichenko, V., Popova, M., Zhuravkov, Benevenuto, O.G., Feinstein, C. & Méndez, M. – *Optical Study of Southern LMXB with High Temporal Resolution: Evidence for Non-thermal Flares*

Harlaftis, E.T. & Charles, P.A. – *More Insight in the Low-Mass X-ray Binary X 1822−371*

Harpaz, A. – *Heating of a Secondary Star in LMXB*

Kalogera, V. – *Study of the Formation of Low-Mass-X-Ray Binaries using Population Synthesis Techniques*

Kanetake, R. & Takeuti, M. – *Vertical Oscillations of Thin Accretion Discs as a Candidate Process for Quasi-periodic Oscillations*

Kolb, U. & King, A.R., – *Implications of Consequential Angular Momentum Loss*

Kunz, M., et al. – *Pulse Phase Dependent Spectra of Her X-1*

Kuulkers, E. & van der Klis, M. – *New Bursts in Two Z-sources*

Lapidus, I., Nobili, L. & Turolla, R. – *Accretion Rates in LMXBs with Expansion in the strongest X-ray Bursts*

Laurent, P., Denis, M., Paul, J., et al. – *New SIGMA Results on GX 1+4*

Martín, E.L., Rebolo, R., Casares, J., Charles, P.A. & Molaro, P. – *A Lithium Search in Companions to Compact Objects*

Naitou, K., Kanetake, R., Takeuti, M. & Dotani, T. – *Time-variation of Pulsars and LMXBs Observed with Ginga*

Navarro, J. – *Quiescent LMXBs and Millisecond Pulsars*

Portegies Zwart, S. – *Period Eccentricity Distribution of Close Binary Evolution Remnants*

Psaltis, D. & van der Klis, M. – *Spectral Models of LMXBs: "Eastern" versus "Western"*

Schandl, S. – *Coronal Winds Producing the Warped Shape of the Accretion Disk in Her X-1*

Shearer, A. – *Observations of Globular Cluster Binaries Using the TRIFFID Camera*

Sheffer, E.K. & Lyutyi – *Optical Studies of the Accretion Disk and Matter Flow Process in the binary system HZ Her/Her X-1*

van der Hooft, F., Kouveliotou, C., Van Paradijs, J., Rubin, B., Finger, M., Harmon, A., van der Klis, M., Lewin, W.H.G. & Norris, J.P. – *Low Frequency QPO in GRO J1719-24*

Vaughan, B., Dieters, S. & van der Klis, M. – *X-ray Time Lags of Sco X-1 and GX 5-1*

Vrtilek, S.D., Charles, P.A., Dennerl, K.O., Hu, E., Kahabka, P., la Dous, C., Marshall, H., Mihara, T., Primini, F.A., Raymond, Rutten, R., Soong, Y., Stull, J., Trümper, J., Voges, W., Wagner, R.M. & Wilson, R. – *Multiwavelength Observations of Her X-1/HZ Herculis*

White, N.E., Zylstra, G., Smale, A., Mitsuda, K. & Corbet, R. – *ASCA Observations of The Accretion Disk Corona Sources X 1822-371 and X 0921-63*

X-ray Observations General

Belloni, T., Mereghetti, S. & Goldwurm, A. – *X-ray Observations of GRS-1758-258*

Brandt, S. & Lund N. – *Monitoring the Activity Variations in Galactic X-ray Sources with WATCH on EURECA*

Cadež, A. & Galičič, M. – *Evidence for phase modulated pulses of the Crab pulsar with a period of $\sim 115\,s$*

Castro-Tirado, A.J., Brandt, S., Lund, N., Lapshov, I.Yu. & Sunyaev, R.A. – *Long Term Observations of X-ray Sources by WATCH*

Grebenev, S., Pavlinsky, M. Sunyaev, R. – *Population of X-ray Sources near the Center of our Galaxy according to ART-P/GRANAT*

Karitskaya, E.A., Cherepashchuk, A.M., Goranskij, V.P., Nadjip, A.E., Savage, A., Shakura, N.I., Sunyaev, R.A. & Volchkov, A.A. – *The Investigation of the Error Boxes of KVANT and GRANAT X-ray Sources in the Region of Galactic Center*

Steshenko, N.V. – *The Spectrum-UV Project*

Sun, X., et al. – *A Revisit of the HEAO1 A-4 All Sky Survey I. Images for Selected Regions*

Zhang, W., Giles, A.B., Jahoda, K., & Swank, J.H. – *The Proportional Counter Array Aboard The X-ray Timing Explorer*

Black Holes

Bao, G. & Østgaard, E. – *X-ray Variability due to Gravitational Lensing by Black Holes and Relativistic Rotation of Accretion Disks*

Bartolini, C., Guarnieri, A., Minarini, R., Piccioni, A., Beskin, G., Mitronova, S., Neizvestny, S., Panferova, I., Plokhotnichenko, V. & Popova, M. – *Optical Studies of the Variability of GRO J0422+32*

Belyanin, A.A. & van Oss, R.F. – *Annihilation Lines from Accreting Black Holes*

Bonnet-Bidaud, J.M. & Mouchet, M. – *The Optical Spectrum of GRO J0422 +32 in Quiescence*

Borozdin, K.N., Alexandrovich, N.L., Arefiev, V.A., Sunyaev, R.A., Skinner, G.K., Patterson, T.G., Willmore, A.P., Brinkman, A.C., Heise, J. & Jager, R. – *Observations of Two X-ray Novae 1993 by KVANT-MIR Module*

Callanan, P., McClintock, J., Garcia, M. & Zhao, P. – *Optical Observations of the X-ray Transient J0422+32: The Outburst and the Decay to Quiescence*

Casares, J. & Charles, P.A. – *The Mass of the Black Hole in GS 2023+338/ V404 Cygni*

Casares, J., Charles, P.A., Harlaftis, E.T., Marsh, T.R., Martin, A.C., Martin, E. & Pavlenko, E.P. – *Doppler Tomography of the X-ray Transient J0422+32 during the Dec 1993 Outburst*

Chakraborty, D.K. & Mishra, K.N. – *General Relativistic Effects on Fluid Disk Rotations around a Black Hole*

Chen, W., Shrader, C. & Livio, M. – *Systematic and Statistical Study of X-ray Nova Light Curves*

Chevalier, C. & Ilovaisky, S.A. – *GRO J0422+32 : Activity and Quiescence*
Dermer, C.D. – *Stochastic Particle Acceleration and High Energy Radiation from AGNs*
Garcia, M.R., Callanan, P., McClintock, J. & Zhao, P. – *Spectroscopy and Photometry of the Black Hole Candidate GRO J0422+32 near Quiescence*
Grebenev, S., Sunyaev, R. & Pavlinsky, M. – *Spectral States of Galactic Black Hole Candidates. Observations with ART-P/GRANAT*
Hadrava, P., Bao, G. & Østgaard, E. – *Reflection by a Relativistic Accretion Disk*
Harmon, B.A., Wilson, C.A., Paciesas, W.S., Pendleton, G.N., Rubin, B.C. & Zhang, S.N. – *The Intensity and Spectral Behavior of GRO J1719−24 = GRS 1716−249 (X-ray Nova Ophiuchi 1993)*
Martin, A.C., CasaresJ., Charles, P.A. & Pavlenko, E.P. – *Spectroscopy of the 6 Hour Variations in the Soft X-ray Transient V404 Cygni*
Oosterbroek, T., et al. – *The "Non-variable Iron Line" in GS 2023+38*
Paciesas, W.S., Pendleton, G.N., Harmon, B.A., Wilson, C.A., Rubin, B.C., Ling, J.C., Skelton, R.T. & Wheaton, W.A. – *The Long-Term Hard X-ray Behavior of Cygnus X-1*
Pavlenko, E.P., Martin, A.C., Casares, J., Charles, P.A. & Ketsaris, N. – *The Optical Light Curve of V404 Cygni: Ellipsoidal Modulation and 6 Hour Variations*
Pavlovski, K. & Vujnović, V. – *The Mass of the Black Hole Candidate in the X-ray Transient GS 2023+338 (V404 Cyg)*
Robinson, E.L., Sanwal, D. & Zhang, E. – *The Infrared Light Curve and Ellipsoidal Variations of the Black Hole Binary V404 Cygni*
Voloshina, I. & Luyty, V. – *The Additional Radiation of the Black Hole Candidate Cyg X-1 at Primary Minimum*
Zakharov, A.F. – *On the Hot Spot near a Kerr Black Hole: Monte Carlo Simulations*

Cataclysmic Variables
Barwig, H., Fiedler, H., Reimers, D. & Bade, N. – *HS 1804+6753 — A New Double Lined Eclipsing Dwarf Nova with High Orbital Inclination*
Billington, I., Marsh, T., Horne, K., Cheng, F., Thomas, G., Bruch, A., O'Donoghue, D. & Eracleous., M. – *An Ultraviolet Dip in the Lightcurve of the Cataclysmic Variable OY Car in Superoutburst*
Chandrasekhar, T., Ashok, N.M. & Ragland, S. – *Near Infrared Coronal Line Emission in Nova Herculis 1991*
Cool, A.M., Grindlay, D.E., Cohn, H.N., Lugger, P.M. & Slavin, S.D. – *Identification of Candidate Cataclysmic Variables in the Post-Core-Collapse Cluster NGC 6397*

Echevarria, J., Tovmassian, G., Tapia, M., Bohigas, J., Shara, M., Gilmozzi, R., Stover, R., Rodriguez, L.F., Martinez, C., Garzon, F., Jones, D.H.P., Costero, R., Barral, J., de Lara, E., Alvarez, M., Wallis, R.E., Roth, M., Lopez, J.A., Vogt, N., Asatrian, N., Zsoldos, E., Mattei, J. & Batteson, F. – *Simultaneous Multiwavelength Observations of Dwarf Novae Outbursts; SU UMa: Minihumps at minioutburst?*

Ercan, E.N., Baykal, A., Esendemir, A., Kızıloğlu, Ü., Ögelman, H., Alpar, M.A. & İkis, G. – *ROSAT observations of TT Ari*

Friedjung, M., Bianchini, A., Cassatella, A. & Selvelli, P.L. – *A wind of the Old Nova V603 Aql*

Hakala, P.J., Piirola, V., Hannikainen, D., Vilhu, O. & Osborne, J. – *Ultimate Polarimetric Variability observed in RE1307*

Hessman, F.V. & Reinsch, K. – *The Mystery of the Emission lines in Eclipsing Cataclysmic Variables*

Hubeny, I. & Lanz, T. – *Modeling the Spectrum of Cataclysmic Binaries*

Ibanoglu, C., Keskin V., Akan, M.C., Evren, S. & Tunca, Z. – *Long-term Luminosity Variations and Period Changes in the White Dwarf Eclipsing Binary V471 Tauri*

Jones, D.H.P., Dhillon, V.D. & Still, M.D. – *The SW Sex Stars: A New Class of Nova-like Variable*

Kjurkchieva, D. & Marchev, D. – *Eclipse Curves of UX UMa in 1992*

Kjurkchieva, D. & Marchev, D. – *R and B Photometry of AM Her during 1993*

Knigge, C. – *The Geometry of Cataclysmic Variable Winds: Constraints from Modelling the C IV Resonance Line in Eclipse Observations of UX UMa*

Kraicheva, Z., Genkov, V. & Popov, V. – *The polar AM Hercules: Photometry in the Time Interval 1988–1993*

Marsh, T., Horne, K. & Cheng, F. – *Ultraviolet Dwarf Nova Oscillations in OY Car*

Mickaelian, A.M. – *New Cataclysmic Variables from the First Byurakan Survey*

O'Donoghue, D., Kilkenny, D., Chen, A.-L., Stobie, B., Koen, C., Warner, B. & Lawson, W. – *EC 15330–1403 and the AM CVn Stars*

Okazaki, A.T. – *Structure of Eccentric Modes in Accretion Disks*

Özkan, M.T., Ak, T., Saygaç, A.T., Esenoğlu, H.H. & Güler, S. – *Orbital Dependence of the UV Spectra of Z Cam Type Dwarf Novae*

Popov, V., Kraicheva, A. & Antov, A. – *Photometry of KR Aurigae 1985–1988*

Pustylnik, I. – *Gas-Eclipsed Binaries*

Schwope, A.D., Mantel, K.-H. & Thomas, H.-C. – *Tomography of the Accretion Stream in the Eclipsing Polar RX J2107.9–0518*

Siarkowski, M. & Pres, P. – *Structure of the AR Lac Corona from ROSAT PSPC All-Sky Observations*
Sion E.M., Cheng, F.H., Long, K.S., Szkody, P., Gilliland, R., Huang, M. & Hubeny, I. – *Hubble Space Telescope FOS Spectrocopy of the Ultra-short Period Compact Binary WZ Sagittae: the Underlying Carbon-rich Degenerate*
Stehle, R. & Kolb, U. – *The Influence of Nova Explosions on the Long-Term Evolution of Cataclysmic Variables*
Suleymanov, V.F. & Andrianov, V. – *The Effect of the Reflecting Radiation on the Spectra of Novalike Stars*
Ulla, A., Mantel, K.-H., Barwig, H., Sabau, L., Goodrich, R.W. & la Dous, C. – *Simultaneous UBVRI high-speed Photometry and Optical Spectropolarimetry of the Peculiar Cataclysmic Variable GP Com*
Ulla, A., Thejll, P. & Sabau, L. – *Search for Late-Type Companions to Hot Subdwarfs using JHK Photometry*
Wheatley, P.J. – *Cataclysmic Variables in the ROSAT WFC Survey*
Wickramasingh, D.T. – *Cyclotron and Zeeman Spectroscopy of White Dwarfs in CVs – Implications for Field Structure*
Wolf, S. & Mantel, K.-H. – *Variable Star Observations with MEKASPEK*
Wood, J.H., Naylor, T., Hassall, B.J.M., Ramseyer, T.F. & Marsh, T.R. – *X-ray Observations of Eclipsing Cataclysmic Variables*
Zwitter, T. & Munari, U. – *CCD Spectrophotometry of CVs. Optical/near-IR Low Resolution Survey of Faint Systems and Echelle High Resolution Study of Emission Line Profiles*

Supersoft Sources and Symbiotic Stars

Hric, L., Skopal, A., Chochol, D., Komžík, R. & Urban, Z. – *Symbiotic Binaries - Basic Results of Six Years Photometric Monitoring*
Meyer-Hofmeister, E. & Meyer, F. – *On the Origin of the Visual Light from Supersoft Sources*
Pakull, M. – *Optical Observations of Supersoft Sources*

Gamma-ray Sources

Cheng, L., Sun, X. & Li, T. – *Position and Proper Motion of Geminga During the COSB Mission*
Li, P., Hurley, K., Sommer, M., Kouveliotou, C., Fishman, G.J., Boer, M., Niel, M., Laros, J. & Cline, T. – *Deep ROSAT Observation of the May 1 1992 Gamma-Ray burst Field*
Li, P., Hurley, K., Kouveliotou, C., Fishman, G.J. & Hartmann, D. – *Flares and Gamma-ray bursts*
Wanajoh, S., Hashimoto, M. & Nomoto, K. – *Gamma Ray Line Emission from Neon Novae*

Various Topics

Aarseth, L.S.J., Anasova, J.P., Orlov, V.V. & Szebehely, V.P. − *The Dynamics of Triple Systems. Close Binary Approaches and Escapes*

Aboelazm, M.S. − *Light Variation of the Variable Star V566 Oph*

Anasova J.P. − *Dynamical Processes of Evolution of Binaries in the Field*

Gorbatsky, V.G. & Prohorov, S.P. − *On the Dynamical Evolution of a Close Binary moving near an AGN*

Hamdy, M.A. − *Light Variation of the Variable Star I-Boo*

Hukeirat, A. − *HDRHD — A Multidimensional Radiative Hydrodynamical Solver for Accretion Flows around Compact Objects*

Kiseleva, L., Eggleton, P., Colin, J. & Orlov, V. − *Stability and Instability of Hierarchical Triple Stars*

Lipunov, V.M., Nazin, S.N., Panchenko, I.E., Postnov, K.A. & Prokhorov, M.E. − *The Gravitational Wave Sky Map*

Lipunov, V.M., Prokhorov, M.E. & Postnov, K.A. − *On the Initial Mass Ratio Distribution of Binary Systems*

Niarchos, P.G. & Pantazis, G. − *A New Approach for the Determination of Gravity Darkening in Close Eclipsing Binaries*

Pogrebenko, S. − *VLBI Detectability of Point Source Image Distortion Caused by a Close Binary Generated Gravity Wave*

Ray, A. & Kluzniak, W. − *Pulsar Timing Residuals & Gravitational Radiation From Binaries*

Tsujimoto, T., Shigeyama, T. & Nomoto, K. − *The Chemodynamical Evolution of Spheroidal Systems*

Ureche, V. & Mioc, V. − *Weighted Bi-polytropic Models for White Dwarfs: Analytic Approach*